Lives and Legacies
An Encyclopedia of People Who Changed the World

Scientists, Mathematicians, and Inventors

Edited by Doris Simonis

Writers

Caroline Hertzenberg
John Luoma
Tami Schuyler
Jonathan Secaur
Doris Simonis
Daniel Steinberg
Valerie Tomaselli
Gayle Weaver
Kelly Wilson

ORYX PRESS
1999

The rare Arabian Oryx is believed to have inspired the myth of the unicorn. This desert antelope became virtually extinct in the early 1960s. At that time several groups of international conservationists arranged to have nine animals sent to the Phoenix Zoo to be the nucleus of a captive breeding herd. Today the Oryx population is over 1,000, and over 500 have been returned to the Middle East.

© 1999 by The Oryx Press
4041 North Central at Indian School Road, Phoenix, Arizona 85012-3397

Produced by The Moschovitis Group, Inc.
95 Madison Avenue, New York, New York 10016

Executive Editor: Valerie Tomaselli
Senior Editor: Hilary Poole
Design and Layout: Annemarie Redmond
Original Illustrations: Lucille Lacey
Copyediting and Proofreading: Carole Campbell, Carol Sternhell, Melanie Rella
Index: AEIOU, Inc.

Published simultaneously in Canada
Printed and Bound in the United States of America

ISBN 1-57356-151-7

Library of Congress Cataloging-in-Publication Data

Simonis, Doris A.
    Scientists, mathematicians, and inventors: lives and legacies: an
encyclopedia of people who changed the world / by Doris Simonis.
        —(Lives and legacies)
    Includes bibliographical references and index.
    ISBN 1-57356-151-7 (alk. paper)
    1. Scientists—Biography—Encyclopedias.   2. Mathematicians—
Biography—Encyclopedias.   3. Inventors—Biography—Encyclopedias.
I. Title.   II. Series.
Q141.S54   1999
509.2'2—dc21
    [B]   98-48484
                                                                              CIP

⊗ The paper used in this publication meets the minimum requirements of American National Standard for Information Science—
Permanence of Paper for Printed Library Materials, ANSI Z39.48, 1984.

Table of Contents

The Biographies

Appendices, Bibliography, and Index

Listing of Biographies

Introduction

This volume of concise biographies of scientists, inventors, and mathematicians is different from most resources about people who influenced their various cultures. Not only have contributions been described in the context of the individual's life, society, and discipline, but also the long-term effects of each individual's work are considered as a global legacy. Unlike traditional encyclopedias, which present facts of individual lives in isolation, this volume makes the thread of interactions between science, technology, and society more apparent in the fabric of human history.

Accordingly, the selection criteria for the 200 exemplars of scientific, mathematical, and inventive thinking were different from those of standard works. Although many famous contributors to knowledge and technology are included, we have also deliberately chosen some people who are not well-known and whose contributions were made by overcoming societal barriers to their productivity. In following this mandate, we have selected and described individuals who fall into one or both of the following categories:

> a) their life and work had significant influence on society in general or on their discipline in particular and their influence extended beyond their lifetime; *or*
>
> b) their work extended beyond the limits imposed on these individuals by contemporary society. This category includes women and minorities whose contributions and legacies were limited because of societal constraints.

We used a flexible selection process based on the two categories above to decide which scientists, inventors, and mathematicians to include in this volume. The attempt to embrace a wider variety of thinkers and doers was sometimes frustrated by the limited information and scholarship available in English about persons of earlier times and various cultures. However, all of those chosen did contribute something new or rare to human society, communicating through their work, ideas, and lives new perspectives on what is possible.

In addition, we used a specific definition of science and related fields to shape our selection process, a definition that emphasized creative thinking as opposed to discovery of existing realities. Science develops from contrarian thinking that produces testable new ideas, rather than the identification of new continents or petrified fossils. Without this distinction, recognition of well-known "greats" in exploration and discovery might have overshadowed recognition of the lesser-known but significant achievements of people who are underrepresented in most historical and biographical works.

In choosing to illustrate a wide variety of contributors, and given the space limitations for this publication, many worthy people have not been included. But we hope that readers will be intrigued by the variety of people, ideas, and inventions noted herein that have contributed to the culture, choices, and comforts available to citizens of Earth.

Practitioners in the field of medicine are included in this volume even though, in the strictest sense, they do not fit our contemporary definition of science. We have done this because of popular notions concerning the nature of medical practice. Indeed, for many readers, the best known "scientist" is their family doctor, the best "invention" would be a cure for cancer or heart disease, and "mathematics" is required to consider how to pay for these resources.

These popular understandings suggest the need for working definitions of science, invention, mathematics, and medicine that highlight their characteristic differences and similarities with other areas of human creativity. Like a timeline, these considerations begin with the oldest of these creative fields, invention of technology.

———◆———

INVENTION is concerned with practical problems requiring new things: devices or processes (technologies) that improve the human condition. Invention is one of the oldest areas of human creativity. Fire, clothing, nets and snares, scrapers and diggers, baskets and bowls, musical instruments and weapons, metallurgy and food preservation all developed long before recorded history—when science or even its prototypes, astrology and alchemy, did not exist.

The practical roots of inventions are evident in the contribution they have made to human survival and culture.

One of the most obvious benefits of many inventions is that they saved time, which could be used for education and intellectual pursuits. Labor-saving devices helped to eliminate slavery, fire and lighting provided both safety and usable hours for after-dark communication and craft development, agriculture made stable settlements possible, and technologies for water purification and vaccinations extended human life wherever they were used.

However, the popular perception of technology as applied science is not supported by most historians. Technology has a much more ancient history. It typically (but not always) precedes science and makes possible new observations and methods of testing scientific ideas. Antoni Van Leeuwenhoek discovered his "invisible world" by using his hand lens, a standard draper's tool, to inspect more than the weave of custom fabrics. Galilei Galileo found that the telescope was invaluable to demonstrate evidence supporting Nicolaus Copernicus's theory that the Earth revolved around the Sun, but it was not invented for that purpose. John Harrison's superb craftsmanship and knowledge of woods were the keys to solving the longitude problem that had stumped theoretical scientists like Isaac Newton and Edmond Halley, his contemporaries. The Wright brothers' airplane was an exercise in practical problem solving, but the Wrights' systematic analyses of wing design provided basic data, which became a foundation for aerodynamics and aeronautical engineering, fields of knowledge that did not even exist before their successful flights.

There is a real difference between an inventor knowing how to solve practical problems, even ones so challenging as making sea voyages of predictable duration and making human flight possible, and a scientist describing nature in ways that support the development of navigational systems or ways that explain how heavier-than-air machines can seem to defy gravity. In these relatively recent examples that difference may be less important than the fact that technology stimulates science and vice versa. Louis Pasteur, who was both an inventor (vaccines, pasteurization) and a scientist (germ theory), once wrote that science and technology are as related as apples are to their tree. In this analogy, which came first historically is much less important than that both remain healthy.

———•———

Mathematicians develop and apply intellectual tools that enable people to describe and identify patterns in nature and events. They use MATHEMATICS, the language of particular sequences of reasoning, which allows recognition, coherent descriptions, and predictions of patterns in time and space. Mathematics allows both practical and imaginative problem solving in real and hypothetical contexts.

Mathematics is another kind of thinking and communicating whose origins pre-date verbal history. Sticks and bones notched for tallies and "sighting stones" that track the movements of the Sun have been found on several continents. These finds have been interpreted as evidence of ways people may have recorded hunting outcomes and anticipated planting times. Some researchers speculate that mathematics as a record keeping function is a natural, universal script older than other written languages.

Measurement and record keeping were essential to planning construction, securing sufficient building materials and food for work crews, and for dividing inheritances. Records of exchanges were vital to the success of traders on every continent as civilizations began to be centered on coastal cities. As writing developed, so did number systems. The best known are the Roman and Hindu-Arabic number systems.

In early recorded history, mathematics that went beyond simple measurement probably began in Mesopotamia (now Iraq) and Egypt many centuries before the Christian era. A millennium of greats living on or near the Mediterranean Sea included the mathematicians Thales, Pythagoras, Euclid, Archimedes, Eratosthenes, and Hypatia. After the destruction of the library in Alexandria (391) and the sack of Rome (455), the intellectual momentum was sustained in Persia where al-Khwarizmi and others preserved the legacy of mathematics through their writings.

In the Roman Empire, trained specialists had to manipulate counting boards and record results in Roman numerals. The tedious tallying of census figures, crop production, mineral wealth, size of armies, etc., was the professional work of reckoning specialists on whom entire economies depended well into the Middle Ages. Leonardo Fibonacci's thirteenth-century introduction to Europe of the more user-friendly Arabic numbers (and the revolutionary concept of zero that came with them) did not change established practice quickly. However, Johannes Gutenberg's invention of the movable-type printing press 200 years later not only institutionalized the use of Arabic numerals but also made mathematical works widely accessible. This was the turning point in the acceptance of numbers now familiar to school children around the world.

Mathematics—and its search for logically consistent patterns in number, space, matter, and human imagination—

has had great influence in contemporary societies. And some contemporary mathematical models (global warming) and theories (game theories, economic projections) have themselves been generated by a series of twentieth-century inventions that speed calculations. Computers have expanded the possible number and kinds of patterns to be explored, an example of technology providing the means for creative investigations that would have been as impossible—as was detection of tiny living animals in a drop of water before the magnifying lens was crafted.

———•———

MEDICINE is primarily a practical art that grew out of knowledge of plant extracts, in particular, that could relieve pain, reduce fever, or speed childbirth. Illness required addition of these natural products and/or removal of bodily fluids believed to be out of balance within the individual patient.

Medicine has a long association with science and invention, growing out of both educational practices and human values. Dissection, one of the most rudimentary tools of medicine, arose from educational practices in a non-medical field. The first universities like the twelfth-century University of Bologna (modern Italy), offered only one specialized advanced degree (in law). The first dissections of corpses were done in law schools to determine whether murder had been committed; otherwise, autopsies were generally forbidden throughout the western world.

By the thirteenth century, the first medical school was organized in Bologna, and stories about Hippocrates and works by Galen, al-Razi, and Ibn Sina were the handwritten bibles from which lecturers preached. Like Galen, instructors typically had a distaste for dissection. Autopsies and invasive surgery were generally avoided, but by the 1530s they were a part of the medical school program in Paris. Andreas Vesalius was probably the first professor there to demonstrate directly (as opposed to delegating dissection to a lowly assistant) the relationships between internal human body systems and was the first to base his anatomical drawings on observations instead of assumptions and inferences.

Many early scientists were physicians who explored a scientific interest during and after formal studies in medicine. Carolus Linnaeus, for example, developed his passion for grouping and classifying plants after being introduced to botanical remedies as part of his professional training.

Starting in the sixteenth century, there were 400 years of inquirers and experimenters who had medical school education. Some examples in this volume are Paracelsus,

Ambroise Paré, William Harvey, Marcello Malpighi, Joseph Black, Edward Jenner, Friedrich Wöhler, Louis and Elizabeth Cary Agassiz, Ignaz Semmelweis, William Morton, Joseph Lister, Robert Koch, Ivan Pavlov, Paul Ehrlich, Sigmund Freud, Karl Landsteiner, and Alexander Fleming. During the twentieth century, however, scientific specializations and advanced degrees in disciplines like biochemistry, genetics, physiology, neurology, bacteriology, and immunology had become the norm for researchers while medical doctors were trained to be practitioners.

The power of a "safe attack" on a human being via a vaccine was not validated until the nineteenth century and widespread use of vaccines is a twentieth-century phenomenon. However, they are still not universally administered (poverty and religious objections are among the reasons) despite the signal success of eliminating at least one global scourge, smallpox, by this means and preventing wholesale epidemics of several others.

Other advances in modern medicine were also made during the nineteenth century. The stethoscope was invented and became an indispensable tool. Ignaz Semmelweis and Florence Nightingale used statistics to develop a rationale for sanitary practices. They invented effective procedures without the rationale of Pasteur's germ theory. Joseph Lister learned of Pasteur's work through translations his wife made for him, and his use of carbolic acid became standard antiseptic procedure for surgery, greatly reducing infections.

The twentieth century saw major advances based on earlier innovations. After Elie Metchnikoff identified phagocytes (cells that engulf foreign materials in the body) and other experimenters of the nineteenth century (Pasteur, Jenner, Koch, Ehrlich) showed the effectiveness of a series of vaccines, other ways to help natural immunity were developed such as blood transfusions, antibiotics, bone marrow transplants, and T-cell infusions in the twentieth century.

Other advances included diagnostic use of X-ray machines that came into use at the beginning of the twentieth century. Sophisticated imaging systems like ultrasound, nuclear magnetic resonance (NMR), and computerized tomography (CT) were all added to the non-invasive options available to physicians by the end of the twentieth century. Surgical strategies were expanded by technology to include microsurgery with specialized tools, and treatment options were multiplied thanks to chemical synthesis of many drugs, enzymes, and hormones that were once available only from living organisms.

Since nearly everyone has direct experience with advances in medical practices, the relationship between invention and quality of life is most often recognized in this area. So, too, the legacies of individuals who make contributions in this field are most easily understood.

—•—

Scientists are more than discoverers of existing realities. SCIENCE is a creative act; it requires new perspectives on what is known, integration of knowledge into new generalizations or subsuming concepts, and development of comprehensive understandings of how the natural universe works, providing direction for future inquiries, forecasts, discoveries, and events. Science generates new ideas about nature and ways to test those ideas. Explorers, adventurers, and voyageurs may "discover" existing land masses, species, or materials not yet recognized by people in their homelands, but they rarely originate powerful ideas, insights, or theories that relate their findings to other knowledge and understandings. (Geographer Alexander von Humboldt is a notable exception)

Another important aspect of science is defining how to study nature. Using instruments to help observe evidence outside of their personal influence and using mathematics to help explain their findings, scientists detect patterns and relationships of universal significance. The key indicator of a tenable scientific generalization is that nature does not provide contradictions to humanly-described patterns. Of course, it is desirable that colleagues agree that a particular explanation (theory) is elegant and satisfying, but the truth of a theory should depend on confirmations from the physical world, not affirmations from human hearts. In this regard, science is quite different from poetry, music, and philosophy.

Science is a way of problem defining and problem solving that provides creative perspectives—seeing things or events in ways they have not been seen before. In other words, at its highest level, science generates new ideas and explanations about how nature works. The hardest part, the one least often achieved, is envisioning things in a way that has not been done before. Knowledge is not enough. Ancient astrologers on several continents had made accurate measurements of the movement of the Sun, moon, planets, and major constellations. But the basic assumption that Earth is a stationary platform around which all heavenly bodies move prevented the most careful and accurate observers from considering any alternatives.

Copernicus's attempt to demonstrate mathematically what the celestial motions would look like to an observer on the Sun required a creative leap to an assumption contrary to common observations. His insight was particularly unwelcome for political and religious reasons. In his time, human worth was measured in large part by creation stories that described Earth as made expressly for people and centrally located in the universe as testimony to their importance. A Sun-centered planetary system undermined the power of that argument and the people who used it.

Some historians claim that without certain technologies, science did not even exist as anything more than philosophical musings about the natural world. Copernicus's revolutionary concept would have been nothing more than an imaginative personal perspective without the extended vision that a telescope provided and Galileo's interpretations of sights it allowed.

Another important break scientists made with the past was the promotion of free and open communication about observations, experiments, and analyses of findings. Modern science broke away from philosophy when Renaissance thinkers not only took a fresh look at nature but also publicly described and defended their viewpoints. Unlike Pythagoreans and alchemists, they did not try to keep their new knowledge secret.

Galileo in the early seventeenth century conducted experiments in a systematic way, using mathematics as well as words. He described his work so logically that most educated people accepted both his methods and his analysis. Later in the same century scientists like Robert Boyle stressed the importance of publishing articles about scientific observations so that they could be replicated by others as tests of their acceptability. Soon after Newton presented his breakthrough concepts to a receptive audience, colleagues in the Royal Society of London. He outlined laws that unified the heavens and Earth into one universal machine, completing a revolution that recognized science as a way of knowing distinct from philosophy.

The scientific world was expanded further by a succession of other original thinkers. By the middle of the nineteenth century, John Dalton had laid the foundations of atomic theory, James Maxwell had developed an inclusive theory uniting electricity, light, and magnetism, and Charles Darwin provided a coherent explanation for the observed diversity of living species with his theory of evolution.

More insights on the material world were developed during the twentieth century. A previously unknown realm of nuclear transformations was revealed by Marie Curie—radioactivity. Her work in radioactivity was the beginning of a better understanding of the internal state of an atomic nucleus. Quantum mechanics led Werner Heisenberg to propose the uncertainty principle, which recognizes limits—in the subatomic realm—to the precise application of classical concepts of location, energy, velocity, and momentum. Albert Einstein developed theories of relativity that recognize relationships between space and time, inertia and gravity, energy and mass. And the concept of the universe as an expanding cluster of galaxies was the brainchild of Edwin Hubble.

New ideas are usually unacceptable to conventional thinkers, especially those who assume that everything important is already known and understood. The most exciting and integrative ideas are often resisted because they require unlearning traditional teachings and/or changing the way of thinking about information in general. Creative scientific work is usually not the result of accumulation of data and searches for accuracy and certainty. It is, instead, a product of independent thinking when looking into the heart of nature. Individuals like those mentioned above exhibited this contrary trait: they proposed new ways of looking at their part of the universe. They did this often to the discomfort of colleagues who "didn't know that they were bound to a model, and you couldn't show them," as Barbara McClintock said when reflecting on the years when her reports of "jumping genes" were either jeered at or ignored. The hidden web of tacit assumptions that limit human thinking has had to be broken again and again to accommodate new observations, unexpected experimental results, and ideas that create relationships among various constituents of the material world.

There is a noteworthy dimension to science that makes it a special kind of creative thought. It is the assumption that the universe is knowable and that relationships observed in one part of it will be true everywhere. This makes science less culture-bound than music, poetry, and other forms of art. More than one person may independently discover the same element, the same physical laws, or the same regularity of events. All humankind shares the reality of the natural world as it exists beyond personal and political orientations. Whether human perceptions of that physical world are determined by external reality, whether we alter that reality in the processes of experimentation and exploration, and whether our understandings merely reflect the way human minds work are all subjects of ongoing debates. Albert Einstein's viewpoint was that:

. . . we are somewhat like a man trying to understand the mechanism of a closed watch. He sees the face and the moving hands, even hears it ticking, but he has no way of opening the case. If he is ingenious, he may form some picture of a mechanism which could be responsible for all the things he observes, but he may never be quite sure his picture is the only one which could explain his observations.

This humble recognition of the tentativeness of scientific knowledge may not appeal to lovers of certainty, but it is a hallmark of modern science.

———◆———

That nearly 75% of the biographies in this volume describe people who lived during the last 200 years should be no surprise. Science is a relatively recent discipline. There are probably more scientists and mathematicians living now than the total of all those in the past. Some of that number is simply a reflection of the unprecedented size of the human population (six billion vs. one billion as recently as 200 years ago), some is a product of access to education by more people than ever before in history, and some is attributable to the positive aspects of technology that have allowed longer, healthier lives with time for study.

The role of literacy and education is especially important to note in the lives of these significant people who influenced our world. The printing press spread new ways of thinking beyond the monastery and the university. The necessity to master Latin as the secret language of knowledge was also removed when publications in people's spoken languages became relatively quick and inexpensive to produce. Skilled translators and interpreters like Mary Somerville became more than helpmates for a single person (as were Lavoisier's and Lister's wives)—their texts were used by generations of new readers.

Obtaining higher education is a recent privilege for most people. Some women (notably Marie Curie, Lise Meitner, Emmy Noether, Harriet Brooks) began their struggles for independence in thought and action by meeting requirements to teach children, one of the few options for self-sufficiency open to women. The societal roadblocks to college education for women were first removed in the United States when some colleges for women were opened that stressed sciences (Mt. Holyoke in 1837; Vassar, where Maria Mitchell taught in 1865) and where new state universties began to accept women in coed classes in the 1860s and 1870s. In 1940 Columbia University awarded its first Ph.D. in medical science to an African American, Charles Drew, and the difficult academic career of Indian physicist Satyendranath Bose is evidence of discrimination beyond U.S. shores. Also, employment opportunities lagged behind educational advancement for both women and minorities.

Another big shift in the professional population began with admission of significant numbers of women into medical schools in the 1970s. Biographies of Elizabeth Blackwell and Jane Cooke Wright are reminders of how difficult it was for women and minorities to get access to medical schools and hospitals. Although the medical profession is no longer exclusively male, some other areas of science are still predominantly single-sex clubs, especially the fields of physics and engineering which were 95% male in the United States in 1994.

Given the above scenarios, it is remarkable that some individual women and people from a variety of nations and cultures have recently contributed to science, invention, and mathematics. Among examples in this volume are: Satyendranath Bose, George Washington Carver, Jane Goodall, Grace Murray Hopper, Percy Julian, Garrett Morgan, Lise Meitner, Hideyo Noguchi, Emmy Noether, C. V. Raman, Bertram Fraser-Reid, Jokichi Takamine, Jane Cooke Wright, Chien-Shiung Wu, Rosalyn Yalow, and Hideki Yukawa.

Although invention is the area of human accomplishment seen as most obviously liberating (by improving health, speeding communication, sustaining agriculture, transporting people and a variety of goods and services), its creators are often unknown or unheralded. During most of human history, originators of new devices or processes were not recorded. The first formal patent process in the world was established in England in 1624. The United States adopted that model in 1790 and has since granted millions of patents. The current (1997) rate is more than 124,000 patents per year in the United States.

The definitions, outlines, and comments above are broad strokes in developing an impressionistic picture of science, invention, medicine, and mathematics rather than a literal chronological history. The biographies in Lives and Legacies aim to illustrate the interconnections among creative people in scientific and technical fields and to foster an appreciation for the diversity of ideas, processes, and mechanical devices that have transformed human civilizations, especially in recent times.

Each biography is divided into two sections: the first describing the individual's life and work and the second explaining the individual's legacy. Space limitations did not allow use of many interesting details about the lives of the creative people selected for this volume. We did, however, try to include facts that suggest positive influences and/or specific challenges to individual achievements. In legacy sections, three time frames—contemporary, immediate, and long-term influence—were consistently considered if that information was available. The timelines help put individual lives and contributions into a historical context. Also we hope readers will explore some of the suggestions for further reading about people, ideas, and inventions that interest them.

———•———

If you can now see science, invention, and mathematics as a sort of unending relay race with a general direction but no fixed route, a journey toward understanding that has no final end, a team effort that any lover of learning may play, you may decide to join in the sport. But if you just enjoy finding out how people think and work, especially those who are "first" to do something in their time or those who have changed the world in which you live, then you will appreciate this book too. If you hope to find a role model or two, a trailblazer to follow, you also will probably not be disappointed. Make your own creative selection. Examples of persistence, curiosity, and inspiration are here!

Doris G. Simonis
July 1998

The
Biographies

Abel, Niels Henrik

Advocate of Rigorous
Mathematical Proofs
1802-1829

Life and Work

During his short life, Norwegian mathematician Niels Abel produced an enormous volume of original work in diverse areas of mathematics and helped to raise the standards for mathematical proof.

Abel was born on August 5, 1802, the son of a poor Lutheran minister in Finnoy, Norway. Poverty and famine were common in Norway and times were difficult for the clergyman and his seven children. At age 13 Abel entered the cathedral school in Christiana (now called Oslo). With his math instructor Bernt Michael Holomboe, Abel studied the works of ISAAC NEWTON, LEONHARD EULER, Joseph-Louis Lagrange and CARL FRIEDRICH GAUSS. He read critically and found gaps in their arguments, quickly becoming a first-rate mathematician.

In 1820 Abel's father died. Although Abel was the second son, at age 18 he took on the responsibility of supporting his mother and his six siblings with money that he earned by taking on private students. Holomboe also helped to find and donate funds so that Abel was able to continue his education at the University of Christiana.

About a year later Abel produced a stunning result on quintic (fifth degree) equations of the form $ax^5 + bx^4 + cx^3 + dx^2 + ex + f = 0$. Solutions of second-degree equations were given by quadratic equations (in the form $ax^2 + bx + c = 0$), and solutions to the third- and fourth-degree equations were also known. Mathematicians had searched for more than 200 years for an algebraic solution to quintic equations. Abel proved that such a solution did not exist. He sent the result to Gauss, who dismissed it without even reading it.

In 1825 the government of Norway sponsored Abel for travel and study in France and Germany. In Berlin he began an association with August Leopold Crelle with whom he founded the first journal of mathematical research. The first three issues of Crelle's *Journal of Pure and Applied Mathematics* contained 22 of Abel's papers. In 1826 Abel submitted to the French Academy of Sciences his paper on transcendental functions (all non-algebraic functions such as exponential, logarithmic, trigonometric functions, etc.). The well-established mathematicians AUGUSTIN-LOUIS CAUCHY and Andrien-Marie Legendre were to review the paper but both forgot about it. Abel realized he was wasting his time and returned to Norway.

The rigorous Abel was unhappy that the proofs of correct theorems rested on shaky reasoning. Among other things, Abel felt that "divergent series are the invention of the devil, and it is a shame to base on them any demonstration whatsoever." He created proofs of his own—most notably he provided the first proof for the binomial theorem.

Abel taught part-time at the University of Christiana and continued to publish papers on the theory of equations and elliptic functions (those that are doubly periodic). Abel died of tuberculosis on April 6, 1829, in Froland, Norway, at the age of 26. In response to a diplomatic inquiry, Cauchy found Abel's paper on transcendental functions in 1830 and it was published in 1841, 12 years after Abel's death.

Legacy

Abel introduced rigor into mathematical proofs, suggested creative methodology for investigating mathematical problems, and left a host of new theorems, proofs, and equations for future mathematicians to learn from.

A pioneer of a wide range of mathematical ideas, Abel helped change the mathematical standards of what constitutes a rigorous proof. He put mathematics on a firmer foundation, one that supported the creative mathematical work of future generations. Abel also put the discipline of mathematics into a professional stance supported by its first journal.

One of his intellectual heirs was GEORGE BOOLE, an elementary school teacher who bought math books to read because they were less expensive than any others available to him. When he studied Abel, he discovered that algebra could be a logical delight, not just a symbolic system for real numbers.

Two of Abel's famous discoveries came from inverting a problem that others had proposed. For example, instead of continuing to search for the formula that would give a solution to the quintic, consider the possibility that none exists, and prove that! Now it is standard practice to consider the inverse or the "upside down" model to get a new perspective for creative work.

Modern math texts reflect the range of Abel's original work by calling his result on transcendental functions "Abel's theorem" and attaching his name to Abelian functions, Abelian equations, and Abelian groups. As HENRI POINCARÉ's teacher, Charles Hermite, claimed, Abel laid out 500 years of work for future mathematicians.

Steinberg

WORLD EVENTS		ABEL'S LIFE
French Revolution	1789	
	1802	Niels Henrik Abel is born
Napoleonic Wars in Europe	1803-15	
	1817	Abel begins work with Bernt Holomboe
	1821	Abel proves that quintic equations have no algebraic solution
	1826	Abel publishes 22 papers in *Journal of Pure and Applied Mathematics*
		Academy of Sciences receives paper on transcendental functions
	1829	Abel dies
	1841	Abel's paper on transcendental functions is published
United States Civil War	1861-65	

For Further Reading:

Bell, E. T. *Men of Mathematics.* New York: Simon & Schuster, 1937.

Calinger, Ronald, ed. *Classics of Mathematics.* Englewood Cliffs, N.J.: Prentice-Hall, 1995.

Kline, Morris, *Mathematical Thought from Ancient to Modern Times.* New York: Oxford University Press, 1972.

Ore, Oystein. *Niels Henrik Abel: Mathematician Extraordinary.* Minneapolis: University of Minnesota Press, 1957.

Agassiz, Louis;
Agassiz, Elizabeth

Originator of Ice Age Theory;
Science Writer
1807-1873; 1822-1907

Life and Work

Louis Agassiz was the first to propose the concept of a past ice age. Elizabeth Cary Agassiz recorded and publicized his work and helped to establish Radcliffe College.

Louis Agassiz was born on May 28, 1807, in Motier, Switzerland. He earned a medical degree from the University of Munich and a Ph.D. in zoology from the University of Erlangen. From 1829 to 1842 he focused on the study of fossil fishes and published studies of the natural history of Brazilian and European fishes.

In 1836 Agassiz turned his attention to the geology of the Swiss Alps, his boyhood home. His interest was piqued by his own investigations of fossils in the region and by recent writings suggesting that the existing glaciers had once been more extensive. He began to study the movement and effects of glaciers. Leading geologists thought that glaciers were static, but Agassiz noted evidence suggesting that glaciers moved, causing striations, or grooves, in rock and depositing massive boulders and debris. From his observations he concluded that in the past much of northern Europe had been covered in ice; he published this theory in 1840.

Agassiz was a creationist and never accepted CHARLES DARWIN's theory of the evolution of species by natural selection (published in 1859). Believing, however, that physical events can result in the disappearance of species, he suggested that an ice age had caused mass extinction.

Obtaining a professorship at Harvard University in 1847, Agassiz moved to the United States and married Elizabeth Cary (born on December 5, 1822, in Boston, Massachusetts) in 1850. She had little science background but learned the principles of natural history from her husband. They traveled together, Elizabeth keeping a detailed account of her husband's observations and research; her notes led to the publication of *A Journey to Brazil* and several shorter articles.

Devoted to innovative teaching methods, Agassiz discouraged reliance on books and encouraged direct contact with nature.

He died on December 14, 1873, in Cambridge, Massachusetts.

After her husband's death, Elizabeth Agassiz focused on women's education. She helped establish Radcliffe College, the sister college of Harvard that gave women access to the university's instructors and educational resources. She served as Radcliffe's president from 1894 to 1903; she died on June 27, 1907, in Arlington Heights, Massachusetts.

Legacy

Louis Agassiz's work formed the basis of research concerning the ice ages and related investigations of climate change. The legacy of Elizabeth Cary Agassiz rests in the publication of accounts that popularized her husband's work and in her influence on higher education for women.

Louis Agassiz's ice age theory was not readily accepted by his contemporaries. A majority of geologists believed that Earth had been undergoing a gradual cooling since its beginning, a theory supported by various pieces of fossil, climatic, and geophysical evidence. A period colder than the present did not fit within this view. Another leading theory held that massive tidal waves, caused by catastrophic earth movements, had swept across Europe in the recent geological past, strewing boulders and debris across the continent.

Geologist Jean de Charpentier developed the concept of an ice age at the same time as Agassiz, but he published after Agassiz and received little recognition. He suggested that a moderate layer of ice had built up and covered Europe after a long period of cold weather, which is closer to the modern view than the sudden, catastrophic climate change envisioned by Agassiz.

The ice age theory began to be accepted in the 1850s. CHARLES LYELL had popularized the idea that gradual physical processes shape Earth's surface features over long periods. In the 1850s John Tyndall and Andrew Ramsay, both of Britain, gathered evidence of the power of glaciers to effect gradual geological change.

Scientists began to search for explanations of climate change extreme enough to cause extensive glaciation. In 1875 James Croll proposed a complex theory based on principles of meteorology and oceanography and involving interglacial periods of moderate climate, for which physical evidence was gathering. In the twentieth century, astronomical data, ocean-bed cores, and radiocarbon-dating techniques have displaced Croll's theory, and now various possible climate-change explanations are debated. It is believed that several ice ages have occurred in the recent as well as distant geological past.

Elizabeth Agassiz preserved Louis's work in her notes and books and, in effect, is responsible for the dissemination and popularization of her husband's ideas. The founding of Radcliffe College at Harvard University, made possible by Elizabeth Agassiz, represented a major leap forward for women's education in the United States. As a part of Harvard, Radcliffe offered women educational resources and opportunities formerly closed to them. It also served as a model as other universities began to open their doors to women.

Schuyler

World Events	Agassizes' Lives
Napoleonic Wars in 1803-15 Europe	
	1807 Louis Agassiz is born
	1822 Elizabeth Cary is born
	1840 Agassiz publishes ice age theory
	1847 Agassiz begins teaching at Harvard
	1850 Agassiz and Cary are married
Germany is united 1871	
	1873 Louis Agassiz dies
	1894 Elizabeth Agassiz helps found Radcliffe College
Spanish-American War 1898	
	1907 Elizabeth Agassiz dies
World War I 1914-18	

For Further Reading:

Agassiz, Elizabeth Cary. *Louis Agassiz, His Life and Correspondence.* Boston: Houghton Mifflin, 1893.

Erickson, Jon. *Ice Ages: Past and Future.* Blue Ridge Summit, Penn.: TAB Books, 1990.

Winsor, Mary P. *Reading the Shape of Nature: Comparative Zoology at the Agassiz Museum.* Chicago: University of Chicago Press, 1991.

Agnesi, Maria

First Important Woman
Mathematician of Modern Times
1718-1799

Life and Work

Maria Agnesi was the first woman of the western world to be recognized as an important mathematician. She used her powerful skills in mathematics and languages to create a comprehensive math text that brought together many great works separated by time and language.

Born in Milan, Italy, in 1718, Agnesi was the eldest of 21 children. Her father held a position as mathematics professor at the University of Bologna. Her early life was steeped in an academic environment. A child prodigy, Agnesi could speak many languages at an early age. Her father encouraged her to participate at the social gatherings of intellectuals held at their home. By age nine, Agnesi had written in Latin a defense of women's education. As she grew older, Agnesi tutored her younger brothers.

In 1738 Agnesi published a collection of essays on natural science in which she also continued to champion education for women. Over the following 10 years, Agnesi began work on a math compendium, which had been started as a text for her younger brothers. Agnesi eventually published it as the four-part

Analytical Institutions in 1748. It included discussions of the work of scientists and mathematicians such as Isaac Newton, René Descartes, and Pierre de Fermat, as well as a comprehensive range of subjects in mathematics such as algebra, geometry, and calculus.

One of the mathematical elements described by Agnesi was a special bell-shaped curve whose equation is $x^2y = a^2 (a-y)$. Her geometric description of the curve, referred to as *versiera* (versed sine curve) of Agnesi (which over time has been incorrectly translated as the "witch of Agnesi"), continues to be of interest to mathematicians.

Her work drew critical acclaim from many in Europe. Pope Benedict XIV thought highly of her accomplishments, and in 1750 he offered her a position as honorary lecturer at the University of Bologna. Two years later, in 1752, she decided to turn her attention to charitable work. Agnesi eventually became the director at a home for the aged in 1771. She remained in this position until her death on January 9, 1799.

Legacy

Maria Agnesi demonstrated to western Europe that a woman could be a successful mathematician. Her compilation of mathematical knowledge reflected her grasp of complex mathematical concepts and provided a source of information for countless others.

Her text, *Analytical Institutions,* was translated from the original Italian into French and English. Her writing made many mathematical subjects—algebra, analytic geometry, the calculus, and differential equations—accessible to a wider audience. By compiling, translating, and organizing widely scattered papers, Agnesi enabled students to spend their time learning about mathematics rather than searching for sources. The work would become an invaluable resource for mathematicians and students of mathematics throughout Europe.

During her lifetime Agnesi advanced the cause of women's education. Her arguments on the matter were widely read and they gained the attention of influential people such as the Empress of Germany and Pope Benedict XIV.

Although Agnesi focused on mathematics for only a third of her life, leaving at the height of her career, she accomplished a great

deal of intellectual work in a relatively short time. Her accomplishments left a model for women mathematicians of succeeding generations to follow.

Wilson

World Events	Agnesi's Life
Peace of Utrecht 1713-15 settles War of Spanish Succession	
	1718 Maria Agnesi born
	1727 Agnesi, at age nine, writes essay on women's education in Latin
	1738 Agnesi's essays on natural science are published
	1748 *Analytical Institutions* is published
	1750 Agnesi named honorary lecturer at University of Bologna
	1771 Agnesi appointed director at home for aged
United States 1776 independence	
French Revolution 1789	
	1799 Agnesi dies
Napoleonic Wars 1803-15 in Europe	

For Further Reading:

Osen, Lynn. *Women in Mathematics.* Cambridge, Mass.: MIT Press, 1974.

Perl, Teri. *Math Equals.* Reading, Mass.: Addison-Wesley, 1978.

Agricola, Georgius

Father of Mineralogy

1494-1555

Life and Work

Georgius Agricola is considered the father of mineralogy. His investigations into the mineral composition of Earth and his writings on the subject began the process of codification of minerals and metallurgy.

Agricola was born Georg Bauer on March 24, 1494, in Clauchau, Germany. As was typical for scholars at that time, he later latinized his German name, which means farmer, to Agricola.

Little is known of his early life. He attended the University of Leipzig from 1514 to 1518, where he focused on the classics and philosophy. In 1523 he went to Italy, where he studied medicine and natural sciences at the universities in Bologna and Padua. Agricola moved to Venice where he completed his medical training and helped to edit a popular edition of writings by the ancient Roman physician GALEN.

For much of his remaining life, Agricola practiced medicine; mineralogy would become an avocation that occupied much of his leisure time. In 1526 he returned to Germany and took up a medical practice in the town of Joachimsthal. The town was situated in a mining region, which offered Agricola opportunities to explore the mineral resources and smelting practices in hopes of finding new drug treatments. This work led Agricola to a rigorous study of mineralogy, on which he wrote prolifically. Over a period of 20 years, Agricola wrote a comprehensive survey of mining and metallurgy, *On Metallurgy* (published posthumously in 1556); it covered historical, scientific, and technical aspects of the subject, and was filled with attractive and instructive woodcuts. In the book Agricola also used his medical expertise to describe miners' diseases and prescribe the use of protective gear for them.

Agricola moved to Chemnitz in 1533, set up his medical practice, and stayed there for the rest of his life. While in Chemnitz, he participated in politics under the sponsorship of Duke Maurice of Saxony. The Duke appointed him mayor in 1546 and his diplomatic representative to Charles V, head of the Holy Roman Empire, which ruled much of Europe during Agricola's life.

His most important book, *On the Nature of Fossils* (the word used at that time for any material taken from the ground), published in 1546, presents a classification system in which he categorized minerals according to "color, taste, odor, place of origin, natural strength and weakness, shape, form, and size." He also differentiated simple elements from compound substances, a notable conclusion given the rudimentary state of chemistry at the time.

Agricola developed his classification system through direct observation and analysis. This methodology contrasted with the mode of scientific inquiry typical at the time, which, inherited from classical philosophers, was based more on informed speculation.

Agricola died on November 21, 1555, in Chemnitz, Germany.

Legacy

Agricola's investigations into the mineral composition of Earth formed the basis for the development of mineralogy. His investigative method avoided the mysticism and alchemy popular at the time; his reliance on direct evidence was a precursor to the scientific methodology that would later be advocated by philosopher Francis Bacon in the early seventeenth century.

Agricola's codification of minerals offered a body of knowledge for future mineralogists and geologists to build on. His *On the Nature of Fossils* was a standard reference on rocks and minerals used for over two centuries. Not until the end of the eighteenth century—following the development of chemical analysis, which lent its tools to mineralogy—would strides be made beyond Agricola's classification system. During the late 1700s Abraham Werner of Germany, René-Just Haüy of France, and William Babington of England each prepared classification plans based on the chemical composition of minerals. Such chemical systems of categorization were later refined and contemporary mineralogists have a wide variety of methods at their disposal including isotopic analysis and X-ray diffraction.

Agricola's *On Metallurgy* was also used as a standard reference on mining practices, problems, and tools for the next two centuries. His advocacy of protective gear helped to improve miners' health in his district. Other problems highlighted by Agricola were solved by future inventors. For instance, the problem of flooding in mines was first addressed in 1698 by Thomas Savery who invented a steam-powered pump to help keep mine shafts dry.

Agricola was one of the first physicians to recognize occupational hazards and diseases, in particular those suffered by miners, and to draw attention to the human costs of economic activity. Today occupational medicine is a respected specialty and mining's effects on people and the environment are no longer ignored.

Tomaselli

WORLD EVENTS		AGRICOLA'S LIFE
Columbus discovers Americas	1492	
	1494	Georgius Agricola is born
	1514-18	Agricola attends University of Leipzig
Reformation begins	1517	
	1523	Agricola travels to Italy; begins studies in Padua and Bologna
	1526	Agricola returns to Germany, begins medical practice in Joachimsthal
	1533	Agricola moves to Chemnitz
	1546	*On the Nature of Fossils* published
	1555	Agricola dies
	1556	*On Metallurgy* is published
Thirty Years' War in Europe	1618-48	

For Further Reading:

Faul, Henry. *It Began With a Stone: A History of Geology from the Stone Age to the Age of Plate Tectonics.* New York: John Wiley, 1983.

Oldroyd, David Roger. *Thinking About the Earth: A History of Ideas in Geology.* Cambridge, Mass.: Harvard University Press, 1996.

Turner, Roland, and Steven Goulden. *Great Engineers and Pioneers in Technology.* New York: St. Martin's Press, 1981.

Aiken, Howard Hathaway

Pioneer in Early Computer Engineering
1900-1973

Life and Work

Howard Aiken was a pioneer in the field of computer science. He designed the Mark I computer, one of the first full-scale, programmable, digital computers in the world.

Aiken was born on March 8, 1900, in Hoboken, New Jersey. He graduated from the University of Wisconsin in 1923, and then worked as chief engineer at Madison Gas for 12 years. Aiken resumed his education at Harvard University in 1935. While there, he investigated the work of CHARLES BABBAGE, who developed an early computer prototype, and circulated a memo in 1937 proposing an "Automatic Calculating Machine."

After receiving his doctorate from Harvard in 1939, Aiken collaborated for five years with a group of IBM engineers on the machine he envisioned at Harvard. It was called the IBM Automatic Sequence Controlled Calculator (ASCC), later referred to as the Mark I. It was one of the first full-scale digital computers that could be controlled by programmed instructions input into the machine. (Konrad Zuse completed a model of the first programmable computer called Z-1 in 1938 in Germany, about the time Aiken was beginning work on the Mark I.) The Mark I had registers for storing 70 23-digit, signed numbers, 60 constants, and could add, subtract, multiply, and divide as well as calculate logarithms, exponentials, sines, and cosines. The programmed instructions were fed to the machine through a hole-punched paper tape.

The size of Mark I was astounding by today's standards: it was 50 feet in length and weighed around five tons. It would have been considered fast if ENIAC, a computer built at the University of Pennsylvania, had not been released a year later. ENIAC was completely electronic and 500 times faster than the electromechanical Mark I, even though the Mark I was more precise.

Following construction of the Mark I, Aiken founded the Harvard Computation Laboratory. There he built three other versions for the Navy and Air Force, including the Mark III in 1950, which was his first electronic machine. Both the Air Force and Navy honored Aiken for his distinguished service. Under Aiken, Harvard became the home for some of the first academic courses in circuit design and the design of components for electronic digital computers.

In 1961 Aiken retired from Harvard. Even in retirement he continued to work. He helped the University of Miami create a computer science department, and he founded Aiken Industries. He died on March 14, 1973, in St. Louis, Missouri.

Legacy

Aiken's Mark I moved computer design in the United States a huge step forward during a period when each new design added a world of further computer engineering possibilities.

The early history of computer design illustrates a common maxim that, when working in a new field, initial efforts and innovations build upon trial and error. Aiken's machines were not theoretical models but working machines. He turned possibilities into reality that others would appreciate and adapt.

As the first fully operational computer governed by a set of programmed instructions in the United States, Aiken's Mark I set the stage for other innovations in post-war Allied countries (those opposing Germany and its allies in World War II). Indeed, the University of Pennsylvania's ENIAC, finished just one year after Mark I, used architecture similar to the Mark I. In May of 1949 Maurice Wilkes at Cambridge University finished the Electronic Delay Storage Automatic Computer (EDSAC). Delivering on the promise of a 1948 prototype built by Max Newman and Freddie C. Williams, it was the first fully operational computer controlled by an internally stored program.

Aiken nurtured the computer engineers with whom he worked; these engineers would make their own independent contributions to the field. In 1952 GRACE MURRAY HOPPER, part of Aiken's Mark I team, was the first person to develop and implement a compiler, a program that translates a high-level programming language into a code understood at the machine level. Aiken also helped to establish computer science as an academic discipline, benefiting not only his students at Harvard, but computer scientists at other universities as well.

Steinberg

WORLD EVENTS	AIKEN'S LIFE
Spanish-American War 1898	
	1900 Howard Aiken is born
World War I 1914-18	
	1923 Aiken graduates from University of Wisconsin
	1938 German Konrad Zuse completes Z-1, model of first programmable computer
World War II 1939-45	
	1944 Mark I, first full-scale programmable computer in U.S., completed
	1949 Maurice Wilkes builds first computer with internally stored program
	1950 Mark III, Aiken's first electronic computer, is completed
Korean War 1950-53	
	1973 Aiken dies

For Further Reading:

Goldstine, Herman. *The Computer from Pascal to von Neumann.* Princeton, N.J.: Princeton University Press, 1972.

Palfreman, Jon, and Doron Swade. *The Dream Machine: Exploring the Computer Age.* London: BBC Books, 1991.

Richie, David. *The Computer Pioneers.* New York: Simon & Schuster, 1986.

Slater, Robert. *Portraits in Silicon.* Cambridge, Mass.: MIT Press, 1987.

Ampère, André

Formulator of Laws of
Electromagnetism
1775-1836

Life and Work

André Ampère formulated the fundamental mathematical laws of electromagnetism.

Ampère was born on January 22, 1775, near Lyons, France. He was a prodigy, achieving a high level of mathematical proficiency by age 12. The French Revolution disrupted his teenage years, and in 1793 his father, a wealthy city official, was guillotined by the French Republican Army.

In 1802 Ampère was appointed to two science professorships, in Bourg and Lyons, during which time his research revolved around the mathematics of probability. He became a mathematics lecturer at the École Polytechnique, Paris, in 1805, and four years later was promoted to professor. In 1808 Napoleon recognized Ampère's talent and appointed him Inspector General of the country's recently overhauled university system.

From 1805 to 1820 Ampère studied psychology, philosophy, physics, and chemistry,

and taught in various disciplines in Paris at the École Polytechnique, the University of Paris, and the Collège de France. His investigations of individual elements led him to suggest a classification system based on the chemical properties of elements, a system similar to the periodic table that would be devised independently by Russian chemist DMITRY MENDELEYEV in the late 1860s. In 1814 Ampère arrived at Avogadro's law (independently discovered by AMEDEO AVOGADRO in 1811), which states that equal volumes of different gases at a given temperature contain equal numbers of molecules.

In 1820 Ampère's attention turned to electricity. That year, the Danish physicist Hans Christian Oersted proved that an electric current generates a magnetic field. Stimulated to conduct his own experiments, Ampère had remarkable results within months. He showed that currents flowing in the same direction in two parallel wires cause the wires to attract each other, while currents flowing in opposite directions create a repulsive force.

In 1827 Ampère published his comprehensive theory of electricity and magnetism, ideas that preceded the discovery of the electrical nature of atoms by 70 years. The treatise included what is now known as Ampère's law, an equation relating the magnetic force between two parallel currents to the product of their currents and to the distance between the conductors. The mathematical relationship between electricity and magnetism was thus established. Ampère called this new field of physics electrodynamics; it is now generally referred to as electromagnetism.

Ampère spent his later years theorizing about the nature of electromagnetism. He died on June 10, 1836, in Marseilles, France.

Legacy

Ampère's contributions to the physics of electricity and magnetism were crucial to our understanding of electricity and its innumerable applications in modern society. The use of his name as the unit of electrical current, the "ampere," honors his legacy.

Oersted's discovery and Ampère's experiments of 1820 inspired a cascade of research related to electromagnetism. That year, French physicist François Arago demonstrated that electric current flowing through copper wire enables the wire to attract iron filings as readily as does a steel magnet. Soon Johann Schweigger, a German physicist, constructed the first galvanometer, a device that uses magnetism to measure the strength of an electric current. In 1823, experimenting with solenoids (coiled wires that act like bar magnets when current flows through them), English physicist William Sturgeon invented the electromagnet by inserting an insulated iron bar into a horseshoe-shaped wire coil; when an electric current flowed through the wire, the device could lift 20 times its own weight. The electromagnet was improved by American physicist Joseph Henry in 1831. Henry coiled insulated wire around an iron bar, enabling more turns of wire to be employed without causing a short circuit, thus allowing a stronger magnetic field to be generated. Henry's device could lift a ton of iron.

Progress continued with English physicist MICHAEL FARADAY's 1831 discovery that placing a changing magnetic field near a wire creates an electric current. This finding had a broad range of effects—it led to the development of the electric motor and generator, to the discovery that light is a type of electromagnetic wave, to the invention of radio communication, and, indeed, to the construction of all electronic devices.

Schuyler

WORLD EVENTS	AMPÈRE'S LIFE
	1775 André Ampère is born
U. S. independence 1776	
	1787 Ampère masters advanced mathematics
French Revolution 1789	
	1793 Ampère's father is guillotined
	1802 Ampère secures first academic post
Napoleonic Wars 1803-15	
	1805 Ampère appointed to mathematics professorship at École Polytechnique
	1808 Napoleon appoints Ampère Inspector General of universities
	1820 Ampère founds the study of electromagnetism
	1827 Ampère introduces Ampère's law
	1836 Ampère dies
Germany is united 1871	

For Further Reading:
Hofmann, James R. *André-Marie Ampère.* New York: Cambridge University Press, 1996.
Meyer, Herbert W. *A History of Electricity and Magnetism.* Cambridge, Mass.: MIT Press, 1971.
Nye, Mary Jo. *Before Big Science: The Pursuit of Modern Chemistry and Physics, 1800-1940.* London: Prentice-Hall International, 1996.
Purrington, Robert D. *Physics in the 19th Century.* New Brunswick, N.J.: Rutgers University Press, 1997.

Appert, Nicolas-François

Developer of Canned Food Process
c.1750-1841

Life and Work

The first to preserve food by sealing and heating jars, Nicolas-François Appert initiated the practice of canning.

About 1750 Appert was born to an innkeeper in Châlons-sur-Marnes, France. Little is known of his childhood except that he gained experience with brewing, pickling, cooking, and baking. He later became a chef and confectioner.

In the late eighteenth century, salting, drying, and smoking were the primary ways of preventing food from rotting and souring, and those practices limited the types of food that could be transported and stored. In 1795, motivated by the desire to offer more diverse meals to its isolated troops, the French government offered a prize for the best method of preserving food for long periods.

Appert heard of the contest and began investigations aimed at identifying an efficient preservation method for a more diverse range of food. In 1765 Italian biologist Lazzaro Spallanzani had shown that lengthy boiling and airtight packaging extends the shelf life of meat broth. Appert drew upon this information, packing food into glass jars capped with airtight lids of cork and wax. He heated the jars in an autoclave, which reaches higher temperatures than boiling water because it heats with pressurized steam (like a modern pressure cooker). Through numerous trials, he refined this process, successfully preserving meats, vegetables, fruits, soups, dairy products, and jams.

In 1810 Appert won the government's 12,000-franc prize and published his findings in *The Art of Preserving All Kinds of Animal and Vegetable Substances for Several Years.* He invested the money in a cannery, the House of Appert, at Massy, France (near Paris). It operated from 1812 to 1933.

Although his invention was highly praised, Appert died in poverty at Massy on June 3, 1841.

Legacy

Appert's achievement, extending the shelf life of prepared food, sparked a culinary revolution whose influence on society stretched far beyond the kitchen.

Canning altered what people cooked and how they thought about food. For the first time, a wide variety of non-local items was readily available, and it was possible to store food for long periods of time without spoilage. Canning made many fruits and vegetables obtainable throughout the year. An international food trade was launched—when foods could endure the long trip between cities and even continents, the variety of choices available in any given country quickly multiplied.

In 1810 French inventor Pierre Durand introduced metal food-packaging receptacles consisting of an alloy of tin and lead. These cans gradually replaced glass jars as the dominant canning medium. Canning factories sprung up around Europe, initiating the modern food industry, which continues as a major economic force in the modern market.

Early metal cans resulted in numerous lethal cases of lead poisoning, but the problem was quickly corrected. Some modern cans still contain small amounts of lead; it is debated whether the amount is harmful.

The work of LOUIS PASTEUR in the 1860s demonstrated the mechanism behind Appert's food-preservation method: the heat kills microorganisms that cause spoilage, and the seal prevents other microorganisms from invading. Pasteur's experiments dislodged the theory of spontaneous generation of life by showing that fermentation and decay are caused by airborne organisms. His results led him to reintroduce and popularize the idea that many diseases are initiated by pathogenic microorganisms, thus launching a revolution in medicine.

Canned food (along with more advanced food-preservation techniques such as chemical additives and freeze-drying) facilitated continued exploration and colonization of distant lands and eventually helped to make space exploration possible. Canned food also shaped the face of warfare by providing a constant, reliable food supply to military troops in combat for prolonged periods of time.

Schuyler

WORLD EVENTS		APPERT'S LIFE
	c.1750	Nicolas-François Appert is born
United States independence	1776	
French Revolution	1789	
	1795	French government sponsors contest for best food preservation method; Appert begins experiments
Napoleonic Wars in Europe	1803-15	
	1810	Appert perfects "canning" process and wins government prize
	1810	Pierre Durand introduces metal cans
	1812	Appert opens canning factory
	1841	Appert dies
Germany is united	1871	
Spanish-American War	1898	
World War I	1914-18	
	1933	Appert's canning factory closes

For Further Reading:

Thorne, Stuart. *The History of Food Preservation.* Totowa, N.J.: Barnes & Noble Books, 1986.

Torgerson, Nancy. *Food Preservation: Before the Mason Jar.* Decatur, Ill.: Glimpse of the Past, 1995.

Archimedes

Inventor; Early Pioneer in Plane
and Solid Geometry
c. 290-c. 212 B.C.E.

Life and Work

Archimedes, one of the most famous mathematicians and inventors in history, devised about 40 labor-saving devices and made important contributions to the geometry of plane and solid figures.

Archimedes was born sometime around 290 B.C.E. in Syracuse, a Greek colony in the Italian province of Sicily. Although Syracuse remained his home, Archimedes spent a considerable amount of time in Alexandria, Egypt, where he studied and worked with other mathematicians and scientists, including ERATOSTHENES and Conon, two of the era's most well-known scholars.

The stories about his inventions have been embellished, yet they portray the ingenuity of this man. One such story tells how King Heiron II of Syracuse suspected that a man hired to make a gold crown had replaced a portion of the gold with silver. Heiron could not prove his suspicions, so he turned the problem over to Archimedes. While bathing one day, Archimedes noticed both how the water his body displaced flowed over the edge and how much lighter he felt in the water. This led him to the insight that the buoyancy of different materials in water might indicate their different

masses. He jumped out of the bath and ran home naked, shouting "Eureka! Eureka!" (I have found it). Thus legend tells us how Archimedes discovered the first law of hydrostatics (Archimedes' Principle), which states that a body wholly or partially immersed in a fluid is pushed upward by a force equal to the weight of the fluid it displaces.

Archimedes is also credited with having invented a device, known as Archimedes' screw, that is still used to raise water for irrigation. Archimedes invented the compound pulley and used it to move a fully loaded ship out to sea with one hand. He also developed a mechanical device that demonstrated the movement of the bodies in the solar system.

One of Archimedes' most important mathematical contributions survives today in *On the Sphere and Cylinder* in which Archimedes explains his discovery of the geometric relationship between a sphere and cylinder. He concludes that the surface area of a sphere equals four times the surface area of its greatest circle and a sphere's volume is two-thirds the volume of the cylinder that inscribes the sphere. Other works that survive include *Measurement of the Circle,* in which Archimedes resolves the value of *pi* to between 3 ¹⁄₇ and 3 ¹⁰⁄₇₁.

Over the course of many years, Archimedes devised weapons of war for Heiron. In 212 B.C.E., Archimedes' catapults hurled blocks of stones and logs against Marcellus' ships during the Roman siege of Syracuse in the Second Punic War. The inventor's cranes were also used to lift enemy ships and let them fall back into the water again, breaking up on impact.

Despite the success of Archimedes' weapons of war, Syracuse was captured by the Roman army in 212 B.C.E.; he died during that capture.

Legacy

Archimedes' practical inventions and contributions to mathematics have had a profound impact on mathematics, science, and technological innovation throughout the centuries.

Archimedes' inventions and mechanical innovations have counterparts throughout history

and some are still used today. The Romans used the catapult extensively and added innovations of their own. Modern screw pumps of similar design are used in sewage treatment plants and in irrigation systems worldwide, and Archimedes' pulley systems can be seen in numerous contemporary applications.

While largely ignored for the rest of antiquity, Archimedes' mathematical work resurfaced to have a lasting impact on mathematicians and scientists in later centuries. Arabic mathematicians of the eighth and ninth centuries were the first to uncover his work, and they produced innovations on Archimedes' ideas in the area of volumes of solids.

Many of his works were translated into Latin in 1558 and again in 1615. The Latin texts made ancient mathematical insights accessible to many mathematicians and scientists who, in turn, profoundly influenced the development of mathematics and science in the following centuries. These scholars included JOHANNES KEPLER (1571-1630), the German astronomer and natural philosopher; GALILEO (1564-1642), the Italian physicist and astronomer; and RENÉ DESCARTES (1596-1650), the French philosopher, scientist, and mathematician. Archimedes' studies of the areas and volumes of curved solid figures and the area of plane figures led to many discoveries in modern science, such as integral calculus.

Weaver

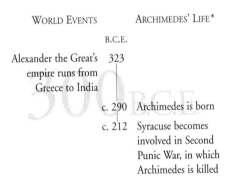

WORLD EVENTS	ARCHIMEDES' LIFE*
	B.C.E.
Alexander the Great's empire runs from Greece to India	323
	c. 290 Archimedes is born
	c. 212 Syracuse becomes involved in Second Punic War, in which Archimedes is killed

* Scholars cannot date the specific events in Archimedes' life with accuracy.

For Further Reading:
Archimedes: New Studies in the History and Philosophy of Science and Technology. Boston: Kluwer Academic, 1996.
Dijksterjuis, E. J. *Archimedes.* New York: The Humanities Press, 1957.
Ronan, C. *Science: Its History and Development Among the World's Cultures.* New York: Facts On File, 1982.

Arrhenius, Svante

Originator of Electrolytic
Dissociation Theory
1859-1927

Life and Work

Svante Arrhenius conducted pioneering work on the conductivity of solutions and introduced the electrolytic dissociation theory.

Arrhenius was born on February 19, 1859, in Vik, Sweden. After excelling in primary and secondary school, he entered the University of Uppsala in 1876 and concentrated on chemistry, physics, and mathematics. He began graduate studies at Uppsala, but transferred in 1881 to the Physical Institute of the Swedish Academy of Sciences, Stockholm, to finish his doctoral research.

Focusing his studies on the electrical conductivity of solutions, Arrhenius began to construct what would become the electrolytic dissociation theory. His theory attempted to explain observed changes in osmotic pressure, vapor pressure, and boiling and freezing points of solutions at various concentrations. He started from the premise that a solution of sodium chloride (salt) dissolved in water conducts an electrical current, but neither pure water nor dry salt does. Arrhenius postulated that when molecules of salt dissolve in water, they break apart, or dissociate, into smaller charged particles (now known as ions).

In 1884 Arrhenius submitted his thesis, which contained the basic principles of electrolytic dissociation theory. His doctoral committee was skeptical of his ideas, questioning in particular that electrically charged particles could exist in water. He was awarded a Ph.D. but given the lowest passing grade.

Arrhenius had his thesis printed and sent to numerous scientists in Germany and the Netherlands who were working on similar problems in the realm of physical chemistry. This group welcomed the electrolytic dissociation theory, and he was offered a job in Germany. Preferring to stay in Sweden, he accepted a lectureship at the University of Uppsala. He stayed only a brief time, though, as he was awarded a traveling scholarship in 1886 to visit the major scientific laboratories throughout continental Europe.

Arrhenius was honored with the Nobel Prize for Chemistry in 1903 for his work on the dissociation of electrolytes. Two years later, he became director of the physical chemistry division at the Swedish Academy of Sciences.

In his later years, Arrhenius's interests broadened. He applied the principles of physical chemistry to biological phenomena, and he theorized that life forms might be transmitted from planet to planet by tiny spores. He also was among the first to postulate the greenhouse effect, which states that carbon dioxide in the atmosphere traps heat radiated from Earth's surface. He hypothesized that reduction in the percentage of carbon dioxide in air caused the ice ages. Arrhenius died on October 2, 1927, in Stockholm, Sweden.

Legacy

The electrolytic dissociation theory was Arrhenius's major scientific contribution. By 1900 the theory had become widely accepted. Further research expanded the theory to include the following principles of electrolytic solutions: dissociation of the solute into free ions takes place even when no current is passed through the solution; conductivity depends on the number and migratory speed of the ions present; in weak electrolytes, degree of dissociation is directly proportional to degree of dilution; and in strong electrolytes, the ions impede each other's migration through the solution, preventing complete dissociation.

The understanding of electrolytic solutions that grew out of Arrhenius's theory paved the way for research into the properties of conductivity and the behavior of ions in solution. This research encompassed investigations of chemical equilibrium (the state of a chemical reaction where the net change in amounts of reactants stabilizes), semi-permeable membranes, osmotic pressure, corrosion, pH measurements, and ion concentration of fluids. Studies of the last were applied to medical problems such as determining the concentration of particular ions in blood.

Several technological developments emerged from the theory of electrolytic dissociation as well. Acid-base batteries were improved and fuel cells were devised. Protection against corrosion became possible with the development of thin-layer electroplating on exposed metal surfaces. In addition, numerous industrial reactions were made possible or more efficient by the increased understanding of chemical kinetics (study of chemical reaction rates).

Schuyler

World Events		Arrhenius's Life
	1859	Svante Arrhenius is born
Germany is united	1871	
	1876	Arrhenius enters University of Uppsala
	1881	Arrhenius transfers to Physical Institute of Swedish Academy of Sciences
	1884	Arrhenius presents electrolytic dissociation theory
	1886	Arrhenius is given traveling scholarship
Spanish-American War	1898	
	1903	Arrhenius is awarded Nobel Prize for Chemistry
World War I 1914-18		
	1927	Arrhenius dies
World War II 1939-45		

For Further Reading:

Farber, Eduard, ed. *Great Chemists*. New York: Interscience, 1961.

James, Laylin K., ed. *Nobel Laureates in Chemistry*. Washington, D.C.: American Chemical Society and the Chemical Heritage Foundation, 1993.

Avogadro, Amedeo

Originator of Concept of Molecules;
Developer of Avogadro's Law
1776-1856

WORLD EVENTS		AVOGADRO'S LIFE
United States independence	1776	Amedeo Avogadro born
French Revolution	1789	
	1800	Avogadro gives up law for science
Napoleonic Wars in Europe	1803-15	
	1811	Avogadro publishes hypothesis of equivalent volumes of gases, later called Avogadro's Law, in *Journal de Physique*
	1820	Avogadro begins teaching at college in Turin
	1856	Avogadro dies
	1858	Stanislao Cannizzaro launches efforts to re-examine and prove Avogadro's Law
United States Civil War	1861-65	

Life and Work

Amedeo Avogadro developed the hypothesis that, under certain conditions, equal volumes of gases contain equal numbers of molecules. This came to be known as Avogadro's Law.

Avogadro was born in Turin, Italy, on August 9, 1776. His father pushed him into law, and the study of law, and Avogadro earned a doctorate by the time he was 20. However, Avogadro found the legal profession profoundly boring; it was chemistry, physics, and mathematics that truly captured his interest. He switched to a career in science in 1800 and had become a physics professor at a small college in Turin by 1820.

Avogadro's work focused on understanding the building blocks of elements and compounds. There was great confusion at the time about the nature of atoms and how they formed different substances. Avogadro proposed the concept (and coined the term) molecule, to indicate the smallest unit of a compound as opposed to an atom, the smallest unit of an element.

The distinction between atoms and molecules helped to clarify the function of atoms. Building on the work of JOHN DALTON and JOSEPH GAY-LUSSAC, Avogadro hypothesized that atoms of the same element could bond together. (This had been deemed impossible because of the similar polarities of the two atoms.) Avogadro correctly stated that certain gases could form diatomic molecules, molecules consisting of two atoms of the same element.

Avogadro's work on molecular behavior in gases formed the basis for his most important contribution: he deduced that equivalent volumes of gases at a specified temperature and pressure would contain an equivalent number of molecules, a hypothesis that later became known as Avogadro's Law. While a brilliant theoretician, Avogadro was weak in laboratory research. Put off by his radical ideas and poor physical evidence, few were willing to listen to his deductions. He published his theory in 1811 in the *Journal de Physique,* but it was largely ignored and lay dormant for years.

Avogadro continued to teach and conduct research in Turin until his death on July 9, 1856.

Legacy

Avogadro made discoveries vital to the understanding of matter, particularly gases. His introduction of the idea of the molecule and his hypothesis about equal volumes of gases containing equal numbers of molecules provided clues about the nature of elements and compounds for later generations of researchers.

While Avogadro was unable to prove his theory concerning equivalent volumes of gases, researchers some 50 years later (in 1858), led by Stanislao Cannizzaro, began to confirm and quantify his ideas—in particular, the specific number of molecules in equivalent volumes of gases. This specific number became known as Avogadro's Number, or Avogadro's Constant, which is 6.02252×10^{23} molecules per 22.4 liters of pure gas. Cannizzaro championed Avogadro's work at conferences and in his own writings; he used Avogadro's work to further his own investigations into the atomic weights of elements in compounds. Cannizzaro established a clear distinction between atomic weight and molecular weight, just as Avogadro established a clear distinction between the atom and the molecule.

Other scientists would use Avogadro's ideas to help further what was known about the elements. Knowing the relative numbers of molecules enabled scientists to determine relative atomic weights. This was influential in DMITRY MENDELEYEV's design of the periodic table.

Avogadro also provided the framework for determining the valences of atoms. Valences helped explain how atoms bonded to one another.

Without Avogadro's grasp of the abstract nature of atoms and molecules, chemists would have continued to be perplexed by the interactions of matter.

Wilson

For Further Reading:

Brock, William. *The Norton History of Chemistry.* New York: Norton, 1992.

Irown, Keith G. *The Romance of Chemistry: From Ancient Alchemy to Nuclear Fission.* New York: Viking, 1959.

Morselli, Mario. *Amedeo Avogadro: A Scientific Biography.* Hingham, Mass.: Kluwer Academic Publishers, 1984.

Babbage, Charles

Grandfather of the Modern
Computer; Designer of the
Analytical Engine
1791-1871

Life and Work

Charles Babbage was a mathematician and inventor who developed a model for an automatic computing machine that anticipated the modern-day digital computer.

Babbage claimed that he was born in London on December 26, 1792, though his biographer lists his birthplace as Devonshire in 1791. He was privately educated until he entered Cambridge University in 1810 where he studied mathematics and became interested in astronomy.

In 1813 Babbage and other mathematicians formed the Analytical Society. Their purpose was to further the study of abstract algebra and to introduce mathematical developments from Europe into Great Britain. Babbage also helped to found the Royal Astronomical Society in 1820.

In 1822 Babbage outlined his ideas concerning a "Difference Engine" intended to mechanize the process of computing astronomical tables. Babbage gained financial support for the project from the British government. It is common today for scientists and inventors to apply for government grants, but this was not the case in his time. The government hoped that accurately calculated star tables would help save lives by improving navigation on the high seas. Unfortunately, Babbage had underestimated the size of the project. He reportedly suffered a breakdown in 1827, ceased working on his machine in 1833, and nine years later the government cancelled its support. A Stockholm printer, Pehr Georg Scheutz, improved Babbage's design and built a working version of it in 1855.

The Difference Engine is not the machine for which we remember Babbage. In 1833 he began working on his Analytical Engine. Babbage adapted elements of the Jacquard weaving loom to use in his calculating machine. The loom employed a set of instructions read from punch cards; they told the loom which operations it had to perform and which color thread to use. Babbage saw that this could be applied to scientific computation. The operations could be input on punch cards and the values on which operations would be performed could be stored in registers. Mathematician Augusta Ada Byron, Countess of LOVELACE, collaborated with Babbage on programming the instructions for the Analytical Engine and helped to fund his endeavor.

Unfortunately, Babbage never built this machine. He died in London in 1871 and his notes were lost until 1937. In 1842 Lovelace translated a paper written by an Italian engineer (who had seen a model of the engine) explicating its design; she added numerous notes and explanations of her own. This paper is the only writing by LOVELACE concerning the Analytical Engine to have survived. In 1991 British scientists used Babbage's specifications to build a working model of the engine; it has been performing accurate, simple calculations since then.

Legacy

Babbage is considered to be the grandfather of the modern computer; his early ideas for mechanical computing machines led to the development, 100 years later, of the first full-scale, automatic devices. More immediately, the punch cards he designed for the Analytical Engine led to other inventions, and he was influential in the development of the field of mathematics in Britain.

Babbage's efforts at the Analytical Society brought GOTTFRIED LEIBNIZ's notation to Britain. After a dispute between the followers of Leibniz and ISAAC NEWTON about who had invented the calculus, the British used Newton's approach for over 100 years. However, Leibniz's approach and notation were more accessible and easier to use. After the Analytical Society successfully introduced Leibniz's notation, the work of other scholars who used Leibniz's notation became accessible to students in England for the first time.

Not just a mathematician, Babbage was an inventor of great imagination, and his inventive use of punch cards in the Analytical Engine resonated in other fields. For instance, musical instrument makers began using punch cards to replace barrel organs. An automatic organ, like a player piano, could generate any tune encoded on an interchangeable stack of cards or replaceable, perforated paper roll.

Babbage's Analytical Engine was the predecessor to the Mark I, a computer built more than 100 years later by HOWARD AIKEN. Aiken knew of Babbage's designs and, in 1944, was the first U.S. engineer to build a full-scale workable computer that could be controlled by a program. Aiken and other computer engineers of his era found inspiration in Babbage's adaptation of punch cards: Aiken used punched paper tape as the method of feeding the program into the Mark I. Many other computers of the mid-1900s, even full-scale electronic ones, would use punch cards and/or tape to input data or instructions.

Steinberg

World Events		Babbage's Life
French Revolution	1789	
	1791	Charles Babbage is born
Napoleonic Wars	1803-15	
	1810	Babbage enters Cambridge University
	1813	Babbage helps found Analytical Society
	1822	Babbage outlines his ideas for Difference Engine and solicits support to build his machine
	1833	Babbage stops work on Difference Engine and starts Analytical Engine
	1842	Countess of Lovelace translates and augments Italian paper on Analytical Engine
Germany is united	1871	Babbage dies
Spanish-American War	1898	

For Further Reading:

Collier, Bruce, and James MacLachlan. *Charles Babbage and the Engines of Perfection.* New York: Oxford University Press, 1998.

Goldstine, Herman. *The Computer from Pascal to von Neumann.* Princeton, N.J.: Princeton University Press, 1972.

Kline, Morris. *Mathematical Thought from Ancient to Modern Times.* New York: Oxford University Press, 1972.

Morrison, Philip, and Emily. *The Strange Life of Charles Babbage.* Scientific American, April 1952.

Palfreman, Jon, and Doron Swade. *The Dream Machine: Exploring the Computer Age.* London: BBC Books, 1991.

Bacon, Francis

Philosopher of Science;
Originator of Scientific Method
1561-1626

Life and Work

In the early seventeenth century, Francis Bacon developed a revolutionary philosophy of science that has since influenced scientific thought and methodology.

Bacon was born on January 22, 1561, in London, England. His father was a member of the royal court, and Bacon remained linked to the courtly elite for most of his life. From 1573 to 1575, he attended Trinity College at Cambridge University and later studied at Gray's Inn, a London-based law school. In 1584, he entered the English Parliament and began writing influential political tracts regarding the state of England's internal and external affairs. His political status steadily rose throughout his career, and he held numerous coveted positions, including lord chancellor and privy councilor. In 1621, however, Bacon was convicted of bribery, impeached, and barred from public office.

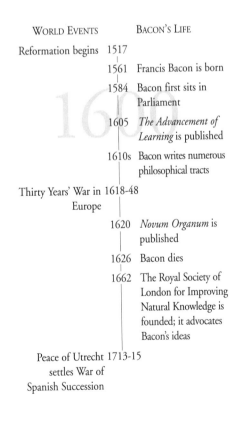

WORLD EVENTS		BACON'S LIFE
Reformation begins	1517	
	1561	Francis Bacon is born
	1584	Bacon first sits in Parliament
	1605	*The Advancement of Learning* is published
	1610s	Bacon writes numerous philosophical tracts
Thirty Years' War in Europe	1618-48	
	1620	*Novum Organum* is published
	1626	Bacon dies
	1662	The Royal Society of London for Improving Natural Knowledge is founded; it advocates Bacon's ideas
Peace of Utrecht settles War of Spanish Succession	1713-15	

In the early 1590s, Bacon's letters and speeches began to reveal his interest in philosophy. In his first lengthy work, *The Advancement of Learning,* published in 1605, he declared that human knowledge was deficient and that society needed to reform the way knowledge was taught and used. In the *Novum Organum* (published in 1620), which outlined his philosophy, Bacon criticized the school of Aristotelian logic, which had dominated philosophical thought in western Europe since ancient times. This traditional philosophy posited that knowledge and truth are revealed most effectively through deductive reasoning, which uses broad, general ideas to draw conclusions about specifics. Bacon introduced a new scientific method, based on inductive reasoning, which attempted to arrive at general truths through the analysis of specific facts.

Retiring to a country estate after his impeachment, Bacon wrote a utopia, *The New Atlantis,* that suggested that science should not be an endeavor of isolated individuals (as it had largely been up to that time), but rather the collective pursuit of many scientists sharing their knowledge. Bacon died on April 9, 1626; several of his philosophical works were published posthumously.

Legacy

Bacon's lasting contributions to science include his theories regarding scientific methodology and his ideas concerning the link between science and society.

In the seventeenth century, science as it is thought of today did not exist in western Europe; there were no institutions of scientific research, little recognition that scientific endeavor can benefit society as a whole, and no standardized experimental methods. Bacon advocated filling this void, thus foreseeing the evolution of the modern scientific community.

It is unclear whether Bacon was greatly influential in his own time, because it is difficult to detect the direct effects of ideas, as opposed to the direct effects of discoveries or inventions. Nevertheless, certain events indicate an indirect Baconian influence. The

scientific communities Bacon detailed in *The New Atlantis* began to appear in Europe soon after his death, with the formation of societies committed to the advancement of natural philosophy. In 1662 the Royal Society of London for Improving Natural Knowledge was founded, and its members advocated Bacon's philosophy of science.

Bacon believed that humans could understand nature and control it, and he claimed that gaining knowledge of the natural world would give humans the power to improve their lives. These assertions suggest that the development of science is integrally connected to all of society, not just to those who practice science. Bacon's ideas have persisted to the present and are reflected in the importance western culture places on science education, on the support of scientific research, and particularly on the development of technology.

The methodology of most modern scientists reflects Bacon's claim that scientific investigation should rely on the empirical evidence gained through observation and experimentation. Because of his insistence on the importance of inductive reasoning, Bacon is credited with having introduced a new scientific method distinct from the type of scientific investigation and thinking typical in his day. In reality, today scientists combine both deductive and inductive reasoning in their ideas and investigations.

Schuyler

For Further Reading:

Eiseley, Loren. *The Man Who Saw Through Time.* New York: Charles Scribner's Sons, 1973.

Mathews, Nieves. *Francis Bacon: The History of a Character Assassination.* Conn.: Yale University Press, 1996.

Peltonen, Markku, ed. *The Cambridge Companion to Bacon.* New York: Cambridge University Press, 1996.

Urbach, Peter. *Francis Bacon's Philosophy of Science.* LaSalle, Ill.: Open Court Publishing, 1987.

Zagorin, Perez. *Francis Bacon.* Princeton, N.J.: Princeton University Press, 1998.

Bardeen, John

Superconductivity Theorist;
Inventor of Transistor
1908-1991

Life and Work

John Bardeen co-invented the transistor and was involved in developing the prevailing theory of superconductivity. He is the only person to have won two Nobel Prizes for Physics.

Bardeen was born on May 23, 1908, in Madison, Wisconsin. His parents introduced him to mathematics and logic at a young age, always encouraging his intellectual development. He received a bachelor's degree in electrical engineering from the University of Wisconsin in 1928 and a doctorate in mathematics and physics from Princeton University in 1936.

Taking a job in 1945 at Bell Telephone Laboratories (now called Bell Labs), Bardeen joined Walter Brattain and William Shockley in the study of semiconductors, materials that conduct electrical current moderately well. Semiconductors have electrical properties that can be altered by change in temperature or amount of incident light (the amount of light striking them) or by the addition of impurities.

In 1948 the team's investigation yielded the transistor, a device that acted both as a rectifier, translating alternating current into direct current, and as an amplifier of current. The transistor was an improvement over the diodes and triodes used from the early 1900s in electronic equipment; it was rugged and small, used little energy, and required neither heating nor vacuum. Transistors revolutionized electronics.

Bardeen, Brattain, and Shockley shared the 1956 Nobel Prize for Physics for their invention.

Bardeen's next project concerned superconductivity, first described in 1911 by Dutch physicist Heike Kamerlingh-Onnes as the ability of some metals to exhibit zero resistance to the flow of electrons at temperatures close to absolute zero or -273°C. Superconductivity offered a potential solution to the energy wasted by electronic devices in overcoming electrical resistance, but it was little understood when Bardeen began to study it in the 1950s.

Bardeen and colleagues Leon Cooper and John Schrieffer formulated a complex mathematical theory of superconductivity in 1957. Their explanation, thereafter called the BCS theory, accounted satisfactorily for all observed phenomena associated with superconducting materials. This breakthrough earned them the 1972 Nobel Prize for Physics.

Retiring in 1975 from the University of Illinois, where he taught from 1959, Bardeen died on January 30, 1991, in Boston, Massachusetts.

Legacy

Bardeen contributed to two of the most revolutionary aspects of twentieth-century technology—the invention of the transistor and the understanding of superconductivity.

The introduction of the transistor was followed by the rapid development of modern electronics. The manageable size and small energy requirement of transistors made them the ideal electrical-control device for electronic machines. Portable radios, electronic telephone switches, televisions, amplified musical instruments, video cameras, smoke detectors, stereos, calculators, and microphones all would have been impossible without the transistor. All modern electronic devices—those that use electricity or magnetism to influence the flow of electrons—contain one or more transistors.

The computer represents perhaps the most significant legacy of the transistor. Computer microchips, the source of computing power, are slices of silicon etched with miniature transistors. Computers and computerized equipment, used to perform an enormous number of diverse tasks, have become one of the most universal and indispensable tools of modern life.

Superconductivity is a focus of advanced technological research in various fields. One area is transportation: engineers are investigating the possibility of constructing a railway train run on rails made of superconducting material. Electronics have also benefited from the BCS theory of superconductivity. In 1962 British physicist Brian Josephson formulated the theory of electron tunneling between superconductors; the phenomenon led to fast-switching devices for computers.

In 1986 Swiss physicist Georg Bednorz discovered materials that exhibit superconductivity at 35 degrees above absolute zero. The following year that temperature was raised to 90 degrees above absolute zero, which makes the materials usable in specialized experimental electronics. In 1993 researchers made a compound of mercury, barium, and copper that becomes superconductive at 133 degrees above absolute zero (which is still 140 degrees below 0°C). The pursuit of higher temperatures continues.

Schuyler

World Events		Bardeen's Life
	1908	John Bardeen is born
	1911	Heike Kamerlingh-Onnes describes superconductivity
World War I	1914-18	
	1936	Bardeen earns Ph.D. in mathematics and physics from Princeton University
World War II	1939-45	
	1948	Bardeen, Walter Brattain, and William Shockley invent transistor
Korean War	1950-53	
	1956	Bardeen, Brattain, and Shockley share Nobel Prize for Physics for transistor
	1957	Bardeen, Leon Cooper, and John Schrieffer formulate theory of superconductivity
	1972	Bardeen, Cooper, and Schrieffer share Nobel Prize for Physics for superconductivity theory
End of Vietnam War	1975	
Dissolution of Soviet Union	1991	Bardeen dies

For Further Reading:

Campbell-Kelly, Martin. *Computer: A History of the Information Machine.* New York: Basic Books, 1996.

Hazen, Robert M. *The Breakthrough: The Race for the Superconductor.* New York: Summit Books, 1988.

Schechter, Bruce. *The Path of No Resistance: The Story of the Revolution in Superconductivity.* New York: Simon & Schuster, 1989.

Simon, Randy, and Andrew Smith. *Superconductors: Conquering Technology's New Frontier.* New York: Plenum Press, 1988.

Barton, Clara

Founder of American Red Cross
1821-1912

Life and Work

Clara Barton founded the American Association of the Red Cross (now called the American National Red Cross) and spent most of her life administering relief aid to wounded soldiers and to victims of peacetime disasters.

Clarissa Harlowe Barton was born on December 25, 1821, in Oxford, Massachusetts. She worked for 18 years as a schoolteacher in Massachusetts and New Jersey, where she founded several free schools. When a throat ailment forced her to give up teaching in 1854, she obtained a clerical post at the U.S. Patent Office in Washington, D.C. At the outbreak of the U.S. Civil War in 1861, she left her job to work as a volunteer organizing the distribution of supplies and caring for wounded soldiers.

In 1865 President Abraham Lincoln authorized Barton to collect records on missing Union soldiers, and she was thus able to identify thousands of the dead. This work was the largest effort of its kind to date and made her famous throughout the country.

Her health failing because of the strain of her work during the war, Barton went to Europe in 1869 to recuperate. She enrolled as a volunteer with the International Red Cross, helping civilian victims of the Franco-Prussian War. She returned to Washington, D.C., in 1873. Four years later, International Red Cross authorities asked Barton to initiate an American affiliate of the Red Cross. The group she organized was incorporated as the American Association of the Red Cross in 1881, and Barton assumed its presidency. She then lobbied the U.S. Congress to vote in favor of the Geneva Convention of 1864, which had adopted Red Cross principles as international law. In 1882 the bill authorizing adoption of the Geneva Convention was passed by Congress.

Barton introduced to the Red Cross a program of aid to victims of peacetime disasters, such as hurricanes, fires, floods, earthquakes, famines, and epidemics. She led expeditions into regions devastated by such disasters. At age 77 she ran a relief operation for U.S. forces in Cuba during the Spanish-American War.

As Barton expanded the mission of the American Red Cross, its reach extended throughout the country. Local chapters began to want less interference from state units and the national organization. This unwieldy growth became more difficult for Barton to manage as she got older. In 1904 she was asked to resign, which she did. She was 83 years old.

Barton wrote *A Story of the Red Cross*, published in 1904, and *Story of My Childhood*, published in 1907. She died in Glen Echo, Maryland, on April 12, 1912.

Legacy

Barton's establishment and leadership of the American Association of the Red Cross set the stage for the organization's irreplaceable medical relief work conducted throughout this century. It is now a highly effective relief agency, extending beyond just medical help,

with millions of members providing various services to those in need.

Following Barton's resignation as president, the Red Cross underwent reorganization. Mabel T. Boardman took over the leadership and aggressively set about raising funds and restructuring the organization. The local chapters were gathered under territorial and state branches, which reported directly to the national headquarters.

During the first two decades of the twentieth century, the Red Cross provided aid to victims of the 1906 San Francisco earthquake and World War I. The war transformed the volunteer organization into a powerful relief program. By the end of the war, there were 31 million members serving all over the world.

The American National Red Cross, as it is now known, continued throughout the twentieth century to provide global relief to victims of natural and human-made disasters. During the Great Depression in the United States, the Red Cross was kept busy by drought, famine, and unemployment, and World War II demanded the organization turn its attention to the armed forces. The International Red Cross conducted relief efforts in many areas during peace and war throughout the century, and has evolved to serve over 50 nations today.

The Red Cross continues to serve the public and maintains the principles that Clara Barton embraced: dedication to promoting peace, humanity, impartiality, neutrality, and voluntary service on a global scale.

Schuyler

WORLD EVENTS		BARTON'S LIFE
Napoleonic Wars in Europe	1803-15	
	1821	Clarissa Barton is born
	1854	Barton works at Patent Office
United States Civil War	1861-65	Barton administers relief aid during Civil War
	1865	Barton gathers missing Union soldier records
	1869	Barton travels to Europe; volunteers with International Red Cross
Germany is united	1871	
	1873	Barton returns to U.S.
	1881	Barton founds American Association of the Red Cross
Spanish-American War	1898	Barton organizes a relief effort during the Spanish-American War
	1904	Barton resigns as president of American Red Cross
	1912	Barton dies
World War I	1914-18	

For Further Reading:

Barton, William Eleazar. *The Life of Clara Barton, Founder of the American Red Cross.* Boston: Houghton Mifflin, 1922.

Burton, David Henry. *Clara Barton: In the Service of Humanity.* Westport, Conn.: Greenwood Press, 1995.

Pryor, Elizabeth Brown. *Clara Barton: Professional Angel.* Philadelphia: University of Pennsylvania Press, 1987.

Bateson, William

Early Mendelian Geneticist;
Originator of Term "Genetics"
1861-1926

Life and Work

William Bateson publicized and helped corroborate the principles of Mendelian inheritance. He raised questions about the relationship between evolution and genetics, which are still being addressed today.

Bateson was born in Whitby, England, on August 8, 1861. He attended Cambridge University, where he developed a passion for zoology, and graduated with first-class honors in 1883. He spent the next two years in the United States investigating the embryological development of echinoderms (a group of various sea animals). Noting that the larval stage of echinoderms possesses nervous-system structures similar to those of chordates (a group of animals having a stiff rod below a single nerve fiber at some point in their life cycle), he proposed that chordates might have evolved from primitive echinoderms, a theory that is now widely accepted.

Bateson staunchly believed in the theory of evolution, but while studying embryology, he began to question the element of CHARLES DARWIN's theory that described evolution as a slow accumulation of small changes. His skepticism arose from evidence suggesting that abrupt and significant changes occur both in individuals and entire species. He began searching for a solution to this puzzle in the laws of heredity and, upon his return to Cambridge, began investigations into breeding and heredity.

At the same time, in 1900, the scientific community rediscovered the research of Austrian botanist GREGOR MENDEL, which demonstrated that heredity is governed by the transmission of certain elements (now called genes) from parent to offspring. Bateson recognized the significance of Mendel's work and published an English translation of it in 1902. The book was an enormous success, elevating Bateson to prominence in academic circles.

Bateson established the application of Mendel's theories to animals with experiments on the inheritance of comb shape in fowl (Mendel had studied plants). He also extended Mendel's pea-plant experiments and discovered that the heredity of some characteristics is controlled by more than one element.

Bateson coined the term "genetics" in 1906 and was appointed the first professor of genetics at Cambridge University in 1908. Bateson did not agree with the growing consensus among scientists that chromosomes are involved in heredity, and he thus fell into academic disrepute in his later years. He died in London on February 8, 1926.

Legacy

Bateson was instrumental in spreading the news of Mendel's remarkable research, launching a flurry of studies and paving the way for a more advanced understanding of the connection between genetics and evolution.

Genetics made rapid progress in the twentieth century, due partly to Bateson's promotion of Mendel's work. When Mendel's papers were first rediscovered, his theories were highly controversial. The success of Bateson's translation and corroborative research inspired further investigations, and soon most biologists agreed that Mendel's theories were sound. Mendel's work is still considered to have laid the foundation of modern genetics, a discipline crucial to all fields of biological science.

Bateson's experiments demonstrated that some characteristics appear to be inherited together, which led to the discovery of an exception to Mendel's laws. According to Mendel's laws, the inheritance of each gene is independent of the inheritance of other genes. In the first decade of the twentieth century, Thomas Hunt Morgan proved that genes reside on chromosomes, and he went on to show that when two genes are close together on a chromosome ("linked"), they are inherited together as a pair. This phenomenon, later called gene linkage, explained Bateson's results.

The question, first posed by Bateson, of whether evolution progresses slowly in small steps, or abruptly in large steps, foreshadowed the theory of punctuated equilibrium. Darwin proposed in the mid-nineteenth century that evolution occurs through the accumulation of many small changes. In the 1970s, two prominent evolutionary biologists, Stephen Jay Gould and Niles Eldredge, suggested that perhaps evolution occurs during short bursts of change which interrupt longer periods of stability. This punctuated-equilibrium theory remains controversial.

Schuyler

WORLD EVENTS		BATESON'S LIFE
	1861	William Bateson is born
Germany is united	1871	
	1881	Bateson graduates from Cambridge University
	1890s	Bateson conducts breeding experiments
Spanish-American War	1898	
	1900	Mendel's work is rediscovered
	1902	Bateson translates Mendel's work
	1906	Bateson coins term "genetics"
	1908	Bateson is appointed first Cambridge professor of genetics
	1910	Morgan proves genes are on chromosomes
World War I	1914-18	
	1926	Bateson dies

For Further Reading:

Bateson, William. *Scientific Papers of William Bateson*, New York: Johnson Reprint, 1971.

Bateson, William. *William Bateson, F.R.S., His Essays & Addresses, with a Memoir by Beatrice Bateson.* New York: Garland, 1984.

Dunn, L. C. *A Short History of Genetics: The Development of Some of the Main Lines of Thought, 1864-1939.* Ames: Iowa State University Press, 1991.

Becquerel, Antoine-Henri

Discoverer of Radioactivity
1852-1908

Life and Work

Antoine-Henri Becquerel was the first to realize that lifeless rocks could emit powerful but completely invisible rays—radioactivity or nuclear radiation—and he proved it through patient, careful experiments.

Becquerel was part of a French scientific family that included several generations of respected scientists. He was born in Paris on December 15, 1852, the son of Alexandre-Edmond Becquerel, who had spent his career studying the rotation of polarized light by materials in magnetic fields. Henri took up that study himself while also working for the Department of Bridges and Highways, for which he was appointed chief engineer in 1894. It had been known that solids and liquids could rotate the direction of polarized light, and the young Becquerel showed for the first time that gases could do the same with light.

When Becquerel became a professor of physics at École Polytechnique in Paris in 1895, his main interest was the phenomenon of phosphorescence, the property of some materials to glow for a while in darkness immediately after they are exposed to light. In particular Becquerel studied the phosphorescence from some uranium compounds, which he assumed needed exposure to sunlight before they would glow.

At the end of 1895 the German scientist WILHELM ROENTGEN discovered that mysterious, invisible rays—which he called X-rays—could be produced in a laboratory. It is now known that they are emitted when fast electrons, accelerated by a high voltage, smash into a glass or metal target. When word of Roentgen's discovery reached him in 1896, Becquerel wondered if phosphorescent materials might also give off X-rays. As a test he placed crystals of uranium compounds in bright sunlight on photographic plates that were well wrapped with dark paper. Sure enough, the plate was exposed, suggesting that an invisible, penetrating radiation was released by the crystals. He repeated the experiment, this time with the crystals and plates in a dark drawer, expecting the plate to stay clear. To his surprise, though, the plates were exposed, just as in the previous experiment. In later experiments he found that pure uranium metal would expose the plates even more strongly than the phosphorescent salts. MARIE CURIE named the phenomenon radioactivity and made it the focus of her research with her husband, Pierre.

Before the year ended, though, ERNEST RUTHERFORD in England showed that at least some kinds of radiation are electrically charged bits of matter. By 1900 Becquerel succeeded in measuring the deflection of beta particles and showed that they were the same as electrons. He also inadvertently discovered the physiological effects of radiation in 1901, when he reported a burn from carrying a sample of radium in his vest pocket.

Becquerel shared the 1903 Nobel Prize for Physics with the Curies. He died in Le Croisic, France, on August 25, 1908.

Legacy

As the first to discover radioactivity and that one component of some radiation, the beta particle, was made of electrons, Becquerel can be considered an ancestor—perhaps the great-grandfather—of the nuclear age.

In Becquerel's lifetime, scientists were regularly producing all kinds of unknown rays from vacuum tubes connected to high voltage. Those rays were clearly the result of the experimenter's action: turn off the power, and the rays stop. In contrast, the rays he discovered emanating from uranium and other minerals came out on their own, from the samples themselves. Figuring out where that energy came from and what it could do spurred the investigations of several generations of scientists, starting with Marie and Pierre Curie. The Curies' studies completed Becquerel's work. They demonstrated "disintegration," the breaking apart of a radioactive atom's nucleus that resulted in the emission of heat and subatomic particles. Other scientists focused quickly on the significance of their findings. Ernest Rutherford characterized the atom as a nucleus surrounded by orbiting electrons in 1911, and NIELS BOHR proposed that electrons circle the nucleus at specific levels of energy in 1913, thus marking the beginning of nuclear physics.

Many applications exist today that are rooted in Becquerel's work. Radiation from radioactive materials is used to sterilize food, to measure thicknesses of metal and plastic sheets in industry, and for treating many kinds of cancer. Radioactive materials themselves are sometimes administered to patients for diagnosis and treatment of some diseases, and radioactive materials have been essential as tracers to decipher complex biological and chemical reactions. Becquerel could not have imagined any of these uses, but they are all descended directly from his work more than 100 years ago.

Secaur

For Further Reading:

Halacy, Daniel S. *Science and Serendipity: Great Discoveries by Accident.* Philadelphia, Penn.: Macrae Smith Co., 1967.

Heathcote, N. H. *Nobel Prize Winners in Physics, 1901-1950.* New York: Schuman, 1953.

Rayner-Canham, M. F., and G. W. *A Devotion to the Science: Pioneer Women of Radioactivity.* Philadelphia, Penn.: Chemical Heritage, 1997.

Bell, Alexander Graham

Inventor of the Telephone
1847-1922

Life and Work

Alexander Graham Bell was born on March 3, 1847, in Edinburgh, Scotland, to a family of speech specialists. He received most of his early education from his parents and, starting at age 15, was trained by his grandfather, a prominent speech tutor in London. In 1867 he became an assistant to his father, who had invented a phonetic visible-speech system for teaching the deaf to speak.

The Bell family moved to Brantford, in Ontario, Canada, in 1870 to help Alexander recuperate from the shock of his two brothers' deaths from tuberculosis. The following year, Bell moved to Boston and began demonstrating his father's speech method to teachers of the deaf. He became professor of vocal physiology at Boston University in 1873.

Bell befriended Thomas Watson, a repair mechanic, and with funds provided by the families of two of Bell's deaf students, Bell and Watson began long nightly research sessions aimed at devising an electrical means of transmitting sound. In 1874 Bell obtained a patent for a telegraph that could send more than one message over a single wire simultaneously.

Watson and Bell continued their efforts to develop electrical-sound transmission. On March 10, 1876, Bell shouted the now-famous words—"Mr. Watson, come here; I want you"—when he spilled acid on his clothes; the words were relayed from Bell's lab transmitter to Watson's receiver in another room. By the end of the year, Bell had sent a spoken message over a distance of 229 kilometers and had been granted a patent for the telephone. The patent covered the method and devices (microphone and loudspeaker) for transmitting vocal and other sounds via electrical undulations similar to the vibrations in air that accompany such sounds. A prototype of Bell's telephone earned the gold medal at the Centennial Exhibition in Philadelphia that same year, and the telephone system spread quickly across the United States and Europe.

Bell spent the rest of his life developing an extremely diverse collection of instruments and conducting varied technological and biological research. He invented a device that locates metallic objects within the human body, an air-cooling system, the forerunner of the iron lung, a card-sorting machine, a hydrofoil boat, and various sound-transmitting instruments that he hoped would help the deaf, including his wife Mabel. His studies and experimental activities also involved sonar detection, flying machines, the physiology of hearing and speech, and sheep breeding.

He received numerous awards during his life, helped found the journal *Science* in 1883, and served as the president of the National Geographic Society from 1898 to 1903. He died on August 2, 1922, on Cape Breton Island, Nova Scotia, Canada.

Legacy

Bell's telephone changed the lives of people in industrialized countries by allowing immediate communication between distant places.

Early telephone systems were limited and required an operator to connect every call. In 1889, Almon B. Strowger invented the automatic-exchange switchboard. In modern systems computers connect calls. Rapid intercontinental communications relied for decades on expensive and short-lived telegraph cables laid on the bottom of the Atlantic Ocean.

Development of durable telecommunications satellites in the 1960s made instant global communications feasible and affordable. Recent technological advances include facsimiles (fax machines), cellular phones, modems, and video conferencing, all of which improved global communication and information exchange.

Bell's lesser-known achievements were also influential. His promotion of the visible-speech method for teaching the deaf to speak led to the method's widespread use. It improved the lives of the deaf by enabling them to communicate more easily with hearing people.

The journal *Science,* founded with Bell's help, has become one of the world's most prominent peer-reviewed scientific journals. It publishes news and research from all fields of science.

As president of the National Geographic Society, Bell guided the transformation of the society's simple pamphlet into an educational journal that illustrates and describes life in distant areas of the world to millions of readers.

Schuyler

WORLD EVENTS		BELL'S LIFE
Napoleonic Wars 1803-15 in Europe		
	1847	Alexander Graham Bell is born
United States 1861-65 Civil War		
	1867	Bell becomes his father's assistant
Germany is united 1871		
	1870	Bell immigrates to Canada
	1871	Bell moves to Boston and demonstrates his father's visible speech method of teaching the deaf
	1873	Bell becomes professor of vocal physiology
	1876	Bell invents telephone
	1883	Bell helps found *Science*
Spanish-American War	1898	Bell becomes president of National Geographic Society
World War I 1914-18		
	1922	Bell dies

For Further Reading:

Eber, Dorothy. *Genius at Work: Images of Alexander Graham Bell.* New York: Viking Press, 1982.

Grosvenor, Edwin S., and Morgan Wesson. *Alexander Graham Bell: The Life and Times of the Man Who Invented the Telephone.* New York: Harry N. Abrams, 1998.

Mackay, James A. *Sounds Out of Silence: A Life of Alexander Graham Bell.* Philadelphia, Penn.: Mainstream, 1997.

Pasachoff, Naomi. *Alexander Graham Bell Making Connections.* New York: Oxford University Press, 1996.

Bernoulli, Jakob
Bernoulli, Johann

Originators of the Calculus of Variations

1654-1705; 1667-1748

World Events		Bernoullis' Lives
Thirty Years' War 1618-48 in Europe		
	1654	Jakob Bernoulli is born
	1667	Johann Bernoulli is born
	1687	Jakob becomes professor of mathematics at University of Basel
	1694	Johann earns medical degree from University of Basel
	1697	Jakob and Johann work on brachistochrone problem, initiating calculus of variations
	1705	Jakob dies
	1713	*The Art of Conjecturing* by Jakob Bernoulli is published
Peace of Utrecht 1713-15 settles War of Spanish Succession		
	1742	*The Work of Johann Bernoulli* is published
	1748	Johann dies
United States independence	1776	

Life and Work

Jakob and Johann Bernoulli, the most prominent members in a family of mathematicians whose work spanned three generations, helped to spread and further the newly invented calculus as well as contributed to the study of probability theory and the calculus of variations.

Jakob (also referred to as James or Jacques), the fifth child of Nikolaus Bernoulli, was born in Basel, Switzerland, on January 6, 1654. Thirteen years later the tenth child, Johann, was born on August 6, 1667. Nikolaus forced Jakob to study theology and hoped that his son would give up his ideas of studying mathematics and astronomy. Jakob won this war of wills.

In 1687 Jakob became professor of mathematics at the University of Basel. He began to study the mathematics of John Wallis and Isaac Barrow, which had inspired Isaac Newton and Gottfried Leibniz to invent the calculus. He taught his brother mathematics, even though Johann's professed field was medicine, for which Johann earned a degree in 1694 from the University of Basel. Johann switched to mathematics, similarly against his father's will. Although the two worked together on many problems, they also were fierce competitors who posed public challenges to each other.

In 1697 the brothers worked on the brachistochrone problem, whose solution described the path that a particle takes to slide down from one point to another point (not directly below) in the least amount of time. Johann's solution—known as the cycloid (a path inscribed by a point on a revolving wheel)—was disputed by Jakob and the dispute led the brothers to develop the calculus of variations, an extension of regular (infinitesimal) calculus that focused on the largest or smallest values of functions that come from curves or other functions, not real numbers.

Jakob's book *The Art of Conjecturing* was published in 1713, eight years after his death. In it he explained his work on probability, of which he was one of the earliest students. The book contained several innovations including an extension of the binomial theorem, the principle known as Bernoulli's theorem. Johann's work was published in 1742 under the title *The Work of Johann Bernoulli*.

Jakob died on August 16, 1705, in Basel. Upon Jakob's death, Johann succeeded him in his position at the University of Basel. Johann died on January 1, 1748 in Basel.

Legacy

It is difficult to separate the achievements of the two Bernoulli brothers. Their legacy rests on their contributions to the early study of the calculus, in particular their origination of the basic principles of the calculus of variations.

In their investigations into the brachistochrone problem, the brothers' formulation of the calculus of variations began. In tackling the problem, Jakob proposed a variational principle: maximize or minimize some quantity subject to some constraint. Such variational principles became the basis of the calculus of variations, and subsequently became helpful in expressing other scientific principles including William Rowan Hamilton's theory of least action, in which the "action integral" must be minimized. The theory of least action is linked to Newton's laws of motion, and other applications of variational principles can be found in the study of elasticity, electromagnetics, and aerodynamics.

Jakob and Johann built on the work of Leibniz, using his notation. The brothers corresponded with Leibniz, and Johann was vocal in his support of Leibniz in the dispute between the Continental mathematicians who believed that Leibniz invented the calculus first and the British mathematicians who attributed that achievement to Newton. Leibniz's notation—easier both to understand and to manipulate in calculations—offered greater utility to mathematicians and scientists; eventually it became the primary system of expressing and using the calculus.

The students and offspring of Jakob and Johann carried on their legacy. Daniel Bernoulli, Johann's son, supported Leonhard Euler throughout his career; Daniel himself was a productive mathematician, focusing on applications in astronomy, physics, and hydrodynamics, a term that he originated. Euler, who studied with Johann and whose father studied with Jakob, was one of the more productive and prolific mathematicians of the eighteenth century. His work included innovations in number theory and algebra, and his textbook on the calculus was used for generations.

Steinberg

For Further Reading:

Bell, E. T. *Men of Mathematics*. New York: Simon & Schuster, 1937.

Dunham, William. *Journey through Genius*. New York: John Wiley, 1990.

Kline, Morris. *Mathematical Thought from Ancient to Modern Times*. New York: Oxford University Press, 1972.

Struik, Dirk J. *A Concise History of Mathematics*. 4th rev. ed. New York: Dover, 1987.

Bessel, Friedrich

Astronomer; Originator of
Bessel Functions
1784-1846

Life and Work

Friedrich Bessel analyzed disturbances in planetary and stellar motion, which enabled him to measure the positions of numerous stars and, for the first time, the distance from Earth to a star other than the Sun. He introduced Bessel functions and showed that Earth was not a perfect sphere.

Bessel was born on July 22, 1784, in Minden, in present-day Germany. He was apprenticed at age 15 to an accountant, but he dreamed of travel and studied navigation at night. From observations of Halley's Comet recorded in 1607, he calculated the comet's orbit and sent his notes to the astronomer Wilhelm Olbers, who published them and sent Bessel to join the Lilienthal Observatory as an assistant.

In 1808 Bessel was asked to oversee construction of a large observatory at Königsberg (now Kaliningrad); two years later he was appointed professor of astronomy at the city's university. He directed the observatory from 1813 until his death.

From 1821 to 1833, Bessel diligently calculated the positions and observed the motions of thousands of the nearest stars, correcting

errors associated with telescope imperfections and atmospheric disturbances.

Bessel's astronomical endeavors required his command of advanced mathematics. In 1824 he formulated new functions, now known as Bessel functions, to assist the understanding of anomalies in planetary motion. Bessel functions represent solutions to certain differential equations.

In 1832 Bessel calculated Earth's ellipticity, the extent to which its shape deviates from that of a perfect sphere, by measuring elements of selected meridian arcs (imaginary arcs inscribed in the sky corresponding to Earth's longitude lines) over East Prussia.

Six years later Bessel accomplished a groundbreaking determination of the distance from Earth to the star Cygni 61. This was the first calculation of the distance from Earth to a star other than the Sun. He found the distance to be about 10.3 light-years, which is within 10% of the current measurement. His method involved stellar parallax, the displacement nearby stars exhibit over time because they are viewed from different locations as Earth moves through its orbit. Bessel's computation was the first use of parallax to make an accurate measurement of stellar distance. He published his results in 1842.

Soon thereafter, Bessel suggested that the wave-like motion of the star Sirius is due to the gravitational pull of an unseen orbiting body. He also predicted the existence of a planet beyond Uranus by observing minute irregularities in its orbit.

Bessel died on March 17, 1846, in Königsberg.

Legacy

Bessel established the scale and framework of the universe, thus lighting the path for following generations of astronomers.

Bessel's later predictions were confirmed soon after his death. In the summer of 1846, astronomers identified the planet Neptune as the body that disturbs the orbit of Uranus. The companion of the star Sirius, dubbed Sirius B, was detected by telescope-lens maker Alvan Clark in 1862. Such two-star systems were found to be common.

Bessel's method of calculating the distance from Earth to the nearest stars led to the best determination of the scale of the universe that

had then been achieved. His star positions allowed the first accurate calculation of distances between stars and led to a means of calculating the size of stars, galaxies, and clusters of galaxies.

Bessel is credited with reforming the approach to astronomical observation and providing the key to astronomical progress with his precise methods of correcting measurement errors.

Bessel functions have applications in pure mathematics and physics. They are useful in studying the distribution and flow of heat and electricity through cylinders and in solving problems related to wave theory, elasticity, and hydrodynamics.

Schuyler

WORLD EVENTS		BESSEL'S LIFE
United States independence	1776	
	1784	Friedrich Bessel is born
French Revolution	1789	
Napoleonic Wars in Europe	1803-15	
	1808	Bessel oversees construction of observatory at Königsberg
	1821	Bessel begins calculating star positions
	1824	Bessel functions are introduced
	1832	Bessel calculates Earth's ellipticity
	1838	Bessel determines distance to Cygni 61
	1842	Bessel publishes results on Cygni 61, including first use of parallax inter-stellar measurements
	1846	Bessel dies
Germany is united	1871	

For Further Reading:

Dragomir, V. C., et al. *Theory of the Earth's Shape.* New York: Elsevier Scientific Publishing, 1981.

Hoskin, Michael, ed. *The Cambridge Illustrated History of Astronomy.* Cambridge: Cambridge University Press, 1997.

Bessemer, Henry

Inventor of Bessemer
Steel-making Process
1813-1898

Life and Work

Henry Bessemer developed and patented an inexpensive method of manufacturing steel known as the Bessemer process. The product obtained was more versatile than the steel previously available.

Bessemer was born on January 19, 1813, in Charlton, England, to an engineer father who recognized and encouraged the boy's mechanical tendencies. His first major invention, completed in 1833, was a typesetting machine for the prevention of government-document forgeries. He also devised an improved graphite pencil, a process for making imitation lace, and a machine to make cheaper "gold powder," a brass-based substance mixed into paints to create a glittery effect.

During the Crimean War (1853-56), demand arose for a new metal to use in making guns. The two available types of iron were flawed: cast iron contained impurities and was brittle, while wrought iron, which was relatively pure, was difficult and time-consuming to manufacture. Steel, defined as iron with less than 2% carbon, was non-pliable and costly to make. In 1855, while attempting to fabricate stronger cast iron, Bessemer developed a quick, inexpensive technique that produced a more valuable steel, an alloy that could withstand considerable stresses.

The Bessemer process, as it came to be known, involves forcing air through a drum of molten crude iron, causing impurities to burn off without the consumption of fuel. Some of the carbon remains in the iron, producing a strong, light, and versatile metal. However, Bessemer had unknowingly used phosphorus-free iron, and other metallurgists, using iron that contained phosphorus, failed to reproduce his results. Phosphorus, which lowers the melting point of iron, allowed more impurities to remain. Such impurities were not removed by the original Bessemer process, whereas they were effectively removed by traditional methods of manufacturing cast and wrought iron. To overcome this obstacle to success, Bessemer moved to northwestern England, where phosphorus-free iron ore was plentiful, and opened a steel foundry.

In his last years, Bessemer engineered a solar furnace, a telescope, and equipment for polishing diamonds. He was knighted in 1879; he died in London on March 15, 1898.

Legacy

Although other innovations outproduced the Bessemer process by the end of the century, Bessemer was responsible for initiating the wide use of steel that continues today.

Sidney Gilchrist Thomas and Percy Gilchrist solved the phosphorus problem of the Bessemer process around 1878. They realized that the weakness lay not in the process itself, but in the ingredients that lined Bessemer's drums. They added limestone to the drum material, which created the necessary environment for the removal of phosphorus from molten iron.

The Bessemer process enormously increased world steel output and transformed many metal industries. Steel became a readily available and highly prized product for numerous construction purposes, including railway lines. Steel replaced wrought iron in ship plate, girders, sheet, rods, wires, rivets, and various other metal components.

Bessemer's success established metallurgy as a distinct discipline on the border between science and technology. Researchers experimented with steel by varying the carbon content and adding ingredients (such as manganese, tungsten, chromium, and vanadium): different types and qualities of steel resulted. The ancient craft of metallurgy thus acquired a scientific framework.

In 1861 Friedrich and Wilhelm Siemens invented the regenerative blast furnace, in which waste heat is efficiently recirculated to preheat the fuel and air entering the furnace. This method achieved higher temperatures with better fuel economy than previous blast furnaces, and it allowed the use of low-grade coal as fuel. Pierre and Emile Martin were the first to modify the regenerative furnace for making steel, yielding larger quantities and allowing tighter quality regulation than the Bessemer process. Later in the nineteenth century, the Bessemer process was overtaken by the Siemens-Martin process. United States industrialist Andrew Carnegie contributed to the demise of the Bessemer process by adopting the Siemens-Martin system for his steel mills.

Schuyler

For Further Reading:

Bessemer, Henry. *Sir Henry Bessemer: An Autobiography.* Brookfield, Vt.: Institute of Metals, 1989.

Cardwell, Donald. *The Norton History of Technology.* New York: Norton, 1995.

Tylecote, R. F. *A History of Metallurgy.* Brookfield, Vt.: Institute of Metals, 1992.

Bjerknes, Vilhelm

Founder of Meteorology
1862-1951

Life and Work

Vilhelm Bjerknes was a founder of meteorology and modern weather forecasting.

Bjerknes was born on March 14, 1862, in Christiana, Norway. He was the son of Karl Anton Bjerknes (1825-1903), a Norwegian scientist who first recognized relationships between hydrodynamics and electrodynamics. As a young boy, Bjerknes learned much by working with his father, who was a professor of mathematical physics at the University of Christiana.

In 1888 Bjerknes received a master of science degree from the University of Christiana, and then branched out on his own, avoiding his father's increasingly secretive behavior. Bjerknes's long academic career began with an appointment to the University of Stockholm in 1895. During his professorship at Stockholm, he sought the support of the Carnegie Foundation in the United States to develop an advanced program of scientific weather forecasting. The foundation awarded him a yearly grant from 1905 to 1941 to pursue this research.

In 1897, early in his tenure at the University of Stockholm, his son Jacob was born. Jacob eventually became a valuable partner in his later years and continued his father's work after Bjerknes's death.

After brief appointments to the University of Christiana (1907-1912) and University of Leipzig (1912-1917), Bjerknes joined the faculty at the Bergen Museum of Natural Science in Bergen, Norway, in 1917, where he established the Bergen School of Meteorology. While there he wrote his most influential publication, *On the Dynamics of the Circular Vortex with Applications to the Atmosphere and Atmospheric Vortex and Wave Motion* (1921). In this work, he presented some of his most significant ideas; he made a direct analogy between hydrodynamics—stream flow, turbulence, and whirlpools in water—to the behavior of air masses. He recognized that the movements of air masses could be better predicted when characteristic vortices (such as cyclones, polar fronts, and squall lines) were understood.

With his son Jacob, Bjerknes established a system of meteorological observation stations in Norway during World War I. The pair also worked together to define weather fronts as interfaces between air masses.

Among Bjerknes's many other publications are *Dynamic Meteorology and Hydrography* (with J. J. Sandstrom) in 1910 and *Kinematics* (with Hesselberg and Devik) in 1911.

Bjerknes joined the faculty of the University of Oslo in 1926, and stayed there until his retirement in 1932. He died on April 9, 1951, in Oslo, Norway.

Legacy

Bjerknes helped to establish the scientific foundation for meteorology, of which he is a founding father.

His application of hydrodynamics to the motion of air masses and his development of the idea of vortex (whirlpool) flow as a factor in determining weather became the foundations for modern concepts in weather forecasting. These vortices today are known as the systems of high and low pressure on weather maps seen on the evening news. Using knowledge of the interaction of weather systems, meteorologists today can help predict the advent of severe weather and therefore help to alleviate potential damage and loss of life.

Bjerknes was the driving force behind the famous Bergen School of Meteorology. The school became known for its systematic training of meteorologists and for its accuracy in weather forecasting. Today the University of Bergen has

expanded its meteorological research beyond the School of Meteorology. The Nansen Environmental and Remote Sensing Center (NERSC), affiliated with the university, is dedicated to understanding regional and global environmental problems through modeling climatic processes, observing marine systems, and monitoring weather-related disasters.

Bjerknes's son Jacob further extended his father's legacy. Jacob settled in the United States in 1940 where he joined the meteorology faculty at the University of California at Los Angeles (UCLA). There, he discovered the jet-stream effect, studied the special weather sub-system called El Niño, and developed the meteorological center at UCLA into an institution of international reputation.

Luoma

For Further Reading:

Bulletin of the American Meteorological Society, 1975.

Friedman, Robert Marc. *Appropriating the Weather: Vilhelm Bjerknes and the Construction of a Modern Meteorology.* Ithaca, N.Y.: Cornell University Press, 1989.

Jewell, Ralph. "The Meteorological Judgment of Vilhelm Bjerknes." *Social Research* 51 (Autumn, 1984): 783-807.

World Events	Bjerknes's Life
United States 1861-65 Civil War	
	1862 Vilhelm Bjerknes is born
Germany is united 1871	
	1888 Bjerknes receives master of science degree from University of Christiana
	1895 Bjerknes is appointed professor at University of Stockholm
	1897 Jacob Bjerknes is born
Spanish-American 1898 War	
	1905 Bjerknes gains Carnegie Foundation annual support for research on scientific weather forecasting
World War I 1914-18	
	1917 Bjerknes becomes professor of meteorology at Bergen Museum of Natural Science
	1921 Bjerknes publishes *On the Dynamics of the Circular Vortex...*
	1926 Bjerknes becomes professor at University of Oslo
	1932 Bjerknes retires
World War II 1939-45	
	1941 Bjerknes's Carnegie grant expires
Korean War 1950-53	
	1951 Bjerknes dies

Black, Joseph

Isolator of Carbon Dioxide;
Definer of Latent Heat and
Heat Capacity
1728-1799

World Events	Black's Life
Peace of Utrecht 1713-15 settles War of Spanish Succession	
	1728 Joseph Black is born
	1754 Black earns medical degree from University of Glasgow
	1756 Black's carbon dioxide work is published; he becomes professor of chemistry and anatomy at University of Glasgow
	c.1760 Black discovers heat capacity
	1762 Black discovers latent heat
	1765 James Watt designs improved steam engine
United States 1776 independence	
French Revolution 1789	
	1799 Black dies
Napoleonic Wars 1803-15 in Europe	

Life and Work

Joseph Black isolated carbon dioxide and, through experimentation, revealed many of its properties. He also introduced the concepts of latent heat and heat capacity.

Black was born on April 16, 1728, in Bordeaux, France. He studied anatomy and medicine at the University of Glasgow, and he received a medical degree from the University of Edinburgh in 1754. Two years later he became professor of chemistry and anatomy at Glasgow.

Black's doctoral work, published in 1756, concerned his investigations of carbon dioxide, which he called "fixed air." He heated limestone, which yielded quicklime and carbon dioxide. Adding water to the quicklime and boiling the mixture with potassium carbonate, he recovered the original weight of limestone. He thus demonstrated that gases could be studied in combination with other chemical substances and that such experiments could be analyzed quantitatively. These and subsequent experiments showed that carbon dioxide acts as an acid, is produced during respiration, and is present in the atmosphere.

Around 1760 Black's research turned to the physical study of heat. He noted that when equal weights of mercury and water were heated over the same flame, the temperature of the mercury rose twice as fast as that of the water. This challenged the assumption that heat fills all substances equally. The experiment showed that substances have differing capacities to capture and hold heat. He called this property heat capacity.

Two years later Black found that when he heated a mixture of ice and water just until the ice melted, the temperature of the mixture did not change. He theorized that heat can cause a change of state in a substance without changing the temperature of the substance. Such heat he called latent, from the Latin for "hidden."

Black died on November 10, 1799, in Edinburgh, Scotland.

Legacy

Black's discoveries of latent heat and heat capacity helped James Watt to revolutionize steam power, and his experiments with carbon dioxide moved gases from the realm of mystery to that of quantitatively understandable chemicals.

In the 1760s, while attempting to improve a single-cylinder Newcomen steam engine, Watt became stuck at a problem involving heat transfer. He had designed an engine with a second cylinder for the condensation of steam. The steam was to be condensed by surrounding this cylinder with cold water. As the heat from the steam warmed the surrounding water, the warm water had to be replaced with cold. Watt wondered how much heat the steam needed to lose to condense and thus how much cold water would be needed to absorb this heat with each stroke of the engine. After discussing the principles of latent heat with Black, Watt performed experiments to determine the latent heat of steam condensation—precisely the information he sought. Armed with experimental data, he completed the design of his steam engine in 1765.

Black's work on carbon dioxide led scientists to study other gases. He inspired Henry Cavendish, in the 1760s, to become the first to study hydrogen in depth and report on its properties, and he encouraged the experiments of his student Daniel Rutherford, who discovered nitrogen in 1772. Black's work also provided the first hint that air is a mixture of chemicals rather than a single element.

Black's discovery of heat capacity enabled researchers to quantify different substances' heat capacities. In 1871 Portuguese scientist J. H. de Magellan coined the term "specific heat" to refer to a particular substance's heat capacity. It is now known as specific heat capacity, the amount of heat required to raise one gram of a substance by one degree Celsius at constant pressure. Substances with higher specific heat capacities, such as water, can absorb or give off large amounts of heat while undergoing a relatively small change in temperature. Specific heat capacity is an important factor in calculating the results of chemical reactions involved in research and numerous industrial proceedings.

Schuyler

For Further Reading:

Crowther, James Gerald. *Scientists of the Industrial Revolution: Joseph Black, James Watt, Joseph Priestly, Henry Cavendish.* London: Cresset Press, 1962.

Donovan, Arthur L. *Philosophical Chemistry in the Scottish Enlightenment: The Doctrines and Discoveries of William Cullen and Joseph Black.* Edinburgh, Scotland: Edinburgh University Press, 1975.

Blackwell, Elizabeth

First American Woman Physician
of Modern Times

1821-1910

Life and Work

Elizabeth Blackwell was the first woman to gain a medical degree in the United States and is considered the first American woman physician of modern times. Her pioneering work helped open the medical profession to women.

Blackwell was born on February 3, 1821, in Bristol, England, to parents whose progressive social and political views shaped her later devotion to social reform. Blackwell and her 11 siblings were educated at home by private tutors. In 1832, her family immigrated to the United States, lived for six years in New York and New Jersey, and then moved to Cincinnati, Ohio. For four years, Blackwell and her sisters ran a boarding school and taught private pupils. In 1842, Blackwell accepted a teaching position at a girls' school in Kentucky, but the work did not appeal to her, and in 1844 she decided to become a medical doctor.

Blackwell's first attempts to gain admission to medical school were unsuccessful. Then, finally, she was accepted at Geneva College in New York. Her acceptance was the result of an accident: admissions authorities at Geneva College thought her application was a hoax from a rival school and decided to accept the application in good humor. When Blackwell arrived, the school honored its invitation. She graduated in 1849 at the top of her class. For the following two years, she trained in Europe.

From 1851 to 1853, Blackwell was blocked from practicing medicine in New York City because of her sex. She gave lectures on hygiene and developed a base of friends and professional connections. In 1853, she opened a dispensary, which in 1868 became the New York Infirmary for Women and Children, the first institution to have an entirely female medical staff. Blackwell developed a more holistic approach to medicine than was typical in her day: she emphasized prevention and the nurturing role of physicians in maintaining a patient's good health. She continued to lecture, both in the U.S. and Europe, advocating the cause of women in medicine. In 1869, Blackwell left the infirmary's operation to her sister Emily, also a physician, moved to England, and ran a flourishing medical practice.

Blackwell died on May 31, 1910, in Hastings, England.

Legacy

Blackwell is a central figure in the history of women in medicine. During the early Renaissance in Europe, women were generally barred from medical training. In the sixteenth century the study of obstetrics was developing, and some women entered the field; Louise Bourgeois, a pupil of the influential physician AMBROISE PARÉ, was among the most prominent. In 1754, the first German female physician earned her degree, prompting public expressions of shock and outrage at the idea of a woman practicing medicine. This attitude was pervasive in Western society when Blackwell applied to medical schools nearly a century later. Blackwell's determined entrance into the medical profession marked the beginning of a change in people's opinions about women's abilities.

Blackwell's work also influenced public health. Her lectures on hygiene prompted people to maintain living conditions that inhibited the spread of disease. Her dispensary and infirmary provided much-needed medical attention to the destitute women and children of New York City. The infirmary became a model for other such humanitarian medical facilities.

Blackwell spoke out in favor of women's medical education, inspiring and encouraging many women to pursue medicine as a career. As women demanded admittance to medical schools, the doors began to open. Most western European countries graduated their first female physician between 1860 and 1900. In the United States, nearly 10% of students at 18 major medical schools were women by the turn of the century.

Schuyler

WORLD EVENTS	BLACKWELL'S LIFE
Napoleonic Wars 1803-15 in Europe	
	1821 Elizabeth Blackwell is born
	1838-44 Blackwell teaches schoolchildren
	1849 Blackwell receives medical degree from Geneva College
	1850-52 Blackwell obtains further medical training in Europe
	1853 Blackwell opens dispensary in New York City
	1860s Blackwell pushes for women's acceptance into medicine
United States 1861-65 Civil War	
	1868 Blackwell's dispensary becomes N.Y. Infirmary for Women and Children
	1869 Blackwell moves to England and practices medicine
Germany is united 1871	
Spanish-American 1898 War	
	1910 Blackwell dies
World War I 1914-18	

For Further Reading:

Baker, Rachel. *The First Woman Doctor: The Story of Elizabeth Blackwell, MD.* New York: J. Messner, 1944.

Blackwell, Elizabeth. *Pioneer Work in Opening the Medical Profession to Women: Autobiographical Sketches.* New York: Schoken, 1977.

Wilson, Dorothy Clarke. *Lone Woman: the Story of Elizabeth Blackwell, the First Woman Doctor.* Boston: Little, Brown, 1970.

Bohr, Niels

Developer of Theory of
Atomic Structure
1885-1962

Life and Work

Niels Bohr developed the modern understanding of atomic structure and contributed greatly to the fields of quantum mechanics and nuclear fission.

Bohr was born on October 7, 1885, in Copenhagen, Denmark, to a family that inspired and encouraged his early interest in science. He received his doctorate from the University of Copenhagen in 1911 and spent a year in England with ERNEST RUTHERFORD and other leaders in theoretical physics working on the structure of the atom.

Bohr developed a planetary model of the atom with electrons in stable orbits around the nucleus to explain why individual elements

emit only certain frequencies of light (atomic spectra) when excited by high temperatures or electronic discharge. Bohr theorized that electrons emit or absorb radiation as they move from one energy level to another. In 1913 this work culminated in a mathematical description of atomic structure that revolutionized classical physics, incorporating both Rutherford's ideas about the atom and MAX PLANCK's quantum theory. Bohr's theory earned him the Nobel Prize for Physics in 1922.

In 1920 Bohr was appointed director of the University of Copenhagen's new Institute for Theoretical Physics. In 1927 he introduced the concept of complementarity, which includes the proposition that atomic phenomena are best described by calculating the probabilities of the various possible results of a given situation, a departure from linear "one-right-answer" thinking.

In the 1930s Bohr developed the liquid-drop model, which proposed that the nucleus consists of neutrons and protons held together strongly, like molecules in a drop of liquid. His understanding of how a liquid drop ruptures enabled him to offer an accurate description of nuclear fission to ALBERT EINSTEIN immediately after LISE MEITNER and Otto Frisch, in 1939, verified that they had split a uranium nucleus.

Bohr, whose mother was Jewish, escaped German-occupied Denmark via Sweden and traveled to the United States in 1943 but only after King Gustav of Sweden promised to grant refuge to Denmark's Jews. In the United States, he joined the Manhattan Project, the government program to develop the atomic bomb. After World War II, Bohr passionately advocated an international open exchange of information regarding nuclear weapons research; he believed this would reduce the possibility of nuclear war. In 1955 Bohr helped organize the first Atoms For Peace conference in Geneva, Switzerland.

Bohr continued as head of the Institute for Theoretical Physics in Denmark until his death on November 18, 1962, in Copenhagen.

Legacy

Bohr's theoretical insight was integral to the development of atomic and nuclear physics during the first half of the twentieth century.

Bohr's theory of atomic structure continues to serve as the foundation for our modern understanding of the structure of atoms. Bohr's contemporaries built upon his model of atomic structure and multiple attempts to use the model to explain atomic behavior proved successful. Bohr's model was later modified several times, but his basic ideas were the foundation for decades of atomic research.

Bohr's complementarity principle was the first systematic formulation of quantum mechanics, an area of physics that seeks to explain the dynamic systems of subatomic particles (electrons, neutrons, protons, etc.). This formulation, known as the Copenhagen Interpretation because it originated with Bohr and his colleagues at the Institute for Theoretical Physics in Copenhagen, has been described as a revolution in scientific thought. It forced physicists to recognize that, at the atomic level, exact laws of causality do not exist. Furthermore, it is impossible to observe atomic behavior without influencing it and without deciding what data to include and what to ignore. In other words, there may be more than one way to conceptualize nature, and when scientists cannot observe phenomena without influencing outcomes, exact cause-effect relationships do not apply.

Bohr's legacy also includes his humanitarian efforts to help save Danish Jews, his advocacy of peace in the Atomic Age, and the encouragement of his son, Aage Bohr, who earned the Nobel Prize for Physics in 1975.

Schuyler

WORLD EVENTS		BOHR'S LIFE
Germany is united	1871	
	1885	Niels Bohr is born
	1913	Bohr publishes theory of atomic structure
World War I	1914-18	
	1920	Bohr becomes director of the Institute for Theoretical Physics in Copenhagen
	1922	Bohr earns Nobel Prize for Physics
	1927	Bohr introduces Copenhagen Interpretation of Quantum Mechanics
	1930s	Bohr develops liquid-drop model of nucleus
	1939	Bohr describes nuclear fission
World War II	1939-45	
	1943	Bohr flees Denmark
	1944	Bohr conducts atomic bomb research in U.S.
Korean War	1950-53	
	1955	Bohr organizes Atoms For Peace conference
	1962	Bohr dies
End of Vietnam War	1975	

For Further Reading:
Albert, David. *Quantum Mechanics and Experience.* Cambridge, Mass.: Harvard University Press, 1992.
Moore, Ruth. *Niels Bohr: The Man, His Science, and the World They Changed.* New York: Knopf, 1966.
Pais, Abraham. *Niels Bohr's Times: in Physics, Philosophy, and Polity.* New York: Oxford University Press, 1991.

Boltzmann, Ludwig

Developer of Statistical Mechanics
1844-1906

Life and Work

Ludwig Boltzmann pioneered the use of statistics in physics, providing the first explanation of the Second Law of Thermodynamics, the physical law that gives direction to time and to the flow of heat.

He was born in Vienna, Austria, on February 20, 1844. Vienna was called the Paris of eastern Europe during the time Boltzmann was growing up. It was the cultural center for opera, dance, and classical music, and he enjoyed them all. He received his doctorate from the University of Vienna in 1866, and then taught mathematics and physics at Vienna and Graz, in Austria, and Munich and Leipzig, in Germany, for the next 40 years.

Unlike many other scientists of his time, Boltzmann was convinced that everything is made of atoms. He believed that the heat in any object was a form of motion, the motion of the atoms in that object. For example, a cup of cocoa feels hot because its atoms move faster than the atoms in our fingertips. As the minutes pass the cocoa always cools until it reaches the same temperature as the room and all the atoms have the same average speed. Critics of the atomic theory asked how it is that the atoms in the cocoa "know" how to slow down over time. Boltzmann's brilliant solution was to apply the laws of probability and statistics. It could happen that the atoms in the cocoa pull energy from the surrounding air, moving faster and growing warmer by themselves, but because of the great number of atoms in the cup, that outcome is overwhelmingly unlikely. Boltzmann found that the cocoa cools because hot objects are simply more likely to cool than to grow warmer.

The Second Law of Thermodynamics, clearly laid out by Rudolf Clausius when Boltzmann was just a boy, suggested that heat flow seemed to have a preferred direction over time. That law also said that the entropy, or disorder, of a system of objects always increases with time. In an 1875 paper Boltzmann demonstrated that disorder is more likely because there are so many more ways it can happen. Boltzmann was the first to explain the Second Law using statistical methods that came to be known as statistical mechanics. He showed that the Second Law was unique among all the laws of physics as it is based on laws of chance.

Later Boltzmann worked with the British physicist James Clerk Maxwell to further develop the techniques of statistical mechanics. With his former teacher Josef Stefan, he also developed the formula describing the rate of heat production of a hot object, such as the filament of a light bulb, or the surface of a star. He published his results on thermal radiation in 1889.

Troubled by bouts of depression, bothered by other scientists' criticisms of his work, suffering from asthma, angina, severe headaches, and very poor eyesight, Boltzmann hanged himself in his hotel room on September 5, 1906, while on vacation in Duino, Italy.

Legacy

Boltzmann's statistical mechanics helped to add credibility to atomic theory and formed the theoretical basis that eventually led scientists to develop quantum theory. His support of Maxwell, along with that of his many students, led to other significant achievements in physics in the first half of the twentieth century.

Boltzmann's statistical mechanics worked so well that most scientists came to accept the atomic theory by the end of the nineteenth century. Albert Einstein used Boltzmann's methods in his 1905 paper on molecular motion that finally settled the issue and proved that atoms must exist.

Although Boltzmann was critical of many lesser scientists, he praised Maxwell and was probably the first scientist outside of England to recognize the importance of Maxwell's new theories of electromagnetism. His interest and unfailing support spread Maxwell's influence throughout Europe. Einstein later recognized the importance of Maxwell's work and used it, in turn, as a foundation of his 1905 theory of relativity.

Ironically, one of Boltzmann's greatest legacies came through a contemporary who strongly disliked him. Max Planck, the great German scientist, was reluctant to use Boltzmann's statistical approach but found it the only way to solve a stubborn problem about the nature of light. That led to the development of quantum theory, the most important advance in physics in the twentieth century.

Boltzmann was a brilliant but demanding teacher and many of Europe's finest young scientists came to work with him. He cared deeply about his students and gave them extraordinary attention. One of his students, Svante Arrhenius, went on to develop the modern theory of acids and bases. His brightest but shyest pupil was Lise Meitner, who discovered and named the process of nuclear fission, the key to nuclear power plants and bombs.

Secaur

World Events		Boltzmann's Life
Napoleonic Wars in Europe	1803-15	
	1844	Ludwig Boltzmann is born
United States Civil War	1861-65	
	1866	Boltzmann receives doctorate from the University of Vienna
Germany is united	1871	
	1875	Boltzmann publishes paper on statistical method of explaining the Second Law of Thermodynamics
	1889	Boltzmann publishes work on thermal radiation
Spanish-American War	1898	
	1905	Albert Einstein uses Boltzmann's methods in his paper on molecular motion
	1906	Boltzmann commits suicide
World War I	1914-18	

For Further Reading:

Broda, Engelbert. *Ludwig Boltzmann: Man, Physicist, Philosopher.* Woodbridge, Conn.: OxBow Press, 1983.

Greenstein, George. *Portraits of Discovery: Profiles in Scientific Genius.* New York: John Wiley, 1998.

Tolstoy, I. *James Clerk Maxwell.* Chicago: University of Chicago Press, 1981.

Boole, George

Co-founder of Symbolic Logic
1815-1864

Life and Work

George Boole was the first to devise an effective system for working out logical arguments using the tools of mathematics.

George Boole was born on November 2, 1815, in Lincoln, England. He realized early that the only way out of poverty was to educate himself. He taught himself Latin and Greek and after his father, a shopkeeper, taught him the mathematics he knew, George took over his own education in this subject as well.

At the age of 16, Boole began teaching to help supplement his family's income. When he was 20 he opened his own school in Lincoln. He continued his education by reading original sources instead of textbooks. He read some of the work of seventeenth-century mathematicians Pierre-Simon Laplace and Joseph-Louis Lagrange and wrote his first original papers about differential equations and invariance. In 1839 Boole began to publish his results in *The Cambridge Mathematical Journal*. In 1844 he published a paper that examined the interplay of techniques in distant algebra and calculus and was awarded a medal by the Royal Society for these contributions.

Boole then began to apply the ideas of abstract algebra to logic. He devised a system in which the logical arguments were written as mathematical expressions. Boole separated his system into quantities, symbolized by letters, and the operations on them. For example he represented those things that were x or y but not both as $x+y$, and those that were both x and y as xy. Given this system, he argued in his 1847 publication *Mathematical Analysis of Logic* that logic is a part of mathematics. As has often happened in the history of mathematics, the same idea was being suggested by someone else at almost the same time. Augusts De Morgan published *Formal Logic* the same year.

Although Boole was almost entirely self-educated, the strength of his publications landed him a professorship in mathematics at Queen's College, Ireland, in 1849. He continued to work on his symbolic logic and published a more careful exposition of these ideas in the 1854 *An Investigation into the Laws of Thought*. He studied logic and probability until his death from pneumonia on December 8, 1864, in Ballintemple, Ireland.

Legacy

Boole's system of formalized, mathematically expressed logic moved the study of logic forward and eventually would have great significance in the development of computer systems in the twentieth century.

More than a century before Boole, GOTTFRIED LEIBNIZ had tried to formalize logic, without success. At the same time as Boole, De Morgan's work paralleled Boole's but stopped short of a comprehensive system of formalized logic. Boole provided a system where rules could be applied to algebraic expressions, and the expressions would represent propositions from which to derive logical conclusions. This system, called Boolean algebra, was later refined and applied to probability and set theory. More important, it helped to form the basis of modern symbolic logic, a formal system of logic represented by symbols, which was later refined by Gottlob Frege in 1879 and furthered in BERTRAND RUSSELL's and Alfred North Whitehead's *Principia Mathematica* in 1913.

Perhaps the most important application of Boolean algebra came in the twentieth century with the development of telephone switching systems and digital computers. Boole's rules are the same as those in a binary system; they can be applied to situations where the values are only 0 and 1, such as a switch that can be either off (0) or on (1), or a test case that can be either false (0) or true (1). This same system, used exclusively in telephone switching and digital computers, has helped to launch a revolution in information processing and communications that will extend into the next century.

Steinberg

WORLD EVENTS	BOOLE'S LIFE
Napoleonic Wars 1803-15 in Europe	
	1815 George Boole is born
	1831 Boole begins teaching
	1835 Boole opens his own school
	1839 Boole's work on differential equations is published in *The Cambridge Mathematical Journal*
	1844 Royal Society awards Boole medal
	1847 Boole publishes *Mathematical Analysis of Logic*
	1854 Boole publishes *An Investigation into the Laws of Thought*, outlining his system of symbolic logic
	1864 Boole dies
Germany is united 1871	

For Further Reading:

Kline, Morris. *Mathematical Thought from Ancient to Modern Times.* New York: Oxford University Press, 1972.

Lewis, Clarence Irving, and Cooper Harold Langford. "History of Symbolic Logic." In *The World of Mathematics,* edited by James R. Newman. New York: Simon & Schuster, 1956.

MacHale, Desmond. *George Boole: His Life and Work.* Dublin: Boole Press, 1985.

McLeish, John. *Number.* New York: Fawcett Columbine, 1992.

Moschkowski, Herbert. *Ways of Thought of Great Mathematicians.* San Francisco: Holden-Day, 1964.

Borlaug, Norman Ernest

Agronomist; Founder of the
Green Revolution
1914-

Life and Work

Norman Borlaug launched the Green Revolution in food production to the level that global food production exceeded population growth.

Born in Cresco, Iowa, on March 25, 1914, Borlaug entered college as the Depression began. He dropped out of school several times to earn tuition for his degrees in forestry and botany. He earned a Ph.D. in plant pathology from the University of Minnesota in 1941.

Borlaug researched the transportation of a particular fungus, called rust, that plagues corn and wheat. He traced movements of rust spores to global harvest cycles, a surprising finding at a time when the jet stream, which carries the spores, was unknown. Results extended Borlaug's interests beyond the Midwest, where "dust bowl" conditions threatened survival of some American farmers. Borlaug noticed that drought had the least effect in the United States wherever high-yield approaches to farming were used. He decided to focus his work on spreading the benefits of high-yield technology to nations where crop failures were recurring, systemic nightmares.

From 1944 to 1960, Borlaug worked in Mexico as a research scientist for the Rockefeller Foundation's Cooperative Mexican Agricultural Program. Later he directed the Inter-American Food Crop Program (1960-1963) and the International Maize and Wheat Improvement Center in Mexico (1964-1979). While working in Mexico he promoted innovations such as shuttle breeding, a technique for developing disease immunity among strains of crops; cereals that could tolerate many climates by eliminating their sensitivity to the number of daylight hours; triticale, a high-protein rye-wheat hybrid; and a dwarf wheat (short plants expend less energy on inedible stalks and leaves in proportion to the grain).

In 1963 Borlaug went to India and Pakistan to promote dwarf wheat. By 1965 famine was so acute that the governments of these two countries agreed to try his methods. A successful first crop weakened official resistance to change. The next harvest doubled yield per acre in Pakistan and, by 1968, that country was self-sufficient in wheat production. India met its domestic needs in cereal production in 1974 and produced some surplus in the mid-1980s.

Borlaug was awarded the Nobel Prize for Peace in 1970 for his contributions to alleviating hunger and poverty. Borlaug began working in Africa in 1981. Since then, he has worked with private organizations to promote high-yield agriculture along with reduction of pesticide use.

Legacy

From 1965 to 1990, malnutrition decreased even though two billion people were added to the world's population. Thanks mainly to Borlaug's efforts, harvesting more food from fewer acres has delayed the prospects of too little food for too many people.

The Green Revolution in high-yield agriculture has been criticized, however, for its dependence on fertilizer and irrigation and for interfering with a natural constraint (famine) on population growth. The counter-argument is that a transition to lower birth rates occurs only after development reduces the need for child labor and provides potential parents with educational and career options that also tend to reduce family size.

Expensive and environmentally risky pesticides have also been used more sparingly since Borlaug and his fellow agronomists began advocating integrated pest management. This system of controls includes spraying only at the most vulnerable stage in the insect's life cycle; use of natural predators; crop rotation; and development of pest-resistant hybrids.

Famine has not been eliminated, but Borlaug's work has bought time and improved the health of people who must now work toward longer-term solutions to world hunger. As his supporters claim, Norman Borlaug has probably already saved more lives than any other person in the history of the world.

Simonis

WORLD EVENTS	BORLAUG'S LIFE
World War I 1914-18	
	1914 Norman Borlaug is born
Word War II 1939-45	
	1941 Borlaug earns Ph.D. in plant pathology at University of Minnesota
	1944 Borlaug begins work in Mexico under Rockefeller Foundation
Korean War 1950-53	
	1960-63 Borlaug directs Inter-American Food Crop Program
	1963 Borlaug begins work in India and Pakistan
	1964-79 Borlaug directs International Maize and Wheat Improvement Center in Mexico
	1970 Borlaug is awarded Nobel Prize for Peace
	1981 Borlaug begins work in Africa
Dissolution of Soviet Union	1991

For Further Reading:

Bickel, Lennard. *Facing Starvation: Norman Borlaug and the Fight Against Hunger.* Pleasantville, N.Y.: Reader's Digest Press, 1974.

Easterbrook, Gregg. "Forgotten Benefactor of Humanity." *The Atlantic Monthly.* January 1997.

Born, Max

Early Pioneer in Quantum Mechanics
1882-1970

Life and Work

Max Born perfected matrix mechanics and developed the probability interpretation of the wave nature of matter; both were crucial steps in the development of quantum mechanics.

Born was born on December 11, 1882, in Breslau, Prussia (now Wroclaw, Poland). His early interests in science and music were fostered by his parents, an anatomy professor and a musician. He enrolled at the University of Breslau in 1901, where his brilliance, eclectic interests, and independent spirit were noted and encouraged by prestigious scientists. In 1907 he received a Ph.D. in physics from the University of Göttingen in Germany.

During World War I, Born began working with German physicists MAX PLANCK and ALBERT EINSTEIN, with whom he maintained life-long friendships.

In 1921 Born returned to Göttingen as director of the university's Physical Institute. He was soon intrigued by Danish physicist NIELS BOHR's newly introduced quantum theory of matter, an attempt to explain the puzzling behavior and wave-like nature of subatomic particles. Born set out to formulate a new set of physical laws based on this subject, which he called "quantum mechanics."

Born's student, WERNER HEISENBERG, developed a mathematical approach to quantum mechanics; in 1925 Born expanded and refined the approach into a system called matrix mechanics. It described the relationship between the position and momentum of an electron within an atom and accounted for other subatomic phenomena.

The following year Austrian physicist Erwin Schrödinger proposed a different system of quantum mathematics called wave mechanics, which was consistent with Born's approach but easier to follow. Wave mechanics included an equation called the wave function that could be used to analyze the properties of subatomic particles.

Schrödinger's wave function was purely mathematical and embodied no physical meaning, but in 1926 Born proposed a physical interpretation that solved this problem. He suggested that the wave function of an electron tells something about the probability of finding the electron at a particular location within the atom. Although difficult to conceptualize, this interpretation was consistent with what was known about quantum mechanics and was widely accepted among scientists.

For this achievement Born received the 1954 Nobel Prize for Physics. The prize was shared with German physicist Walther Bothe, in recognition of unrelated work on cosmic rays.

Forced by the Nazi regime to leave Germany because of his Jewish heritage, Born taught in England during and after World War II. He retired in 1953 and died in Göttingen on January 5, 1970.

Legacy

Born helped establish the modern understanding of quantum mechanics, which represents a revolutionary view of matter and is a theoretical foundation of twentieth-century physics.

Born joined the investigation of the subatomic realm just as the central formulations of the new quantum mechanics were being made. In 1900 Planck discovered that energy is emitted and absorbed in discrete packets, or quanta; this concept formed the foundation of quantum theory. Planck and Einstein soon showed that waves have particle-like characteristics, and in 1924 French physicist Louis de Broglie demonstrated that particles have wave-like characteristics. The mathematics of Schrödinger and Heisenberg provided the basis for Born's probability interpretation, which was the foundation upon which Bohr finally developed the Copenhagen Interpretation of quantum mechanics (presented in 1927), the leading modern approach to subatomic physics.

Further developments followed Born's treatment of quantum mechanics. In 1927 Heisenberg introduced the uncertainty principle, which states that it is impossible to know both the position and momentum of a particle at a given time because when one is measured the other is unavoidably altered. This concept, which for the first time removed the absolute determinacy inherent in classical physics, deeply troubled some physicists (including Einstein), but has become widely accepted.

The principles of quantum mechanics were also applied to chemistry. In 1927 the properties of the chemical bond between the atoms of a hydrogen molecule were explained by quantum mathematics and later extrapolated to other molecules. During the next decade, quantum mechanics helped build the modern model of atomic and subatomic structure. The model is continually refined as more particles are discovered and as quantum theory becomes more sophisticated.

Quantum mechanics, originally an attempt to explain the results of experiments with subatomic phenomena, forced scientists to alter their views of nature and to accept theories that are not intuitively comprehensible. The development of quantum theory launched a revolution in scientific thought that continues to unfold.

Schuyler

WORLD EVENTS		BORN'S LIFE
Germany is united	1871	
	1882	Max Born is born
World War I	1914-18	
	1923	Born begins studying quantum theory
	1925	Born refines Werner Heisenberg's matrix mechanics
	1926	Erwin Schrödinger formulates wave mechanics
	1926	Born introduces probability interpretation
	1927	Copenhagen Interpretation of quantum mechanics presented
World War II	1939-45	
	1954	Born wins Nobel Prize for Physics
	1970	Born dies
End of Vietnam War	1975	

For Further Reading:
Albert, David. *Quantum Mechanics and Experience*. Cambridge, Mass.: Harvard University Press, 1992.
Ballentine, Leslie E. *Quantum Mechanics*. Englewood Cliffs, N.J.: Prentice-Hall, 1990.
Born, Irene, trans. *The Bohr-Einstein Letters*. New York: Walker, 1971.
Zukav, Gary. *The Dancing Wu Li Masters: An Overview of the New Physics*. New York: Bantam Books, 1980.

Bose, Satyendranath

Pioneer in Quantum Mechanics
1894-1974

Life and Work

Satyendranath Bose was the first person to derive the fundamental idea of quantum theory without having to start from assumptions based on older theories. He also developed the mathematical rules that describe the boson, a type of elementary particle, named in his honor.

Bose was born in Calcutta, India, on January 1, 1894. India was for most of his life a colony of Great Britain, and the teachers and professors who were sent from Great Britain to India often felt Indian students were inferior. Such discriminatory attitudes were reflected in the college physics laboratories that Bose used, which were often poorly equipped and run by professors who didn't take the students seriously. Nevertheless, Bose earned a doctorate from the University of Calcutta.

In 1914 a new science department opened at the University of Calcutta, the oldest and largest university in India. It was funded by wealthy contributors in India who required the teachers to be Indian rather than European. Bose and his friend M. N. Saha were the first two Indian scientists appointed to the new University College of Science. Together they published several papers, including, in 1919, the first English translation of ALBERT EINSTEIN's papers on relativity.

Five years later, in June of 1924, Bose sent Einstein a paper in which he demonstrated mathematically that light energy must be transferred in tiny bundles, which we now call quanta. More than 20 years earlier, German scientist MAX PLANCK had first made such a calculation, which established the basis for quantum theory but used some assumptions and methods from nineteenth-century classical electromagnetic theory. Bose was the first to see how the fundamental ideas of the new theory could be derived from basic properties of space and energy, without using older theories. Einstein liked the paper, translated it into German, and had it published for Bose, thus beginning a lifelong collaboration between the two scientists.

Bose's purpose in his 1924 paper was to simplify and purify the foundation of quantum theory. But Einstein realized that Bose's work accurately described the behavior of an ideal gas at the atomic level in terms of quantum theory. At very low temperatures, Bose and Einstein predicted, atoms in a gas would link together in a mysterious way and act as one great atom. This phenomenon was later referred to as a Bose-Einstein Condensate. Bose also introduced what is now referred to as Bose-Einstein Statistics in his 1924 paper. These mathematical rules developed by Bose and Einstein describe the behavior of bosons, elementary subatomic particles characterized by their spin.

Bose traveled in Europe, meeting many of the leading scientists of the time, but he was quiet and shy and was never able to think of himself as their intellectual equal. He returned to India in 1926 and became professor of physics at the university in Dacca, where he taught until he returned to the University of Calcutta to teach in 1945. He retired from teaching in 1956 and died in Calcutta on February 4, 1974.

Legacy

Bose's work augmented the theoretical underpinnings of particle physics and quantum theory, and helped to develop quantum mechanics, a system of mechanics describing how elementary particles behave.

Bose-Einstein Statistics provided one step in the ongoing development of quantum theory and particle physics. Another set of statistics called Fermi-Dirac Statistics, also developed in the mid-1920s by ENRICO FERMI and P. A. M. Dirac, described the behavior of the fermion, an elementary particle having a type of spin distinct from the boson. Since the mid-1920s other elementary particles have been discovered, including the pi- and mu-meson predicted by HIDEKI YUKAWA in 1935, and "strange particles" (lighter than protons but heavier than mesons) discovered by British physicists Clifford Butler and George Rochester in 1947. They were so named because of their strange behavior in relationship to nuclear matter. In 1964 MURRAY GELL-MANN and American scientist George Zweig independently proposed that these var-

ious elementary particles were made of smaller components that Gell-Mann later termed quarks, the word still used today.

The Bose-Einstein Condensate, in which atoms in a gas condense at very low temperatures, has contemporary ramifications. In 1995 two American scientists succeeded in cooling a few million rubidium atoms down to less than a degree above absolute zero: the atoms merged, the first example of a Bose-Einstein Condensate in a laboratory. Electrons in superconductors, metals that have no electrical resistance at low temperatures, can pair up and act just as Bose and Einstein predicted. While practical applications for the Bose-Einstein Condensate have not yet been developed, the model of behavior they provide—even about superconductors—is invaluable. Superconductors are already useful in making powerful and energy-efficient magnets.

Secaur

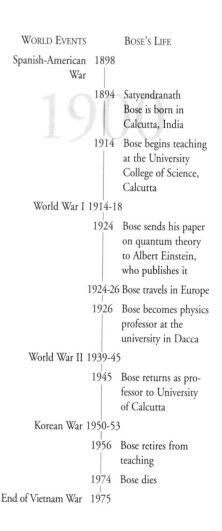

WORLD EVENTS		BOSE'S LIFE
Spanish-American War	1898	
	1894	Satyendranath Bose is born in Calcutta, India
	1914	Bose begins teaching at the University College of Science, Calcutta
World War I	1914-18	
	1924	Bose sends his paper on quantum theory to Albert Einstein, who publishes it
	1924-26	Bose travels in Europe
	1926	Bose becomes physics professor at the university in Dacca
World War II	1939-45	
	1945	Bose returns as professor to University of Calcutta
Korean War	1950-53	
	1956	Bose retires from teaching
	1974	Bose dies
End of Vietnam War	1975	

For Further Reading:

Blanpied, W. A. "Satyendranath Bose: Co-founder of Quantum Statistics." *American Journal of Physics.* (September 1972).

Keller, A. *The Infancy of Atomic Physics: Hercules in His Cradle.* New York: Oxford University Press, 1983.

Newman, Harvey B., and Thomas Ypsilantis, eds. *History of Original Ideas and Basic Discoveries in Particle Physics.* New York: Plenum Press, 1996.

Boyle, Robert

Founder of Chemistry

1627-1691

Life and Work

Robert Boyle established the field of chemistry as a science distinct from alchemy and medicine, and founded modern experimental chemical analysis and chemical theory.

Boyle was born on January 25, 1627, to a wealthy aristocratic family in Lismore Castle, Ireland. He attended schools in Switzerland and Italy, where he developed an enthusiasm for science while reading GALILEO's accounts of his astronomical discoveries. From 1644 to 1654, he retired to his family estate at Stalbridge in Dorsetshire, England, and there he began to investigate pneumatics, the study of the properties of gases.

WORLD EVENTS	BOYLE'S LIFE
Thirty Years' War 1618-48 in Europe	
1627	Robert Boyle is born
1640s	Boyle reads Galileo's works
1644-54	Boyle resides at family estate and studies pneumatics
1654	Boyle moves to Oxford and works with Robert Hooke
1661	Boyle postulates particulate matter theory in *The Sceptical Chemist*
1662	Boyle articulates Boyle's Law on gas pressures and co-founds Royal Society for Improving Natural Knowledge
1691	Boyle dies
French Revolution 1789	

In 1654, Boyle moved to Oxford, where he began his scientific work in earnest. Enlisting the help of English physicist ROBERT HOOKE, he constructed an air pump with which he demonstrated the role of air in combustion, respiration, and sound transmission. Boyle joined a group of scholars who advocated FRANCIS BACON's philosophy that scientific investigation should rely on the empirical evidence gained through observation and experimentation. In 1662 the group became the Royal Society for Improving Natural Knowledge, also known as the Royal Society and the Royal Society of London.

Boyle attacked the traditional Aristotelian theory that all matter consists of four elements: earth, air, fire, and water. In *The Sceptical Chemist* (1661), he proposed that matter is made up of primary particles that coalesce into corpuscles, and that different types of matter are distinguished by the number, position, and motion of the corpuscles they contain.

Boyle made numerous other discoveries. He was the first person to suggest that heat was a result of the motion of the corpuscles (molecules) and that air is a mixture of gases. He was also the first chemist to collect a gas and to devise equipment to test the properties of gases. In 1662, he synthesized the results of his experiments into what is known as Boyle's Law, which states that pressure times the volume of a gas will be constant if the temperature is constant. For example, doubling the pressure on a gas squeezes it into half its original volume if the temperature remains the same.

In 1668, Boyle moved to London, where he lived with his sister until his death on December 30, 1691.

Legacy

Boyle's innovative experiments and theories set the stage for chemists in the eighteenth and nineteenth centuries to develop the theories and methodologies of modern chemistry.

Boyle is known as the "father of modern chemistry"; he is thought to be the first person to have engaged in chemistry as a pure science. Although Boyle maintained some alchemical views regarding transmutation, he saw chemistry as a science distinct from both alchemy

and medicine. Alchemy was a medieval art whose aims were to achieve the transmutation of base metals into gold or silver, to discover a universal cure for disease, and to find the means to prolong life indefinitely. Boyle's extensive experimentation and rigorous application of scientific procedures distinguished his work from the alchemists of the day, who were more interested in the practical goals of transmutation and curing diseases.

Although Boyle stopped short of postulating different kinds of primary elements, his corpuscular theory was significant because it was among the first to break from the dominant four-element theory. Throughout the eighteenth century, scientists discovered new elements, as the concept of different kinds of primary elements took root. By 1800, 34 elements had been identified. Boyle's ideas also moved chemistry toward a theory of elemental particles (atoms) and congregations of elemental particles (molecules).

Boyle's Law is among the most important mathematical relationships in chemistry. The law holds for most gases, except at extremely high pressures, and thus facilitates quantitative analyses of experiments with gases. The most common application of Boyle's Law is to calculate the volume of gases consumed or produced by a chemical reaction.

One of Boyle's most significant contributions to chemistry was the importance he placed on experimentation and accurate observation.

Schuyler

For Further Reading:

Hunter, Michael, ed. *Robert Boyle Reconsidered.* New York: Cambridge University Press, 1994.

More, Louis Trenchard. *The Life and Works of the Honorable Robert Boyle.* New York: Oxford University Press, 1944.

Brahe, Tycho

Discoverer of Supernova
1546-1601

Life and Work

Tycho Brahe performed the most accurate observations of any pre-telescope astronomer. He corrected extensive errors in previous astronomical tables, paving the way for future studies of the solar system.

Brahe was born on December 14, 1546, in Knudstrup, Denmark (now in southern Sweden), and was raised by a wealthy uncle. In 1560 he witnessed a total eclipse of the Sun; awestruck, he determined to continue astronomical observations. He attended the Universities of Copenhagen and Leipzig to study law, in accordance with his uncle's wishes, but at night he examined and charted the star-studded sky.

In 1563 Brahe observed and recorded an overlapping of Jupiter and Saturn with the aid only of crude compasses and globes. When he investigated existing astronomical almanacs, he found that the measurements compiled by COPERNICUS were inaccurate. With this revelation, he decided to devote himself to correcting faulty astronomical data through careful observation and calculation.

Brahe made the groundbreaking observation of a supernova, an exploding star, in the constellation Cassiopeia in 1572. Because the star had been previously invisible to the naked eye, the event caused an uproar among intellectuals who believed in the unchanging nature of the heavens. After publishing his finding in 1573, Brahe became a celebrated astronomer.

Denmark's King Frederick II offered Brahe funds for an astronomical observatory to be constructed on the island of Hven (now Ven) in 1576. Until 1597 the artfully adorned and well-equipped Uraniborg observatory served as a major center for astronomical research. Brahe determined the correct positions of more than 700 fixed stars and improved the measurements of planetary positions using a huge quadrant, righting nearly every known astronomical inaccuracy. Without a telescope, he calculated the length of the year from his own measurements with less than a one second error. Yet Brahe never fully accepted Copernicus's sun-centered solar system, instead proposing that the planets orbit the Sun, which itself orbits a stationary Earth.

Funds for Uraniborg began to dwindle following the death of Frederick II in 1588, and in 1597 Brahe moved to Prague. Brahe acquired astronomer JOHANNES KEPLER as an assistant in 1600; he taught him the importance of making multiple observations over long periods to get complete records of planetary cycles. The two worked together until Brahe's death on October 24, 1601.

Legacy

Brahe's observations and corrections provided a framework for the developments in astronomy and physics that took place after the introduction of the telescope in the early seventeenth century.

Brahe's supernova discovery helped dislodge ancient philosophical doctrine. Many intellectuals of the time believed that the future was secured by the harmonious continuity of the whole world and that such harmony was ruled by the perfect and unchanging stars. The foundations of this belief system, established by Aristotle in the fourth century B.C.E., were shaken by the supernova appearing where no star had previously been visible. Brahe's observations, along with the eventual acceptance of the Sun-centered view of the solar system, caused a revolution in seventeenth-century thought.

Brahe's extensive data contributed to Kepler's formulation of the laws of planetary motion, which accurately describe the planets' paths, speeds, and periods (the time it takes to complete a cycle of motion). In 1627 Kepler published the Rudolphine Tables, improved tables of planetary motion resulting from the short collaboration between himself and Brahe. The tables, upon which modern ones are still based, allowed astronomers to calculate planetary positions at any time in the past or future.

Kepler's laws led ISAAC NEWTON to his theory of gravitational force, which governs objects on Earth's surface as well as the orbits of the moon and planets.

Schuyler

WORLD EVENTS		BRAHE'S LIFE
Reformation begins	1517	
	1546	Tycho Brahe is born
	1560	Brahe witnesses solar eclipse
	1563	Brahe observes overlapping of Jupiter and Saturn
	1572	Brahe discovers a supernova
	1576	Construction on Uraniborg observatory begins
	1588	Brahe's benefactor King Frederick II dies
	1597	Brahe leaves Denmark for Prague
	1600	Johannes Kepler joins Brahe in Prague as his assistant
	1601	Brahe dies
Thirty Years' War 1618-48 in Europe		

For Further Reading:

Chapman, Allan. *Astronomical Instruments and Their Users: Tycho Brahe to William Lassell.* Brookfield, Vt.: Variorum, 1996.

Thoren, Victor E. *The Lord of Uraniborg: A Biography of Tycho Brahe.* New York: Cambridge University Press, 1990.

Braille, Louis

Inventor of Braille Alphabet
1809-1852

World Events	Braille's Life
Napoleonic Wars 1803-15 in Europe	
	1809 Louis Braille is born
	1812 Braille is blinded
	1819 Braille attends National Institute for the Blind
	1825 Braille introduces dot-based alphabet
	1826 Braille begins teaching at National Institute for the Blind
	1829 Braille writes first treatise on his alphabet for the blind
	1837 Braille completes second treatise on his alphabet
	1852 Braille dies
United States 1861-65 Civil War	
Germany is united 1871	
	1878 Braille alphabet is presented at International Congress in Paris

Life and Work

Louis Braille invented the written alphabet of raised dots, which is known by his name and used by blind people in many countries around the world.

Braille was born at Coupray, France, on January 4, 1809. At age three, he was blinded by an accident in his father's harness shop; an awl slipped and entered the boy's eye. The injury eventually caused sympathetic blindness in the uninjured eye. Braille attended the village school and forged ahead of his classmates despite his handicap.

In 1819, he traveled to Paris to continue his education at the National Institute for the Blind, and he began teaching there in 1826. He also became renowned in Paris for his accomplishments as an organist and cellist.

While a student, Braille became concerned about the limited reading and writing opportunities of blind people; he and his peers had access to only a few bulky books with enlarged, embossed letters. Braille began searching for a new printing and reading technique. He was inspired by Charles Barbier, a French army officer who had developed a system of raised dots and dashes for written communication on battlefields at night. Braille recognized that such a system had potential as a method of touch reading because it was simple and would fit within the span of the fingertips.

In 1825 Braille introduced a new alphabet based on the "Braille cell," a six-point unit similar to the dots of a domino. He assigned a different spatial combination of the six dots to each of the 26 letters of the Roman alphabet. He also designated dot values to numbers and musical notation. The Braille alphabet enabled students to read and also to write, by using a simple sliding metal rule with windowlike openings for punching dots out of thick paper. Braille wrote treatises about his alphabet in 1829 and 1837.

Braille became ill with tuberculosis and died on January 6, 1852, in Paris.

Legacy

The Braille alphabet enabled blind people to read and write in an efficient manner—improving their education, opportunities, and acceptance in society.

Braille's system appeared more than three decades after the idea of writing in relief for the blind was first introduced. In 1793 Valentin Haèuy, the founder of the National Institute for the Blind, designed an enlarged italic embossed print. It was used, with some modification, until Braille introduced his dot-based alphabet.

Braille's system opened new doors for the blind. It enabled them to read and write and to rely on a standardized, widespread system of reading and writing. Blind people were able to study extensively, perform complex mathematical calculations, transcribe musical notation, and correspond with each other through the mail. The Braille alphabet allowed blind people to live more normal lives and to pursue careers and activities that would have been much more difficult without an education involving reading and writing.

In the 1830s Englishman James H. Frere improved Braille's alphabet by proposing the return-line system, in which one line reads from left to right, and the next follows from right to left, so that the finger does not lose its place between lines.

The Braille alphabet came into widespread use outside of France after it was presented at the 1878 International Congress in Paris.

In the twentieth century, technological progress was harnessed to produce a more extensive and more convenient library of books for the blind. The early books printed in Braille were so large and bulky that their practical use was limited. The modern Braille library is stored on a computer, and when needed, Braille text is transcribed from the computer onto tactile paper by a special printer.

Schuyler

For Further Reading:
Bickel, Lennard. *Triumph Over Darkness: The Life of Louis Braille.* Sydney, Australia: Allen & Unwin Australia, 1988.
Kuglemass, J. Alvin. *Louis Braille: Windows for the Blind.* New York: Messner, 1951.

Bramah, Joseph

Inventor of Hydraulic Press
and Bramah Lock
1748-1814

Life and Work

Joseph Bramah invented the hydraulic press, the thief-proof Bramah lock, and other useful items. He was also indirectly responsible for the training of a whole generation of engineers in the craft of precision design and manufacturing.

Bramah was born on April 13, 1748, in Stainborough, England. At age 16 he was injured and made lame in an accident while working on his father's farm. After completing an apprenticeship to a carpenter, he relocated to London, opened his own business, and began to build mechanical instruments and tools. Among his first well-known inventions was the flushable toilet, introduced in 1778.

In 1784 Bramah patented the design for a safety lock, subsequently called the Bramah lock, which involved a hollow key with grooves of unequal lengths. The grooves pressed on corresponding blades within the lock. Bramah had taken advantage of his mastery of high-precision engineering and his shop of high-quality machine tools to construct this superior lock. He advertised an award for whomever could pick the lock, but it was not until 1851, long after his death, that a

mechanic was able to open the lock after 51 hours of trying.

Bramah had hired an extremely talented assistant named Henry Maudslay, who was crucial to the successful production of both the Bramah lock and the next invention, the hydraulic press. Devised in 1795, this instrument was a practical application of Pascal's Law, which states that the pressure exerted on a confined liquid is constant in all directions. In Bramah's hydraulic press, a small amount of force was applied to a confined amount of liquid, which in turn amplified the force by pressing on a larger surface.

Bramah also invented a paper-making machine, a device for numbering bank notes, and a pump for extracting beer from cellars. He died on December 9, 1814, in London, England.

Legacy

Bramah's inventiveness and success at realizing his ideas in precise detail laid the foundations for the style of engineering that made the Industrial Revolution possible. He was responsible also for advances in hydraulics and lock design.

Bramah's assistant Maudslay, who constructed many of Bramah's machine tools and used them to build the devices that Bramah patented and marketed, developed a reputation as a skilled mechanic. After working for Bramah, he set up business on his own and invented and manufactured an improved lathe with an extremely accurate lead screw. Joseph Clement, who later developed the first self-regulating lathe, was among Maudslay's students; Clement himself became well respected and was hired by Charles Babbage, who designed a model of one of the earliest precursors of the computer. Many other inventors of the early nineteenth century were educated and inspired by Bramah and Maudslay to execute precise, accurate engineering practices.

The use of hydraulics traces its origin to the development of Bramah's press. The press was used immediately in the lifting of massive iron and steel parts for large construction projects. Robert Stephenson, son of steam-locomotive pioneer George Stephenson, used the hydraulic press to position the enormous tubular spans of the Menai Straits bridge,

erected in 1850, which carried the London to Holyhead railway and was of political and economic importance in England.

Hydraulic technology has produced hydraulic jacks, mechanical excavators, and presses for consolidating and binding waste paper and metal. Automobile brakes embody perhaps the most far-reaching impact of hydraulic systems. Hydraulic brakes, which ensure that the braking action occurs smoothly and operates evenly on all the wheels, made cars and other vehicles safer.

The technology of locks also advanced by the end of the nineteenth century. The Yale lock was an improvement that allowed an infinite number of combinations, and timing systems were added to bank safes so that locks could be opened only at specific times.

Schuyler

World Events		Bramah's Life	
Peace of Utrecht settles War of Spanish Succession	1713-15		
		1748	Joseph Bramah is born
United States independence	1776		
		1778	Bramah introduces flushable toilet
		1784	Bramah designs pick-proof lock; offers award to anyone who can pick his lock
French Revolution	1789		
		1795	Bramah builds hydraulic press
		1814	Bramah dies
		1850	Menai Straits bridge is completed using hydraulic press
		1851	Mechanic picks Bramah's lock after 51 hours of trying
Germany is united	1871		

For Further Reading:

McNeil, Ian. *Joseph Bramah: A Century of Invention, 1749-1851.* New York: A. M. Kelley, 1968.

Roper, C. A. *The Complete Book of Locks and Locksmithing.* Blue Ridge Summit, Penn.: Tab Books, 1983.

Brooks, Harriet

Pioneering Researcher in
Radioactivity
1876-1933

Life and Work

Harriet Brooks was a pioneer in studying radioactivity. She contributed to the understanding of the nature of radon and the radioactive decay series (the transmutation of elements from thorium to lead in particular).

Brooks was born in Ontario, Canada, on July 2, 1876. Her father was a salesman who moved his large family frequently during Brooks's school years. Without encouragement from her family, Brooks decided to go to McGill University, Montreal, in 1894 to gain a teaching diploma. This unconventional decision was one of the few ways she could leave home without marrying and gain some financial independence.

McGill University did admit women but provided separate classrooms and instructors for them during the first two years of undergraduate coursework. Brooks graduated with honors, obtaining her teaching diploma in 1898. She then joined the research group of physicist ERNEST RUTHERFORD, who discovered the atomic nucleus, at McGill; she was his first graduate student. She completed research on electricity and magnetism that earned her a master's degree in physics in 1901.

Brooks began work with radioactive materials in 1899. She identified one of the products of thorium's decay as a heavy gas (radon), leading Rutherford and Frederick Soddy to the realization that transmutation of that element had occurred.

Brooks continued to explore radioactivity while she held a series of short-term positions at Bryn Mawr College (1901-1902), the Cavendish Laboratory, England, under J. J. Thompson (1902-1903), McGill University (1903-1904), Barnard College (1904-1906), and Laboratoire Curie, Paris (1906-1907) with Marie Curie.

Brooks made the first measurements of the half-life of thorium, and she tracked small electric charges that radioactive materials induced. She also demonstrated that radiation emitted from several natural substances was composed of identical high energy beta-particles (electrons) regardless of their source.

In 1904 she published two major papers in nuclear physics. "A Volatile Product from Radium" described her observation that when radium decays to radon and then to polonium, walls of the testing vessel become radioactive. She was the first to record this recoil effect of nuclear radiation. The second paper, "Decay of Excited Radioactivity from Thorium, Radium, and Actinium," demonstrated that at least two successive transmutations are typical of the decay of the named elements before they reach a stable state (lead).

Brooks lost her teaching position at Barnard College in 1906 when she announced her engagement. "The dignity of women's place in the home demands that your marriage shall be a resignation," Dean Gill wrote. When she married Frank Pitcher of Montreal in 1907, Brooks abandoned her search for professional employment.

Brooks died in Montreal of leukemia on April 17, 1933.

Legacy

Brooks's research was fundamental to understanding the sequence of events in radioactivity.

Her contributions focused on radon gas and the elements naturally following from its decay. Brooks's identification of radon as a separate element, not a volatile form of thorium, was the pioneering step that led to Rutherford and Soddy's startling assumption in 1902 that transmutation of one element into another lighter element must have occurred. In his papers and public lectures, Rutherford repeatedly gave credit to Brooks for her vital work. He was recognized as a Nobel laureate in 1908 for the breakthrough in the understanding of matter that her studies in radioactivity made possible.

Her demonstration that radiation emitted the same beta-particles from several different substances pointed to the existence of electrons. This type of evidence fired NIELS BOHR's imagination in the 1910s as he developed a new model of atomic structure.

Brooks's other legacy rests in her position as a woman in science. Marriage or career at lowest wages in academia were the only obvious choices for a woman without independent income, yet Brooks made important contributions in the positions she was able to secure.

Simonis

Timeline

WORLD EVENTS		BROOKS'S LIFE
United States Civil War	1861-65	
Germany is united	1871	
	1876	Harriet Brooks is born
	1894	Brooks enters McGill University
Spanish-American War	1898	Brooks becomes Ernest Rutherford's graduate assistant
	1899	Brooks starts radioactivity investigations and identifies radon
	1901	Brooks earns master's degree in physics
	1902-03	Brooks works with J. J. Thompson in England
	1904	Brooks publishes two major papers on nuclear physics
	1906-07	Brooks works with Marie Curie in Paris
	1907	Brooks marries Frank Pitcher and is forced to abandon teaching
World War I	1914-18	
	1933	Brooks dies
World War II	1939-45	

For Further Reading:

Rayner-Canham, M., and G. Rayner-Canham. *A Devotion to Their Science: Pioneer Women of Radioactivity*. Philadelphia Penn.: Chemical Heritage Foundation, 1997.

———. *Harriet Brooks: Pioneer Nuclear Scientist*. Montreal: Queen's University Press, 1992.

Brown, Robert

Botanist; Discoverer of
Brownian Motion
1773-1858

Life and Work

Robert Brown discovered Brownian motion, identified the difference between gymnosperms and angiosperms, recognized the nucleus as a constant cellular element, and refined plant classification.

Brown was born on December 21, 1773, in Montrose, Scotland. He studied medicine at the University of Edinburgh, but never completed a degree.

Upon meeting the noted botanist Joseph Banks in 1798, Brown had his first opportunity to exercise his interest in plants. Banks helped him obtain the position of naturalist aboard the ship *Investigator,* which set sail to explore Australia in 1801.

Returning to England in 1805, Brown painstakingly classified the approximately 4,000 botanical specimens (most of which were unknown) that he had collected during the voyage to Australia. He used and improved the classification system of French botanist Antoine-Laurent de Jussieu, which took into account more of a plant's physical, microscopic characteristics than the prevailing Linnaean system.

Brown was librarian at Banks's residence from 1810 to 1820. In 1827 the extensive library and botanical collection were transferred to the British Museum, where Brown then became curator.

In 1827 Brown observed under a microscope that pollen grains suspended in water inexplicably moved about randomly. The water itself was still, and he wondered if the movement was due to some kernel of life within the grains. He tested this idea by suspending inorganic dye particles in water; they too exhibited random motion. The following year he published these findings, and the phenomenon has since been called Brownian motion.

During his detailed investigations of plant morphology (the study of the size, shape, and structure of plants), Brown became the first to recognize the fundamental difference between gymnosperms (conifers and related plants) and angiosperms (flowering plants): angiosperm seeds develop within closed chambers, or ovaries, and gymnosperms lack these chambers.

In 1831 Brown observed that all plant cells contain a body, previously noticed but ignored by others, that he suspected might be an important biological element. He named it nucleus for the Latin "little nut."

He died in London on June 10, 1858.

Legacy

Brown contributed directly to the development of botany, but his legacy was felt across disciplines: the molecular mechanism of Brownian motion, first explained in the twentieth century, influenced fundamental theories in physics.

Brown's revision of traditional schemes of plant classification and his identification of hundreds of unknown plants represented progress in both theoretical and practical botany. With Brown's work as an example, biologists became more attentive to microscopic studies and the potential revelations to be garnered from morphology.

The cell was not considered important at the time Brown identified the nucleus; thus he was unaware that he had named one of the most significant elements of living things. In 1839 Matthias Schleiden and Theodor Schwann discovered that cells are the building blocks of all life; research on cells then blossomed. But not until the late nineteenth century did biologists recognize the nucleus as the director of cell activity.

In the twentieth century, the mystery of Brownian motion was solved. During Brown's lifetime and in the decades that followed, scientists observed Brownian motion and postulated numerous explanations. In 1905 ALBERT EINSTEIN formulated the first mathematically sound model for the phenomenon, and within 10 years others had confirmed this model experimentally. Brownian motion, defined as the constant movement of particles suspended in a solvent (liquid or gas), occurs because the particles are continually and randomly buffeted by the molecules that make up the solvent. This explanation provided both a convincing confirmation of the existence of molecules and strong evidence for the kinetic theory of matter, which states that the molecules of all matter are in a constant state of motion.

Schuyler

World Events		Brown's Life
United States independence	1776	
	1773	Robert Brown is born
French Revolution	1789	
	1801	Brown joins Australian voyage as naturalist
Napoleonic Wars in Europe	1803-15	
	1805	Brown begins classifying Australian plant specimens
	1810	Brown becomes Joseph Banks's librarian, supervising Banks's botanical collection
	1827	Brown becomes curator at the British Museum to which Banks's collection is transferred
		Brown discovers Brownian motion
	1831	Brown names the cell nucleus
	1858	Brown dies
United States Civil War	1861-65	

For Further Reading:

Borodin, A. N. *Handbook of Brownian Motion: Facts and Formulae.* Boston: Birkhauser-Verlag, 1996.

Chung, Kai Lai. *From Brownian Motion to Schrödinger's Equation.* New York: Springer-Verlag, 1995.

Buffon, Georges

Developer of
Survey of Natural World
1707-1788

Life and Work

Georges Buffon wrote the widely read and acclaimed eighteenth-century classic *Natural History*, which represented the first attempt at a comprehensive survey of the natural world.

Buffon was born on September 7, 1707, in Montbard, France, to an aristocratic family. He completed a law degree in 1726 and then studied medicine, botany, and mathematics at Angers. When his mother died in 1732, he returned to his family estate in Montbard, where he lived for the rest of his life.

Buffon's generous inheritance allowed him to pursue his interest in the sciences, particularly mathematics and botany. He conducted studies

WORLD EVENTS	BUFFON'S LIFE
Thirty Years' War 1618-48 in Europe	
	1707 Georges Buffon is born
Peace of Utrecht 1713-15 settles War of Spanish Succession	
	1726 Buffon earns law degree
	1732 Buffon receives inheritance and begins scientific work
	1735 Buffon translates *Vegetable Staticks*
	1739 Buffon becomes keeper of the Jardin du Roi
	1740 Buffon translates Isaac Newton's *Fluxions*
	1749 First volumes of Buffon's *Natural History* appear
	1788 Buffon dies
French Revolution	1789 Final part of *Natural History* is published

in probability theory, and he also translated several works into French: a botanical treatise, *Vegetable Staticks* by Stephen Hale, in 1735; and a mathematical work called *Fluxions* by ISAAC NEWTON in 1740.

In 1739 he was appointed keeper of Jardin du Roi, France's royal botanical and zoological gardens, and was charged with creating a catalog of the garden's collection. He developed the project into an attempt to describe the whole of nature. The result was *Natural History*, a series of 36 volumes published from 1749 through 1789. The work contained careful, detailed descriptions of biological and geological history accompanied by artistic illustrations.

Natural History included Buffon's non-theological views on the history of Earth and its inhabitants. He posited that Earth and the other planets were formed out of stellar fragments broken off the Sun by a comet's collision. He claimed that Earth began as a molten body and gradually cooled to its present state, a process that he estimated to have taken at least 70,000 years.

Buffon noted that variation exists within species, and he described the vestigial features of some animals. He later wrote that related animals have common ancestors, but he stopped short of positing that species evolve.

Buffon became ill in 1785 and died in Paris on April 16, 1788.

Legacy

Buffon's *Natural History* represented the first attempt to catalogue all of nature in a single publication, and the work provoked controversy as well as widespread interest in natural history. It also influenced the development of evolutionary theory.

Natural History caused an uproar among theologians. Immediately following the book's publication, religious authorities attacked Buffon for his non-biblical views. The Bible, in the book of Genesis, states that Earth was created by God and implies the event occurred about 6,000 years ago, whereas Buffon posited that a natural cosmic collision formed Earth and reasoned that Earth must be at least 70,000 years old. While his estimate of Earth's age now seems vastly incorrect, compared to current

estimates that date back to roughly 3.9 billion years ago (the age of the oldest rocks known to scientists), his estimate was one of the first to scientifically consider Earth's history apart from calculations based on biblical considerations. The controversy was part of the gulf developing between science and religion in Europe during the Enlightenment.

Natural History became one of the most widely read works of the eighteenth century. Its engaging style and skillful illustrations captivated readers, and the ensuing popular interest in natural history carried through to the beginning of the twentieth century.

Buffon's ideas helped shape evolutionary theory. He understood that changes in Earth's surface features have an impact on the distribution and adaptation of living forms. And although he did not conceive of evolution, his discussion of variation within species, vestigial features, and common ancestors marked the beginning of uncertainty about species immutability.

Buffon's work contributed to the development of natural history as a field of study. *Natural History* brought together zoology, botany, geology, mineralogy, cosmology, and anthropology into one discipline: the study of Earth, its physical history, and its inhabitants. These had previously been studied as distinct subjects. Uniting them made it possible to explore connections among the various elements and thus to develop new theories.

Schuyler

For Further Reading:

Bajema, Carl Jay, ed. *Natural Selection Theory: From the Speculations of the Greeks to the Quantitative Measurements of the Biometricians*. New York: Van Nostrand Reinhold, 1983.

Fellows, Otis, and Stephen F. Miller. *Buffon*. New York: Twayne Publishers, 1972.

Bunsen, Robert

Developer of Chemical
Spectral Analysis
1811-1899

Life and Work

Robert Bunsen pioneered the use of spectral lines (wavelengths associated with the light and other energy emitted from atoms) to analyze chemical elements. He also discovered new elements and invented several important pieces of laboratory equipment, including the spectroscope.

Bunsen was born on March 31, 1811, in Göttingen, Germany. He received a doctorate in chemistry from the University of Göttingen in 1830 and subsequently taught at three other German universities. In 1852 he was appointed professor of experimental chemistry at the University of Heidelberg, where he was extremely popular with students and built a highly regarded department of chemistry.

Bunsen worked in several diverse fields of chemistry. He discovered an antidote to arsenic poisoning in 1834: freshly precipitated hydrated ferric oxide. In 1837 he began to work with cacodyl compounds, dangerous arsenic-containing substances. He lost the use of one eye in a laboratory explosion, nearly died of arsenic poisoning, and eventually banned the study of organic chemistry from his laboratory.

Bunsen constructed a carbon-zinc electric cell and used it to produce a bright electric arc light in 1841. To measure the light's brightness, he invented a grease-spot photometer in 1844. Using the carbon-zinc cell, he became the first to isolate metallic magnesium and to show that it emits high-intensity light when burned in air. He also prepared the first pure sample of the metal lithium.

In the 1850s, Bunsen studied the heat lost with the waste gases of blast furnaces and presented his results in 1857 in *Gasometric Methods*, his only major publication.

Bunsen began a partnership with Gustav Kirchhoff, which led to their invention of the spectroscope in 1859. The two chemists observed that each element gives off light of a distinct wavelength, represented by a dark band within a spectrum of light. In 1860 they discovered two new elements, cesium and rubidium, when they recognized two unknown bands in the light spectra of certain mineral compounds.

In his later years, Bunsen designed and built a filter pump, an ice calorimeter, and a vapor calorimeter. Although the Bunsen burner bears his name, Bunsen did not invent it. It was constructed by his technician, Peter Desdega, based on a design by MICHAEL FARADAY.

Bunsen died in Heidelberg, Germany, on August 16, 1899.

Legacy

Bunsen's experimental endeavors left a significant mark on the development of various fields of chemistry.

Working with Kirchhoff, Bunsen recognized that every element when heated gives off a spectrum of electromagnetic radiation that is the "fingerprint" unique to that element. Bunsen and Kirchhoff's technique of spectral analysis enabled other scientists to discover several unknown elements after 1860. Heavy elements proved to have very complicated spectra, and scientists could not fully explain all the spectral components that appeared. This called for attempts to determine the origin and structure of spectra, the results of which led to revisions in the fundamental tenets of chemistry and physics.

Bunsen and Kirchhoff's spectral-analysis research was soon used to determine the chemical components of heavenly bodies, and the field of astrophysics was born. Spectral analysis proved especially valuable in characterizing solar flares. Applying spectral analysis to other stars, scientists found that stars did not all have the same spectra, as had been thought previously. Identification of the various stellar spectra led to the modern system of spectrum-based stellar classification.

Bunsen's work in organic chemistry stimulated one of his students, Edward Frankland, who became a successful experimental chemist. Frankland prepared and analyzed the first known organometallics, the zinc dialkyls. Organometallics encompass a wide range of substances, including many catalysts, all of which include at least one carbon-metal band.

This work led him to the theory of valence, which states that there is a limit to the number of bonds two atoms or groups of atoms can create while forming a compound. Valence theory set the foundations for the emerging field of structural chemistry.

Schuyler

WORLD EVENTS		BUNSEN'S LIFE
Napoleonic Wars in Europe	1803-15	
	1811	Robert Bunsen is born
	1834	Bunsen discovers arsenic antidote
	1837	Bunsen begins organic chemistry research
	1841	Bunsen constructs carbon-zinc electric cell
	1844	Bunsen invents grease-spot photometer
	1857	Bunsen's *Gasometric Methods* is published
	1859	Bunsen and Gustav Kirchhoff invent the spectroscope
	1860	Bunsen and Kirchhoff discover rubidium and cesium
Germany is united	1871	
Spanish-American War	1898	
	1899	Bunsen dies

For Further Reading:

Hollas, J. Michael. *Modern Spectroscopy.* New York: John Wiley, 1992.

Moore, F. J., and W. T. Hall. *History of Chemistry.* New York: McGraw-Hill, 1939.

Burbank, Luther

Pioneer Breeder of
New Plant Varieties
1849-1926

World Events		Burbank's Life
	1849	Luther Burbank is born
	1870	Burbank buys first plot of land
Germany is united	1871	
	1873	Burbank introduces improved potato (known as Burbank or Idaho)
	1875	Burbank moves to California and builds experimental nursery and farm
Spanish-American War	1898	
World War I 1914-18		
	1915	Burbank finishes 12-volume treatise on his work called *Luther Burbank, His Methods and Discoveries and Their Practical Applications*
	1921	*How Plants Are Trained to Work for Man* is published
	1926	Burbank dies
	1927	Burbank's autobiography, *Harvest of the Years,* is published
World War II 1939-45		

Life and Work

Luther Burbank, a pioneer plant breeder, developed more than 200 improved fruit, vegetable, and ornamental plant varieties, and triggered widespread interest in horticulture.

Born in Lancaster, Massachusetts, on March 7, 1849, Burbank was raised on a farm and received little formal education beyond high school. In 1870 he bought a 17-acre plot of land near Lunenburg, Massachusetts, and began plant-breeding experiments. He was deeply influenced by Charles Darwin's book, *The Variation of Animals and Plants Under Domestication,* and recognized that selective breeding would produce improved plants.

The development of a large, hardy potato (now called the Burbank or Idaho potato) in 1873 was his first success. Normally potato crops are cultivated by replanting parts of the tuber, which prevents the introduction of variation into the new crop, because vegetative propagation replicates genes of the original plant only. Burbank chose to plant potato seeds, the creation of which involves sexual recombination (the re-ordering of genes following sexual reproduction), and he thus created numerous new varieties of potato. He selected the one that was most resistant to spoilage, and it quickly made its way into the national produce market.

In 1875 Burbank used the profit from his improved potato to travel to California and establish a nursery, greenhouse, and experimental farm near Santa Rosa, north of San Francisco. He crossed foreign strains with native varieties and employed grafting to gauge the success of hybrids. Devoted to the development of improved plants, he expanded the farm until it was capable of sustaining more than 3,000 experiments simultaneously.

Although Burbank used little scientific methodology, he was skilled at recognizing desirable characteristics. He introduced numerous improved (with respect to taste and profit potential) varieties of fruits, vegetables, and decorative flowers. He believed his success supported the Lamarckian theory of heredity, which posited that organisms can pass to their offspring the characteristics they acquire during their lifetime. The development of modern genetics showed this theory to be false.

Burbank completed a 12-volume treatise, *Luther Burbank, His Methods and Discoveries*

and Their Practical Applications, in 1915, and he published his eight-volume *How Plants Are Trained to Work for Man* in 1921. With the help of author Wilbur Hall, he also wrote an autobiography, *Harvest of the Years,* which was published in 1927. He died in Santa Rosa on April 11, 1926.

Legacy

Burbank's work, which was widely acclaimed during his lifetime, greatly enhanced food-plant variety and established experimental plant-breeding as a foundation of both agriculture and genetics research.

Burbank produced about 800 new plant varieties, including more than 200 that proved to be valuable for culinary or decorative purposes. He was particularly successful with plums, apricots, berries, lilies, and roses; many of these are still commercially lucrative. Profits from the Burbank potato exceeded a billion dollars during the first 50 years of its cultivation.

As a result of Burbank's successes, plant breeding flourished in the early twentieth century, both as a hobby and as an agricultural endeavor. Since then, advances in understanding genetics have allowed botanists and agronomists to manipulate the outcomes of cross-breeding experiments more efficiently and to value the characteristics of food plants that are more insect, disease, and drought resistant than many cultivated varieties.

Schuyler

For Further Reading:

Dreyer, Peter. *A Gardener Touched with Genius: The Life of Luther Burbank.* Santa Rosa, Calif.: Luther Burbank Home and Gardens, 1993 (originally published by University of California Press, 1985).

Cannon, Annie Jump

Developer of
Star Classification System
1863-1941

Life and Work

Astronomer Annie Jump Cannon identified and classified more stars than anyone else in the world. She also developed the classification of stars by their spectral types, and extensively rearranged the original Henry Draper classification of stars to make it more flexible, expandable, and orderly.

Cannon was born in 1863 in Dover, Delaware, to a prominent family. Her father was a shipbuilder and lieutenant governor of the state; her mother, who had taken a course in astronomy, interested her daughter in the stars. As a young girl, Cannon observed the stars in her family's makeshift attic observatory and recorded her observations by candlelight. She was also intrigued by the rainbow colors of light passing through the glass prisms in the candelabra in her home, an interest in the behavior of light that lasted throughout her life.

Although higher education for women was uncommon at that time, Cannon's teachers and parents agreed that Annie had unusual abilities, and her father agreed to send her to Wellesley College. She worked under Professor Sarah Whiting, the first woman to study physics at The Massachusetts Institute of Technology. Graduating in 1884, Cannon went home to Delaware and did not pursue science until the death of her mother 12 years later. While still in mourning, Cannon returned to the hobby she had shared with her mother and considered new questions about what she saw.

In 1896 Cannon started taking graduate courses at Radcliffe College where she met Edward Pickering, director of the Harvard Observatory. At age 34 she became his assistant and began a career in astronomy. Her job was to classify stars according to their spectra, the ribbons of light crossed from top to bottom by telltale lines and bands (spectra) that result when starlight passes through a prism at the end of a telescope. These spectra (nature's bar codes) tell a trained observer the star's composition, temperature, speed, and size. Not only did Cannon become expert at this type of analysis, she also devised an improved classification scheme for stellar information, the "Harvard" system, ranking stars from hottest to coolest. The system was adopted as the international standard by the International Astronomical Union in 1913.

Harvard began recording Cannon's systematic records of the stars in nine volumes known as the Henry Draper Catalogue, so named by Draper's widow who subsidized the project in his honor. Cannon was in her sixties by the time this work was published, but she was not finished. She continued to look for and catalog thousands of fainter stars and discovered 300 variable stars and five novae (exploding stars). She classified 350,000 stars in all. Using photographs taken through a telescope, she participated in the first scientific mapping of the sky. Cannon became curator of the Harvard College Observatory in 1911. She enjoyed travel and spent six months working at the Harvard Branch Observatory in Peru.

Cannon was recognized as one of the greatest living experts in her field. She was awarded six honorary degrees, including an honorary doctorate from Oxford University in 1925, the first ever granted to a woman. The National Academy of Sciences honored her in 1931 with the Draper Medal. Harvard University, however, did not make her a faculty member until 1938, shortly before her death. Cannon died in 1941 at age 78.

Legacy

Annie Jump Cannon may be considered one of the world's first astrophysicists, as she developed the Harvard classification of stars by their spectral types as well as contributed enormously to stellar taxonomy.

The seven star classes that Cannon differentiated (O, B, A, F, G, K, and M with O being the hottest and largest and M being the coolest and smallest) was the standard until 1998. That year J. Davy Kirkpatrick of the California Institute of Technology and his colleagues found examples of dwarf stars only one-third the temperature of the Sun. They expanded Cannon's series by adding a new L class for these previously undetected bodies whose distinctive spectral lines indicate atypical atmospheres rich in metal hydrides.

Developed, refined, and organized by Cannon, the Henry Draper Catalog of stellar spectra laid the foundation for modern stellar spectroscopy. Whenever a new observatory is established on any continent, this catalogue is purchased as essential equipment. These studies made by a single observer are a benchmark for Earth-based astronomy, yet they were hidden behind the name of a Harvard-approved man. For many intellectual women of her time, invisibility was a small price to pay for the privilege of doing challenging work and making valuable contributions to science.

As a scholar of international reputation, Cannon set an example for other astronomers. Her enthusiasm and tenacity in photographing stars, especially faint ones, and her skill in detecting their distinctive behaviors as coded in starlight received on Earth set a standard of fruitful observations unlikely to be equaled by any other individual. Her longevity and intellectual initiative also contradicted the "common knowledge" that education would destroy a healthy woman's mind.

Hertzenberg

WORLD EVENTS	CANNON'S LIFE
	1863 Annie Jump Cannon is born
Germany is united 1871	
	1896 Cannon begins graduate courses at Radcliffe College
	1911 Cannon becomes curator at Harvard College Observatory
	1913 International Astronomical Union adopts Cannon's system of star classification
World War I 1914-18	
	1925 Cannon becomes first woman to receive honorary degree from Oxford University
	1931 Cannon receives Draper Medal of National Academy of Sciences
	1938 Cannon joins faculty at Harvard
World War II 1939-45	
	1941 Cannon dies
Korean War 1950-53	

For Further Reading:

Greenstein, George. *Portraits of Discovery.* New York: John Wiley, 1998.

Mack, Pamela E. "Straying from Their Orbits: Women in Astronomy in America." In *Women of Science: Righting the Record.* Edited by G. Kass-Simon and Patricia Farnes. Bloomington: Indiana University Press, 1990.

Ogilvie, Marilyn Bailey. *Women in Science: Antiquity through the Nineteenth Century.* Boston: MIT Press, 1986.

Carnot, Sadi

(in full Nicholas-Leonard-Sadi)

Developer of Heat Engine Theory
1796-1832

Life and Work

Sadi Carnot proposed the theory of how heat engines operate. He mathematically connected the efficiency of heat engines to measures of work, disorder, probability, and information.

Carnot was born in Paris, France, in 1796 to a prominent family; his father Lazare was a well-known leader in the French Revolution, and his nephew, also named Sadi, became president of the French Republic. When he graduated from the elite École Polytechnique in Paris in 1814, Great Britain was a long-term enemy of France and the greatest power in the world. Carnot realized that the steam engine, invented and improved in Great Britain, was one of the most important sources of that country's strength. Carnot committed to study how steam engines

WORLD EVENTS		CARNOT'S LIFE
French Revolution	1789	
	1796	Sadi Carnot is born
Napoleonic Wars in Europe	1803-15	
	1814	Carnot graduates from school and begins his engineering career
	1824	*Reflections on the Motive Power of Fire*, in which Carnot presents his ideas on heat engines, is published
	1832	Carnot dies
	1849	Carnot's work is discovered by Lord Kelvin
Germany is united	1871	

operate in hopes that France might catch up with, or even surpass, Great Britain.

Carnot began by comparing the flow of heat through a steam engine with the flow of water through a waterwheel. He assumed that heat was a material substance, commonly called caloric by scientists of his time. (We know now that heat is a form of vibrational motion and not a separate substance. Remarkably, though, Carnot's logical description of heat flow was so thorough that his theory was easily corrected after his death.) His great discovery was that the operation of a heat engine, a device that converts heat into motion just as water turns a waterwheel, did not depend on what fluid was used; any gas (a form of fluid) would follow the same laws as steam. His work was later interpreted to show that there is a limit to how much work any heat engine can do, even if it had perfect bearings and parts with no friction.

In 1824 he published the results of his study in a book, *Reflections on the Motive Power of Fire*. Six hundred copies were printed at his own expense and went practically unnoticed. Eight years later he contracted cholera as an epidemic swept Paris, and he died on August 24, 1832, only 36 years old.

Legacy

Carnot's ideas became the starting point for thermodynamics, the field dealing with mathematical relationships that connect the efficiency of heat engines to measurements of work available from their fuel.

A year after Carnot's death, another French engineer, Émile Clapeyron, found and read Carnot's book. Clapeyron simplified the main ideas and published them in 1834, after several journals had refused to print the article. Finally in 1849 the great British scientist William Thomson (later known as LORD KELVIN) read Clapeyron's article, became aware of Carnot's work, and recognized its importance.

Carnot's work was eventually widely understood and appreciated; it became the basis of the Second Law of Thermodynamics, which states that heat will never flow on its own from a colder object to a warmer one. Carnot also came close to developing the First Law, which asserts that the amount of energy in the universe is constant, but he died before he could

get his notes in order and present his work. His surviving brother did not publish his notes until 1878, well after other scientists had understood the nature of heat energy.

Many aspects of contemporary technology employ applications of Carnot's discoveries. Steam engines are still in use today, although they have become less visible than in Carnot's time. All electric power plants that burn coal, oil, or gas, along with all nuclear power plants, are actually steam engines. Those fuels supply heat to boil vast quantities of water into steam and force it through giant pinwheels, called turbines. The turbines spin and force enormous generators to turn and produce electricity. The engines that power modern cars, buses, trains, and planes all use the same principles that Carnot discovered, but with hot, expanding gases as the working fluid rather than steam.

Unfortunately, the inherent problems that Carnot discovered in the theory of heat engines remain. Carnot's work was interpreted to show that the absolute maximum efficiency of any heat engine, whether fueled by steam, gasoline, or anything else, is limited by the difference in temperature between the source of heat entering the engine and its exhaust. The exhaust from gasoline engines is very hot and that represents wasted heat. Only about one-third of the fuel is actually put to use, and we can not improve that figure very much without running into the theoretical limits discovered by Carnot.

Secaur

For Further Reading:

Commoner, Barry. "Thermodynamics, the Science of Energy." In *The Poverty of Power*. New York: Knopf, 1976.

Segrè, Emilio. *From Falling Bodies to Radio Waves*. New York: Freeman, 1984.

Carothers, Wallace

Inventor of Nylon
1896-1937

Life and Work

Wallace Carothers formulated a theory governing the synthesis of polymers, introduced a synthetic rubber-like material (neoprene), and invented nylon, the first synthetic fiber.

Carothers was born on April 27, 1896, in Burlington, Iowa. As a child he gained a deep appreciation for music from his mother and a sense of the importance of education from his father, a teacher. Recognized as an outstanding student at Tarkio College, Missouri, he was appointed to replace the director of the chemistry department during World War I. He completed a bachelor's degree in 1920 and continued at the University of Illinois, earning a Ph.D. in chemistry in 1924.

Carothers accepted a position at Harvard University in 1926. In addition to teaching chemistry, he researched the properties of large polymers, molecules consisting of numerous repeating sequences of atoms strung together.

In 1928 Carothers joined the Experimental Station of the Du Pont Company in Delaware as director of a new program aimed at investigating the emerging field of synthetic materials. Beginning with molecules in the acetylene family, he devised experimental methods based on the work of Belgian-born American chemist and botanist Julius Arthur Nieuwland. This research yielded a rubber-like substance formed from the combination of vinylacetylene and chlorine compounds. In 1931 the synthetic substance was marketed as neoprene.

Carothers then focused on the creation of synthetic fibers with properties similar to those of silk and cotton. The task called for a chemical reaction that would cause certain groups of chemicals to bind together into a polymer. Working with families of compounds called diamines and dicarboxylic acids, Carothers developed a theory explaining the factors involved in the polymerization (the process of polymer formation) of these chemicals. About 1935 this work culminated in the creation of the first synthetic fiber, later called nylon; it was similar to silk but stronger and more elastic.

In January 1937 the death of his sister Isobel plunged Carothers into despair. Having long suffered fits of severe depression, he committed suicide on April 29, 1937, in Philadelphia, Pennsylvania.

Legacy

Carothers's work laid the foundation for the commercial polymer industry.

Neoprene, Carothers's first marketable invention, replaced natural rubber to a large extent. At the beginning of the twentieth century the automobile industry was rapidly expanding, increasing the demand for rubber for tires. One of Du Pont's research aims was to prevent a future shortage of rubber by supplying a synthetic alternative. Although not ideal for tires, neoprene was, and still is, used for making wire, cable, hose, machinery belts, molded goods, soles, heels, and adhesives.

The Du Pont Company started marketing nylon in 1940; it met with immediate success in the production of toothbrushes and stockings. Toothbrush bristles had previously been made from the stiff hair of animals such as hogs, horses, and badgers. Nylon was softer, cheaper, and resistant to the growth of bacteria. Synthetic toothbrushes made effective tooth-cleaning devices available to more people, ushering in the modern era of dental care.

Prior to the fabrication of nylon stockings, most stockings were made from silk. The debut of nylon stockings in 1940 caused great excitement in the United States following a campaign that advertised the synthetic stockings as virtually indestructible. Within a few years delicate silk stockings were obsolete. Today nylon is used in some fashion in nearly all industries.

Carothers's success with neoprene and nylon followed the introduction of the first plastics and launched the development of other synthetic polymers. Celluloid and Bakelite, in regular use by the early 1920s, were followed by cellophane, acetate, vinyl, plexiglass, acrylic, styrene, formica, and polyester. In 1945 Du Pont chemist Earl Tupper produced polyethylene (soon commercialized as Tupperware), one of the most versatile synthetic polymers. Modern synthetic polymers, the result of more than a half century of industrial experimentation, include various combinations of chemicals engineered to produce products with specific properties.

Convenient, lightweight, airtight, cheap, and durable, synthetic polymers revolutionized modern life and helped conserve some natural resources. However they also yielded a mass of commercial waste, created by the continuous production of disposable but non-degradable items.

Schuyler

World Events		Carothers's Life
	1896	Wallace Carothers is born
Spanish-American War	1898	
World War I	1914-18	
	1924	Carothers earns Ph.D. in Chemistry from University of Illinois
	1926	Carothers begins teaching at Harvard University
	1928	Carothers joins Du Pont
	1931	Neoprene, invented by Carothers, is marketed in place of rubber
	c. 1935	Carothers invents nylon
	1937	Carothers commits suicide
World War II	1939-45	
	1940	Du Pont markets nylon
Korean War	1950-53	

For Further Reading:
Bowden, Mary Ellen. "Wallace Carothers." In *Chemical Achievers*. Philadelphia, Penn.: Chemical Heritage Foundation, 1997.
Hermes, Matthew E. *Enough for One Lifetime: Wallace Carothers, Inventor of Nylon.*
Washington D.C.: American Chemical Society and the Chemical Heritage Foundation, 1996.

Carson, Rachel

Early Advocate of
Environmental Movement
1907-1964

Life and Work

Rachel Carson was an acclaimed marine biologist and nature writer whose 1962 book *Silent Spring* aroused worldwide concern about human-caused environmental damage.

Carson was born on May 27, 1907, and grew up on an isolated farm in Springdale, Pennsylvania. Her writing talent was evident early; when she was 10 her first published story appeared in *St. Nicholas* magazine. She earned a full scholarship to attend the Pennsylvania College for Women (now Chatham College) in Pittsburgh and graduated with a zoology degree in 1929.

Carson spent a summer at the Woods Hole Marine Biology Laboratory in Massachusetts, where she saw the ocean for the first time and

developed a passion for aquatic biology. In 1932 she earned a master's degree from Johns Hopkins University in Baltimore while also teaching zoology at the University of Maryland. Subsequently joining the Bureau of Fisheries (later the Fish and Wildlife Service) in 1936, she wrote various government publications and radio scripts. In 1941 her first book, *Under the Sea-Wind*, appeared and received positive reviews.

Carson's next book, *The Sea Around Us*, was published in 1951 and included information from previously classified government research (including oceanographic discoveries) and her own recent diving experiences. The book was a success, received various awards, and established Carson's reputation as a science writer. In 1953 the RKO film based on this book won an Oscar for best documentary. *The Edge of the Sea* was published in 1955 and won more honors.

Carson had long harbored concern about the increasing worldwide use of pesticides, particularly the chemical dichlorodiphenyltrichloroethane (DDT). Enlisting the aid of numerous scientific experts, she gathered data for her next book, *Silent Spring*, which was published in 1962. In this book she argued that pesticides threaten environmental health and that human efforts to control nature had begun to destroy ecological systems. Chemical corporations attacked the book and attempted to discredit Carson by questioning her scientific credentials. Some reviewers resorted to sexist comments, calling her "a hysterical woman." She had anticipated these negative comments but had the courage to speak out publicly about the widespread effects of pesticides. Her calm demeanor and extensive knowledge displayed during lectures earned people's respect and raised their concern. Subsequent public outcry led to government investigations of pesticide contamination of food and the environment; bills to regulate pesticides were quickly passed.

Carson's health had been failing for several years, and she died of cancer on April 14, 1964, in Silver Spring, Maryland. In 1980 President Jimmy Carter posthumously awarded Carson a Presidential Medal of Freedom, the highest honor a civilian can earn from the United States government.

Legacy

Carson's *Silent Spring* brought national attention to the issue of protecting the environment from pollution, causing controversies that are still alive today. She built the framework for the environmental movement; she emphasized working with natural processes instead of fighting against nature.

Carson inspired a public campaign to influence environmental legislation. Immediately a presidential advisory committee, to which Carson presented influential testimony, was formed, and by the end of 1962, state legislatures had introduced more than 42 bills to curb pesticide use. The campaign led eventually to the formation of the Environmental Protection Agency, a government group that monitors and regulates biological conservation.

Raising concern about pesticide damage and the squandering of natural resources, Carson helped initiate a public movement, a strong social and political force, aimed at preserving the environment. A worldwide reduction in agricultural chemical output (including that of pesticides such as DDT) was one of the results of efforts to decrease the environmental impact of human activities. The movement has also altered the way many people live and think by introducing the concept of bio-accumulation, the process of residual pesticides accumulating in the diet.

Her lyrical writing about life in water, soil, and air, and her joy in sharing that natural world with her readers are testimony to the sense of wonder she promoted and celebrated in her life and works. Carson's descriptions of the interdependencies of living species are models of beauty and subtle logic that inspire people to care about the environment. This is the core of her enduring legacy.

Schuyler

WORLD EVENTS	CARSON'S LIFE
	1907 Rachel Carson is born
World War I 1914-18	
	1929 Carson earns zoology degree from Pennsylvania College for Women
	1932 Carson receives master's degree from Johns Hopkins University
	1936 Carson begins working for U.S. Bureau of Fisheries
World War II 1939-45	
	1941 Carson's first book, *Under the Sea-Wind*, is published
Korean War 1950-53	
	1951 *The Sea Around Us* is published
	1955 *The Edge of the Sea* is published
	1962 *Silent Spring* causes public concern for environment; government investigation into pesticide use begins
	1964 Carson dies
End of Vietnam War 1975	

For Further Reading:

Brooks, Paul. "The Courage of Rachel Carson." *Audubon* (January 1987).

Carson, Rachel. *Silent Spring*. Boston: Houghton Mifflin, 1962.

Lear, Linda. *Rachel Carson: Witness for Nature*. New York: Henry Holt, 1997.

McKay, Mary A. *Rachel Carson*. New York: Twayne, 1993.

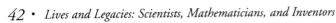

Carver, George Washington

Originator of Crop
Rotation Practices

c.1860-1943

Life and Work

George Washington Carver developed scientific agriculture in the southern United States by introducing crop rotation and devising hundreds of new uses for peanuts and sweet potatoes.

About 1860, Carver was born in Newton County, Missouri, to African-American slaves held by farmers Susan and Moses Carver. His mother died when he was a boy, and he was brought up on the Carvers' farm, where he indulged a deep curiosity about nature by collecting frogs and maintaining a plant nursery. Even as a child, he was known as the "plant doctor."

In 1877 he left Missouri to seek a high school education, a goal he accomplished in the 1880s after years of traveling and subsisting on odd jobs. He entered Simpson College in Indianola, Iowa, in 1890 and transferred the following year to Iowa State Agricultural College in Ames; he received a master's of science degree in 1896. As a graduate student, he was appointed to the agricultural experiment station where he collected fungi and identified several new species, which were named for him.

Carver became director of the agriculture department at Tuskegee Institute in Alabama in 1897. He devoted much of his time to research aimed at helping southern farmers, especially black sharecroppers, who were facing a crisis—years of cotton cultivation had depleted and eroded the soil. Carver spent years studying soil chemistry and crop rotation and directing an experimental farm. In 1914 he made his results public and urged the introduction of peanuts, soybeans, and sweet potatoes to vary the nutrients delivered to and removed from the soil. With innovative laboratory research, he developed hundreds of new uses for these products, including foods, inks, dyes, plastics, soaps, oils, cosmetics, and glues.

Carver received numerous awards in the early 1900s for his contributions to agriculture and for his lectures aimed at increasing interracial respect. In 1921 he spoke before the United States House Ways and Means Committee, and his expert agricultural testimony led to the institution of a tariff to protect U.S. peanut crops from foreign competition. In 1940 he donated his life's savings to help establish a new agricultural research foundation at Tuskegee. He died in Alabama on January 5, 1943.

Legacy

Carver's experimental, educational, and political work revolutionized agriculture and revitalized the economy of the southern United States in the early twentieth century.

The introduction of new crops to the South improved many people's lives. In Carver's time, agriculture was the single largest occupation in the United States, and many farmers in the South relied almost exclusively on cotton cultivation. The success of peanuts, sweet potatoes, and soybeans liberated farmers from their dependence on cotton.

The development of the peanut as a major cash crop boosted the economy of the South. When Carver arrived in Alabama in 1896, the peanut was not even recognized as a crop. Over the next 50 years, the peanut became one of the six largest crops in the United States and the second-largest crop in the South. Five million acres of peanut cropland were allotted to farmers by the government in 1942.

Carver's outreach programs helped uneducated farmers gain an advantage. He encouraged improved farming methods and instituted practices of soil conservation and crop diversification. The institute Carver helped found at Tuskegee continues to contribute to the development of innovations in agriculture today.

Basic tenets of scientific agricultural practice, which persist in modern times, were established by Carver's pioneering efforts. Crop rotation is a widely recognized method of maintaining healthy soil-nutrient balance and avoiding erosion. The development of varied uses for agricultural products is also an ongoing practice.

In addition to his scientific contributions, Carver's journey from his beginnings in slavery to his status as an internationally renowned scientific expert has made him a hero in the history of the United States.

Schuyler

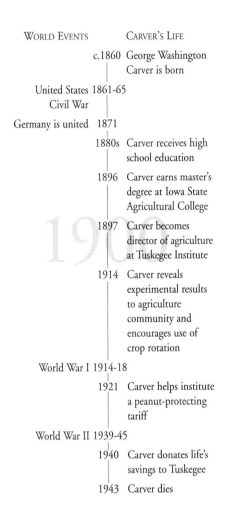

WORLD EVENTS	CARVER'S LIFE
	c.1860 George Washington Carver is born
United States 1861-65 Civil War	
Germany is united 1871	
	1880s Carver receives high school education
	1896 Carver earns master's degree at Iowa State Agricultural College
	1897 Carver becomes director of agriculture at Tuskegee Institute
	1914 Carver reveals experimental results to agriculture community and encourages use of crop rotation
World War I 1914-18	
	1921 Carver helps institute a peanut-protecting tariff
World War II 1939-45	
	1940 Carver donates life's savings to Tuskegee
	1943 Carver dies

For Further Reading:

McMurry, Linda O. *George Washington Carver: Scientist and Symbol*. New York: Oxford University Press, 1981.

Kremer, Gary R., ed. *George Washington Carver: In His Own Words*. Columbia: University of Missouri Press, 1987.

Cauchy, Augustin-Louis

Early Developer of the Calculus
1789-1857

Life and Work

Augustin-Louis Cauchy gave the calculus a logically acceptable algebraic foundation.

Cauchy was born in Paris on August 21, 1789, during the height of the French Revolution. His father moved the family to Arcueil where he provided for them as best he could and wrote textbooks to educate his children. Cauchy's early malnutrition affected his health for the rest of his life.

One of the family's neighbors in Arcueil was the well-known mathematician Pierre Laplace; he met with young Cauchy and encouraged him in mathematics. When Cauchy was 11 his father was appointed Secretary of the Senate and it was in his office that Cauchy met another famous mathematician, Joseph Lagrange. Lagrange, who also recognized Cauchy's mathematical talents, advised Cauchy's father on the boy's education, suggesting a well-rounded course of study. Cauchy was very successful in school, winning prizes in Latin and Greek, and at 16 entered the École Polytechnique in Paris. Here his principled and obstinate character was revealed in his public observance of Catholicism when quiet religious observance might have been an easier path to follow. Following two years of training in engineering he was given a commission in Napoleon's army and spent the next three years in Cherbourg as an engineer. In addition to his military work, he continued to study and teach mathematics.

Cauchy returned to Paris in 1813 and began to write at a fantastic pace, producing papers on several mathematical topics including the relationship between the number of sides, edges, and vertices of a polyhedron, and a solution to a problem on polygonal numbers posed by PIERRE DE FERMAT. This prolific writing continued throughout his life; he produced a total of 789 papers and seven books during his lifetime. Within three years he became a professor at the Polytechnique, won the Grand Prize of a contest sponsored by the Institute de France for his theory of waves, and was appointed to France's Academy of Sciences.

In the 1820s Cauchy conducted his most important work. He focused on the calculus—also known as complex analysis—in order to simplify, clarify, and systematize its rules. He compiled his lessons and lectures on the subject in three treatises published in the 1820s in which he established the concept of limits and proposed theories of convergence, continuity, derivatives, and integrals. During the same period he developed theoretical details of the functions of complex variables, those using a multiple of the square root of minus one.

In 1830 King Charles X was exiled. Cauchy had been loyal to King Charles and so, rather than swear allegiance to the new King, he exiled himself and ended up teaching in Turin, Italy. Cauchy returned to the Polytechnique in 1838, when an oath of allegiance to the state was no longer a requirement for serving on the faculty. Ten years later he also joined the faculty of the Sorbonne. In 1852 Napoleon III reinstated the requirement of the oath but excused Cauchy from having to take it. Cauchy's response was to donate his salary to the poor of his home town. Cauchy died on May 23, 1857, in Sceaux, France.

Legacy

Cauchy made contributions in various fields of mathematics, but his legacy rests primarily on his systemization of the calculus and his refinement of the theory of complex variables.

In developing and systematizing the concepts, symbols, and theorems of the calculus, Cauchy gave mathematics a powerful tool for solving problems in science and engineering. NIELS HENRIK ABEL reacted as if he'd had a religious conversion after reading Cauchy's published lecture notes. He used Cauchy's method in his own mathematical work on infinite series. Bernhard Riemann was also directly affected and, with Karl Weierstrass, extended Cauchy's work to complex function theory. Later Georg Cantor, Richard Dedekind, Eduard Heine, and Weierstrass developed the modern theory of real numbers. The favorable intellectual climate to make all this possible was created by Cauchy, the sometimes aloof professor who threw away seminal papers sent to him by young admirers like Abel.

For the next hundred years calculus textbooks were almost exclusively based on Cauchy's method and proofs. No other mathematician has had so many concepts and theories bearing his name, but perhaps no other ever examined the foundations of his discipline so rigorously and attempted such a comprehensive and creative reorganization of it.

His theory on complex variables is now invaluable in the several fields of applied mathematics, including applications in physics and aeronautics.

Steinberg

WORLD EVENTS		CAUCHY'S LIFE
United States independence	1776	
French Revolution	1789	Augustin-Louis Cauchy is born
	1800	Cauchy's father becomes Secretary of the Senate in France
Napoleonic Wars in Europe	1803-15	
	1810	Cauchy joins Napoleon's army as an engineer
	1813	Cauchy returns to Paris
	1816	Cauchy becomes full professor at École Polytechnique and wins Grand Prize of Institute de France
	1820s	Cauchy compiles and publishes lectures on the calculus
	1830	Cauchy leaves Paris in support of exiled King Charles X
	1857	Cauchy dies
United States Civil War	1861-65	

For Further Reading:

Bell, E. T. *Men of Mathematics.* New York: Simon & Schuster, 1937.

Grabiner, Judith V. *The Origins of Cauchy's Rigorous Calculus.* Cambridge, Mass.: MIT Press, 1981.

Kline, Morris. *Mathematical Thought from Ancient to Modern Times.* New York: Oxford University Press, 1972.

Cavendish, Henry

Investigator of Hydrogen;
Calculator of Earth's
Gravitational Constant
1731-1810

Life and Work

Henry Cavendish made many contributions to physical science, the most important of which were his calculation of Earth's gravitational constant and extensive experimentation with gases, including hydrogen.

Cavendish was born on October 10, 1731, to a wealthy English family in Nice, France. He was sent to Dr. Newcome's Academy in London and entered Cambridge University in 1749. Cavendish left the school four years later without a degree, an occurrence that was not uncommon in his day. His father provided him with a laboratory and generous allowance. Cavendish became a member of the Royal Society in 1760 and attended the meetings regularly. While at Society meetings, he rarely spoke but listened avidly to the debates. His shy demeanor developed into increasingly reclusive behavior as he grew older.

Many of Cavendish's significant discoveries were in chemistry. One major finding was that the composition of the atmosphere remained constant regardless of location. He also experimented with atmospheric gases and studied what he called inflammable air. This gas was later named hydrogen by ANTOINE LAVOISIER. In a 1766 paper he discussed many of hydrogen's properties, noting that it was much lighter than other gases. He experimented with hydrogen during 1784 and combined it with ordinary air. When the hydrogen burned, water was produced. This conclusively demonstrated that water was not an element as previously believed, but rather a compound. Cavendish isolated another gas, nitrogen, but did not publish his results and therefore was not credited with its discovery. Cavendish used nitrogen to create nitric acid in 1795.

Although he experimented extensively with electricity and wrote personal notes on its properties and uses during the 1770s, he only published a few of his findings.

His last major work was a calculation of Earth's gravitational force. ISAAC NEWTON had postulated a way to find this constant value but could not do the exacting experimentation. Cavendish devised a system of lightweight balls suspended on a flexible rod. The gravitational attraction between the objects caused the rod to twist. Using this minute measurement of attraction and knowing the mass of the objects, Cavendish was able to find the gravitational force between the two objects. From that information he was able to calculate the gravitational force for Earth. He used this value to eventually find the density and mass of Earth within 10% of current estimates. In 1798, Cavendish published his findings.

In his later years Cavendish became increasingly withdrawn. He shunned the company of others, particularly women. He died at his London home on February 24, in 1810.

Legacy

Much of what Cavendish discovered during his solitary life was not shared with the scientific community for many years. What he did share provided important information for understanding gravity and how gases interact.

The isolation and study of hydrogen provided scientists with an understanding of a known gas that could be used for comparison with other gases. Cavendish discovered that hydrogen was the lightest known gas, and, as such, it became the basis for chemist JOHN DALTON's comparative atomic weights, which in turn were the precursor to the periodic table. As the lowest density gas, hydrogen was useful for a number of things. Its lighter-than-air properties made it very useful in the construction of dirigibles. These flying airships became multipurpose modes of transportation. The use of hydrogen in water experiments allowed Cavendish to show that water was not an element and paved the way to understanding the difference between elements and compounds. This break from the Greek idea about the four components of matter (fire, earth, air, and water) opened new paths of thinking among scientists.

The calculation of the constant gravitational force on Earth not only allowed Cavendish to estimate the mass of Earth, it

also provided the key to understanding other gravitational systems. This constant has been used to explain tides, the orbits of the moon and planets, and the paths of falling objects. It is a factor in projectile motion. Satellites used for communications and our understanding of planetary systems are dependent upon this constant.

Perhaps the greatest contribution of Cavendish came after his death. In 1871 his family funded the creation of a laboratory at Cambridge in his honor. The prestigious Cavendish Laboratory is well known for its experimental physics research facilities.

While Cavendish did not share many of his findings, JAMES MAXWELL collected and published Cavendish's notebooks in 1879, nearly seven decades after his death. In these writings Cavendish described many of the principles of electricity that are credited to later scientists, for example, the concept of voltage and the inverse square law of attraction. Historians have speculated that Cavendish's work, had it been disseminated during his life, could have precipitated some of the great leaps in scientific thought that occurred many years later.

Wilson

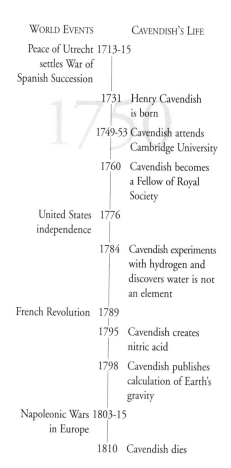

WORLD EVENTS	CAVENDISH'S LIFE
Peace of Utrecht 1713-15 settles War of Spanish Succession	
	1731 Henry Cavendish is born
	1749-53 Cavendish attends Cambridge University
	1760 Cavendish becomes a Fellow of Royal Society
United States 1776 independence	
	1784 Cavendish experiments with hydrogen and discovers water is not an element
French Revolution 1789	
	1795 Cavendish creates nitric acid
	1798 Cavendish publishes calculation of Earth's gravity
Napoleonic Wars 1803-15 in Europe	
	1810 Cavendish dies

For Further Reading:

Brock, William. *The Norton History of Chemistry.* New York: Norton, 1992.

Travers, B., ed. *World of Scientific Discovery.* Detroit, Mich.: Gale Research, 1994.

Wilson, George. *The Life of the Honorable Henry Cavendish.* New York: Arno Press, 1975.

Chadwick, James

Discoverer of the Neutron
1891-1974

Life and Work

James Chadwick isolated and identified the neutron, a basic nuclear particle, in 1932.

Chadwick was born in Manchester, England, on October 20, 1891, and graduated from Manchester University in 1911. He stayed there to work with physicist ERNEST RUTHERFORD following his graduation. He received a scholarship in 1913 to study in Germany, which put him in that country at the start of World War I; Chadwick was detained in a civilian prisoner of war camp. By 1919 he was back in England, conducting research at Cambridge University, where he was appointed assistant director of research of the Cavendish Laboratory in 1923.

Chadwick returned to England eight years after Rutherford had discovered that atoms have tiny, dense nuclei. It was known that the nucleus held a positive charge, but only in multiples of the charge of a hydrogen nucleus. For example, helium had exactly twice the charge of hydrogen, but the helium nucleus weighed almost four times as much as the hydrogen nucleus. Scientists were wondering where this extra mass was hiding.

Chadwick helped to answer this question in 1932 with his identification of the neutron, a

particle in the nucleus with no electrical charge. As early as 1920, Rutherford had guessed that a neutral particle of some sort might exist in the nucleus. Experiments first conducted by Rutherford, then by Chadwick and other scientists, involved the bombardment of elements with alpha particles, positively charged particles discharged by radioactive material. This allowed scientists to analyze the effect of the release of nuclear particles and determine the mass of particles from the effects of the release.

Chadwick devised an experiment in 1932 that answered the question about the unidentified source of mass in the nucleus. He smashed alpha particles into beryllium and allowed the radiation that was released to strike another target, paraffin wax. The radiation from the beryllium hit hydrogen atoms in the wax and knocked them into a detecting chamber. Only a particle with about the same mass as a hydrogen atom could kick out the hydrogen that way. Chadwick had shown that the collisions with beryllium atoms were releasing massive neutral particles, which he called neutrons because of their lack of charge. His identification of neutrons provided the explanation for hidden mass in atoms like helium and established the atomic weight as the combined mass of protons and neutrons.

Chadwick won the Nobel Prize for Physics in 1935 for his discovery. He worked on the British atomic bomb project during World War II; he was a scientific advisor to the atomic bomb policy team led by J. ROBERT OPPENHEIMER, chief of the United States' Manhattan Project, which produced the first atomic bomb. He was knighted in 1945. He died in Cambridge on July 24, 1974.

Legacy

Chadwick's discovery of the neutron solved the puzzle of the weights of atoms and formed the foundation for investigating important questions in nuclear physics concerning the nature of the nucleus and the forces held within it.

Nuclear physicists around the world soon began incorporating Chadwick's discovery into new models of the nucleus. In 1935 Japanese physicist HIDEKI YUKAWA reasoned that, if the nucleus contains only neutrons and positively charged protons, the protons, all of the same

positive charge, should repel each other and the nucleus would break apart. To answer this question, he suggested that some other particle must exist with a negative charge to help hold the nucleus together. This theory became known as the meson theory, named for the meson particles that Yukawa and other scientists eventually identified in the nucleus. Later modified and refined, the meson theory led other scientists to discover other types of nuclear particles; in 1965, the American scientists MURRAY GELL-MANN and George Zweig independently suggested that tinier building blocks, later termed quarks, composed the newly discovered particles within the nucleus.

The discovery of the neutron helped scientists answer puzzling questions concerning radioactive isotopes, versions of chemical elements with varying atomic masses whose nuclei disperse extra energy through radiation. The analysis of the atomic weight of radioactive substances would help scientists identify the isotope forms of elements, and by 1935, the major isotopes were known for nearly all of the elements.

Chadwick's discovery was also important for experimental work. Because neutrons have no electrical charge, neutrons fired from a source can penetrate deeply into thick layers of materials, reaching into the nuclei of target atoms. After Chadwick's discovery, scientists around the world began bombarding all kinds of materials with neutrons. They discovered that when uranium is the target, nuclear fission is possible, which is the basis for nuclear power plants and nuclear weapons.

Secaur

WORLD EVENTS	CHADWICK'S LIFE
	1891 James Chadwick is born
Spanish-American War 1898	
World War I 1914-18	
	1919 Chadwick goes to the Cavendish Laboratory, Cambridge, England
	1932 Chadwick discovers the neutron
	1935 Chadwick receives the Nobel Prize for Physics
World War II 1939-45	Chadwick works on British atomic bomb project
	1945 Chadwick is knighted
Korean War 1950-53	
	1974 Chadwick dies
End of Vietnam War 1975	

For Further Reading:

Brown, Andrew P. *The Neutron and the Bomb: A Biography of Sir James Chadwick*. New York: Oxford University Press, 1997.

Heathcote, N. H. *Nobel Prize Winners in Physics, 1901-1950*. Freeport, N.Y.: Books for Libraries Press, 1971.

History of Original Ideas and Basic Discoveries in Particle Physics. Newman, Harvey B., and Thomas Ypsilantis, eds. New York: Plenum Press, 1996.

Snow, C. P. *The Physicists*. Boston: Little, Brown, 1981.

Clausius, Rudolf

Founder of Thermodynamics
1822-1888

Life and Work

Rudolf Clausius formalized the laws of thermodynamics and helped to establish thermodynamics as a field of study.

Born on January 2, 1822, in Köslin, Prussia, Rudolf Clausius was the son of a pastor who also served as principal of the local school. At the University of Berlin, Clausius was at first attracted to history, but gradually became more interested in science. Because of financial difficulties, he had to work as a teacher in Berlin while he finished his own education, receiving his doctorate in 1848. In 1850 he took a professorship at the Royal Artillery School and in that same year presented his first paper on thermodynamics, the study of how energy moves and transforms from one form to another. He taught at several other universities, finally settling at the University of Bonn, Germany, in 1869, where he remained for the rest of his life.

Many scientists had studied heat during the first half of the nineteenth century and had reached some important conclusions. For example, SADI CARNOT described the way heat flows in a steam engine, and JAMES JOULE showed that heat is equivalent to mechanical energy. Clausius clarified their terminology, improved their mathematics,

and unified their ideas into the two great laws of thermodynamics.

Clausius tackled a problem in Carnot's work on heat engines: Carnot had considered only idealized, perfectly reversible processes. In fact, all of classical physics assumed that laws of nature work equally well running forward or backward, like watching a movie in reverse. Clausius found a term in the mathematical equations used to describe heat engines that increases with time, making reversing the process impossible. He developed the concept concerning this new term by 1851, and in 1865 finally introduced it by name. He called it entropy, from the Greek word trope, meaning "transformation." Entropy is a measure of disorder or randomness in a system. In Clausius's new terms, the two laws of thermodynamics are: 1) the energy of the universe is constant 2) the entropy of the universe tends toward a maximum.

The British scientist William Thomson (later named LORD KELVIN) developed the same laws at about the same time. Clausius stated them more simply and succinctly, and he was the one who conceived of the name, entropy, for the fundamental change that occurs in a system over time.

Tragedy struck Clausius in 1875 when his wife died delivering their sixth child. He spent most of the rest of his life raising his family and died in Bonn on August 24, 1888.

Legacy

Rudolf Clausius's two laws of thermodynamics created the groundwork for the field of thermodynamics. He analyzed the work of Joule and Carnot, synthesized their contributions, and created a universal theory that benefited future generations.

Immediate applications of his laws resounded in all fields of science—from the study of gases under real conditions, to electrochemistry, to boiling, melting, and vaporizing, to capillary action, and especially to industrial chemistry. The laws took chemists, in particular, from trial and error methods to understanding theoretically what can and can not be accomplished.

Clausius lived to see his own work taken to yet another level. In 1875 the great American scientist J. Willard Gibbs used Clausius's two laws of thermodynamics as the opening words

in his paper on the theory of chemical equilibrium, which states that a point exists in the progress of a reversible chemical reaction where no net change occurs in the amounts of original or resulting chemicals. Chemical equilibrium proved to be a tremendously useful tool for chemists in both explaining chemical processes and predicting their outcomes.

The laws that Clausius developed had extensive impact on fundamental questions regarding the universe. They began one of the longest-lasting controversies in physics: how will the universe end? The first law claims that energy, whatever it is, is a special, conserved quantity in the universe; however much energy there was at the beginning of time, that energy is still around us today and always will be, in one form or another. The second law sets limits on the transformation of energy: heat must flow from hotter places to cooler ones, and if all the parts of a system are at the same temperature, no energy transformations can happen. These laws suggest the theory of the "heat death" of the universe: the universe will end in continual expansion, as the stars burn out one by one. It is proposed, however, that this won't happen for dozens of billions of years. Our Sun alone has enough fuel to last for five billion years or more.

Secaur

For Further Reading:

Cardwell, D. S. L. *From Watt to Clausius: The Rise of Thermodynamics in the Early Industrial Age.* Ames: Iowa State University Press, 1989.

Hecht, E. *Physics in Perspective.* Addison-Wesley, 1979.

Segrè, Emilio. *From Falling Bodies to Radio Waves.* New York: Freeman, 1984.

WORLD EVENTS		CLAUSIUS'S LIFE
Napoleonic Wars in Europe	1803-15	
	1822	Rudolf Clausius is born
	1848	Clausius receives doctorate from University of Berlin
	1850	Clausius presents first paper on thermodynamics
	1851	Clausius develops laws of thermodynamics
	1865	Clausius coins the term entropy and applies it to second law
	1869	Clausius takes a professorship at University of Bonn
Germany is united	1871	
	1888	Clausius dies
Spanish-American War	1898	

Comstock, Anna Botsford

Entomologist;
Pioneer in Science Education
1854-1930

World Events	Comstock's Life
	1854 Anna Botsford is born
Germany is united 1871	
	1878 Botsford marries John Henry Comstock
	1885 Anna Comstock awarded B.S. from Cornell University
	1895 Nature Study Movement is founded; Comstock becomes its spokesperson
	1899 Comstock appointed assistant professor at Cornell
	1900 Comstock demoted to lecturer at Cornell
	1911 Comstock publishes *Handbook of Nature Study*
	1913 Comstock again becomes assistant professor at Cornell
World War I 1914-18	
	1917-23 Comstock is editor of *Nature Study Review*
	1920 Comstock promoted to full professor at Cornell
	1930 Comstock dies
World War II 1939-45	

Life and Work

At a time when few women participated in science, Anna Botsford Comstock worked as an entomologist and contributed significantly to science education, notably in promoting popular knowledge and appreciation of nature study.

Anna Botsford was born in 1854 in New York State. She was the only child of Marvin and Phoebe Botsford, a prosperous Quaker farming couple. When she became a college student at Cornell University in Ithaca, New York, in 1874, Botsford intended to study English. However, she enrolled in a class in invertebrate zoology taught by her future husband, John Henry Comstock, and became fascinated by entomology, the study of insects.

After marrying in 1878, the Comstocks moved to Washington, D.C., where Henry Comstock was appointed chief entomologist at the U.S. Department of Agriculture. During their two-year stay in Washington, Comstock did clerical, editorial, and laboratory work for her husband, and illustrated his books on insects; her drawings were widely acclaimed. She later mastered wood engraving to illustrate the entomological books they published together, under their own company called Comstock Publishing Company. The company later became part of Cornell University Press.

On their return to Cornell, Comstock completed her degree in natural history in 1885; she subsequently worked in the field as a spokesperson for the Nature Study Movement, founded in 1895 to educate children about the relationship between farming and nature study. She became active in science education all over New York State, lectured at a number of universities and teachers' colleges, and produced numerous leaflets for classroom use.

In 1899 she was appointed assistant professor at Cornell, becoming the first woman to reach professorial rank at Cornell. After some conservative trustees objected, her rank was reduced to lecturer in 1900. In 1913 she was once more appointed to an assistant professorship, and in 1920 she became a full professor.

She authored and co-authored numerous books, and contributed to many other publications as author and editor. From 1917 to 1923, she was the editor for the *Nature Study Review;* her most important work, *Handbook of Nature Study,* was published in 1911.

In her work, Comstock appears to have received constant support from her husband; from all accounts, he was as proud of her achievements as she was of his. She achieved national recognition, and in 1923 a poll named her one of America's 12 greatest living women.

Comstock retired from full-time teaching in 1922, but continued teaching part-time until two weeks before she died at age 75 in August 1930.

Legacy

Anna Comstock inspired public interest in science, notably in entomology and nature study. Through her many activities and work in the nature study movement, she influenced science education policy.

Her work in promoting nature study in schools was very effective, and her publications were important in the popularization of natural history, both among her contemporaries and in succeeding generations. Her *Handbook of Nature Study,* in particular, became a bible for teachers, garden club members, farmers, scoutmasters, and others interested in natural cycles and processes. By 1939 this bestseller was in its twenty-fifth edition in the United States and had been translated into eight languages.

As the first woman to achieve professorial status at Cornell University, Comstock, along with other early women pioneers in universities, helped to break down the societal boundaries that barred women from careers in higher education.

Hertzenberg

For Further Reading:

Bailey, Martha J. *American Women in Science.* Denver: ABC-CLIO, 1994.

James, Edward T., ed. *Notable American Women 1607-1950: A Biographical Dictionary.* Cambridge, Mass.: Harvard University Press, 1971.

Ogilvie, Marilyn Bailey. *Women in Science: Antiquity Through the Nineteenth Century.* Boston: MIT Press, 1986.

Copernicus, Nicolaus

Father of Modern Astronomy
1473-1543

Life and Work

Nicolaus Copernicus proposed the revolutionary theory that the Earth rotates on its axis and revolves around the Sun. He is considered to be the father of modern astronomy,

Copernicus was born on February 19, 1473, in Toran, Prussia (now Poland). His father, a wholesaler of copper, died when Nicholas was only 10. Copernicus's uncle, Lucas Watzelrode, the Bishop of Ermeland, guided his education, which prepared Copernicus for a role in the Catholic church. His education at University of Cracow (1491-94) included humanities, law, canon law, medicine, theology, and astronomy.

In January of 1497, Copernicus was appointed to a canonry at the Cathedral of Frauenburg, but he immediately took a leave to continue his studies. He remained abroad, finishing his studies, until 1506. Then Copernicus returned to live with his uncle, assisting in administration of the diocese while continuing his studies and writing on astronomy.

Copernicus studied at a time when most people believed the ideas of the great astronomer, Ptolemy of Alexandria (c. 100-

170). The planetary system that Ptolemy proposed put Earth at the center of the universe with the other planets, Sun, and moon moving around it. This system, which would eventually be replaced by Copernican theory, is known as geocentric (Earth-centered). Copernicus rejected such an idea. In 1510 he began work on a manuscript that proposed a heliocentric (Sun-centered) astronomy and outlined seven revolutionary hypotheses about planetary motion.

Copernicus was reluctant to publish his manuscript when it was completed around 1530. This reluctance may have stemmed from a belief that many would attack such a revolutionary idea. However, he shared his ideas with some colleagues and church experts. Many of them encouraged Copernicus to publish his theories; one church official, Cardinal Schoenberg, even offered to cover the costs of printing. However, it was not until 1542 that Copernicus agreed to release his manuscript. *Revolutions of the Heavenly Spheres* was published in 1543. It is recorded that Copernicus glanced over his completed book just a few hours before he died on May 24, 1543, in Frauenburg.

Legacy

Copernicus's heliocentric model of the Earth, Sun, and planets caused a cataclysmic change in our understanding of our place in the universe. For at least 14 centuries people had believed that the universe revolved around Earth and humankind. Heliocentrism displaced humans from that central role. His new model formed the basis for future astronomical investigations for centuries to come.

Copernicus's ideas made sense in that they addressed many discrepancies in the Ptolemaic system; however, proof of the motions of the planets or Earth's rotation was not immediately attainable in this era before telescopes. Therefore, by its very nature, the Copernican theory was open to controversy. Many religious leaders opposed his theories; indeed, the book remained on the Catholic church's Index of Prohibited Works for 150 years.

However, Copernicus started a revolution that would not be ignored. Many followers,

including Johannes Kepler (1571-1630) and Galileo Galilei (1564-1642), further developed his theories and used them as springboards for revolutionary scientific insights. Kepler expanded Copernican theory by formulating the three laws of planetary motion. Galileo supported heliocentrism after observing phases of Venus through a telescope and openly opposed the Roman Catholic church's restrictions on Copernican theories.

By stepping outside conventional ways of viewing the world, Copernicus was able to break free of many misconceptions of his time. His insights eventually stimulated a wave of new understandings of the universe and of the nature of science itself.

Weaver

World Events		Copernicus's Life
Ottoman Empire captures Byzantine capital, Constantinople	1453	
	1473	Nicolaus Copernicus is born
	1483	Copernicus's father dies; Lucas Watzelrode adopts Copernicus
	1491-94	Copernicus studies at University of Cracow
Columbus discovers Americas	1492	
	1497	Copernicus appointed canon of Cathedral of Frauenburg
	c. 1510	Copernicus outlines heliocentric system in manuscript
Reformation begins	1517	
	1543	*Revolutions of the Heavenly Spheres*, which explains Copernicus's heliocentric model, is published

Copernicus dies |
| Thirty Years' War in Europe | 1618-48 | |

For Further Reading:

Goodstein, David. *Feynman's Lost Lecture: The Motion of Planets and the Sun*. New York: Norton, 1996.

Hoyle, Fred. *Nicolaus Copernicus: An Essay on His Life and Work*. New York: Harper & Row, 1973.

Koestler, Arthur. *The Sleepwalkers*. New York: Macmillan, 1976.

Rosen, Edward. *Copernicus and His Successors*. London: Hambledon Press, 1995.

Cousteau, Jacques-Yves

Pioneer Oceanographer;
Developer of Scuba Equipment
1910-1997

Life and Work

Jacques-Yves Cousteau developed scuba-diving and filmmaking equipment, produced highly acclaimed oceanographic documentaries, and aroused popular concern about threats to marine life.

Cousteau was born on June 11, 1910, in St. André-de-Cubzac, France. As a teenager he became intrigued by film and bought a movie camera. In 1929 he joined the French naval academy with dreams of world travel, and during his first year abroad he made movies of everything he encountered.

World Events		Cousteau's Life
	1910	Jacques Cousteau is born
World War I 1914-18		
	1936	Cousteau involved in near-fatal accident
World War II 1939-45		
	1943	Cousteau co-develops Aqua-Lung
Korean War 1950-53		
	1953	*The Silent World* (book) is published
	1956	*The Silent World* (film) wins prize at Cannes Film Festival
	1957	Cousteau becomes director of Monaco's Oceanographic Institute
	1964	Cousteau's film *World Without Sun* is produced
	1968	Television program, *Undersea World of Jacques Cousteau,* begins
End of Vietnam War	1975	
Dissolution of Soviet Union	1991	
	1997	Cousteau dies

After several years in China, Cousteau returned to France and joined the aviation academy. He was nearly killed in a car accident in 1936 and subsequently began a swimming regimen to rehabilitate his mangled left arm. Wearing goggles, he marveled at the beauty of underwater life in the Mediterranean.

During World War II, Cousteau investigated new techniques for improving underwater filmmaking. His first undersea documentary, *Sixty Feet Down,* appeared in 1942 and won praise at the Cannes Film Festival.

Frustrated by his restrictive, heavy diving equipment, Cousteau collaborated in 1943 with Emile Gagnan to develop the Aqua-Lung, an apparatus with an automatically regulated compressed-air system. He used the lung while exploring a sunken ship, collecting footage for another movie, *Wreck,* which earned him a position with the French Navy's Underwater Research Group.

Cousteau sailed for the Red Sea in 1951. He and his team identified several unknown species, discovered volcanic basins on the sea floor, and, on their way home, discovered a treasure-filled Roman shipwreck. In 1953 Cousteau's daily logs were published as *The Silent World,* which enjoyed immense popularity. A second journey yielded a film version of his daily logs, which won a prize at Cannes in 1956.

In 1957 Cousteau became director of the Oceanographic Institute and Museum of Monaco, filling its aquariums with rare organisms collected during his expeditions. That year he also initiated the Conshelf Saturation Dive program, which involved researchers living underwater for extended periods. The program led to his successful 1964 film, *World Without Sun.*

Cousteau produced the acclaimed eight-year television series, *Undersea World of Jacques Cousteau,* starting in 1968. The equally popular series *Cousteau Odyssey* followed.

In the 1980s, Cousteau produced many oceanographic television documentaries that focused on his growing environmental concerns. He also continued to develop improved oceanographic equipment.

He died in Paris on June 25, 1997.

Legacy

Cousteau's technological contributions yielded numerous oceanographic discoveries, and his films helped to develop a worldwide concern for the marine environment.

Diving equipment engineered by Cousteau led to expanded diving opportunities and the discovery of the complex ecosystems hidden in the deep sea, opening new areas of biological research. The Conshelf program allowed biologists to observe daily changes in natural marine environments over extended periods.

The principles of Gagnan's and Cousteau's Aqua-Lung can be found in modern scuba diving equipment. Innovations based on Cousteau's principles, such as decompression schedules and carefully calculated gas mixtures, now allow safer and deeper dives, and dives in adverse conditions.

Other oceanographic applications of Cousteau's enterprise include the Sea Spider, used for biochemical analysis of seawater, and the Turbosail, used to reduce fuel consumption in ocean-going vessels. Industries such as offshore oil drilling were boosted by adopting Cousteau's innovations.

His lyric writing style and engaging film presence drew an enthusiastic audience and spread concern about the damage human activity wreaks on the ocean environment. His films, books, and television programs revealed the diversity of marine life and led to efforts to reduce marine pollution. In the 1970s the Cousteau Society, an environmental group focusing on both marine issues and world peace, was formed.

His eldest son Jean-Michel Cousteau, a prominent oceanographer, has been involved in his father's projects since his childhood and carries on his father's technological and environmental legacy.

Schuyler

For Further Reading:
Cousteau, Jacques. *The Silent World.* New York: Harper, 1953.
Madsen, Axel. *Cousteau: An Unauthorized Biography.* New York: Beaufort Books, 1986.
Munson, Richard. *Cousteau: The Captain and His World.* New York: Morrow, 1989.

Crick, Francis; Watson, James

1916- ; 1928-

Discoverers of DNA Structure

Life and Work

Francis Crick and James Watson proposed the double-helix model for the genetic material deoxyribonucleic acid (DNA). For this accomplishment they shared (with Maurice Wilkins) the 1962 Nobel Prize for Physiology or Medicine.

Crick was born on June 8, 1916, in Northampton, England. He interrupted his doctoral work in physics to develop mine technology during World War II. Thereafter he focused on biology and in 1949 became involved in efforts to determine the structure of large molecules using X-ray crystallography at the Cavendish Laboratories of Cambridge University.

Watson, born on April 6, 1928, in Chicago, Illinois, entered the University of Chicago at age 15; he graduated with a degree in zoology after four years. He received a Ph.D. from Indiana University in 1950 and proceeded to Copenhagen to study bacterial metabolism. Convinced that genetics could be fully understood only after the molecular structure of DNA had been described in detail, he joined Crick at the Cavendish Laboratories in 1951.

When Crick and Watson began collaborating, it was known that genes, organized into DNA molecules, carry hereditary information. The basic units of DNA had also been identified: nucleotides, each containing a phosphate, a sugar, and one of the four nitrogen bases—cytosine, guanine, adenine, and thymine. Crick and Watson combined biochemical evidence and physical clues from molecular models and X-ray diffraction images (created by ROSALIND FRANKLIN and Maurice Wilkins) to determine the three-dimensional structure of the DNA molecule: two sugar-phosphate strands twisted into a double helix and linked by bonds between complementary bases (adenine and thymine; cytosine and guanine). They announced this discovery in 1953.

In 1961 Crick helped show that the sequence of nucleotides in DNA represents a three-part code that designates the sequence of amino acids in proteins. In 1977 he became professor at the Salk Institute for Biological Studies, and in the 1980s and 1990s he proposed theories concerning the origin of life on Earth and the nature of consciousness.

In 1955 Watson joined the faculty of Harvard University and in 1968 became director of the Cold Spring Harbor Laboratory of Quantitative Biology on Long Island, New York. In 1988 he began coordinating the Human Genome Research project, a plan to map the sequence of all human genes; he resigned in 1993 in opposition to the government's plan to patent genetic information gleaned from the project.

Legacy

Watson and Crick's description of the structure of DNA overcame a major hurdle in the development of biological science and laid the groundwork for modern molecular genetics.

The model of DNA structure proposed by Crick and Watson suggested a mechanism for the construction of new DNA within the cell. Confirmed experimentally in 1958 by Matthew Meselson and Franklin Stahl, DNA replication involves the uncoiling of the two strands of a DNA helix, which each then serve as a template for the synthesis of a new DNA strand.

The next step was to determine how DNA directs the production of proteins. Over the next few years researchers found that DNA is first transcribed into messenger ribonucleic acid (mRNA), and that mRNA is translated into the sequence of amino acids in a protein. Crick had shown by 1961 that translation involves a three-nucleotide code. In the mid-1960s Marshal Nirenberg and Severo Ochoa, working independently, cracked the code and demonstrated which nucleotide triplets designate which amino acids.

Molecular biology has continued to be a productive area of research. The discovery of methods to determine the base sequence of DNA molecules in the 1970s initiated the era of recombinant DNA technology, separating genes and recombining them with genes from original donors or others. This technology has had profound implications in all areas of biological research, as well as in medicine, pharmaceuticals, and agriculture. Advances include DNA sequencing, DNA fingerprinting, DNA cloning, the search for genetic causes of diseases, gene therapy, genetic engineering of plants and animals, and the Human Genome Project (in which researchers are determining the blueprint of the complete human genome). Scientists are also conducting an enormous variety of pure research aimed at understanding the complexity of genetics, the origin of life, and the details of evolution.

The implications of these advances are profound, as was seen in 1996 when a team of British scientists led by Ian Wilmut cloned an adult sheep using DNA from one of its organs; the result was the now-famous ewe called Dolly. Both excitement and controversy erupted over the possible positive applications of cloning (for example, cloning of top-quality livestock and regenerating endangered species) and the potential misuse of cloning if developed for humans.

Schuyler

WORLD EVENTS	CRICK'S AND WATSON'S LIVES
World War I 1914-18	
	1916 Francis Crick is born
	1928 James Watson is born
World War II 1939-45	
	1951 Crick and Watson begin collaboration
Korean War 1950-53	
	1953 Structure of DNA is announced by Crick and Wilkins
	1958 Matthew Meselson and Franklin Stahl confirm process of DNA replication
	1961 Crick helps show the triplet nature of genetic code
	1962 Crick, Watson, and Wilkins share Nobel Prize for Physiology or Medicine
	1968 Watson begins directing Cold Spring Harbor Laboratory
End of Vietnam War 1975	
	1977 Crick joins Salk Institute
	1988 Watson directs Human Genome Research project
Dissolution of Soviet Union 1991	
	1993 Watson resigns from Genome Project

For Further Reading:

Crick, Francis. *What Mad Pursuit: A Personal View of Scientific Discovery.* New York: Basic Books, 1988.

Elseth, Gerald D. *Principles of Modern Genetics.* Minneapolis-St. Paul, Minn.: West Publishing, 1995.

Watson, James D. *The Double Helix: A Personal Account of the Discovery of the Structure of DNA.* London: Weidenfeld & Nicholson, 1983.

Curie, Marie

Founder of Radiation Chemistry

1867-1934

Life and Work

Marie Curie pioneered the field of radiation chemistry, coining the term "radioactivity" to describe the spontaneous emission of energy from unstable elements. Her many years of arduous research earned her two Nobel Prizes and a reputation as a scientist of rare insight and dedication.

Curie was born Marya Sklodowska on November 7, 1867, in Warsaw, Poland, which was then under Russian domination.

In 1891, after attending an underground school run by Polish nationalists, she enrolled at the Sorbonne in Paris, where she earned degrees in physics and mathematics. In 1895 she married Pierre Curie, a French physicist, with whom she collaborated on much of her subsequent research.

In the late 1890s, Curie became interested in the 1896 discovery, by ANTOINE-HENRI BECQUEREL, that the element uranium emits energy, a phenomenon that no one could explain. She began testing other elements for energy emission, and at first concluded that only uranium and thorium had this property, which she called "radioactivity." In 1898, after completing the tedious analysis of numerous ore samples, Curie isolated an unknown radioactive element and named it polonium after her homeland. Six months later, she discovered another and called it radium.

Curie became the first woman to earn a Nobel Prize when, in 1903, the prize for physics was awarded jointly to her and her husband and to Becquerel for their discovery of radioactivity. In 1906 Pierre Curie died, and Curie took over her husband's teaching position at the Sorbonne. In 1909 she was appointed co-director of a newly founded branch of the Sorbonne, the Institut du Radium, dedicated to the study of radioactivity. In 1911 she was awarded a second Nobel Prize, this time in the field of chemistry, for her discovery of radium and polonium.

By 1920 Curie's health was deteriorating because of leukemia, more than likely caused by exposure to radiation, and on July 4, 1934, she died at a nursing home in the French Alps.

Legacy

Curie founded the study of radioactivity, which played a key role in the advancement of nuclear physics and led to the development of valuable technological applications.

Curie entered the field of chemistry at a time when her colleagues were making swift progress toward a description of the properties and structure of the atom. In the last decade of the nineteenth century, scientists relinquished the long-held belief that atoms were indivisible units and began to suggest that atoms might consist of smaller particles. The discovery of radioactivity by Becquerel and the Curies was one of the key experimental results that confirmed this suggestion. Their studies told them that the nucleus of a radioactive atom emits heat and subatomic particles when it breaks apart, transforming the atom into different components. Curie coined the term "disintegration" to describe this loss of particles, and the term "transmutation" to describe the transformation. Other discoveries and theories soon followed. In 1911 ERNEST RUTHERFORD described the atom for the first time as a heavy nucleus circled by electrons. In 1913 NIELS BOHR used quantum theory (introduced in 1900 by MAX PLANCK) to theorize that electrons orbit the nucleus in distinct energy levels. The field of nuclear physics was thus born.

Radioactivity gave scientists the opportunity to explore the potential benefits of a naturally occurring phenomenon and to harness its energy. Physicians use radioactivity to destroy cancer cells and sterilize medical equipment; genetics researchers use radioactivity to identify genes. Carbon dating, which is used to determine the age of ancient organic remains, relies on the radioactive decay of carbon-14, and radioactive uranium fuels nuclear power plants.

Curie also left a scientific legacy in her daughter, Irène Joliot-Curie, who produced and identified artificial radioactivity and studied its by-products. She received the Nobel Prize for Chemistry jointly with her husband Frédéric Joliot in 1935.

Schuyler

World Events		Curie's Life
	1867	Marya Sklodowska is born
Germany is united	1871	
	1891	Sklodowska enters the Sorbonne
	1895	Sklodowska marries Pierre Curie
	1896	Antoine-Henri Becquerel discovers that uranium emits energy
Spanish-American War	1898	M. Curie conducts experiments in radioactivity and discovers polonium and radium
	1903	M. Curie, P. Curie, and Becquerel receive the Nobel Prize for Physics for their research into radioactivity
	1906	P. Curie dies; M. Curie is appointed professor at the Sorbonne
	1911	M. Curie is awarded a second Nobel Prize for Chemistry
World War I 1914-18		
	1934	M. Curie dies
World War II 1939-45		

For Further Reading:

Curie, Eve. *Madame Curie: A Biography by Eve Curie.* Vincent Sheean, trans. New York: Doubleday, 1937.

Pasachoff, Naomi. *Marie Curie and the Science of Radioactivity.* New York: Oxford University Press, 1996.

Pflaum, Rosalynd. *Grand Obsession: Marie Curie and Her World.* New York: Doubleday, 1989.

Cuvier, Georges

Early Pioneer in Comparative Anatomy and Paleontology
1769-1832

Life and Work

Georges Cuvier helped lay the foundations of modern comparative anatomy and paleontology.

Cuvier was born on August 23, 1769, in Montbéliard, France. He studied comparative anatomy at the University of Stuttgart, Germany, and returned to France to tutor children in the northern coastal province of Normandy. His employer had a good library and the nearby beaches were strewn with marine invertebrates that he studied in his spare time. He sent his observations to Étienne Geoffroy Saint-Hilaire, a zoologist at the Museum of Natural History in Paris, who hired him in 1795 to teach at the museum. In 1800, he joined the faculty of the Collège de France, also in Paris.

Cuvier was a popular lecturer and became an authority on animal anatomy, physiology, and classification. He introduced the idea that an animal's anatomy is determined by its environment, disagreeing with Saint-Hilaire, who believed that an animal's anatomy requires it to adopt a particular lifestyle. Cuvier contended that animals of past ages differed widely from those currently alive. He reconstructed two skeletons of major extinct plant-eating animals from fragments found in a local quarry, and he was first to identify the bird-like dinosaur called the pterodactyl.

Fossil excavations led Cuvier to ponder geology. He believed that Earth's life span was short (tens of thousands of years) and was therefore impressed by the immense geological changes that had occurred. He revitalized the catastrophe theory, which states that a series of sudden, massive geological events, such as floods and land upheavals, had both extinguished entire species and sculpted the surface of the planet.

Most of Cuvier's contemporaries—including Saint-Hilaire and Jean Baptist Lamarck, who developed an early theory of evolution—believed that animal species are organized in a hierarchical series from the simplest to the most advanced (humans). Cuvier suggested instead that animals be classified non-linearly into four groups, based on four distinct body plans: vertebrates, articulates (e.g., insects, crabs), radiates (e.g., snails, clams, oysters), and mollusks (e.g., corals, anemones).

In 1830 Cuvier and Saint-Hilaire publicly debated the amount of anatomical uniformity among animals. Cuvier argued that the four body plans represent distinct animal types, while his opponent countered that all animals belong to one anatomical type. Cuvier also argued that the function of organs, not just structural resemblances, be considered in classification. Their disagreement sprung from the lack of a satisfactory explanation for the similarities and differences among species, a problem that would be solved in 1859 with CHARLES DARWIN'S theory of natural selection.

Cuvier died in Paris on May 13, 1832.

Legacy

Cuvier's achievements mark the transition from pre- to post-evolutionary thought. Though many of his theories were shown to be incorrect, he helped lay the foundations of modern comparative anatomy and paleontology (the study of prehistoric life forms through examination and analysis of fossilized remains) by focusing on the relationship between structure and function and by recognizing that many fossil animals were very different from other species.

Maintaining that species were created independently of each other and were immutable, Cuvier differed from other prominent biologists of his time. Saint-Hilaire believed in evolution, as did Lamarck, who in 1809 published a theory positing that species evolve through the inheritance of acquired characteristics. This idea dominated evolutionary theory until Darwin's theory of natural selection took hold in the late nineteenth century. Natural selection posits that certain individuals in a population have characteristics that give them reproductive advantages over other individuals, allowing the slow accumulation of those favorable characteristics and the eventual transformation of the population into a new species.

Cuvier's four-type classification system dislodged the idea of a linear progression of animals from simple to advanced, and it dominated zoological doctrine until the theory of evolution was widely accepted. Evolutionary relationships explained the similarities and differences among animals and indicated new categories of classification.

Cuvier's talent for the reconstruction of animals from limited fossil fragments provided a basis for the specialty of paleontology. He worked primarily with the fossils of reptiles and mammals, exhibiting interpretive powers that surprised his colleagues and adding extensively to the vertebrate collection of the Museum of Natural History in Paris.

Schuyler

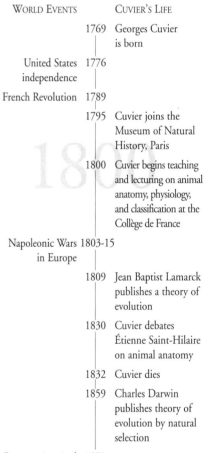

WORLD EVENTS		CUVIER'S LIFE
	1769	Georges Cuvier is born
United States independence	1776	
French Revolution	1789	
	1795	Cuvier joins the Museum of Natural History, Paris
	1800	Cuvier begins teaching and lecturing on animal anatomy, physiology, and classification at the Collège de France
Napoleonic Wars in Europe	1803-15	
	1809	Jean Baptist Lamarck publishes a theory of evolution
	1830	Cuvier debates Étienne Saint-Hilaire on animal anatomy
	1832	Cuvier dies
	1859	Charles Darwin publishes theory of evolution by natural selection
Germany is united	1871	

For Further Reading:

Appel, Toby A. *The Cuvier-Geoffroy Debate: French Biology in the Decades Before Darwin.* New York: Oxford University Press, 1987.

Smith, Jan C. *Georges Cuvier: An Annotated Bibliography of His Published Works.* Washington, D.C.: Smithsonian Institution Press, 1993.

Daguerre, Louis

Inventor of the First Photograph

1789-1851

Life and Work

Louis Daguerre invented the daguerreotype, the first practical photograph.

Daguerre was born on November 18, 1789, in Cormeilles, France. His elementary school-ing was disrupted by the French Revolution, as political upheaval caused a shortage of funds for public education. His talent for drawing became apparent at an early age; it led him to an apprenticeship with an architect at age 13, however, he was more interested in painting landscapes and portraits, and he moved to Paris in 1804 to become an artist.

After studying under a stage designer, Daguerre began painting landscapes in 1807 but received no encouragement from art critics for his work. Turning to the theater, he obtained a design post at the Théatre Ambigu-Comique, where his innovative use of color and light was admired.

In 1822 Daguerre introduced the Diorama, a technique in which sets were painted on large translucent linen screens and light was shone through them to create illusions of depth and movement. The basic sketches for the Diorama were often made using a *camera obscura,* a box with a small lens at one end and a screen at the other, and Daguerre began investigating ways to capture and fix camera-obscura images to increase the efficiency of the work.

Daguerre met French inventor Joseph Nicéphore Niépce, who was developing the science of photography, in 1826. They formed a partnership in 1830 and conducted experiments aimed at fixing an image on a surface treated with light-sensitive chemicals.

Niépce died in 1833, but Daguerre continued his work with great success: using silver salts he found that mercury vapor could develop an image on a copper plate coated with silver iodide. Developing the image, dubbed a "daguerreotype," took only 20 minutes, while Niépce's process had taken eight hours and produced a fainter image.

The French government gave Daguerre a lifetime pension in exchange for the rights to the invention; the daguerreotype was introduced to the public in 1839.

Daguerre retired in 1840 and died on July 10, 1851, in Bry-sur-Marne, France.

Legacy

The technology behind Daguerre's invention, soon called a "photograph," was rapidly improved and formed the basis for the various applications of modern photography.

Daguerre's photographic process had several flaws. The equipment was bulky and exposure required 10 to 15 minutes in strong sunlight; the image was laterally reversed, and it could not be replicated. In 1840 exposure time was reduced by adding silver bromide to the silver iodide coating, and in 1841 it was further reduced by the addition of an improved camera lens. In the 1840s photography studios opened in Europe and North America.

In 1841 the photographic negative was introduced by English inventor William Talbot. The negative, in which light and dark regions of the image were reversed, could be used to produce multiple copies of the original; it also solved the problem of lateral reversal. Negatives led to the first books illustrated with photographs.

American chemist John Draper, a pioneer of scientific photography, produced a photograph of the moon in 1840. To illustrate a book on physiology published in 1850, he made the first microphotographs, photographs requiring a magnifying instrument for proper viewing.

In 1851 English inventor Frederick Archer devised a photographic process involving collodion (a solution of nitrocellulose in ether) that was more sensitive than previous solutions and reduced exposure time to between two and 20 seconds.

The photographic dry plate made its debut in 1871. While experimenting with photographic solutions, English chemist Joseph Swan found that heat increased their sensitivity. Even drying the solutions on the plates produced better images, and the resulting dry plates were easier to use.

American inventor GEORGE EASTMAN made photography portable and popular with the introduction of flexible film and the small Kodak camera in 1888. Twentieth-century technology advanced to include movies, color film, Polaroid "instant" prints, three-dimensional holograms, and many sophisticated cameras and specialized film.

Schuyler

WORLD EVENTS		DAGUERRE'S LIFE
French Revolution	1789	Louis Daguerre is born
Napoleonic Wars in Europe	1803-15	
	1804	Daguerre moves to Paris
	1822	Daguerre invents the Diorama
	1826	Daguerre meets Joseph Niépce
	1830	Daguerre and Niépce form partnership to develop early photographic techniques
	1833	Niépce dies
	1839	Daguerreotype is introduced
	1840	Daguerre retires
	1841	Photographic negative is invented
	1851	Daguerre dies
Germany is united	1871	Dry plates are invented
	1888	Kodak camera and film introduced
Spanish-American War	1898	

For Further Reading:

Davenport, Alma. *The History of Photography: An Overview.* Boston: Focal Press, 1991.

Gernsheim, Helmut and Alison. *L. J. M. Daguerre: The History of the Diorama and the Daguerrotype.* New York: Dover, 1968.

Lovell, Ronald. *Two Centuries of Shadow Catchers: A Compact History of Photography.* Albany, N.Y.: Delmar Publishers, 1996.

Dalton, John

Originator of Atomic Theory

1766-1844

Life and Work

John Dalton is credited with unifying a vast amount of information into a theory about the structure of matter. His atomic theory provided a framework for understanding how matter is composed of atoms and how those atoms combine.

Born to a Quaker family on September 6, 1766, in Eaglesfield, England, Dalton attended school until he was 12. At this young age, he became a teacher and taught for the rest of his life. By accident he discovered he was colorblind. This phenomenon fascinated him. In 1794 he published his first scientific paper on the subject. Colorblindness came to be called Daltonism because of his research.

A friend introduced him to meteorology, sparking his interest in the natural sciences. In 1787 Dalton began a daily record of weather observations that would eventually include over 200,000 items. As he became more involved with these observations, Dalton wondered about the atmosphere and its components.

Dalton recognized patterns in the interactions of atmospheric gases. In 1803 he published his law of partial pressures, now called Dalton's Law. This law states that when different gases are combined the resulting pressure is the sum of what each gas exerts individually. He explored other characteristics of gases including solubility in water and how temperature affects volume.

These studies led Dalton to infer that gases must be made of small particles. Dalton expanded on the Greek Democritus' idea that atoms could not be split and created a new atomic theory. He stated that all matter is composed of tiny, indivisible parts and that these atoms are indestructible and cannot be changed. Dalton also asserted that each element has a unique type of atom. For example, all gold atoms are alike but different from those of other elements. Additionally, each atom has a measurable weight that can be experimentally determined. He included in his 1808 publication, *New System of Chemical Philosophy*, a list of weights for certain elements. Dalton used the lowest density gas, hydrogen, as a basis for the comparative weights. In conjunction with his theory, he also discussed how atoms form compounds. Although he did not use the term molecule, Dalton hypothesized that atoms combine in certain proportions to become compound substances. Even though several of his measurements later proved incorrect, Dalton's basic theory of matter and atoms was a tremendous achievement. He was awarded a Royal Medal in 1826 for his groundbreaking work.

Throughout his life Dalton supported himself by teaching; in later years, he spent less time on his scientific research. He continued to be recognized for his earlier work until he died on July 27, 1844, at his home in England.

Legacy

Dalton's atomic theory provided scientists with new ways of viewing the physical world. The concepts of atomic weights and fixed ratios of atoms within compounds challenged researchers to explore the chemical composition of matter further.

Dalton's assertion that atomic weights could be experimentally determined provided scientists with a new key to studying various substances. New information on the atomic weights of elements later enabled Russian chemist DMITRY MENDELEYEV to create the periodic table of elements. This, in turn, sparked the discovery of new elements. Dalton also proposed symbols for the elements that could be used to represent combinations of atoms in compounds.

These were later changed to the abbreviations used today, but the initial idea for a consistent system was Dalton's. And while Dalton's theory was well-accepted at the time, it also raised many new questions about the interactions of atoms. New ideas about the structure of atoms and how they interact helped to refine atomic theory.

During Dalton's time, the division between physics and chemistry was just emerging. His theory helped clarify both disciplines. The atom is a cornerstone for understanding forces and matter in physics and chemistry, in particular in quantitative analysis, an aspect of chemistry that measures how much of a particular substance is present in a sample. Dalton's physical view of the atom was later refined through the work of ERNEST RUTHERFORD, NIELS BOHR, and others. Nuclear energy and fusion research today are the progeny of Dalton's theory. In chemistry Dalton's work provides a basis for stoichiometry (the study of the quantitative relationships between substances in chemical reactions), structural theory and how the arrangement of atoms influences properties, and other fields. His early models of compounds evolved into an understanding of the three-dimensional quality of molecules and how structure affects properties. While Dalton's theory contained inaccuracies, his atomic theory provided a foundation for later generations of scientists—a foundation that didn't exist before Dalton.

Wilson

WORLD EVENTS	DALTON'S LIFE
Peace of Utrecht 1713-15 settles War of Spanish Succession	
	1766 John Dalton is born
United States 1776 independence	
French Revolution 1789	
	1803 Dalton publishes law of partial pressures
Napoleonic Wars 1803-15 in Europe	
	1808 Dalton publishes *New System of Chemical Philosophy*, which presents his atomic theory
	1826 Dalton is awarded medal by Royal Society
	1844 Dalton dies
United States 1861-65 Civil War	

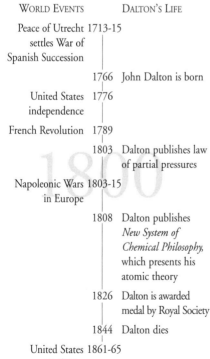

For Further Reading:

Asimov, Isaac. *A Short History of Chemistry.* Westport, Conn.: Greenwood Press, 1965.

McDonnell, John J. *The Concept of an Atom from Democritus to John Dalton.* Lewiston, N.Y.: Edwin Mellen Press, 1992.

Travers, B., ed. *World of Scientific Discovery.* Detroit, Mich.: Gale Research, 1994.

Darby, Abraham

Originator of the Use of Coke for
Iron Smelting

c. 1678-1717

Life and Work

Abraham Darby was the first to use coke as
the fuel for iron smelting; he demonstrated that coke is superior in cost and efficiency to charcoal, the traditional fuel.

World Events	Darby's Life
Thirty Years' War 1618-48 in Europe	
	c.1678 Abraham Darby is born
	1708 Darby founds the Bristol Iron Company
	1709 Darby uses coke to smelt iron at Coalbrookdale
	1712 Thomas Newcomen invents first practical steam engine
Peace of Utrecht 1713-15 settles War of Spanish Succession	
	1717 Darby dies
	1758 By this time, coke process produces more than 100 six-ton cylinders for Newcomen engines
	1760 Carron Ironworks opens
	1779 Abraham Darby III, Darby's grandson, erects cast-iron bridge
French Revolution 1789	
	1802 Coalbrookdale builds steam locomotive

Little information exists about Darby's life. He was born about 1678 near Dudley, England, and became an ironworks master at Bristol, where coke was already utilized for smelting copper. In 1708 Darby founded the Bristol Iron Company and began investigating the possibility of smelting iron with coke. He acquired an iron foundry at Coalbrookdale, near ample deposits of iron ore and good coking coal. He successfully used coke to smelt iron in 1709.

In seventeenth-century England, technical problems limited the development of the iron industry. To smelt iron, a source of fuel is required to react with oxygen, producing heat to melt the iron ore. Charcoal was the traditional fuel, but the scarcity of wood, from which charcoal was made, combined with an increased demand for charcoal had raised its price significantly. Attempts to increase the iron yield by increasing the size of the furnaces were unsuccessful: charcoal is too weak to support a large column of iron ore. Coal could not be used in place of charcoal because coal contains sulfur, which spoils the quality of the iron. Darby took the logical next step and replaced charcoal with coke, which contains no sulfur. (Coke is the purified residue left after coal has undergone destructive distillation, a purification process.)

Coke's strength allowed the use of larger furnaces, increasing the rate at which iron could be smelted. Larger furnaces meant more draft and hotter fires, further improving the process. The high quality of his coke-smelted iron permitted Darby to manufacture thin castings for producing pots and other cookware that were competitive with the heavy brass items then available.

Darby did not patent his coke-smelting invention. He died on March 8, 1717, at Madeley Court, England.

Legacy

Darby's direct legacy takes the form of his son's and grandson's significant contributions to the iron industry. In a broader sense, the introduction of coke to the iron-smelting process was an essential step along the path toward industrialization.

Only a half dozen coke furnaces for iron smelting were built in the 50 years following Darby's invention; by some estimates, charcoal was still cheaper to use than coke, perhaps accounting for the slow adoption of coke smelting.

However, Darby's Coalbrookdale foundry was kept busy, particularly by THOMAS NEWCOMEN's 1712 invention of the first practical steam engine. Newcomen engines required six-ton iron cylinders, and by 1758, the coke process had produced more than 100 of them. Charcoal-fueled iron smelting could not have produced such huge iron cylinders.

Darby's eldest son, Abraham Darby, Jr., succeeded his father as head of the Coalbrookdale iron foundry. He helped it retain its prominence within the industry.

Darby's grandson, Abraham Darby III, designed, cast, and erected one of the world's first cast-iron bridges, near Coalbrookdale, in 1779. The Coalbrookdale foundry also built, in 1802, the first high-pressure steam railway locomotive, for engineer Richard Trevithick.

As charcoal prices skyrocketed in the mid-eighteenth century, coke rapidly became the primary fuel in iron smelting. In 1760 the enormous Carron Ironworks, the largest iron plant in Great Britain, opened in Scotland. Coalbrookdale supplied its equipment and operators. Great Britain was soon producing the cheapest and strongest iron and steel in the world. As these metals were invaluable to numerous industries, the country led the Industrial Revolution. Coke remains the predominant fuel in modern iron smelting.

Using coke for fuel in iron processing made the metal less expensive and more readily available for new products: water pipes to and from homes (replacing leaky wooden ones), machinery parts, and cast-iron rails were just some of the new products the process made possible. Iron implements became widely available, which helped to improve health, safety, and efficiency at home and in the factory.

Schuyler

For Further Reading:

Burke, James. *Connections.* Boston: Little, Brown, 1995.

Cardwell, Donald. *The Norton History of Technology.* New York: Norton, 1995.

Carr, J. C., and W. Taplin. *History of the British Steel Industry.* Cambridge, Mass.: Harvard University Press, 1962.

Darwin, Charles

Founder of Evolutionary Theory
1809-1882

Life and Work

Charles Darwin developed the theory of the evolution of species by natural selection, which gave a new coherence to the study of biology and profoundly influenced modern thought.

Darwin was born in Shrewsbury, England, on February 12, 1809. At first an undistinguished student, Darwin developed a passion for the natural sciences at Cambridge University, from which he graduated in 1831. That year he joined a scientific tour of North Wales and learned geologic fieldwork methods. He was then invited to serve as an unpaid naturalist on a voyage of the HMS *Beagle*, departing in December 1831.

Darwin took meticulous notes and collected biological and geological specimens during the *Beagle's* survey of the coasts of South America. He noticed that fossils of some extinct species were similar to some living species and that there were distinct variations of certain species living on separate, but nearby, islands. He began to question the conventional view that species were immutable.

Upon Darwin's return to England in 1836, he investigated the immutability question and soon became convinced that species do change over time. However, he was reluctant to reveal such a revolutionary view until he could explain how such

evolution occurs. Darwin was led to such an explanation in 1838 when he read an essay by economist Thomas Malthus, who argued that a limited food supply checks human population growth. Darwin reasoned that every form of life must struggle for existence. This idea led Darwin to his theory of natural selection, which states that variation among individuals within a population exists, and that some traits allow certain individuals to be more successful survivors than others. Over time, the population transforms into a new species as individuals with those advantageous traits multiply.

Darwin secretly elaborated his theory until another naturalist, Alfred Russel Wallace, sent him an outline of the very same theory in 1858. A joint paper by Darwin and Wallace, summarizing the theory of evolution by natural selection, was made public later that year. In 1859, Darwin published *On the Origin of Species by Means of Natural Selection*. The book's first printing sold out immediately. In 1871, Darwin published *The Descent of Man*, which proposed the same principle of natural selection for human biology.

Darwin studied and wrote until he died at Down, England, on April 19, 1882.

Legacy

Darwin's introduction of evolutionary theory kindled a major revolution in biology in the nineteenth century and has influenced biological study and modern thought ever since.

The revolutionary nature of his theories caused an immediate uproar in the religious community. Christian leaders vehemently opposed Darwin's theory because it described a grand biological design governed by laws of science, not by God. These controversies are still alive today.

The scientific impact of Darwin's work was immediate also. Research into anatomy and paleontology expanded, as scientists searched for evidence supporting or refuting Darwin's views. Most members of the scientific community quickly accepted the theory of evolution upon the publication of *On the Origin of Species*. Many scientists had been discussing the possibility of evolution for years, but Darwin and Wallace were the first to propose the idea of natural selection.

Within Darwin's lifetime, proponents of evolution split into two conflicting camps, Darwinian and Lamarckian, which dispute the source of variation within populations. Darwin

and his followers could not explain how variation occurred. Jean Lamarck, a French naturalist, had proposed in the early 1800s that organisms acquire traits during their lifetimes as they adapt to their environment and that their offspring inherit those traits. Lamarckians believed that those traits were the source of variation in populations. It was not until 1900, when the work of GREGOR MENDEL was rediscovered, that Lamarck was proved wrong and scientists began to understand that genetic mutations cause variation among individuals.

Darwin's ideas permeated nearly all fields of study. His theory of natural selection was formulated into social Darwinism in the late 1800s. Social Darwinists, most notably British sociologist and philosopher Herbert Spencer, believed that competition for survival governed human society. Epitomized by the term "survival of the fittest," coined by Spencer, social Darwinism was used as a scientific and philosophical justification for laissez-faire economics, in which individuals are left on their own to compete for resources and achieve whatever they can. While social Darwinism has fallen out of favor in the fields of philosophy and sociology, its influence on economics and popular thought holds firm.

Schuyler

For Further Reading:
Bowler, Peter J. *Charles Darwin: The Man & His Influence.* Cambridge: Cambridge University Press, 1996.
De Beer, Gavin. *Charles Darwin.* Westport, Conn.: Greenwood Press, 1976.
Keynes, R. D., ed. *Charles Darwin's Beagle Diary.* Cambridge: Cambridge University Press, 1988.
Stefoff, Rebecca. *Charles Darwin and the Evolution Revolution.* New York: Oxford University Press, 1996.

WORLD EVENTS	DARWIN'S LIFE
Napoleonic Wars 1803-15 in Europe	
1809	Charles Darwin is born
1831	Darwin graduates from Cambridge University
	Darwin sets sail aboard the HMS *Beagle* as a naturalist
1836	Darwin returns from voyage
1830-50s	Darwin privately elaborates his theory of evolution
1858	Darwin and Alfred Wallace present the theory of evolution by natural selection
1859	*On the Origin of Species...* is published
Germany is united 1871	*The Descent of Man* is published
1882	Darwin dies
Spanish-American War 1898	
1900	Mendel's heredity research is rediscovered

Davy, Humphry

Discoverer of Elements;
Pioneer in Electrolysis
1778-1829

WORLD EVENTS		DAVY'S LIFE
United States independence	1776	
	1778	Humphry Davy is born
French Revolution	1789	
	1800	Davy publishes findings on anesthetic properties of nitrous oxide
	1801	Davy invited to lecture at Royal Institute
Napoleonic Wars in Europe	1803-15	
	1806	Davy publishes "On Some Chemical Agencies of Electricity," explaining use of electrolysis in chemical analysis
	1807	Davy isolates potassium and sodium
	1808	Davy discovers borium, strontium, and calcium
	1815	Davy invents mining lamp
	1829	Davy dies
United States Civil War	1861-65	

Life and Work

Humphry Davy isolated many new elements, pioneered the use of electrolysis in chemistry, and invented a miner's safety lamp.

Davy was born on December 17, 1778, to a woodcarver's family in Cornwall, England. When his father died, Davy was forced to find a job to help support the family. He apprenticed to an apothecary, or druggist, who exposed him to the world of chemistry. Later Davy worked for a physician who studied the effects of various gases on patients. During this time he taught himself chemistry by reading texts by contemporary scientists such as ANTOINE LAVOISIER.

In 1798 Davy was appointed chemical superintendent at the Pneumatic Institute in Clifton, England, which conducted investigations into the practical medical applications of gases. There Davy studied the make-up of all kinds of gases and inhaled them to test their effects. He discovered the exhilaration of nitrous oxide ("laughing gas"), which he suggested could be used as an anesthetic. He published *Researches, Chemical and Philosophical,* a report of his findings, in 1800 and was invited to give lectures at the Royal Institute in 1801.

Davy's work in chemistry expanded to include experiments with elements, acids, and metals. Applying the work of Luigi Galvani and ALLESANDRO VOLTA, he pioneered the use of electrolysis to separate compounds into their respective elements. He published a paper on the properties of electrolysis, "On Some Chemical Agencies of Electricity," in 1806. Davy isolated several elements that had yet to be named, including potassium and sodium in 1807, and borium, strontium, and calcium in 1808. He challenged his idol, Lavoisier, on the idea that all acids contain oxygen by demonstrating that hydrochloric acid (then named oxy-muriatic acid) was a compound of only two elements: hydrogen and chlorine.

During his life Davy was committed to using science to help society. He designed a safety lamp in 1815 that would not ignite the dangerous methane gases present in coal mines. He also worked on using chemistry to improve agricultural practices.

Davy became influential in the scientific community, especially at the Royal Institute, where he held an honorary professorship. He interviewed MICHAEL FARADAY and was instrumental in his appointment as a laboratory assistant at the Institute in 1813.

During the later years of his life Davy's health declined, possibly due to exposure to dangerous chemicals. He traveled in Europe, seeking new challenges and better health. Davy died in Switzerland on May 29, 1829.

Legacy

Davy's rigorous investigative methods, discoveries of new elements, and work with electrolysis expanded the base of knowledge available to future chemists and led to important practical applications still in use today.

Davy's work had an immediate impact in the field of science. His encouragement of other working-class men, especially Faraday, to become scientists helped to weaken the image of science as only a rich man's game.

In the research arena Davy challenged the idea that all acids contain oxygen. He used precise experimental means to support his argument. Although it was difficult for many fellow scientists to change their positions, Davy's irrefutable evidence eventually swayed them. His discovery of potassium, sodium, chlorine, calcium, magnesium, and boron fueled other research as scientists became aware of the new elements and their properties. Nitrous oxide is still used as an anesthetic in dentistry today.

Davy's experiments with electrolysis led to numerous applications. One of the most important practical applications included its use in determining the electrical affinities of elements. This eventually led to an understanding of bonds and of valence, the property of an element that indicates how many atoms an individual atom can bond with. Electrolysis is used today to separate salt from seawater, a process called desalination. This is an invaluable technology in many areas of the world where fresh water is a scarce resource.

In the public sector his mining lamp saved hundreds, perhaps thousands, of lives. It revitalized the dying coal mining industry and made available an important fuel resource. He also worked with the agriculture and tanning industries to find better chemical methods to improve their products.

Wilson

For Further Reading:

Brock, William. *The Norton History of Chemistry.* New York: Norton, 1992.

DeFries, Amelia. *Pioneers of Science: Seven Pictures of Struggle and Victory.* Freeport, N. Y.: Books for Libraries Press, 1970.

Descartes, René

Founder of Analytic Geometry
1596-1650

Life and Work

A philosopher, scientist, and mathematician, René Descartes formulated analytic geometry (the use of algebra to represent geometrical shapes), thus founding modern mathematics.

Descartes was born on March 31, 1596, in La Haye, France. His mother died when he was an infant, leaving him with sufficient funds to live comfortably for the rest of his life. His father, a judge, sent him to a school run by Jesuits at La Flèche, where he received a rigorous secondary education in languages, humanities, logic, philosophy, physics, and mathematics. Mathematics, with its deductive precision, was the only subject he truly enjoyed.

At age 16 Descartes went to Paris and became reacquainted with Father Marin Mersenne, an advocate of scholarship whom he had known at school. Together they spent two years investigating mathematics in quiet retirement. Descartes returned to school, earning a law degree from the University of Poitiers in 1616.

Deciding to experience the world firsthand, from 1617 to 1628 Descartes served as an army volunteer under Dutch and Bavarian command. He did not participate in battle; his time in the army was spent mostly in studies of philosophy and mathematics.

In 1628 Descartes quit the army, moved to Holland, and began 20 years of meditation and writing. The first six years were spent compiling a physics treatise that he did not publish because he feared the Catholic Church would condemn him for believing in the Earth's movement around the Sun (he knew the church had forced GALILEO to recant the same view in 1632).

In 1637 Descartes published *Discourse on Method,* a philosophical work containing three scientific appendixes. The first two, "Dioptrics" and "Meteorology," address the properties of light and atmospheric phenomena. The third, "Geometry," represents the foundation of analytic geometry.

With "Geometry," Descartes became the first to represent geometric shapes on a coordinate system of two axes. He defined the positions of points in a plane by their distance from two fixed axes, one horizontal and one crossing diagonally (as opposed to the perpendicular axes used today); he was thus able to describe lines and curves with algebraic equations. He introduced the notation of x, y, and z to signify variables (unknown quantities) and a, b, and c to designate constants. He systematized the use of negative and imaginary roots and of exponential variables. Finally, he addressed numerous problems in algebra, particularly those relating to polynomial equations.

Descartes continued to write, and his *Principles of Philosophy*, appearing in 1644, included some general scientific theory. He advocated the search for a single mechanical description of all natural phenomena (within the fields of physics, chemistry, and physiology), to be based on mathematical methods.

In 1649 Descartes accepted an invitation to live in Stockholm, Sweden, to teach philosophy to Queen Christina. Unaccustomed to the icy climate and the early-morning lessons demanded by the queen, he caught pneumonia and died on February 11, 1650.

Legacy

Descartes was among the most influential figures in the history of western science. He initiated both the modern approach to mathematics and the systematic application of mathematical methods to other fields of science.

Descartes's *Discourse on Method,* in which "Geometry" appeared, was written in French (rather than Latin) and thus had popular appeal.

It was widely read during the author's lifetime and was responsible for his fame as a philosopher and mathematician.

Descartes's analytic geometry provided the tools for a revolution in mathematics. Analytic geometry made it possible to apply algebraic calculation—whose properties were fairly well understood—to geometrical shapes such as lines and curves. Descartes's two-axis system, named the Cartesian coordinate system by GOTTFRIED WILHELM LEIBNIZ in 1692, was the key to analytic geometry. The coordinate system is used in most branches of modern mathematics; it is the basis of all graphs and an important element in cartography. Descartes's notation for constants and variables also survives to this day.

Analytic geometry's main offshoot was the calculus, the use of algebra to study changing quantities. It was invented independently by ISAAC NEWTON in England and by Leibniz in Germany in the 1660s and 1670s. Descartes's "Geometry" was the spark that began Newton's interest in mathematics and it continued to inspire him as he worked out elements of the calculus. The calculus provided invaluable methods for solving problems in physics and astronomy and survives as a basic tool in higher mathematics and physics.

Descartes's influence on Newton is also evident in the latter's *Principia*, published in 1687. The work was a quintessential example of Descartes's idea that nature should be studied by applying mathematical laws to physical phenomena. *Principia* laid the foundations of modern celestial mechanics and the mathematical physics of the following two centuries.

Schuyler

WORLD EVENTS		DESCARTES'S LIFE
Reformation begins	1517	
	1596	René Descartes is born
	1612	Descartes moves to Paris
	1617	Descartes joins army
Thirty Years' War in Europe	1618-48	
	1628	Descartes leaves army
	1637	Descartes's *Discourse on Method* is published; it introduces his system of analytic geometry
	1649	Descartes is summoned to Sweden to teach Queen Christina
	1650	Descartes dies
England's Glorious Revolution	1688	

For Further Reading:

Cole, John R. *The Olympian Dreams and Youthful Rebellion of René Descartes*. Urbana: University of Illinois Press, 1992.

Hollingdale, Stuart. *Makers of Mathematics*. New York: Penguin Books, 1989.

Muir, Jane. *Of Men and Numbers*. New York: Dodd, Mead, 1961.

Shea, William R. *The Magic of Numbers and Motion: The Scientific Career of René Descartes*. Canton, Mass.: Science History Publications, 1991.

Diesel, Rudolf

Inventor of Diesel Engine
1858-1913

World Events		Diesel's Life
Napoleonic Wars in Europe	1803-15	
	1858	Rudolf Diesel is born
	1870	Diesel begins to obtain engineering education
Germany is united	1871	
	1880	Diesel works in refrigeration and studies engines
	1890	Diesel decides to build heat engine
	1893	Diesel builds efficient compression-ignition engine
Spanish-American War	1898	
	1899	Diesel founds engine-manufacturing company
	1913	Diesel dies
World War I	1914-18	

Life and Work

Rudolf Diesel invented the four-stroke internal-combustion Diesel engine, which powers all types of heavy-duty transport equipment and vehicles.

Diesel was born on March 18, 1858, in Paris, to German parents. During his early years in Paris he frequented the Museum of Arts and Crafts and was fascinated by the display of Joseph Cugnot's steam-propelled carriage, designed in 1769.

In 1870 Diesel's family moved to London to escape the Franco-German War, and he was sent to Augsburg, Germany, to continue his education. Excelling in engineering, he obtained a scholarship to attend Munich's technical college, where he studied under Carl von Linde, a pioneer in refrigeration technology. From von Linde he learned thermodynamics and the theory of heat engines.

From 1880 to 1890 Diesel worked at von Linde's refrigeration firm in Paris; his first two patents were for ice-making machines. He spent his spare hours researching engines. In 1890 he decided to try to design a highly efficient internal-combustion engine by applying the principle of thermodynamics stating that work is performed as heat flows from a high to a low temperature state. Such an engine would have to operate in a wide temperature range, which would require that the air within the cylinder be greatly compressed (because the temperature of air depends on its pressure).

Diesel built his first engine in 1893. It utilized the basic four-stroke cycle introduced by German inventor Nikolaus August Otto in 1876, but differed from the Otto engine in not requiring a spark for ignition. In Diesel's engine, air is drawn into the cylinder and compressed, raising its temperature to nearly 600°C. Then a fine spray of fuel is injected into the cylinder, the fuel ignites because of the high temperature of the air, and the ignition powers the resulting power stroke. A final exhaust stroke removes the burned fuel, and then the cycle repeats. The engine ran on a type of refined mineral oil.

In 1899 Diesel established a company in Augsburg to manufacture and market his engines, which were an immediate commercial success.

On September 29, 1913, Diesel apparently drowned after vanishing from a steamer en route to London. His body was never found.

Legacy

The Diesel engine proved to be an energy-efficient alternative to other internal-combustion engines and continues to be used in a variety of vehicles and power generators.

Rudolf Diesel's invention embodied both advantages and disadvantages of other successful engines of the time. It could run on lower-cost fuels and was more efficient, and its lack of electrical ignition combined with its non-explosive fuel reduced the risk of accidental fire. However it was heavy, because it had to be made with heavy-gauge metals to withstand high pressures, and it was noisy. It was also relatively expensive to manufacture.

The development of the Diesel engine and its applications continued throughout the twentieth century. It was used to power submarines during World War I after undergoing improvements introduced primarily by German engineer Karl Bosch. Locomotives, streetcars, and large ships were soon running on Diesel engines. Further design alterations during the 1930s adapted it for use in small boats, and later contributions by British designers Cedric Dicksee and Harry Ricardo moved it into trucks, buses, and tractors. By the 1950s, most ships and many railways were Diesel-powered, and by the 1980s, some automobiles had Diesel engines. Also, at mid-century, many coal-burning boilers and water-flow alternators had been replaced by Diesel engines in the generation of electricity.

Schuyler

For Further Reading:

Dark, Harris E. *Auto Engines of Tomorrow.* Bloomington: Indiana University Press, 1975.

Moon, John Frederick. *Rudolf Diesel and the Diesel Engine.* London: Priory Press, 1974.

Nitske, Robert W. *Rudolf Diesel: Pioneer of the Age of Power.* Norman: University of Oklahoma Press, 1965.

Drew, Charles

Developer of Technique
for Storing Blood Plasma
1904–1950

Life and Work

Charles Drew devised effective techniques of storing blood plasma for long periods, contributed to blood-transfusion efforts during World War II, and spoke out against the practice of segregating stored blood according to race.

Drew was born on June 3, 1904, in Washington, D.C. As a teenager he excelled in both academics and athletics, winning a scholarship to Amherst College in western Massachusetts. After graduating in 1926, he taught chemistry and biology at Morgan State College in Baltimore, Maryland, to pay off his student loans and to save money for medical school. In 1928 he enrolled at McGill University in Montreal, Canada, where he worked on problems of blood storage. He earned a medical degree in 1933.

By that time blood transfusions had become safe because of KARL LANDSTEINER's discovery of the four major blood groups; however, whole blood spoiled quickly and could be stored safely only for seven days. Returning to Washington, D.C., after finishing at McGill, Drew investigated the use of blood plasma instead of whole blood in transfusions. Blood plasma, he reasoned, would be better for emergencies because its lack of red blood cells (which contain the factors determining blood type) makes a match between donor and recipient unnecessary. In 1939, while working at Columbia University in New York City, he developed a procedure for processing and preserving blood plasma so that it could be transported great distances.

In 1940 Columbia University awarded Drew a Ph.D. in medical science. He was the first African-American to earn this degree from Columbia. (In 1908, Columbia had awarded a black man from England named Travis Johnson a medical degree.) In his dissertation he demonstrated that blood plasma can be stored far longer than whole blood.

During World War II Drew became medical supervisor of a blood-transfusion program in England and introduced the use of "bloodmobiles," trucks equipped with refrigerators, to transport blood to victims of bombings. Upon his return from England in 1941, he was recruited to direct the American Red Cross Blood Bank in New York City. His first assignment was to organize a massive blood drive for the military, but he resigned in protest when the government backed the military's request that the collected blood be categorized according to the donor's race. He argued that there was no scientific evidence showing that blood differs with race, statements that were later confirmed by others' experiments.

Later that year Drew became professor of surgery at Howard University and chief surgeon at Freedmen's Hospital. In 1944 he earned a Spingarn Medal from the National Association for the Advancement of Colored People. On March 31, 1950, he was killed in a car crash near Burlington, North Carolina, en route to a medical meeting at the Tuskegee Institute in Alabama.

Legacy

Drew's pioneering work in blood preservation saved thousands of lives and opened the way for future innovations in blood preservation.

The replacement of whole blood with blood plasma in transfusions expanded the opportunities to perform life-saving transfusions, both because donor–recipient matching is not necessary with plasma and because it can be stored for longer periods of time without going bad.

Drew is credited with saving the lives of thousands of European civilians and soldiers during World War II. The preservation, storage, and use of blood plasma for emergencies and the introduction of bloodmobiles increased the life expectancy of the badly wounded. Ironically Drew himself died of blood loss during a car accident. (Controversy currently exists over the belief held by some historians that Drew was denied treatment because he was an African-American, and died as a result.)

Drew was also an inspirational teacher and role model for many African-American students. His insistence on drawing attention to racial biases in medical research—when he resigned as director of New York City's American Red Cross Blood Bank—encouraged others to disprove the assumption that there is a difference in blood types of different races.

More than two decades after his death, government agencies recognized Drew's achievements. In 1977 the American Red Cross headquarters in Washington, D.C., was renamed the Charles R. Drew Blood Center in his honor, and his face graced a U.S. postage stamp in a 1980 postal series of "Great Americans."

Schuyler

WORLD EVENTS	DREW'S LIFE
	1904 Charles Drew is born
World War I 1914–18	
	1928 Drew enrolls in McGill University in Montreal, Canada
	1933 Drew earns medical degree
	1939 Drew develops blood-plasma preservation methods
World War II 1939–45	
	1940 Drew earns Ph.D. from Columbia University, the first African-American to do so
	Drew supervises blood-transfusion program in war-torn Britain
	1941 Drew directs Red Cross blood drive; resigns because of racial discrimination
	1944 Drew receives Spingarn Medal from National Association for the Advancement of Colored People
	1950 Drew dies
Korean War 1950-53	

For Further Reading:

Hartwick, Richard. *Charles Richard Drew: Pioneer in Blood Research.* New York: Scribner's, 1967.

Sammons, Vivian Ovelton. *Blacks in Science and Medicine.* New York: Hemisphere Publishing, 1990.

Wynes, Charles. *Charles Richard Drew: The Man and the Myth.* Urbana: University of Illinois Press, 1988.

Eastman, George

Inventor of Hand-held Camera
1854–1932

Life and Work

George Eastman popularized photography with the invention of the simple and affordable hand-held camera.

Eastman was born in Waterville, New York, on July 12, 1854. He dropped out of school at age 14 to help support his family and later worked briefly at an insurance company and a bank. He began photography as a hobby, and in 1879 devised an improved method of making dry photographic plates. He began to manufacture his invention, but his business remained modest because of the small number of amateur and professional photographers.

Eastman recognized the potential of making photography available to a wider market if a substitute could be found for the heavy, fragile glass plates used in cameras of his day. In 1884 he introduced flexible film that could be wound on a spool. Four years later he developed his first camera, a lightweight box camera loaded with paper-stripping film. He named his invention Kodak and backed it with aggressive advertising. He sold more than 100,000 cameras in two years. These first Kodaks were sold for $25, contained 100 exposures, and had to be sent to Rochester for developing and reloading.

In 1892 Eastman organized the Eastman Kodak Company in Rochester, New York, for the manufacture of his cameras. He introduced further improvements to the camera and its film, making them more affordable and eliminating the need to send the whole camera to Rochester for reloading.

Eastman spent large sums of money on industrial research, advertising, and marketing. He was among the first to establish health services, retirement plans, and profit-sharing plans for his employees. He also shared his fortune generously, giving a total of $75 million to various institutes in his later years.

Eastman took his own life in Rochester on March 14, 1932.

Legacy

Eastman's innovations marked the beginning of modern photography.

The Eastman Kodak Company held a monopoly over the photography industry for several decades. Kodak became a household name, synonymous with cameras.

Eastman transformed photography from an expensive, exclusive, labor-intensive trade into a popular hobby. Prior to the introduction of Eastman's Kodak camera, the virtual darkroom they had to transport to each shooting location limited photographers.

World Events		Eastman's Life
	1854	George Eastman is born
Germany is united	1871	
	1879	Eastman devises improved photographic-plate-making method
	1884	Eastman introduces flexible film
	1888	Eastman introduces the Kodak camera; sells more than 100,000 cameras in first two years
	1892	Eastman launches the Eastman Kodak Company
Spanish-American War	1898	
World War I 1914-18		
	1932	Eastman commits suicide
World War II 1939-45		

Kodak cameras brought photography to the general public, who enthusiastically embraced it as a hobby and as an art.

Further photographic improvements quickly followed the Kodak camera. In the 1890s, new emulsions and gelatin plates, faster lenses, and flash powder appeared. Social documentation was one immediate effect of the new photographic capabilities. Social realities were captured and presented with immediacy and impact. Photographs brought viewers into contact with conditions they would never otherwise have seen, such as the slums of New York as documented by Jacob Riis.

Eastman's philanthropic contributions enabled several institutions to remain afloat or to expand. He gave to the Mechanics Institute (later renamed the Rochester Institute of Technology), the Massachusetts Institute of Technology, the University of Rochester, Hampton Institute, and Tuskegee Institute. He provided funds for the Eastman School of Music and the schools of medicine and dentistry at the University of Rochester. He also made possible the establishment of dental dispensaries in many countries.

Photography's popularity spurred development of related technologies. The eight-millimeter movie camera appeared in 1932, and color film, panchromatic film, and automatic-exposure controls soon followed. The "instant" Polaroid camera was first demonstrated by Edwin Land in 1947, and patents for filmless, digital cameras generating electronic images date back to 1973.

Schuyler

For Further Reading:

Brayer, Elizabeth. *George Eastman: A Biography.* Baltimore, Md.: Johns Hopkins University Press, 1996.

Newhall, Beaumont. *Focus: Memoirs of a Life in Photography.* Boston: Little Brown, 1993.

Edison, Thomas

Prolific Inventor;
Creator of the Light Bulb
1847-1931

Life and Work

Thomas Edison invented the incandescent light bulb, the phonograph, a motion-picture machine, and numerous other items that resulted in more than 1,300 patents.

Edison was born on February 11, 1847, in Milan, Ohio. After only a few months of formal schooling at age eight, he received the rest of his education at home from his mother. Fascinated by chemistry, he set up a laboratory in the basement at age 10 and began a life of innovation and experimentation. At age 12 Edison worked aboard a train selling magazines and snacks and spent his free time reading and working in his laboratory. About this time he began to lose his hearing, and he remained nearly deaf for the rest of his life.

He settled in Boston in 1867, where his spare moments were spent inventing. Reading about the electrical experiments of MICHAEL FARADAY, he was inspired in 1869 to construct a device for recording votes electrically, an invention that attracted little attention.

Edison moved to New York City later that year, where after brief but lucrative employment as a technical manager, he formed his own instrument-design company. He invented a stock-ticking machine, which was bought by a telegraphic company for $40,000. With the money Edison built an "invention factory" in Newark, New Jersey, which produced about 200 minor devices in seven years.

In 1876 he opened a larger plant at Menlo Park, New Jersey, where he improved ALEXANDER GRAHAM BELL's recently patented telephone by adding a carbon granule microphone that increased the volume of the transmission considerably.

Edison patented the first phonograph in 1877. It consisted of a steel stylus that transferred sound vibrations to a cylinder wrapped in tin foil; the machine could record and reproduce sounds. The phonograph clinched Edison's fame.

The following year Edison began experiments to develop an incandescent light bulb. English physicist Joseph Swan had been investigating the plausibility since the 1850s. He knew that when an electrical current passes through a thin wire, the wire heats up because of the resistance met by the current, and the wire glows (reaches incandescence) when it becomes hot enough. But he had not overcome the problem that wires oxidize, or burn up, when they get hot. Creating a vacuum inside the bulb would have prevented oxidation, but in Swan's time vacuum technology was insufficiently advanced. Edison worked tirelessly to solve these problems. In 1879, using cotton thread as the filament, he was able to produce light inside an evacuated bulb (vacuum technology had improved) for 45 hours, and after trying 6,000 types of material, he found a bamboo fiber that could produce 1,000 hours of light. He also drastically improved the necessary generating, switching, and transmitting devices to power a number of light bulbs simultaneously and established the world's first power station in New York City in 1882.

While experimenting with electric light, Edison noted the phenomenon, now known as the Edison effect, in which a small electric current flows from a heated filament to a nearby electrode, or wire.

About 1887 Edison invented the kinetograph, an early form of the motion picture, involving a series of photographs mounted on a strip of celluloid film run through a separate projector. In 1903 he produced *The Great Train Robbery* with this technique.

During his later years Edison made contributions to the technology behind the lead storage battery, the mimeograph machine, the dictating machine, the fluoroscope, and the production of cement. The recipient of numerous awards and honors, he died on October 18, 1931, in West Orange, New Jersey.

Legacy

Edison is widely recognized as one of the most influential and prolific inventors of the Western world. With the introduction of electric illumination and various elements of communication and entertainment, he influenced the technological and cultural shape of twentieth-century society.

Edison's incandescent light bulb revolutionized domestic electric lighting, which created demand for public electricity supply. This supply was delivered by Serbian-born American inventor NIKOLA TESLA, who introduced alternating-current generators. Although Edison built the first power station, his system utilized direct current, which was limited to local use. Tesla's alternating-current generators allowed strong electrical currents to travel long distances and were therefore ideal for large-scale power transmission. Alternating current eventually became the primary method of supplying cities with electricity throughout the world.

The discovery of the Edison effect was prerequisite to the development of electronic instrumentation. In 1904 English inventor John Fleming built upon the Edison effect to devise the electronic vacuum tube, later called a diode, a device that could amplify electric audio signals and that could generate, amplify, and detect radio waves. The diode made long-distance telephone connections possible and was at the heart of early radio and television broadcasting. It was also important in the development of the computer.

Although Edison did not fully develop the phonograph and kinetograph (seeing little commercial future for those devices), his inventions provided the prototypes and inspiration for modern music and film production.

Schuyler

World Events		Edison's Life
	1847	Thomas Edison is born
United States Civil War	1861-65	
	1869	Edison designs and sells vote-counting machine
Germany is united	1871	
	1876	Edison improves Alexander Graham Bell's telephone
	1877	Edison invents phonograph
	1879	Edison develops incandescent light bulb
	1882	Edison builds world's first power station
	1887	Edison invents kinetograph
Spanish-American War	1898	
	1904	John Fleming invents diode
World War I	1914-18	
	1931	Edison dies
World War II	1939-45	

For Further Reading:

Baldwin, Neil. *Inventing the Century*. New York: Hyperion, 1995.

Melosi, Martin V. *Thomas A. Edison and the Modernization of America*. Glenview, Ill.: Scott, Foresman/Little Brown Higher Education, 1990.

Wachhorst, Wyn. *Thomas Alva Edison: An American Myth*. Cambridge, Mass.: MIT Press, 1981.

Ehrlich, Paul

Immunologist;
Developer of Cure for Syphilis
1854–1915

Life and Work

Paul Ehrlich discovered a cure for syphilis and conducted pioneering experimental research in the fields of hematology (study of blood), immunology (study of body's defenses against foreign organisms), and chemotherapy (treatment of illness with chemical substances).

Ehrlich was born on March 14, 1854, in Strehlen, Prussia (now Strzelin, Poland). As a young student, he had trouble writing and memorizing, and developed a habit of illustrating his ideas with diagrams instead of words.

His interest in chemistry came early. At age eight he had cough drops made to his own recipe by the town pharmacist. But he was considered a poor student and he had to change medical schools three times before finally graduating in 1878 from the University of Leipzig.

Ehrlich was fascinated with color and his early experiments, begun while a student, were on the effects of various dyes on living tissues. By 1883 he had made significant progress in cell staining, including a technique for identifying tuberculosis bacteria that he showed to ROBERT KOCH, the German bacteriologist who first isolated the tuberculosis bacteria in 1882.

As he continued his investigation, Ehrlich noted that some dyes seem to be quite selective in their behavior, targeting disease-causing bacteria. This prompted his idea that the right mixture of chemicals could be a "magic bullet" to destroy the invaders and make the patient immune to future attacks by that particular disease. The problem, then, was to identify chemicals that would behave so specifically as to attack the bacteria only and not damage human cells and their ability to produce antitoxins (now called antibodies). His much-debated hypotheses were the beginning of a century of research in the field of immunology.

In 1890 Ehrlich helped Emil Behring devise an antitoxin serum effective against diphtheria. Later experiments showed that some diseases (particularly those caused by protozoa) do not respond to antitoxins, and he began to seek substances that kill parasites without harming the body.

Tests in 1910 indicated that an arsenic-containing substance (later dubbed Salvarsan), which Ehrlich had synthesized, was highly effective against syphilis and several other diseases. Ehrlich collaborated with a chemical manufacturer to distribute 65,000 samples of Salvarsan to physicians around the globe. A few harmful side effects occurred, but the drug was soon widely used to treat syphilis.

In addition to numerous awards and honorary degrees, Ehrlich received the 1908 Nobel Prize for Physiology or Medicine (shared with ÉLIE METCHNIKOFF, who conducted independent work in immunology) in recognition of his immunological research. He died in Hamburg, Germany, on August 20, 1915.

Legacy

Ehrlich's various achievements created the foundations for subsequent research into infectious diseases, the immune response, and chemotherapy.

Ehrlich's staining innovations aided in the development of the understanding, diagnosis, and early treatment of infectious diseases such as tuberculosis, diphtheria, and typhoid. The ability to identify the agents of disease by applying dyes to tissues and cells allowed scientists to isolate and study the causes of diseases much more readily. It also illustrated where in the body specific agents congregate.

With his search for synthesized drug treatments, Ehrlich founded chemotherapy. Salvarsan was the first chemotherapeutic drug, and Neosalvarsan, a more stable and less toxic substance, soon followed. The success of these two drugs in relieving the suffering caused by syphilis launched a flurry of attempts to synthesize parasite-killing chemical compounds.

The distribution of free samples of Ehrlich's Salvarsan foreshadowed the modern collaboration between researchers and the pharmaceutical industry. The distribution was unprecedented at the time. It is commonplace now: researchers develop new drugs for pharmaceutical companies, which finance the manufacture of the product. Free samples are commonly sent to physicians so they can introduce the product to the public and gain support for its eventual paid use.

Schuyler

For Further Reading:
De Kruif, Paul. *Microbe Hunters*. San Diego: Harcourt Brace, 1996.
Bäumler, Ernst. *Paul Ehrlich: Scientist for Life*. New York: Holmes & Meier, 1984.

Einstein, Albert

Developer of Relativity and
Quantum Theories
1879-1955

Life and Work

The most important scientist in the first half of the twentieth century, Albert Einstein developed radical new theories of relativity and advanced the new quantum theory of light.

Einstein was born on March 14, 1879, in Ulm, Germany, and was a quiet child. When he did speak he would say things twice, first softly to himself, as if he were practicing or testing the sounds, then out loud. An ordinary student, he did not impress his teachers in the strict German schools; one said that he would never amount to anything. He entered the Swiss Polytechnic School on his second try. After graduation he was unsuccessful obtaining a teaching position there, so he took a job as a junior official in the Patent Office in Bern.

The work at the patent office was easy, so he continued his own work in physics and mathematics. In March of 1905 his first paper was published, demonstrating that light energy was transferred in separate, tiny bursts or shocks, called quanta, an idea first suggested by MAX PLANCK and published in 1900. Two months later came his paper on the motion of dust particles suspended in a liquid, confirming mathematically that atoms are real and are not merely a theoretical model.

In June 1905 he published his revolutionary

paper on the special theory of relativity, explaining that measures of distance, time, and mass are not absolute but that they depend on the motion of the observer. As unbelievable as it seems, the faster an object travels in space, the slower an object travels in time. Another implication of the theory is that matter is frozen or crystalline energy, as represented by $E = mc^2$, perhaps the most famous equation of all time.

The quality of Einstein's work earned him significant attention, and from 1909 to 1913 he taught in Bern, Zurich, Prague, and then Berlin as better positions continually came to him. During his 20 years in Berlin he published his next great paper, in 1915, a general version of the relativity theory. Its remarkable claim is that space tells matter how to move, and matter tells space how to curve. In 1919, only four years after the general theory was announced, scientists observing a total eclipse of the Sun confirmed that light from distant stars bent slightly in the curved space around the Sun, just as Einstein had predicted.

In 1921 he won the Nobel Prize for Physics for his work on quantum theory, although his work on relativity was overlooked; the Nobel committee thought it was too impractical. During the 1920s he traveled and taught extensively; in 1933 Hitler's rise to power forced him to relocate to the United States. He lived in Princeton, New Jersey, and worked at the Institute for Advanced Study there for the rest of his life.

Einstein was a fine writer, able to clearly communicate his ideas to readers who have only a little background in science. After moving to the United States in 1938, he completed *The Evolution of Physics,* one of his most famous volumes, written with his longtime friend Leopold Infeld.

When he died on April 18, 1955, he was the best known scientist in the world.

Legacy

While Einstein was not responsible for the practical development of nuclear power or nuclear weapons, he made the essential theoretical discovery behind them, that mass is really stored energy. He also helped to develop the theoretical foundation for quantum theory, one of the most important scientific theories of the twentieth century.

Physicists in the nineteenth century had neatly categorized the universe and had segregated space from time, matter from energy, and particles from

waves. In every case Einstein showed that what was previously thought to be separate was actually intertwined. Relativity makes it clear that moving rapidly through space slows the passing of time, and that matter is crystalline energy in storage. Quantum theory shows that waves of light are grainy and lumpy, while electrons are diffuse and wavy. Our view of the universe and the tools scientists used to study it were profoundly changed by his work.

Nuclear weapons and nuclear energy would not have been possible without Einstein's work. The atomic bombs detonated over Japan in World War II each released the energy stored in just over one gram of matter, equal to the mass of about one paper clip. And the potential energy possible from chain reactions created in a nuclear reactor, using a relatively small core of uranium, can light entire cities.

Einstein's work on quantum theory has revolutionized our most basic understanding of nuclei, atoms, and the relationship between cause and effect. Quantum theory has led to practical discoveries as well. In 1948 scientists at Bell Laboratories were testing quantum theory by investigating the motion of electrons in the element germanium. Their work led directly to the development of the transistor and eventually to the integrated circuits used in all computers. Ironically, Einstein himself never fully appreciated the range and power of quantum theory.

Secaur

For Further Reading:
Clark, Ronald W. *Einstein: The Life and Times.* New York: World Publishing, 1971.
Fritzsch, Harald. *An Equation that Changed the World.* Chicago: University of Chicago Press, 1994.
Pais, Abraham. *Subtle Is the Lord: The Science and Life of Albert Einstein.* New York: Oxford University Press, 1982.

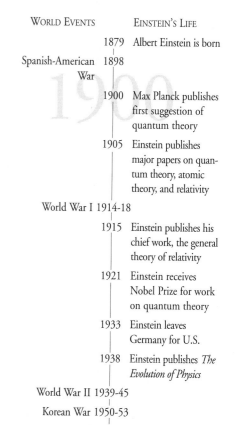

World Events		Einstein's Life
	1879	Albert Einstein is born
Spanish-American War	1898	
	1900	Max Planck publishes first suggestion of quantum theory
	1905	Einstein publishes major papers on quantum theory, atomic theory, and relativity
World War I 1914-18		
	1915	Einstein publishes his chief work, the general theory of relativity
	1921	Einstein receives Nobel Prize for work on quantum theory
	1933	Einstein leaves Germany for U.S.
	1938	Einstein publishes *The Evolution of Physics*
World War II 1939-45		
Korean War 1950-53		
	1955	Einstein dies

Eratosthenes

Mathematician; Geographer;
Librarian at Alexandria
c. 284-192 B.C.E.

Life and Work

The foremost geographer of antiquity, Eratosthenes was the first person to calculate the circumference of Earth with reasonable accuracy. He was also a gifted mathematician who invented a way of determining prime numbers.

World Events	Eratosthenes' Life*
	B.C.E.
Alexander the Great's empire runs from Greece to India	323
	c. 284 Eratosthenes is born
	c. 240 *Geography* outlines procedure for calculating circumference of Earth
	c. 192 Eratosthenes dies

Scholars find it difficult to date the specific events in Eratosthenes' life with accuracy.

Eratosthenes was born about 284 B.C.E. in Cyrene (now Shahhat in Libya) and studied in Athens at Plato's Academy. Ptolemy invited him to Alexandria (present-day Egypt) where he worked as the director of the famous library there. While in Alexandria, Eratosthenes had a specific interest in collecting and furthering geographical knowledge.

His book *Geography* (c. 240 B.C.E.) was the first treatise to give a mathematical basis for measuring Earth as a sphere measured along latitudinal lines and meridians. Eratosthenes used the fact that, on the first day of summer, the sun was directly over the city of Syene (now Aswan in Egypt) but that it made a small angle with a vertical pole at Alexandria. Then, using an estimate of the distance between the two cities and a theorem from EUCLID's *Elements*, Eratosthenes calculated the circumference of Earth as being, in modern terms, about 24,662 miles. Our modern estimate is 24,860 miles: Eratosthenes' calculation is remarkably close.

Eratosthenes invented a method for doubling the cube as well as a way of finding prime numbers, now known as the Sieve of Eratosthenes. A prime number is divisible only by one and itself. His construction eliminates all whole numbers that are products of other whole numbers, so only the prime numbers fall through the "sieve." He does this incrementally by keeping the number one, and then eliminating every second number (multiples of two), then every third number (the multiples of three, which is the next remaining number), then every fifth number (the next remaining number), and so on. While the method is slow and repetitious, it is foolproof in its accuracy.

It is thought that, in about 192 B.C.E., unable to continue as a scholar because of failing eyesight, Eratosthenes starved himself to death.

Legacy

As the first mathematician and geographer to reasonably calculate the circumference of Earth, Eratosthenes is considered to be the father of geodesy, the study of Earth's size and shape. His mathematical insights and constructions remain with us today, and his position as librarian at Alexandria influenced scholars of his and succeeding generations.

Eratosthenes' vast knowledge, acquired not just through study but through his years of directing the library at Alexandria, earned him the label of "the second Plato" from his colleagues. The extent of his influence over his contemporaries, as director of the library, cannot be overestimated; he functioned as one of the organizers and gatekeepers of the knowledge accumulated from the vast holdings of Greek civilization, then centered in Alexandria.

Pappus of Alexandria, one of the last great mathematicians of antiquity, preserved Eratosthenes' work, as well as that of other mathematicians, in his *Synagoge,* finished in 340 C.E. Pappus' efforts did much to keep mathematics alive; however, in the late Roman Empire (235 to 641 C.E.), the contributions of Eratosthenes and other Greek mathematicians fell by the wayside.

Eratosthenes' famous calculation of Earth's circumference is considered to be the first step in the early field of geodesy. Several centuries before, the Greek mathematician PYTHAGORAS (c. 580-500 B.C.E.) concluded that Earth is spherical, and Aristotle and other Greek scholars concluded the same. Eratosthenes' efforts, however, marked the first effort to apply a mathematical technique to calculating Earth's circumference.

Eratosthenes' calculation remained the most accurate measurement until late in the seventeenth century. At the end of the 1500s, the Danish astronomer TYCHO BRAHE introduced the concept of triangulation, which uses a chain of triangles rather than Eratosthenes' single triangle, to measure an arc. Soon after, Dutch mathematician Willebrord van Roijen Snell used triangulation in calculating the circumference of Earth, and, by 1684, when Jean Picard used a telescope to help measure successive triangles, a more accurate measurement supplanted Eratosthenes' calculation.

Eratosthenes' method of determining prime numbers survives today. At an early stage in students' mathematical careers, they are inevitably exposed to the Sieve of Eratosthenes. Not only does it enable students to arrive at a list of prime numbers, it also—and more importantly—helps explain the concept of prime numbers in a practical and unambiguous way.

Steinberg

For Further Reading:
Bell, E. T. *Men of Mathematics.* New York: Simon & Schuster, 1937.
Dunham, William. *Journey through Genius.* New York: John Wiley, 1990.
Kline, Morris. *Mathematical Thought from Ancient to Modern Times.* New York: Oxford University Press, 1972.

Euclid

Organizer of Geometry;
Author of Landmark *Elements*
c. 330-c. 275 B.C.E.

Life and Work

Euclid's text, *Elements,* presented geometry in such a complete and logical format that much of it is still being studied and taught more than 2,000 years later.

Little is known about the specific events in Euclid's life, but ancient historians suggest that he was born around 330 B.C.E. near Athens, Greece, where he was educated.

Euclid moved to Alexandria shortly after 300 B.C.E., when Alexander the Great's successor, Ptolemy Soter, established the famous library and school there. Euclid began teaching mathematics and soon became recognized as one of the leading mathematicians in Alexandria.

Around 280 B.C.E., he completed *Elements,* a 13-volume synthesis of Greek mathematics including the geometry usually associated with his name, Euclidean geometry. In it Euclid listed all the axioms and postulates, those "common notions" that were assumed without proof, at the beginning of his work. He also clearly laid out important definitions and organized theorems in a logical order.

Other works of Euclid include a geometry text, *Data,* and a book on astronomy and the geometry of the sphere, called *Phenomena.*

Euclid's work was not restricted to geometry. Even *Elements* contains results from other areas of mathematics including number theory. His work in number theory includes the classic and beautiful proof that infinite numbers of prime numbers exist.

He also wrote the book *Optics,* an important treatise on mathematical representations of optical phenomena.

Euclid died around 275 B.C.E.

Legacy

Euclid's influence on mathematics was immediate and lasting. In *Elements,* he introduced a logical structure to mathematical exposition that continues to influence mathematicians and the way in which they write mathematics.

Euclid was not the first author of a geometry book. In fact much of the mathematics included in *Elements* had appeared before. Also Euclid was not the first to compile the known geometry into a single text. The geometry text used at Plato's Academy was written by Theudius; this is the book that Aristotle presumably used. However the influence of Euclid's *Elements* lies in its logical structure and comprehensiveness, features that quickly distinguished it from other existing works in mathematics. Euclid's choice of what needed to be assumed and the logical way he structured the subject were the reasons that *Elements* quickly replaced texts that preceded it and why most of it is still used today.

Elements contains some mistakes that have offered challenges to mathematicians throughout the centuries. These mistakes have been noted and, where possible, corrected.

Proving Euclid's parallel postulate preoccupied mathematicians for centuries. The postulate states that given a line *L* and a point *P* not on the line, there is exactly one line through *P* that is parallel to *L*. The postulate preoccupied even Euclid; he delayed using the parallel postulate until he had proved all that he could without it. After many false proofs, future mathematicians realized that Euclid's geometry needed the parallel postulate as an assumption that could not be proved.

Replacing the parallel postulate with others led to various non-Euclidean geometries. For instance, Russian mathematician Nikolay Lobachevsky established a theory of non-Euclidean geometry in 1840 that insisted that the parallel postulate did not hold. He started a revolution in conceptualizing continuous space that led others (including Georg Riemann and ALBERT EINSTEIN) to try to describe three-dimensional space in ways that accommodate how light and matter behave.

Steinberg

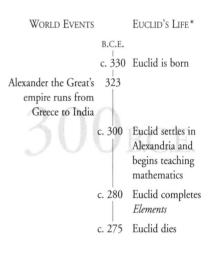

WORLD EVENTS	EUCLID'S LIFE*
	B.C.E.
	c. 330 Euclid is born
Alexander the Great's	323
empire runs from	
Greece to India	
	c. 300 Euclid settles in Alexandria and begins teaching mathematics
	c. 280 Euclid completes *Elements*
	c. 275 Euclid dies

* Scholars cannot date the specific events in Euclid's life with accuracy.

For Further Reading:

Allman, George Johnston. *Greek Geometry from Thales to Euclid.* New York: Arno Press, 1976.

Heath, Thomas L. "Introduction" in *Euclid: The Thirteen Books of the Elements.* New York: Dover, 1956.

Kline, Morris. *Mathematical Thought from Ancient to Modern Times.* New York: Oxford University Press, 1972.

Euler, Leonhard

Pioneer in Algebra
and Number Theory
1707-1783

Life and Work

Leonhard Euler's prolific work in mathematics includes contributions in algebra, geometry, trigonometry, and number theory. He also applied pure mathematics to important concepts in mechanics and astronomy.

Leonhard Euler was born in Basel, Switzerland, on April 15, 1707. His father, a Calvinist clergyman, had studied mathematics with JAKOB BERNOULLI, one of nine members of a multigenerational family of distinguished mathematicians and physicists. While a teenager, Euler studied mathematics at the University of Basel with Jakob's brother, JOHANN BERNOULLI. Euler received his master's degree when he was 17 and, with the Bernoulli family's encouragement, he changed his career from the ministry to mathematics.

In 1727 Johann Bernoulli's son Daniel helped Euler obtain a modest stipend in Russia, teaching in the medical section of the St. Petersburg Academy. When Daniel returned to Switzerland six years later, Euler was chosen to replace him as the Academy's leading mathematician. He produced ideas, calculations, and proofs at an amazing pace. In 1735 he lost the sight in one eye after he worked nonstop for three days to obtain a result that his peers estimated would require months to calculate and years to evaluate. In 1736 he wrote *Mechanics,* the first book entirely devoted to that subject.

Euler returned to Germany, where he was to remain for 25 years, in 1741 at the invitation of King Frederick the Great. There, he joined the Berlin Academy. In 1744 he identified the numerous solutions of polynomial algebraic equations consisting of powers of x. These algebraic solutions included whole numbers, fractions, negatives, irrationals, and complex numbers. Beyond them, however, he described transcendental numbers that are never outcomes of an algebraic equation: exponents, logarithms, trigonometric functions, etc. He also introduced standardized mathematical symbols including summation notation and e for the base of the natural logarithm.

Euler contributed significantly to number theory. In particular, he solved and extended several theorems presented by sixteenth century mathematician PIERRE DE FERMAT.

Euler applied mathematical techniques to the theory of lunar motion and what is called the three-body problem, which refers to an imperfect understanding of the interaction of the movement of the Sun, moon, and Earth. Using ISAAC NEWTON'S gravitational theory, Euler was able to provide an approximation for the three bodies.

After 25 productive years in Germany, Euler accepted a more rewarding and lucrative position in Russia in 1766. Even though he was completely blind by then and had to rely on his students and his children to write his papers for him, he spent the last 17 years of his life thinking and composing ideas about mathematics. The more than 70 volumes of *Opera Omnia* contains nearly 900 smaller books and original papers. Euler worked until he died suddenly on September 18, 1783, in St. Petersburg, Russia.

Legacy

Euler's legacy pervades nearly every area of mathematics—from number theory and geometry to algebra and trigonometry—and his practical application of mathematics had direct influence on navigation at sea.

His work on proving Fermat's last theorem (there are no whole number solutions for the equation $x^n + y^n = z^n$ when n is greater than 2) provided a major clue to that problem for mathematicians of his generation and aided in its eventual solution over 200 years later. Euler's work was advanced in the nineteenth century by KARL GAUSS, who is considered to be the father of modern number-theory methodology.

Evidence of the lasting effects of his originality infuse every area of mathematics. Euler's many contributions are memorialized in mathematics texts by references to Euler's theorem, Euler's coefficients, Euler's proofs, Euler's constant, Euler's integrals, Eulerian circuits, Euler's transcendental functions, and Euler lines and angles.

Euler's immediate influence on mathematics extended beyond what he invented. He wrote textbooks on calculus that were used by generations of students and professionals alike. The clarity of his explanations continues to inspire and educate students of mathematics. Euler's standardization of mathematical notations and his introduction of symbols have been adopted worldwide.

Although most of Euler's work was theoretical, some of it had practical applications. Most importantly, his application of Newton's gravitational theory to describe mutual effects of the motion of the Sun, moon, and Earth led to lunar tables that the British Navy used to help determine latitudinal position of ships at sea.

Steinberg

WORLD EVENTS		EULER'S LIFE
Thirty Years' War in Europe	1618-48	
	1707	Leonhard Euler is born
Peace of Utrecht settles War of Spanish Succession	1713-15	
	1724	Euler earns master's degree from University of Basel
	1733	Euler obtains mathematics chair at St. Petersburg
	1741	Euler joins the Berlin Academy
	1744	Euler identifies many solutions of polynomial equations containing powers of x
	1766	Euler returns to St. Petersburg
United States independence	1776	
	1783	Euler dies
French Revolution	1789	

For Further Reading:
Dunham, William. *Journey through Genius.* New York: John Wiley, 1990.
———. *The Mathematical Universe: An Alphabetical Journey Through the Great Proofs, Problems, and Personalities.* New York: John Wiley, 1990.
Turnbull, Herbert. "The Great Mathematicians." In *The World of Mathematics.* Edited by James R. Newman. New York: Simon & Schuster, 1956.

Everson, Carrie J.

Inventor of Ore
Concentration Process

c.1844-1913

Life and Work

Carrie J. Everson invented a method of separating heavy metals, such as gold and silver, from surrounding rock. This process enabled miners to secure higher yields of precious metals and commensurately greater profits.

Details of Carrie Everson's life are sparse and often inaccurately reported. Born in Sharon, Massachusetts, Carrie (family name unknown) graduated from high school and teacher training college. Later she married a Dr. Everson, who shared with her his knowledge of medicine and chemistry. When the family moved west to pursue mining ventures in the gold and silver fields of Colorado and California, she applied what she knew to mineralogy.

Everson improved upon the then-current method of separating ore from the surrounding rock matrix, the rock material in which ore is embedded. In 1885 she found that the addition of acid to the oil and water used to extract metals from pulverized rocks produced far more effective results than oil and water alone. What she had done manually, crushing and treating four ounces of ore at a time, would have to be done on a large scale with heavy machinery.

Recognizing that her procedure required expensive, specialized equipment, Everson concentrated at first on the more valuable metals such as silver, gold, and copper, so that the investment in making the machinery would pay off. Everson was awarded a patent in 1886 for this oil flotation process and she was given two more patents (with Charles Hebron) in 1892 for further refinements in the process, including a dry flotation technique.

Everson was never able to gain stable financial backing for her inventions. After several failed attempts, she turned to nursing to support herself and her family. Ironically, during the time she patented these processes, ore was plentiful, so most people could not see a need for Everson's work. It was when the lodes began to dwindle that her methods of extraction became a boon to prospectors.

This newfound interest in the oil flotation process led to a search for its inventor; a newspaper tracked down Everson's son, who was able to identify his mother as the inventor of the method. But by the time her process came into wide use, Everson's patent had run out and she received no money.

She died in California in 1913, two years before the search for her began.

Legacy

Everson's oil flotation and acid wash processes helped thousands of miners reap extra ore (and wealth) from the rock surrounding the valuable ore. Without this invention, untold amounts of gold and silver would have remained in the matrix.

During her lifetime the process was not widely used in the United States. The frenzied searching of the American gold rush days was for mother lodes. Little interest was paid to the smaller amounts of ore laced throughout the rocks. Only as the major veins became smaller (following successive exploitation) and the mines less productive did Everson's technique gain wide use. However, the concentration of ores through the use of petroleum distillation residues and acid wash was used in DeBeers diamond mines in South Africa by 1900.

To be commercially viable, Everson's processes for concentrating valuable ores had to be applied to tons of crushed rock. Her methods were known by mining engineers who experimented and applied them. There was a big problem: extracting the metal-bearing mineral from the thick oil used to attract and float desirable ore while leaving the rest of the material to settle out. The problem was solved by using a specialized large-scale centrifuge.

By the time the search was made for the process's obscure inventor, concentration of ores by selective action of oil and acid on certain minerals was standard procedure in mining operations around the world.

Wilson

WORLD EVENTS	EVERSON'S LIFE
	c.1844 Carrie Everson (family name unknown) is born
United States 1861-65 Civil War	
	1864 Carrie marries Dr. Everson
Germany is united 1871	
	1885 Everson improves extraction of ore with addition of acid
	1886 Everson patents process of oil flotation separation
	1892 Everson is awarded patent for dry flotation process
Spanish-American 1898 War	
	1913 Everson dies
World War I 1914-18	

For Further Reading:

Altman, Linda Jacobs. *Women Inventors.* New York: Facts On File, 1997.

Macdonald, Anne. *Feminine Ingenuity.* New York: Ballantine Books, 1992.

Faraday, Michael

Discoverer of
Electromagnetic Induction
1791-1867

Life and Work

Michael Faraday discovered how to generate electricity. He was the first person to realize that a changing magnetic field could create an electric field to move electrons and make them do useful work, like lighting lamps.

Michael Faraday was born on September 22, 1791, near London, England. Faraday was so poor as a child that his mother would give him one loaf of bread to last a week; he would make 14 marks in it so he could measure it out and be sure of having at least something to eat twice a day. At 13 Faraday began working, selling newspapers and binding books, spending his spare time reading pages from the books he was binding. Two books he worked on in the shop included a section of an encyclopedia

about electricity and a book on chemistry. His life work sprung directly from reading them.

One of the customers at the book shop was so impressed with Faraday's hard work that he offered to take him to a series of lectures by HUMPHRY DAVY, one of the most famous scientists of the time. Faraday took detailed notes and sent them to Davy, hoping to get Davy's attention. It worked, and the scientist hired him as his assistant and sponsored Faraday's admission to the Royal Institution, a science school and laboratory. The quality of his work was so good that in 1813 Davy asked Faraday to accompany him on a trip to France, Germany, and Italy. There the young Faraday met—and learned from—the greatest scientists in Europe. Faraday apprenticed with Davy until 1820.

Some scientists knew how to make electricity from chemical reactions, the beginnings of the batteries used today. But Faraday became convinced that there was a connection between electricity and magnetism, and discovered that when he thrust a magnet into a coil of wire, a trickle of electricity emerged. By 1831 he had developed a generator, called a disk dynamo, the prototype of the giant electric power plants that generate electricity today.

Faraday's innovations were numerous, including the discovery of benzene in 1825, a pioneering method for liquefying gases, and the production of stainless steel. He also used optical glass to rotate the plane of a polarized light passing through a strong magnetic field, demonstrating a relationship between two different forms of energy. With his mentor Davy, Faraday also thoroughly investigated the element chlorine.

Faraday was a genius in the lab, conducting experimental investigations with thoroughness and rigor. He was also a gifted teacher; beginning in 1826, he communicated the beauty of science to large and appreciative audiences. He died peacefully in his study chair on August 25, 1867.

Legacy

Michael Faraday was the first person to convert mechanical energy into electricity, laying the foundation for great strides in future theoretical and practical work concerning electricity and magnetism.

Within a generation a scientist in neighboring Scotland, JAMES CLERK MAXWELL, would build on Faraday's discovery and use it for the cornerstone of his comprehensive laws on electricity and magnetism, published in 1873.

Other scientists in many fields were directly affected by Faraday's methodology. His improved experimental methods—clearly stating the problem, carefully carrying out experimental procedures, and completely recording results—provided a model that others would follow.

Faraday's work led to the development of transformers, coils of wire wound on a metal frame that could concentrate electrical energy and increase voltage in a circuit, or spread out energy to make a lower voltage. Transformers are everywhere today, from power packs for computer games to large canisters on utility poles and huge fenced switchyards at power plants and substations. Faraday's insight not only provided the means to generate electricity but also to distribute it over long distances.

In chemistry Faraday provided an improved understanding of how ions in solution conduct electricity, leading to a way to count and measure atoms and to our modern technique of electroplating. Chlorine, which he investigated, would become one of the most important industrial chemicals in the twentieth century.

Faraday's friend and teacher, Davy, was once asked late in his life about his greatest achievement. He answered that his greatest discovery was Michael Faraday.

Secaur

World Events		Faraday's Life
French Revolution	1789	
	1791	Michael Faraday is born
Napoleonic Wars in Europe	1803-15	
	1812-20	Faraday apprentices with Sir Humphry Davy
	1824	Faraday elected as Fellow of the Royal Society
	1825	Faraday isolates benzene
	1826	Faraday begins famous public lectures
	1831	Faraday makes first dynamo and discovers electromagnetic induction
United States Civil War	1861-65	
	1867	Faraday dies
Germany is united	1871	

For Further Reading:

Gooding, D., and Frank James. *Faraday Rediscovered: Essays on the Life and Work of Michael Faraday.* New York: Stockton Press, 1985.

Tricker, R. A. R. *The Contributions of Faraday and Maxwell to Electrical Science.* New York: Pergamon Press, 1996.

Williams, L. Pearce. *Michael Faraday: A Biography.* New York: DaCapo Press, 1987.

Fermat, Pierre de

Pioneer in the Calculus
and Number Theory
1601-1665

Life and Work

Pierre de Fermat, a mathematician only by hobby, carried out pioneering work in a variety of mathematical fields including the calculus and number theory.

Fermat was born in August 1601, in Beaumont de Lomagne, France. Almost nothing is known of his early years except that he attended a local Franciscan school. He studied law at Toulouse, gaining a degree from the University of Orléans in 1631. A life-long career in law and parliamentary service followed.

Fermat pursued advanced mathematics in his spare time, corresponding with another amateur mathematician, Father M. Mersenne who was a friend of RENÉ DESCARTES and BLAISE PASCAL. In the early 1630s Fermat devised a coordinate system of two axes. He used it for the representation of such algebraic equations as lines and curves, a topic now known as analytic geometry. Simultaneously and independently Descartes also developed the use of coordinate axes; he published a description of his system and was recognized as the founder of analytic geometry. Fermat was reluctant to publish his work, most of which was recorded in disordered manuscripts, in letters to friends, and as notes jotted in the margins of books.

In 1637 Fermat completed work that represented the beginnings of the calculus (broadly defined as the use of algebra to calculate the properties of changing quantities). He developed a method of finding the tangent to a point on a curve; the tangent's slope represents the instantaneous rate of change of the curve at that point. This operation is equivalent to differentiation in the modern calculus.

At the same time Fermat worked on his favorite mathematical topic, number theory. He formulated numerous theorems (a theorem is a mathematical formula or statement deduced through other formulas to be true), often sending them, unproved, to friends as a challenge. His most famous theorem, called Fermat's last theorem, states that there are no whole-number solutions to the equation $x^n + y^n = z^n$ when n is an integer greater than 2. He wrote the theorem in the margin of a book (probably in 1637) but never recorded the proof he claimed to have formulated for it. A satisfactory solution was not found until 1995.

In the 1650s Fermat corresponded with French mathematician Blaise Pascal concerning probabilities. Through their letters, they established the foundations of probability theory, the mathematical prediction of outcomes to events.

In his later years Fermat investigated optics. Using mathematics he found that light travels more slowly through denser media. He also suggested that light travels by the quickest path possible, now known as Fermat's principle.

Fermat died in Castres, France, on January 12, 1665.

Legacy

Fermat contributed greatly to various topics in the development of modern mathematics.

Because of his refusal to publish, Fermat was not widely known during his lifetime. Only his friends and correspondents, mostly other mathematicians, understood his contributions. After his death the extent and significance of his work began to be recognized.

Fermat's work on the relation of tangents to curves was important in the evolution of ISAAC NEWTON's ideas concerning the calculus. Newton's calculus, developed in the 1660s, addressed instantaneous rates of change, maximum and minimum values of functions, the length of a specified portion of a curve, and the area under or enclosed by a curve. (GOTTFRIED WILHELM LEIBNIZ independently developed the calculus without borrowing from Fermat.) The calculus proved to be an important tool for advancing solutions to problems in seventeenth- and eighteenth-century physics and celestial mechanics, and it is invaluable in all fields of modern physics.

In the century following Fermat's death, there was little interest in number theory, but in the eighteenth century LEONHARD EULER revived Fermat's discoveries in that field. Euler solved, generalized, and built upon several theorems proposed by Fermat but not since proved by anyone else. Euler's work was advanced in the nineteenth century by CARL FRIEDRICH GAUSS, who is credited as the father of modern number-theory methodology.

Probability theory, founded by Fermat and Pascal, evolved continuously through the work of numerous mathematicians, including JAKOB BERNOULLI, Pierre Laplace, Siméon-Denis Poisson, and Andrei Kolmogorov. Modern probability theory is applied to many scientific fields, including particle physics, quantum mechanics, population biology, and epidemiology.

Fermat's last theorem has likely generated more work than any other problem in the history of mathematics. Thousands of incorrect proofs to the theorem have been proposed. By 1976 centuries of work had shown the theorem to be true for all values of n up to 125,000. Finally, in 1995, Andrew Wiles of Princeton University completed the proof after more than nine years of work; it was published in the May 1995 issue of *Annals of Mathematics*. Some experts now suspect that Fermat thought he had formulated a correct proof but actually had not.

Schuyler

For Further Reading:

Hollingdale, Stuart. *Makers of Mathematics.* New York: Penguin Books, 1989.

Mahoney, Michael S. *The Mathematical Career of Pierre de Fermat (1601-1665).* Princeton, N.J.: Princeton University Press, 1973.

Singh, Simon. *Fermat's Enigma: The Quest to Solve the World's Greatest Mathematical Problem.* New York: Walker, 1997.

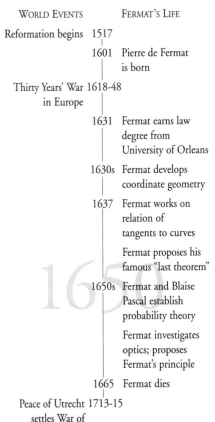

WORLD EVENTS		FERMAT'S LIFE
Reformation begins	1517	
	1601	Pierre de Fermat is born
Thirty Years' War in Europe	1618-48	
	1631	Fermat earns law degree from University of Orleans
	1630s	Fermat develops coordinate geometry
	1637	Fermat works on relation of tangents to curves
		Fermat proposes his famous "last theorem"
	1650s	Fermat and Blaise Pascal establish probability theory
		Fermat investigates optics; proposes Fermat's principle
	1665	Fermat dies
Peace of Utrecht settles War of Spanish Succession	1713-15	

Fermi, Enrico

Designer of First Nuclear Reactor
1901-1954

Life and Work

Enrico Fermi confirmed fundamental theory about the behavior of subatomic particles and built the first controlled nuclear reactor.

Born in Rome on September 29, 1901, and educated there, Fermi received his doctorate from the University of Pisa when he was only 21. In 1926 he became a professor of physics at the University of Rome. That same year he developed the mathematics underlying the behavior of electrons, verifying what is called the "exclusion principle," which says that no two electrons in an atom can exist in identical states. That is, something must be different about each electron—the orbital in which it lies around the nucleus, the shape of the cloud that it forms, or the direction of its spin.

In the early 1930s, exciting discoveries drew Fermi into experimental physics. In 1932 JAMES CHADWICK discovered that atomic nuclei contain neutral particles called neutrons, and two years later Irene Curie (daughter of scientists Pierre and MARIE CURIE) and her husband Frédéric Joliot found that they could produce new, radioactive

isotopes by bombarding stable, naturally occurring atoms with alpha particles—positively charged particles discharged from radioactive material. Fermi realized that neutrons, with no charge, could penetrate atoms more easily than positively charged alpha particles and might be more effective in promoting nuclear reactions. His team of young physicists attacked the elements of the periodic table with zeal, producing scores of radioactive isotopes that had never been seen before. When Fermi bombarded uranium, the heaviest known element, he had surprising results. He and his team thought they were seeing a new, heavier element, but in 1939 LISE MEITNER and OTTO HAHN explained the results as the fission or splitting of uranium, leading to the development of nuclear weapons.

Fermi and his wife Laura escaped from the Fascist government of Italy in 1938, the year he received the Nobel Prize for Physics for his work on radioactivity and neutron bombardment. He came to the University of Chicago in 1939 and three years later joined the Manhattan Project, sponsored by the U.S. government to build the first nuclear weapons. On December 2, 1942, a team under his direction pulled the control rods from the first experimental nuclear reactor and released nuclear energy as planned, demonstrating for the first time that a sustainable, controlled series of fission reactions was possible.

Element number 100 is fermium, named for him, and an annual Enrico Fermi award is given in the United States for advances in nuclear theory and peaceful uses of nuclear energy. He died of cancer in Chicago on November 28, 1954.

Legacy

Enrico Fermi made groundbreaking advances in nuclear theory and its application, innovations that profoundly affected science and the world.

Fermi's theoretical work alone leaves a tremendous legacy; it has come to explain fundamental phenomena in physics and chemistry. Particles that make up the material universe—protons, neutrons, electrons, and the quarks that make up protons and neutrons—have the value of a property called spin, which makes them subject to the rules that Fermi described. All of those particles are called fermions. In any bound system, such as an atom, no two fermions can ever be identical. This explains the organization of electrons within atoms, the patterns of chemical behavior, the durability of matter, and matter's ability to take up space.

Fermi's leadership on the Manhattan Project was invaluable; he was described as the most versatile scientist on the project and was respected by all of the scientists working on it. While his work on the nuclear reactor in 1942 was instrumental in the creation of the atomic bombs dropped on Japan and ultimately led to the development of the United States' nuclear arsenal, he urged caution. With other colleagues, including J. ROBERT OPPENHEIMER and ALBERT EINSTEIN, he helped to form the moral imperative against further development of nuclear weapons, in particular the more powerful hydrogen bomb, following the first demonstrations of their destructive capacity.

The technique Fermi developed for creating and analyzing nuclear isotopes is used today in neutron activation analysis, one of the most precise analytical methods ever developed. A sample to be analyzed is exposed to a powerful beam of neutrons, making many of its component atoms radioactive. Each emits radiation characteristic of that new isotope, giving technicians a unique fingerprint of that element's presence. Chemical detection methods always require trillions of atoms or molecules to make a detectable reaction, but equipment to measure radiation detects the decay of single atoms, making it possible to determine the tiniest quantities of materials. A sample of Napoleon's hair was analyzed with this method, revealing chemically undetectable traces of arsenic, suggesting that he may have been poisoned.

Secaur

WORLD EVENTS		FERMI'S LIFE
Spanish-American War	1898	
	1901	Enrico Fermi is born
World War I	1914-18	
	1922	Fermi receives doctorate from the University of Pisa
	1926	Fermi develops mathematical theory of electron exclusion
	1938	Fermi receives Nobel Prize for Physics
World War II	1939-45	
	1942	Fermi achieves the first sustainable nuclear fission chain reaction
	1942-45	Fermi works on Manhattan Project to build the nuclear bomb
Korean War	1950-53	
	1954	Fermi dies

For Further Reading:

Cooper, Dan. *Enrico Fermi and the Revolution of Modern Physics.* New York: Oxford University Press, 1998.

Fermi, Laura. *Atoms in the Family.* Chicago: University of Chicago Press, 1954.

Segrè, E. *Enrico Fermi: Physicist.* Chicago: University of Chicago Press, 1970.

Snow, C. P. *The Physicists.* Boston: Little, Brown, 1981.

Feynman, Richard

Developer of Quantum
Electrodynamics
1918-1988

Life and Work

Richard Feynman was one of the most colorful physicists in the second half of the twentieth century. A gifted theoretician and a first-rate teacher, he developed the influential theory of quantum electrodynamics (QED).

Born in New York City on May 11, 1918, the son of Jewish immigrants, Feynman studied physics at the Massachusetts Institute of Technology and Princeton University, where he received his doctorate in 1942. During World War II he worked on the atomic bomb project at Princeton University and then at the government's secret facility in Los Alamos, New Mexico, from 1943 to 1945. Feynman was the youngest of the section leaders in the project, heading a group of scientists using mechanical adding-machine-type calculators, slide rules, and paper to perform the vast numbers of calculations required to design and build the bomb. In his spare time at Los Alamos, he enjoyed figuring out how to crack the military officers' safes and mailing coded nonsense messages to his wife to pester the security police that checked all the mail.

After the war Feynman taught at Cornell University from 1945 to 1950; he became professor of theoretical physics at California Institute of Technology (Caltech) in 1950, and remained there for the rest of his career. He shared the Nobel Prize for Physics in 1965 for his work on quantum electrodynamics, the theory of the interaction between light and matter; the two scientists he shared the prize with, Julian S. Schwinger and Tomonaga Shin'ichiro, developed similar theories, which ultimately were not as far-reaching as Feynman's. In the nineteenth-century theory of electromagnetism formulated by JAMES CLERK MAXWELL, one charged particle affected another some distance away through the action of electric and magnetic fields. That theory works well for larger objects, such as two socks tumbling in a clothes dryer, or even two pieces of lint. It does not work well,

though, for atoms or the particles within them. To Feynman, electromagnetic forces were not attributable to mysterious action at a distance but to the exchange of particles of light, the quanta of the electric field. Feynman's quantum electrodynamics explained the interaction between charged particles as a result of the exchange of force-carrying particles, called bosons.

That the theory is nicknamed QED is part of Feynman's sense of humor. Those letters are the first letters of three Latin words that together could be construed as meaning "Thus ends the demonstration." Medieval scholars would write QED at the bottom of a page of a mathematical or logical proof. Calling his theory QED suggests, in a joking way, that the theory is the ultimate end of all physics theories.

Feynman perfected the mathematics of QED and improved it by introducing simple drawings, now called Feynman diagrams, to keep track of the interactions between particles in the theory. He could solve the most difficult mathematical problems of a theory and also explain a theory in a simple model—few scientists could work at both extremes with such ease.

Feynman served as part of the Presidential Commission to investigate the explosion of the Challenger space shuttle in 1986. He will be remembered for dipping a rubber O-ring into his glass of ice water to demonstrate how low temperatures on the day of the launch may have affected rubber seals in the booster rockets. On February 15, 1988, Feynman died of cancer in Los Angeles, California.

Legacy

Feynman's development of quantum electrodynamics was one of the most important advances in quantum mechanics, the system that explains how subatomic waves and particles behave.

Feynman's approach in QED has been used to unravel the operation of other forces in nature. Other theoreticians have applied Feynman's ideas to the strong nuclear force that binds quarks together to form protons and neutrons. That situation is more complex but the theory, called quantum chromodynamics (QCD), has been completely successful

in explaining the strong force. Another similar application of the theory explains the operation of the weak nuclear force involved in radioactivity and nuclear reactions as attributable to the exchange of other particles.

Feynman was a brilliant lecturer and a superb teacher who could teach a lesson equally well with the full power of calculus or with no mathematics at all. His lectures at Caltech from 1961 to 1963 were recorded, edited, and published as *The Feynman Lectures on Physics*. They are widely read even today for Feynman's clear explanations and witty presentations.

Secaur

WORLD EVENTS	FEYNMAN'S LIFE
World War I 1914-18	
	1918 Richard Feynman is born
World War II 1939-45	
	1942 Feynman receives Ph.D. from Princeton University
	1943-45 Feynman works on atomic bomb project at Los Alamos
	1945-50 Feynman teaches at Cornell University
	1950 Feynman begins teaching at California Institute of Technology
Korean War 1950-53	
	1965 Feynman shares Nobel Prize for Physics for his work on quantum electrodynamics
End of Vietnam War 1975	
	1986 Feynman serves on Presidential Commission investigating the Challenger explosion
	1988 Feynman dies
Dissolution of 1991 Soviet Union	

For Further Reading:
Feynman, Richard. P. *Surely You're Joking, Mr. Feynman.* New York: Bantam, 1985.
———. *What Do You Care What Other People Think?* New York: Bantam, 1986.
Gleick, James. *Genius: The Life and Science of Richard Feynman.* New York: Pantheon, 1992.

Fibonacci, Leonardo

Popularizer of Arabic Numerals;
Pioneer in Algebra and Geometry
c. 1170-c. 1250

Life and Work

Leonardo Fibonacci (also known as Leonardo of Pisa) popularized the use of Arabic numerals in the West and extended the frontiers of algebra and geometry.

Fibonacci was born about 1170 in Pisa, Italy. He spent some of his youth in Algeria, where his father, a government official of Pisa, worked at a customs house. There he received mathematical training from North African scholars. Little is known of his early adult years except that he traveled extensively through several Mediterranean regions, observing various mathematical methods of commerce and consequently becoming fluent in the use of Arabic numerals (the familiar numerals used today: 0, 1, 2, 3, 4. . .).

Settling in Pisa about 1200, Fibonacci began writing his first work, the *Liber Abaci* ("Book of Counting"), which appeared in 1202. In it, Fibonacci described the Arabic numeral system, explained how to use it, and presented numerous examples demonstrating its superiority over the traditional system of Roman numerals. The book also contained studies of algebra, fractions, prime numbers, multiplication tables, commercial applications of mathematics, and series. The section on series included what is now known as the Fibonacci sequence, in which each term equals the sum of the previous two terms: (1, 1, 2, 3, 5, 8, 13, 21 . . .).

Fibonacci's next work, *Practica Geometriae* ("Practice of Geometry"), was published in 1220. It addressed arithmetic and geometric problems with algebraic methods.

Fibonacci's writing attracted the attention of Holy Roman Emperor Frederick II, who summoned the mathematician to his court and presented several problems to test his skill. One problem demanded the solution to the equations: $x^2 + 5 = u^2$ and $x^2 - 5 = v^2$. Fibonacci delivered an answer: $x = 4\frac{1}{2}$, $u = 4\frac{3}{12}$, $v = 3\frac{1}{2}$. Problems of this type are called diophantine equations, defined as indeterminate equations (those with more than one solution to unknowns) for which rational solutions are sought. Such equations take their name from third-century Greek mathematician Diophantus. In 1225 Fibonacci's *Liber Quadratorum* ("Book of Squares"), which dealt exclusively with diophantine equations, was published under the patronage of the emperor.

Fibonacci died about 1250 in Pisa.

Legacy

Fibonacci's most lasting legacy was the spread of Arabic numerals throughout the West. His work also influenced Renaissance mathematicians.

Although Fibonacci was not the first to introduce Arabic numerals to Europe, his *Liber Abaci* was responsible for causing their wide adoption. In Fibonacci's time calculations were done on counting boards, and the results were recorded as Roman numerals. Fibonacci's

use of Arabic numerals, which demonstrated the enormous simplification they afforded, attracted proponents as copies of the *Liber Abaci* circulated. Moveable-type print, introduced in the 1450s, standardized the form of Arabic numerals, which therefore look today nearly the same as they did 500 years ago. However, opposition to change was strong, and common use of Arabic numerals had to wait until the sixteenth century. The zero symbol was likely the most difficult aspect of the system for people to accept, as it had no traditional counterpart.

The *Liber Abaci* was used as a standard textbook of arithmetic, algebra, and geometry for two centuries after its publication. However, mathematics fell into academic disrepute in Europe during that time, so Fibonacci's work was largely disregarded. In the sixteenth century, the work of several mathematicians revived Fibonacci's achievements. Nicolo Tartaglia, Girolamo Cardano, and Raphael Bombelli, all of Italy, were encouraged by Fibonacci's writing to work on problems in algebra.

The Fibonacci sequence, so named by French number theorist Edouard Lucas in the nineteenth century, was important in drawing attention to the study of series. With several curious mathematical properties, the Fibonacci sequence pointed the way toward using series to approach approximations of quantities and to construct cryptographic codes.

Schuyler

World Events	Fibonacci's Life
First Crusade 1095	
Settling of Timbuktu, c.1100 present-day Mali	
	c.1170 Leonardo Fibonacci is born
	1202 Fibonacci's *Liber Abaci* appears; it popularizes use of Arabic numerals
	1220 Fibonacci's *Practica Geometriae* appears
	1225 Fibonacci's *Liber Quadratorum* appears
	c. 1250 Fibonacci dies
Hapsburg dynasty 1273 begins dominance in Holy Roman Empire	

For Further Reading:
Dunham, William. *The Mathematical Universe: An Alphabetical Journey Through the Great Proofs, Problems, and Personalities.* New York: John Wiley, 1994.
Hollingdale, Stuart. *Makers of Mathematics.* New York: Penguin Books, 1989.

Fleming, Alexander

Bacteriologist;
Discoverer of Penicillin
1881-1955

Life and Work

Alexander Fleming discovered the bacteria-fighting substance penicillin, which kindled a major revolution in medical science by initiating the development of antibiotics.

Fleming was born on August 6, 1881, to a farming family in Lochfield, Scotland. Walking four miles through the countryside each day to attend school, he developed an early interest in the natural world. He ranked consistently at the top of his class, but poverty prevented him from going to college immediately after high school. He worked as a shipping clerk for four years and, at age 20, he began medical studies.

Fleming earned a medical degree at St. Mary's Medical School in 1906 and began research at London University. He studied bacteriology at that institution for the rest of his career, excepting several years service in the British military medical corps during World War I. His goal was to find a safe and effective substance for treating diseases caused by infectious organisms.

About 1921 Fleming cultured a sample of his own nasal mucus and noted that it destroyed a colony of bacteria growing on the same dish. He dubbed the bacteria-killing substance in the mucus "lysozyme" and subsequently found that it is present also in blood serum, tears, and saliva. Lysozyme dissolves bacterial cell walls, thus preventing some infections, but it is not effective against most disease-causing organisms. However, the evidence that natural secretions could kill germs inspired him to search for other plant or animal substances that might help the human body fight infectious diseases more effectively.

Fleming made his major breakthrough in 1928, observing that a mysterious mold had contaminated a plate of the bacterium *Staphyloccocus aureus,* an infectious agent. The mold appeared to have inhibited bacterial growth. Fleming isolated the antibacterial substance in the mold (identified as *Penicillium notatum)* and called it "penicillin." Further investigations showed that penicillin kills many infectious bacteria and leaves human tissue unharmed.

The circumstances that led to Fleming's discovery were extremely improbable. The mold and the bacteria require particular conditions to grow, and it was only because of unusually damp weather in London in 1928 that the organisms ended up inhabiting the same plate at the same time. But Fleming's trained eye noticed the anomaly and his persistence in studying the effects of common bread mold not only resulted in discovery of a natural healer but also confirmed his idea that antidotes to infection are produced by many species.

Fleming found penicillin unstable and difficult to purify. Fleming's discovery remained obscure until 1940, when chemists Ernst Chain and Howard Florey developed methods to purify, preserve, and mass-produce penicillin. The medical community soon recognized the potential of this new drug for fighting disease. Fleming was knighted in 1944 and, in the following year, Fleming, Chain, and Florey shared the Nobel Prize for Physiology or Medicine. Fleming retired in 1948 and died in London on March 11, 1955.

Legacy

By introducing antibiotics to medicine, Fleming's discovery significantly reduced the spread of infectious diseases in the twentieth century. His ideas also prompted others to look for antibacterial, antifungal, and antiviral agents that would improve human health. Streptomycin, chloromycetin, and tetracycline are some of the successes.

Prior to the discovery of penicillin, only a few substances were known to have antibacterial properties, and they produced toxic effects within the human body. After Chain and Florey devised a means of purifying and concentrating penicillin, large-scale clinical trials were begun to test the new drug's efficacy. By 1942, the trials had been declared enormously successful, and penicillin had been established as a wonder drug for many infectious diseases. Penicillin was immediately—and still is—used to treat pneumonia, scarlet fever, gonorrhea, diphtheria, meningitis, strep throat, and various wound infections.

For the treatment of many diseases, antibiotics represent the most effective tools available. However, many antibiotic-resistant strains of pathogenic bacteria have recently developed, because of the overuse and misuse of penicillin and other antibiotics. The search is on for new ways to help people resist and overcome common diseases by promoting growth of naturally occurring antidotes. These modern endeavors are rooted in Fleming's work.

Although Fleming is remembered mostly for his discovery of penicillin, his earlier work has also been influential. The enzyme lysozyme has become a valuable research material used to break apart bacteria for chemical analysis.

Schuyler

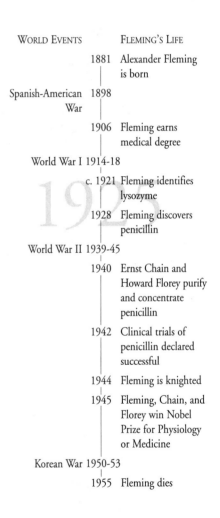

WORLD EVENTS	FLEMING'S LIFE
	1881 Alexander Fleming is born
Spanish-American War 1898	
	1906 Fleming earns medical degree
World War I 1914-18	
	c. 1921 Fleming identifies lysozyme
	1928 Fleming discovers penicillin
World War II 1939-45	
	1940 Ernst Chain and Howard Florey purify and concentrate penicillin
	1942 Clinical trials of penicillin declared successful
	1944 Fleming is knighted
	1945 Fleming, Chain, and Florey win Nobel Prize for Physiology or Medicine
Korean War 1950-53	
	1955 Fleming dies

For Further Reading:

Hughes, W. Howard. *Alexander Fleming and Penicillin.* London: Proiry Press, 1974.

MacFarlane, Gwyn. *Alexander Fleming, the Man and the Myth.* Cambridge, Mass.: Harvard University Press, 1984.

Ford, Henry

Automobile Engineer;
Inventor of the Model T
1863-1947

Life and Work

Henry Ford was an automobile engineer and producer who introduced the Model T, an inexpensive car that transformed the automobile from a luxury into an everyday necessity. He also developed the modern assembly line and interchangeable parts—practices that shaped business and industry in the twentieth century.

Ford was born on July 30, 1863, on a farm near Dearborn, Michigan. From 1879 to 1882 he worked at machine shops in Detroit, where he was exposed to the internal-combustion engine for the first time. In the 1880s he worked as an engineer and in his spare time pursued his dream of building an automobile.

Ford built his first gasoline-powered engine in 1893 and his first "horseless carriage" (as automobiles were called) in 1896. In the seven years that followed he built several racing cars that broke speed records. During this period he continually disappointed financiers who invested in his innovations: they wanted an automobile to market to the consumer, but Ford insisted that each new automobile he built needed incremental improvements before it would be ready for consumer use.

Ford founded the Ford Motor Company in 1903, with the goal of making automobiles safe and affordable for the general public. In 1908 Ford realized his goal: he introduced the Model T, of which more than 15 million were sold during 19 years of its production. The Model T, a 20-horsepower vehicle with two forward gears and one reverse, was built to be sturdy and easy to drive and maintain. The early models were started by crank, but in 1912 Ford offered electric starting as an option.

By 1913 Ford had introduced standardized interchangeable parts and the assembly line, with highly specialized divisions of labor, into his factories. In production the automobile moved down an assembly line along which each worker carried out one specific task, adding a component or solidifying the construction in some way. These practices, rare in industry at that time, lowered costs and improved output. In 1914 Ford increased his employees' wages to $5 per day, double the industry standard, and shortened their workday from nine to eight hours. These benefits curtailed employee-turnover rates and made Ford a national celebrity.

By 1927 Ford had begun utilizing a cross-state manufacturing assembly line. He owned all the means of production of the Model T, including coal mines, foundries, stamping mills, glassworks, and automobile factories.

When the Model T's popularity started to wane, Ford introduced the Model A in 1927, but by 1932, the Ford Motor Company was being outsold by General Motors and Chrysler.

Ford died in Dearborn on April 7, 1947, and his vast holdings went to the Ford Foundation, a non-profit organization he had set up to retain family control over the company.

Legacy

Ford was a key player in initiating the movement of people in the United States from rural to urban surroundings and in creating the car-dominated culture of the modern world. His introduction of the assembly line helped to revolutionize industrial practices in the United States.

The success of the Model T made the automobile a commonly owned means of transportation. It was affordable to the general public and enjoyed wide sales. It also led other manufacturers to focus on the middle-class market, which resulted in General Motors's Chevrolet and Chrysler's Plymouth.

The spreading popularity of automobiles changed the face of agriculture and increased the size of cities. The horse-drawn carriage quickly disappeared, and land that had formerly been used to grow hay was available for other uses. Cars also made people more mobile, allowing them to travel to new places and settle in distant cities, where they were exposed to new ideas and learned new skills.

Automobiles led to demand for paved roads and highway systems. They also helped to shape many modern U.S. cities, which have stretched out in vast urban sprawls in recent decades.

The success of the automobile industry has also led to a rapid consumption of fossil-fuel resources and an increase in pollution. Oil reserves have been exploited to keep up with the demands of transportation, and automobile emissions represent a large proportion of the air pollution that damages the environment and human health.

Henry Ford's use of the assembly line, interchangeable parts, and division of labor spread quickly to other industries and helped to increase productivity and efficiency of the U.S. manufacturing processes. Today, the assembly line is still used even though modern management techniques try to emphasize a team approach to manufacturing, especially in the auto industry.

Schuyler

World Events		Ford's Life
United States 1861-65 Civil War		
	1863	Henry Ford is born
Germany is united	1871	
	1893	Ford builds his first gas-powered engine
Spanish-American War	1898	
	1903	Ford Motor Company is founded
	1908	Model T is introduced
	1913	Ford uses interchangeable parts and assembly lines
	1914	Ford increases wages and shortens work day
World War I 1914-18		
	1927	Ford's Model A is introduced
	1932	Ford Motor Company is outsold by Chrysler and General Motors
World War II 1939-45		
	1947	Ford dies

For Further Reading:

Bryan, Ford R. *Beyond the Model T: The Other Ventures of Henry Ford*. Detroit, Mich.: Wayne State University Press, 1990.

Gelderman, Carol W. *Henry Ford: The Wayward Capitalist*. New York: St. Martin's Press, 1989.

Franklin, Benjamin

Prolific Inventor;
Investigator of Electricity
1706-1790

Life and Work

Benjamin Franklin investigated the nature of electricity, and invented a safer and more efficient stove and bifocal lenses, among other devices. Aside from his technological and scientific accomplishments, he was a statesman and helped to lead the movement for United States independence.

Franklin was born on September 17, 1706, in Boston, Massachusetts, the fifteenth of 17 children. At age 10, after only two years of formal education, he entered his father's tallow and candlemaking business, and three years later he became an apprentice at his brother's printing shop. As a teenager he read extensively and published many essays under a pseudonym.

Franklin moved to Philadelphia in 1723, but left for London the next year to complete his training. He spent two years there, during which he became a master printer, and returned to Philadelphia in 1726, where he soon established a printing business. That year he organized a club for the discussion of politics and philosophy, which was later reorganized as the American Philosophical Society. Ideas discussed at the meetings led to the introduction of a police force, a public library, and a volunteer fire department.

In the 1740s Franklin devised improved methods of street paving and lighting, and helped establish the institution that later became the University of Pennsylvania. He constructed a stove in 1744 with greater fuel efficiency and heating capability than open fires; it also dramatically reduced the number of deaths from burning, a major occupational hazard for women who did most of the cooking in Colonial times.

In 1747 Franklin began his experiments with electricity. He theorized that an "electrical fluid" flows between materials, such as water and metal, and that an electrical charge builds between a collection and an absence of electrical fluid. To prove that lightning is caused by electricity, he devised his famous 1752 experiment: in a thunderstorm he flew a kite with a metal key at the end of the string and received an electrical shock. He concluded that thunderclouds collect electrical fluid, and that lightning represents a discharge of electricity between the cloud and the ground. He invented the lightning conductor, reasoning that a metal rod (connected by wire to the ground) would attract and direct the electrical discharge and thus prevent bolts of lightning from harming people or property.

In 1784 Franklin invented bifocal lenses, so he could read and look up without removing his glasses. Bifocals have since solved the vision difficulties of millions of people.

The second half of Franklin's life was devoted to politics. He played a leading role in events leading to the independence of the United States and was a member of the committee, along with John Adams and Thomas Jefferson, assigned to draft the Declaration of Independence, which was signed in 1776. Franklin was also involved in the development of the Constitution and the support he lent to it helped in its ratification. He developed an international reputation as a skilled diplomat by the time he died, on April 17, 1790, in Philadelphia.

Legacy

Franklin's various achievements made him one of the most revered men in the United States and one of the most respected Americans in the world. His scientific legacy includes his civic improvements, his invention of the Franklin stove, and his contributions to the understanding of electricity.

Franklin's improved pavement and lighting methods for Philadelphia streets decreased accident rates and increased the safety and aesthetic appeal of the city. His introduction of a free library, fire department, and police force provided models for other cities.

The Franklin stove, as his invention came to be known, had a major impact on people's everyday lives. The stove's efficiency decreased the cost of heating and cooking, and its capacity to produce great amounts of heat made the frigid winters of the Northeast more tolerable. Improved versions of the Franklin stove were soon used across the country and are still manufactured today.

Franklin was instrumental in advancing the understanding of electricity, one of the most significant elements of the scientific and Industrial Revolutions and of modern technological progress. Others in both Europe and the American colonies had begun to experiment with electricity in the eighteenth century, but Franklin was the first to formulate a theory that explained nearly all observed electrical phenomena. Electrical research advanced rapidly, and electricity has become an integral part of the modern world, providing innumerable conveniences.

Schuyler

World Events		Franklin's Life
	1706	Benjamin Franklin is born
Peace of Utrecht settles War of Spanish Succession	1713-15	
	1719	Franklin apprentices as a printer
	1723	Franklin moves to Philadelphia
	1726	Franklin starts printing company
	1730s	Franklin active in civic service
	1744	Franklin invents efficient stove
	1752	Franklin investigates electricity and performs kite experiment
United States independence	1776	Franklin helps to draft U.S. Declaration of Independence
	1784	Franklin invents bifocals
French Revolution	1789	
	1790	Franklin dies

For Further Reading:

Bowen, Catherine. *The Most Dangerous Man in America: Scenes from the Life of Benjamin Franklin.* Boston: Little, Brown, 1986.

McKnight, John. *Benjamin Franklin.* New York: Oxford University Press, 1998.

Wright, Esmond. *Franklin of Philadelphia.* Cambridge, Mass.: Harvard University Press, 1986.

Wright, Esmond, ed. *Benjamin Franklin: His Life As He Wrote It.* Cambridge, Mass.: Harvard University Press, 1990.

Franklin, Rosalind

**Pioneer of X-Ray Diffraction
for DNA Imaging
1920-1958**

Life and Work

Rosalind Franklin created the X-ray diffraction images that FRANCIS CRICK and JAMES WATSON used to elucidate the double-helical structure of deoxyribonucleic acid (DNA).

Franklin was born on July 25, 1920, to a Jewish family in London, England. She attended one of the few primary schools in London that offered physics and chemistry to

WORLD EVENTS	FRANKLIN'S LIFE
World War I 1914-18	
	1920 Rosalind Franklin is born
World War II 1939-45	
	1941 Franklin graduates from Cambridge University
	1951 Franklin begins X-ray diffraction of DNA
Korean War 1950-53	
	1953 James Watson and Francis Crick publish the structure of DNA
	1958 Franklin dies
	1962 Watson, Crick, and Wilkins receive Nobel Prize for Physiology or Medicine
	1968 Watson's *The Double Helix* is published
	1975 Anne Sayre's *Rosalind Franklin and DNA* is published
End of Vietnam War 1975	

girls, and by age 15, she had decided to pursue a career in science. In 1941, she graduated from Cambridge University with a degree in chemistry. In the late 1940s, Franklin became proficient at the X-ray diffraction technique, a method that scientists had begun to use to determine the structures of large molecules.

In 1951, Franklin took a position at King's College, London, where she was assigned to use her X-ray diffraction expertise to help reveal the structure of DNA. Franklin made quick progress, producing numerous X-ray diffraction images of the A and B forms of DNA. She concentrated on the A form, which appeared to give the most decipherable images, but which unfortunately exhibits the double-helical structure less clearly than does the B form. Meanwhile, at Cambridge University, JAMES WATSON and FRANCIS CRICK were working on the same problem; however, they were focusing on the technique of molecular-model building. In 1953, Maurice Wilkins, one of Franklin's coworkers, showed Franklin's X-ray diffraction images to Watson and Crick, without Franklin's permission or knowledge. In those images, Watson and Crick found the final bit of evidence they needed to claim confidently that DNA consisted of a two-stranded helix. In April 1953, Watson and Crick published their theory and became famous as the discoverers of DNA's structure.

In 1954, Franklin joined a laboratory at Birbeck College in London where she worked on the structures of several viruses. In 1956, she was diagnosed with cancer. This news did not curb her research, but she died at the age of 37 in London on April 16, 1958.

Legacy

Franklin was a pioneer in the use of X-ray diffraction, and her work was crucial to one of the most significant discoveries of the twentieth century, the structure of DNA.

Watson and Crick's model of DNA unleashed a flurry of enthusiasm among molecular biologists and laid the groundwork for modern molecular genetics. Their model proposed that the DNA molecule consists of two polynucleotide strands held together by hydrogen bonds between complementary

bases: adenine and thymine; cytosine and guanine. This model offered a satisfying explanation of how the DNA molecule replicates. One strand of the helix acts as a template, then a new strand of DNA is synthesized as the four bases bind to their complements on the original strand. The new model also suggested a possible mechanism for the production of proteins. As was shown in the 1960s, the sequence of bases in a DNA molecule acts as a code that dictates the order of amino acids in protein molecules.

Many science historians believe that Franklin's legacy also includes her success as a woman scientist. They claim she remained a devoted researcher—a model for other women scientists—despite working among men who were reluctant to accept her as their equal. The first publicly available suggestion of this reluctance appeared within the pages of *The Double Helix,* Watson's autobiographical account of his and Crick's discovery, published in 1968. The book paints an unflattering portrait of Franklin that has been declared by Franklin's biographers to be inaccurate, cruel, and the product of sexism.

In 1962, Watson, Crick, and Wilkins received the Nobel Prize for Physiology or Medicine for the discovery of DNA's structure. Nobel Prizes are not awarded posthumously, so it will never be known whether Franklin would have shared this honor had she been alive. Franklin's role in the determination of the structure of DNA has only recently been widely recognized.

Schuyler

For Further Reading:

Sayre, Anne. *Rosalind Franklin and DNA.* New York: Norton, 1975.

Watson, James D. *The Double Helix.* New York: Atheneum, 1968.

Fraser-Reid, Bertram

Synthesizer of
Organic Compounds
1934-

Life and Work

Bertram Fraser-Reid works today to find new ways to synthesize complex organic compounds. His discoveries have numerous applications in areas such as medicine and organic chemistry.

Fraser-Reid was born on the island of Jamaica in 1934. His father, principal of a local school, encouraged him to pursue a career in education. Fraser-Reid taught for five years before meeting a new teacher who introduced him to chemistry. Fraser-Reid was enthralled with the science. In 1956 he decided to leave the island to attend Queen's College in Ontario, Canada. Quickly attaining a bachelor's degree in science, Fraser-Reid joined the research lab of a professor at the school. Here he explored the chemistry of carbohydrates. To complete his doctoral work, Fraser-Reid worked in another lab at the University of Alberta that used nuclear magnetic resonance (NMR) to study the structure of complex carbohydrates.

After a stint in England Fraser-Reid took a position at the University of Waterloo in Canada in 1966. There he began to break new ground in the science of carbohydrates.

His various research experiences led him to look for and find connections between sugars and other carbon compounds. He eventually was able to manufacture non-carbohydrate compounds from simple carbohydrates (sugars). This was a real breakthrough because it reduced the reliance on petroleum as the raw material for modern chemistry. He published his findings in 1975. Fraser-Reid also developed chemicals that were tested for their medical benefits and investigated beetle pheromones looking for a use in pest control.

Fraser-Reid moved to the United States in 1980 to work at the University of Maryland. In 1983 he accepted a research professor chair at Duke University where he continued to seek ways to create compounds from complex carbohydrates.

In 1988, he invented a process to manufacture a complex carbohydrate that regulates certain reactions in human cells. Its usefulness is still being explored.

Fraser-Reid has researched complex carbohydrates, called oligosaccharides, that occur on viruses and human cell membranes; in 1988 he patented a process to manufacture them. He continues his research in organic chemistry, making unique contributions to the technology of renewable materials.

Legacy

Fraser-Reid's methods of altering complex carbohydrates to synthesize other organic compounds have applications in modern pharmaceuticals, plastics, paints, fabrics, and many other products of chemical industries. His work promotes sugarcane and sugar beets as renewable resources that may supplement or replace the non-renewable chemical resources (coal, oil) on which we now depend.

Another benefit of Fraser-Reid's work is that uniform, pure batches of chemicals can be predictably synthesized. This is particularly true of mirror forms of organic compounds. Even simple carbon compounds (like the sugars glucose and fructose) having the same molecular composition may occur in two forms because some of their atoms are positioned differently, like mirror images or thumbs on hands. One form of such a compound may be useful while the other is inert or even toxic. Often both forms are produced when petroleum is the raw material, and then they must be separated; such impurities do not result when sugar is the base for synthesis.

His recent work on complex carbohydrates occurring on viruses and human cell membranes may lead to techniques for interfering with undesirable interactions at the cellular level (such as AIDS and cancer).

Although it is too soon to assess Fraser-Reid's full legacy, it is not difficult to foresee that sugar will have increasing value as a renewable resource as petroleum stocks dwindle. His provocative work on the physiology of cell membranes may also lead others to health-related discoveries in carbohydrate chemistry. Fraser-Reid's 250 research publications are major contributions to the field and inspiration for the young scientists he has tried to challenge and encourage.

Wilson

WORLD EVENTS		FRASER-REID'S LIFE
	1934	Bertram Fraser-Reid born
World War II 1939-45		
Korean War 1950-53		
	1956	Fraser-Reid enrolls in Queen's College in Canada to study chemistry
	1961	Fraser-Reid begins using nuclear magnetic resonance to study carbohydrates
	1966	Fraser-Reid joins faculty at University of Waterloo in Canada
End of Vietnam War	1975	Fraser-Reid publishes method to synthesize non-carbohydrate compounds
	1983	Fraser-Reid joins Duke University chemistry department
	1988	Fraser-Reid patents process to make oligosaccharides
Dissolution of Soviet Union	1991	

For Further Reading:
Kessler, J. H., J. S. Kidd, and K. A. Morin. *Distinguished African-American Scientists of the 20th Century.* Phoenix, Ariz.: Oryx Press, 1996.
Yount, Lisa. *Black Scientists.* New York: Facts On File, 1991.

Freud, Sigmund

Founder of Psychoanalysis
1856–1939

Life and Work

Sigmund Freud developed many of the theories upon which the principles of psychoanalysis are based.

Freud was born to a Jewish family on May 6, 1856, in Freiberg, Habsburg Empire (now Příbor, Czech Republic). The Freuds moved first to Leipzig to escape anti-Semitic riots, and then to Vienna in 1860. In his third year of medicine at the University of Vienna, he began to investigate the central nervous system, a subject he found so engrossing that his other courses suffered and he graduated three years late in 1881.

In 1885 Freud went to Paris to study under the neurologist Jean Charcot, who was treating hysteria using hypnosis. Experience with Charcot steered Freud's interests toward psychopathology, the study and treatment of the abnormal functioning of the mind. In 1886, upon his return to Vienna, he opened a private practice specializing in nervous disorders. In 1896 he coined the term "psychoanalysis" for the clinical study of mental states.

Freud believed that the mind has mechanisms of repression and resistance: repression makes inaccessible the memory of painful or threatening events, and resistance makes the mind unaware of this repression. His first psychoanalytic work, *Studies on Hysteria*, published in 1895, described hysteria as the manifestation of undischarged emotional energy associated with repressed psychic trauma. Freud used hypnosis to help a patient recall a traumatic experience and thereby release the damaging energy. However, he soon abandoned the use of hypnosis and adopted instead the technique of free association, in which the patient is encouraged to recount the flow of spontaneous thought.

From 1895 to 1900 Freud analyzed patients' dreams in free-association sessions. His analyses led him to posit that a person's sexuality begins at infancy and develops in stages throughout life. He formulated the idea of the Oedipus complex: the development of sexual attachment to the parent of the other sex and hostile feelings toward the parent of the same sex. Expanding these claims, he developed the theory of transference, stating that the emotional attitudes children harbor toward their parents are transferred in later life to other people. In 1900 Freud completed *The Interpretation of Dreams*, a work outlining the fundamental concepts underlying psychoanalytic theory.

Freud was appointed full professor at the University of Vienna in 1902. However, the medical community was antagonistic toward his work, and when his next two treatises appeared—*The Psychopathology of Everyday Life* in 1904, and *Three Contributions to Sexual Theory* in 1905—this hostility increased.

Freud's final model of the mind, presented in 1923, involved the components he called "Id," "Ego," and "Superego." The Id is the unconscious mind, the source of passions and instincts. The Ego is the agency that controls behavior by considering both the immediate social situation and the instinctual impulse of the Id. The Superego, formed by the Oedipus complex, inhibits Oedipal desires and enforces moral behavior.

In 1938 Nazi forces occupied Austria, and Freud escaped with his wife and children to London, where he died on September 23, 1939.

Legacy

Freud created a new approach to studying and understanding the human mind, and because of his work, Western culture has embraced theories about human psychology and interpersonal and social interactions that pre-Freudian culture lacked.

Despite the hostility to psychoanalysis of many of his contemporaries, Freud attracted numerous followers who became influential thinkers in their own right. Two of his disciples, Alfred Adler and Carl

Jung, eventually each developed their own theoretical basis for psychiatric inquiry and launched the first neo-Freudian schools of psychoanalysis.

Freud's emphasis on the unconscious mind affected the development of medical psychology. Slowly physicians began to accept that the state of people's unconscious can have a major impact on their physical health, and that some illnesses—neurological and physiological—have psychological origins.

Freud's clinical practice inaugurated the fields of psychiatry and therapy, and he formulated a psychoanalytical doctrine that remains prevalent in modern society. Clinical psychiatric analysis and therapeutic counseling are commonly used services. Many therapists use Freud's free-association technique, and many therapists adhere to his theories of repression, resistance, and transference.

The models of feeling, thought, and behavior described by Freud have influenced many other disciplines and much of contemporary Western culture. Freud's concern with analysis of dreams, for instance, fortified the surrealist trend in modern art and his ideas have even found their way into the popular film industry—from serious psycho-thrillers such as *Spellbound* (1945) and *Psycho* (1960) to the light-hearted contemporary films of Woody Allen.

Many contemporary critics take issue with some of Freud's ideas. For instance, feminist scholar Teresa de Lauretis critiques Freud's ideas of feminine sexuality in *The Practice of Love*. Nevertheless, Freud's work is of such seminal importance to modern Western thought that cultural theorists and historians cannot avoid or ignore his influence, and Freudian scholarship will likely continue to be conducted by both those who applaud him and those who challenge his conclusions.

Schuyler

World Events		Freud's Life
	1856	Sigmund Freud is born
Germany is united	1871	
	1886	Freud opens private psychiatric practice
	1895	*Studies on Hysteria* is published
Spanish-American War	1898	
	1900	*The Interpretation of Dreams*, outlining psychoanalytic theory, is completed
	1902	Freud is appointed full professor at University of Vienna
World War I 1914-18		
	1923	Freud presents theory of Id, Ego and Superego
	1938	Freud escapes to England
	1939	Freud dies
World War II 1939-45		

For Further Reading:
Jones, Ernest. *The Life and Work of Sigmund Freud.* New York: Basic Books, 1981.
Neimark, Anne E. *Sigmund Freud: The World Within.* New York: Oxford University Press, 1997.

Fuchs, Leonhard

Developer of Foundations
of Botany
1501-1566

Life and Work

Leonhard Fuchs wrote the influential botanical work, *The Natural History of Plants,* which contains accurate drawings and detailed descriptions of about 500 plants, their pharmaceutical uses, and a glossary of botanical terms.

Fuchs was born at Wemding, Germany, on January 17, 1501. Little is known about his childhood. He attended the University of Erfurt, received a bachelor's degree, joined the faculty, and opened a private school at Wemding, all while he was still in his teens. In 1519 he entered the University of Ingolstadt, soon earning a master's degree in classics and, in 1524, a degree in medicine.

After receiving his medical degree, Fuchs moved to Munich, where the political environment was favorable for Protestants (Fuchs was brought up Catholic, but had become Protestant under the influence of Martin Luther). In Munich he practiced as a physician, but, by 1526, he had returned to Ingolstadt to teach medicine.

Fuchs became private physician to the Margrave of Brandenburg in 1528. Finally, in 1535, he began teaching medicine at the University of Tübingen, which would remain his residence for the rest of his life.

At Tübingen Fuchs began studying botany in earnest, and, in 1542, his masterpiece, *The Natural History of Plants,* was published. The work, which was written as a medicinal handbook, includes precise descriptions of more than 500 native German and foreign plants, many illustrated with woodcuts. Fuchs organized the plants alphabetically and gave for each a description of its form, habitat, powers of healing, and the season in which it is best collected. He also included a glossary of botanical terms, an attempt to establish systematic nomenclature for the science of botany.

The book made Fuchs famous throughout Europe; in 1548, the powerful Duke Cosimo de Medici of Florence asked him to become director of the botanical gardens at Pisa. Fuchs refused the invitation.

Fuchs died at Tübingen on May 10, 1566.

Legacy

Fuchs's illustrated herbal text played a key role in the introduction of accurate drawings and detailed descriptions to the study of plants. The work also helped lay the foundation for the emergence of botany as a science distinct from the field of medicine.

The Natural History of Plants included unique elements that set the standard for further botanical studies. Botanists began to include in their works the type of organized presentation, accuracy, and beautiful illustrations found in Fuchs's book. The book was among numerous botanical handbooks that led to the acceptance of botany as a field of study in its own right. Although many of those books were intended as field guides for the collection of medicinal herbs, they were immensely popular among non-physicians. In the seventeenth century, botanical gardens and specialized lecturers in botany emerged.

Fuchs continued a conservative, descriptive trend in the study of plants, not introducing any new, revolutionary biological discoveries. Similarly, for the next century, botanists focused primarily on the detailed description of plant anatomy and spent relatively little time on other aspects of plant biology. Unlike Fuchs, many botanists did, however, identify and describe newly discovered plants, particularly those from the the Americas, recently discovered by European explorers.

Along with Otto Brunfels and Hieronymous Bock, two other prominent sixteenth-century botanists, many modern scholars have designated Fuchs one of the "German Fathers of Botany."

Fuchs's legacy also endures in the name of a flowering-plant genus, Fuchsia. Members of Fuschia belong to the evening primrose family and usually sport flowers of deep pinks, reds, and purples.

Schuyler

WORLD EVENTS		FUCHS'S LIFE
Columbus discovers Americas	1492	
	1501	Leonhard Fuchs is born
Reformation begins	1517	
	1519	Fuchs enters University of Ingolstadt
	1524	Fuchs earns medical degree
	1526	Fuchs becomes professor of medicine at Ingolstadt
	1535	Fuchs begins teaching medicine at University of Tübingen
	1542	*The Natural History of Plants* is published
	1548	Fuchs invited to be director of Pisa's botanical gardens, but stays in Germany
	1566	Fuchs dies
Thirty Years' War in Europe	1618-48	

For Further Reading:
Ronan, Colin A. "From Renaissance to Scientific Revolution." In *Science: Its History and Development Among the World Cultures.* New York: Facts On File, 1982.
Sachs, Julius. *History of Botany, 1530-1860.* Authorized translation by Henry E. F. Garnsey. New York: Russell & Russell, 1967.

Galen of Pergamum

Leading Physician of Antiquity
129–c.200

Life and Work

Galen of Pergamum was the most accomplished physician of the Roman Empire, and his ideas formed the basis of Western medical practice for over a millennium. He conducted extensive experimental research into the anatomy and physiology of animals and wrote prolifically about his medical methods and beliefs.

Galen was born in 129 to Greek parents living in Pergamum, Asia Minor (now in western Turkey). Pergamum had a shrine to Asclepius, the god of healing, and Galen was educated at the adjoining medical school. The high priest at Pergamum kept a troop of gladiators, which provided students with the opportunity to examine wounds and observe diseases.

In the 150s, Galen traveled to Smyrna, Corinth, and Alexandria, where he practiced animal dissection and met with numerous physicians. He dissected mostly goats, pigs, and monkeys, and from these he made inferences concerning human anatomy.

In 157, he returned to Pergamum and became the chief physician for the gladiators. The position enabled him to test wound treatments and increase his knowledge of anatomy.

Galen made great leaps in the study of animal biology. He identified cranial nerves, described the valves of the heart, and distinguished structurally between arteries and veins. He experimented on the spinal cord and showed how injury at various levels affects the nervous system. Through vivisection experiments, he demonstrated kidney and bladder functions and showed that the arteries carry blood. He noted the role of the liver in the vascular system and described the structure of bones and muscles. He believed the Hippocratic doctrine stating that health requires a proper balance among the four bodily humors, or bodily fluids: phlegm, black bile, yellow bile, and blood. While this doctrine was later proven incorrect, Galen's modification of it—in particular, his suggestion that imbalances in the humors can focus on specific organs—permitted more targeted diagnoses and cures.

In 161, Galen relocated to Rome, where he developed a reputation as a successful physician. He held lectures and performed public dissections, awing the Romans and creating envious colleagues. His fame earned him an appointment to the court of Emperors Marcus Aurelius and Lucius Verus.

He practiced medicine and wrote books until his death in about 200.

Legacy

Galen greatly advanced the understanding of human anatomy and physiology, providing a base for the evolution of modern medicine. His opinions and experimental results dominated medical curricula in Europe for 15 centuries.

Galen's immediate impact consisted of his physiological experimentation, which was without equal among his contemporaries. He opened new avenues of research and treatment by demonstrating the functions and internal structures of many parts of the body. Because of the fame his experiments brought, his medical theory and practice set the standard for Roman medicine.

Galen's medical theories were taught at the intellectual centers of Alexandria and Byzantium during the first few centuries of the Christian era. Galen supported many ideas from the Hippocratic tradition, and with his dominance in medical education, Hippocratic dogma traveled forward into modern European thought.

Galen's legacy survived in the treatises he produced. By the seventh century, Byzantine Christian influence in Persia had assured that many Greek manuscripts remained intact, translated into the Syriac language. This trend continued into the ninth century, when Arab scholars collected and translated Galen's works. In the following two centuries, Europeans translated Galen into Latin. During the fifteenth and sixteenth centuries, as Europeans hosted a revival of Greek and Roman culture and knowledge, industrious physicians attempted to repeat Galen's experiments and observe his results for themselves.

Galen advocated the erroneous theory of the four bodily humors, derived from the Hippocratic Collection (a compilation of HIPPOCRATES' teachings), as well as misconceptions of his own, such as his idea that the blood carried pneuma, or life spirit. These misconceptions limited medical understanding and practice through the many years of his dominance, and it was not until the 1500s, with the work of such physicians and scientists as ANDREAS VESALIUS and PARACELSUS, that such misconceptions began to be corrected.

Schuyler

World Events		Galen's Life
Roman Empire conquers Armenia and Mesopotamia	117	
	129	Galen of Pergamum is born
	150s	Galen travels and learns dissection technique
	157	Galen becomes chief physician of Pergamum gladiators
	160s	Galen is appointed to courts of two Roman Emperors
	161	Galen moves to Rome
	c. 200	Galen dies
Last Severan emperor of Rome is killed; disorder ensues	235	

For Further Reading:

Brain, Peter. *Galen on Bloodletting.* Cambridge: Cambridge University Press, 1986.

Gilbreath, Allan. *Galen.* Memphis, Tenn.: Ronin Enterprises, 1997.

Sarton, George. *Galen of Pergamum.* Lawrence, Kans.: University of Kansas Press, 1954.

Galilei, Galileo

Pioneer in Laws of Motion and
Early Telescopy
1564-1642

Life and Work

In Galileo's time most scientists believed that the Earth and the sky were different and had their own laws and patterns. Galileo made great discoveries in both areas and helped show that the same laws apply throughout the universe.

Born in Pisa, Italy, on February 15, 1564, Galileo was the son of a talented and well-respected musician. At 17 he enrolled at the University of Pisa to study medicine but soon took a greater interest in mathematics and physics. One day during a service in the Pisa cathedral he used his pulse to show that the time it took the great chandeliers hanging from the ceiling to swing back and forth was always the same. His careful experiments with pendulums of lead pellets attached to strings confirmed that the time it takes an ideal pendulum to swing depends on its length, not on its mass or the size of the arc through which it moves.

By age 25 he was appointed to a mathematics professorship at Pisa. At about that time he began experiments to refute the work of Aristotle, the ancient Greek philosopher whose ideas still influenced science in Galileo's time. Aristotle wrote that objects fall because their natural place is in the Earth, and the heavier they were, the faster they would fall. Galileo realized that anything he dropped would fall too quickly

to measure, so he conducted his experiments by rolling balls down long, gently sloping ramps. He reasoned, correctly, that the ramp would offset the effects of gravity without changing the overall result. He was not surprised to find that lightweight balls rolled just as fast as heavy ones. His greatest discovery, he felt, was that however far any ball rolled in one second, it would move three times farther in the next second, five times farther in the next, and so on. He was the first to prove that objects accelerate as they fall.

By 1609 he was ready to publish his results on motion when his work was interrupted: a Dutch optician had combined two lenses in a way that made distant objects appear larger. Galileo quickly improved the design and apparently was the first to turn his telescope to the night sky. Everywhere he looked he saw thousands of stars that no one had ever seen before. "But the greatest marvel of all," he wrote in *The Starry Messenger* in 1610, "is the discovery of four new planets. I have observed that they move around the Sun." He was referring to four large moons that orbit Jupiter (now called the Galilean moons), which he discovered in 1610. His observations convinced him that COPERNICUS's Sun-centered vision of the solar system was correct, that the Earth was another planet in motion around the Sun.

His conclusions conflicted with the official view supported by the Roman Catholic Church, and in 1616 Galileo was forbidden to teach or promote the Copernican theory. In 1632 he broke his silence and published a book discussing the two conflicting theories of the solar system, for which he was punished by house arrest in 1633 until his death. In 1638 Galileo smuggled out of the house the manuscript for a new book on the mechanics of motion, summarizing his work from before 1609 and after his arrest. In it he described his discovery of the constancy of pendulum motion and claimed, for the first time, that moving objects tend to keep moving unless an outside force acts on them.

Galileo died on January 8, 1642, in Artreci, Italy.

Legacy

Galileo's legacy extends beyond his great discoveries. He changed forever the way science is practiced. Until his time natural philosophers focused on why things happen. Instead Galileo considered how things happen. He strove

to uncover the simple patterns that describe the motion of falling objects. Ever since Galileo, scientists have separated "how" from "why," making hypotheses to explain why something happens only after they can accurately describe how it happens.

Galileo's work on the mechanics of motion was taken up after his death by the Dutch mathematician CHRISTIAAN HUYGENS. Improving and refining Galileo's work, Huygens developed the first practical pendulum clock in 1659. Galileo's conclusion regarding the tendency of objects to keep moving unless altered by an outside force came to be known as the law of inertia, and was adopted by ISAAC NEWTON as the first law of motion and incorporated into his comprehensive theory of motion outlined in *Mathematical Principles of Natural Philosophy* (1687), also known as the *Principia*.

Galileo was the first scientist to realize the value of abstracting from nature to the laboratory. Aristotle and all others before Galileo insisted on studying falling motion by looking at actual objects in nature, such as leaves and snowflakes. Galileo's genius was to realize that those motions are complicated and altered by friction, and that more would be learned about the essential process by working in the lab to reduce the effects of friction.

Secaur

WORLD EVENTS		GALILEO'S LIFE
Reformation begins	1517	
	1564	Galileo Galilei is born
	1609	Galileo formulates laws of pendulum motion and falling motion
	1610	Galileo discovers moons of Jupiter, and makes other telescopic discoveries
	1616	Galileo ordered by Roman Catholic Church not to support Copernican theory
Thirty Years' War in Europe	1618-48	
	1632	Galileo publishes book on two theories of solar system
	1633	Galileo tried by church, sentenced to house arrest for life
	1638	Galileo publishes book on motion and mechanics after smuggling it out of Italy
	1642	Galileo dies

For Further Reading:

Gindikin, S. *Tales of Physicists and Mathematicians*. Cambridge, Mass.: Birkhauser, 1988.

Kaplan, Joseph. *Great Men of Physics: The Humanistic Element of Scientific Work*. Alhambra, Calif.: Tinnon-Brown, 1969.

Sis, Peter. *Galileo*. New York: Knopf, 1996.

Gates, Bill

Computer Programmer;
Founder of Microsoft Corporation
1955-

WORLD EVENTS		GATES'S LIFE
Korean War 1950-53		
	1955	William Gates is born
End of Vietnam War	1975	Gates quits Harvard and co-founds Microsoft Corporation with Paul Allen
	1980	Gates designs operating system called MS-DOS
	1984	Use of MS-DOS has become standard
	1980s	Gates develops application software
	1986	Gates offers Microsoft stock to the public and becomes billionaire
	1990	Microsoft introduces Windows 3.0
Dissolution of Soviet Union	1991	
	1995	Windows 95 is released
	1997	Microsoft is world's largest computer software company

Life and Work

William Henry Gates III, popularly known as Bill Gates, co-founded Microsoft Corporation with Paul Allen and developed personal computer software that revolutionized computing capabilities and shaped modern personal computer systems.

Gates was born in Seattle, Washington, on October 28, 1955. As a teenager he became interested in the new field of computer science. In high school he developed proficiency in computer programming, helped computerize local power grids, and established Traf-O-Data, a company that used computers to analyze local traffic patterns.

While attending Harvard, Gates read a magazine article about the development of a microcomputer called the Altair and decided he wanted to be at the forefront of computer software development. Dropping out of Harvard in 1975, he and friend Paul Allen founded Microsoft Corporation in Redmond, Washington. They devised an interpreter that enabled the Altair and other early microcomputers to run the computer language BASIC. In 1980 International Business Machines (IBM) asked Gates to design the operating system for its new personal computer; he formulated the Microsoft Disk Operating System (MS-DOS).

Gates subsequently convinced IBM to release the specifications of its personal computer (PC) so others could write programs for the new machine, and he also firmly established MS-DOS as the standard operating system. As a result, by 1984 more than two million copies of MS-DOS had been sold, making it the leading PC operating system in the country.

In the mid-1980s Gates designed application software, such as word processing and financial spreadsheet programs. He also began to promote CD-ROMs, PC-compatible, "read-only" disks with tremendous storage capacity.

Microsoft introduced Windows 3.0 in 1990, enabling PCs to be operated by a handheld "mouse" and on-screen symbols rather than keyboard commands. By 1993 Gates's company was selling one million copies of Windows per month.

In August of 1995 Microsoft released Windows 95, an updated version of the Windows operating system. Within four days of its release, it had sold one million copies. Except for Macintosh personal computers, which employ their own proprietary operating system, nearly all personal computers now come installed with the Microsoft Windows operating system.

In 1997 and 1998 Microsoft has undergone increased scrutiny from the United States Department of Justice (DOJ) concerning alleged monopolistic practices relating to the licensing of its software to hardware manufacturers. In particular, the DOJ has tried to put legal limitations on Microsoft's effort to tie the use of its Windows operating system to that of its Internet browser Microsoft Explorer.

By offering Microsoft stock to the public in 1986, Gates became a billionaire at age 31. Gates is chairman and chief executive officer of Microsoft Corporation, which in 1997 was the largest computer software supplier in the world.

Legacy

When Gates and Allen founded Microsoft Corporation, they hoped to put "a computer on every desk and in every home." Although Gates has not yet seen this goal fulfilled, the widespread use of PCs attests to his software engineering ingenuity and acute business sense. PCs have revolutionized the information-related aspects of business, education, finance, industry, and science.

Gates left a lasting legacy when he developed MS-DOS and convinced computer companies to standardize it as the operating system for PCs. As a result, most PC software is interchangeable among a wide variety of machines, facilitating the exchange of information. This compatibility also opens the use of PCs to more people.

Microsoft's applications dominate word processing and spreadsheet software. The company's Microsoft Word and Excel lead yearly sales of such programs. The Windows operating system enjoys widespread use both in the United States and abroad.

Gates's company has become the giant of computer software companies, setting standards and shaping the development of the industry.

Schuyler

For Further Reading:

Gates, Bill. *The Road Ahead.* New York: Penguin Books, 1996.

Ichbiah, David, and Susan L. Knepper. *The Making of Microsoft.* Rocklin, Calif.: Prima, 1991.

Manes, Stephen, and Paul Andrews. *Gates.* New York: Doubleday, 1993.

Wallace, James, and Jim Erickson. *Hard Drive.* New York: John Wiley, 1992.

Gauss, Carl Friedrich

Multi-faceted Mathematician;
Developer of
Complex Number System
1777-1855

Life and Work

Carl Friedrich Gauss was one of the greatest mathematicians of the nineteenth century. While other mathematicians were beginning to specialize, Gauss's interests spanned all of mathematics, both theoretical and applied, and his work is described as the transition from eighteenth century mathematics to that of the present day.

Gauss, the son of a mason, was born on April 30, 1777, in Brunswick, Germany. At age 10 Gauss summed 100 terms in an arithmetic series moments after his instructor wrote the problem on the board. His instructor, realizing the boy's abilities, brought him to a tutor, Johann Bartels, who befriended Gauss, and together they studied mathematics.

Bartels introduced Gauss to Ferdinand, the Duke of Brunswick. Until his death in 1806 the Duke supported Gauss, making possible Gauss's single-minded pursuit of mathematics. At age 15, in 1792, Gauss entered Brunswick's Caroline College. While there, he completed, but did not publish, theorems in the calculus that later would be independently constructed by AUGUSTIN-LOUIS CAUCHY and Karl Weierstrass. He also discovered non-Euclidean geometry, but buried his ideas for decades in a private notebook.

Gauss entered the University of Göttingen in 1795 and concentrated on mathematics. He discovered that a regular 17-sided polygon could be constructed using only a straight edge and compass, and devised formulas for constructing such polygons. He was the first mathematician to further EUCLID's work on this problem.

Gauss's doctoral thesis, completed through the University of Helmstadt in 1799, contained a proof of the fundamental theorem of algebra (he published others later), which said that any equation containing complex coefficients must have a complex solution. His first major book, *Disquisitiones Arithmeticae* (1801), is one of the most important in all of mathematics: it discusses concepts in number theory, including the properties of integers and complex numbers. Complex numbers contain both real and imaginary numbers, real numbers being values that can be expressed as a terminating or non-terminating decimal, and imaginary numbers being values that are the square roots of negative numbers.

Gauss's work on number theory was extensive. He tried to expand the real number system. He argued for including imaginary and negative quantities, and he showed how they could be plotted on a graph with x and y coordinates. When real and imaginary numbers converge to determine a point, Gauss combined them as separate complex numbers. All other numbers can be derived from Gauss's complex numbers.

In 1807 Gauss became director of the Göttingen Observatory, where he remained for life. He became involved with problems in astronomy and geodesy, the study of the size and curvature of Earth's surface. He determined the orbit of the asteroid Ceres, explaining his methods in *Theory of Motion of Heavenly Bodies* (1809). His interests in mapmaking and shortest paths led to work in differential geometry, including publication of *General Investigations of Curved Surfaces* (1827). In 1833 he invented an electric telegraph, using it to communicate with collaborating scientist Wilhelm Weber.

Gauss died at age 78 on February 23, 1855, in Göttingen, Germany.

Legacy

The scope and usefulness of mathematics became fully realized in the nineteenth century, and Gauss led the way. He closed gaps in each of its traditional branches: geometry, algebra,

the calculus, and number theory.

One of Gauss's most important contributions was his interpretation of complex numbers: it made the number system sufficient to solve any algebraic equation. Completion of the number system parallels the development of the periodic table in chemistry; it was a landmark achievement that alone would have made Gauss famous and the work of future mathematicians easier.

Gauss's statistical work was adopted by biologists, physicists, astronomers, and social scientists. For instance, Lambert Quetelet (1796-1874) applied the Gaussian method to census data of 1829 to support his inference that better social and economic conditions would reduce crime. With minor modifications Gauss's method of least squares has been adapted to modern computers, and the normal distribution of measurements is called "Gaussian."

Gauss challenged two succeeding generations of mathematicians. He personally selected the doctoral topic for Bernhard Riemann, who proposed a non-Euclidean geometry of curved space that Einstein later claimed as a physical reality. Other Gaussian heirs were Augustin Cauchy, Karl Weierstrass, and his student Georg Cantor. These mathematicians helped systematize contemporary mathematics.

Steinberg

WORLD EVENTS		GAUSS'S LIFE
	1777	Carl Friedrich Gauss is born
French Revolution	1789	
	1792	Gauss enters Caroline College in Brunswick
	1795	Gauss enters University of Göttingen
	1796	Gauss proves 17-sided polygon can be constructed using straight edge and compass
	1801	Gauss publishes *Disquisitiones Arithmeticae*
Napoleonic Wars in Europe	1803-15	
	1807	Gauss appointed director of Göttingen Observatory
	1809	Gauss publishes *Theory of Motion of Heavenly Bodies*
	1827	Gauss publishes *General Investigations of Curved Surfaces*
	1833	Gauss and Weber invent electric telegraph
	1855	Gauss dies

For Further Reading:
Bell, E. T. "The Prince of Mathematics." In *The World of Mathematics,* vol. I. New York: Simon & Schuster, 1956.
Dunham, William. *Journey through Genius.* New York: John Wiley, 1990.
Kline, Morris. *Mathematical Thought from Ancient to Modern Times.* New York: Oxford University Press, 1972.
Muir, Jane. *Of Men and Numbers.* New York: Dodd, Mead, 1961.

Gay-Lussac, Joseph-Louis

Discoverer of Law of
Combining Volumes
1778-1850

Life and Work

Joseph-Louis Gay-Lussac excelled as both a chemist and a physicist. His most important contribution was his articulation of key laws on the combination and expansion of gases.

Born in France on December 6, 1778, Gay-Lussac had the advantage of a good education. His family sent him to the École Polytechnique at age 14. Gay-Lussac came to the attention of the famous chemist Claude Berthollet in 1800 and went to work in his lab. Throughout his career Gay-Lussac collaborated with many different scientists. He experimented both in the laboratory and in the field and worked as a professor of chemistry and physics at various universities in France.

Gay-Lussac's work drew attention almost immediately. His high altitude hot-air balloon flights in 1804, conducted to study the composition of the atmosphere and the Earth's magnetic field, became famous.

World Events		Gay-Lussac's Life
United States independence	1776	
	1778	Joseph-Louis Gay-Lussac is born
French Revolution	1789	
	1802	Gay-Lussac publishes gas expansion law
Napoleonic Wars in Europe	1803-15	
	1804	Gay-Lussac conducts experiments on atmospheric gases in hot-air balloon ascents
	1808	Gay-Lussac proposes Law of Combining Volumes; isolates boron with Louis-Jacques Thenard
	1815	Gay-Lussac discovers cyanogen
	1850	Gay-Lussac dies
United States Civil War	1861-65	

Gay-Lussac's hallmark was his quantitative studies of gases. His volumetric analysis techniques enabled him to make accurate measurements of gases. In 1802 he published a law of gas expansion, which stated that gases expand at a constant rate at the same temperature. (The law is now more commonly called Charles's Law since French chemist Jacques Charles discovered the relationship first, but published it after Gay-Lussac.) In 1808 he proposed another law concerning the behavior of gases, stating that gases combine in simple whole number proportions. This Law of Combining Volumes is the basis for understanding how molecules form and gases interact. For example, Gay-Lussac determined that one part carbon would readily interact with two parts oxygen to form carbon dioxide. Gay-Lussac also studied various elements, including chlorine, iodine, and oxygen; his work with chlorine involved an analysis of how light affected reactions of that element with hydrogen. With Louis-Jacques Thenard, he isolated boron in 1808. Gay-Lussac discovered cyanogen in 1815.

Much of his research attempted to solve practical problems. Gay-Lussac developed an accurate method to determine the purity of silver using a titration system. One of his inventions, an absorption tower, removed toxic gases from factories producing sulfuric acid. He also created a new type of candle and various types of laboratory equipment and methodologies including ways of eliminating water vapor from experiments.

As Gay-Lussac grew older, he focused more of his time and efforts on industrial uses of science. This benefited the citizens of France and better supported his growing family. Gay-Lussac died in Paris on May 9, 1850.

Legacy

Gay-Lussac's analytical work with gases and articulation of two key laws on the combination and expansion of gases provided scientists with insights into matter and its characteristics.

During his lifetime, Gay-Lussac mentored many students. These young scientists, such as the German chemist Justus von Liebig, went on to disseminate his techniques throughout Europe and made their own discoveries. Gay-Lussac's first gas law provided the groundwork for AMEDEO AVOGADRO to determine that all gases contain the same number of molecules at the same temperature and pressure, a principle now known as Avogadro's Law. It also helped LORD KELVIN develop the temperature scale that starts at absolute zero (the theoretically lowest possible temperature), which is now called the Kelvin scale. Gay-Lussac's discovery of cyanogen opened a new field of study, the cyanides.

Gay-Lussac's work immediately influenced the practical world. In 1832, his investigations of silver resulted in a standardization of silver purity used by the French Mint for the next 50 years. His new candle assisted the general populace on a daily basis by providing additional light and durability in warm weather.

Although many of his inventions and discoveries furthered the field of science during his generation, Gay-Lussac's contributions have endured into the twentieth century. His absorption towers that captured nitrogen oxides during the production of sulfuric acid were in use well into the 1900s. These devices prevented untold amounts of toxic fumes from entering the atmosphere. His method of titration and of eliminating water vapor from experiments are still in use today.

He also helped to provide a theoretical basis for important aspects of contemporary chemistry and physics. In particular his laws of gas expansion and combining volumes remain part of the fundamental theoretical basis in chemistry. Also, Gay-Lussac's measurements of how light affects the reaction of chlorine and hydrogen demonstrated the role of light energy in chemical reactions. This provided a foundation for photochemistry, which helped lead to a quantum theory of the transformation of energy and matter.

Wilson

For Further Reading:

Brock, William H. *The Norton History of Chemistry.* New York: Norton, 1992.

Crosland, Maurice. *Gay-Lussac: Scientist and Bourgeois.* New York: Cambridge University Press, 1978.

Gell-Mann, Murray

Developer of the Quark Model
1929-

Life and Work

Murray Gell-Mann, one of the greatest physicists of the late twentieth century, was the first to realize that protons and neutrons are made of smaller particles, which he called quarks.

Born in New York City on September 15, 1929, Gell-Mann entered Yale University at age 15 and earned his doctorate from the Massachusetts Institute of Technology in 1951. He taught briefly and worked with the great physicist ENRICO FERMI at the University of Chicago before becoming a professor of theoretical physics at the California Institute of Technology in 1955.

When Gell-Mann was young, most physicists believed that the proton, neutron, and electron were fundamental particles, unable to be reduced to something smaller. But during the 1950s scientists working at powerful new particle accelerators discovered more than 100 other particles, most with very short life spans, but with properties much like protons and neutrons.

Organizing all the new particles and making patterns of them was the challenge Gell-Mann took on. He discovered that all known subatomic particles could be grouped into eight families, with the proton and neutron as the least massive particles in one of those groups. He called the pattern the Eightfold Way, after the eight attributes of right living in Buddhist philosophy. Gell-Mann realized that there were holes in the pattern, some particles that had yet to be discovered. He predicted the properties and decay mechanisms of one of them, called the omega particle, and in 1964 it was discovered and shown to act in just the way he had described. He received the 1969 Nobel Prize for Physics for his discovery of these patterns that organize subatomic particles.

More importantly, in 1963 he suggested that the whole pattern could be reduced to just three new particles, which he called quarks, joined in combinations of twos and threes. A proton, for example, is made of two up quarks and one down quark, while a neutron is one up and two downs. The other relative in that same family is the sideways quark, which has since been renamed the strange quark. The names are all whimsical choices that reveal something of Gell-Mann's sense of humor. He went on to develop quantum chromodynamics, the theory of how the quarks interact and link together, through other particles called gluons.

Gell-Mann is as brilliant at language as he is at physics; he speaks more than a dozen languages fluently and often corrects the grammar and pronunciation of other speakers. He is an avid environmentalist and bird watcher; one colleague described him as "having his head in the clouds and his feet in the mud."

His current major interest is the interconnection between the simple and the complex in nature. In 1984 he co-founded the Santa Fe Institute dedicated to interdisciplinary research on complex adaptive systems, including advanced computers and living organisms.

Legacy

Gell-Mann's legacy rests on his innovative conception and model of the subatomic particles he called quarks, ideas that have influenced much of the current research in particle physics.

Gell-Mann's quark model seemed ridiculous at first, as it required particles with fractional charges. The up quark carries a charge of $+\frac{2}{3}$, as do the more recently discovered relatives, the charm and top quarks. The down, sideways (or strange), and bottom quarks each have a $-\frac{1}{3}$ charge.

The quark model quickly became accepted as the standard for research in particle physics. Scientists now seem certain that quarks exist, as recent discoveries have continued to confirm and expand Gell-Mann's original model. In 1974 scientists at the Stanford Linear Accelerator Center and the Brookhaven National Laboratory found a particle that could only be explained as containing a fourth quark, called charm, a massive relative of the up quark. Just three years later the Fermi National Accelerator Laboratory (Fermilab) was the site of the discovery of a fifth quark, called bottom, a heavy version of the strange and down quarks. Seventeen years of intensive research finally produced the ultra-massive top quark at Fermilab in 1994 and proved that it was a heavy version of the up and charm quarks.

Will experiments of the future continue to find more and more kinds of quarks? Perhaps, but experimental evidence suggests that there are only six, although current theory cannot explain why. Another problem relates to the masses of the quarks: a single top quark has as much mass as an entire atom of silver. Scientists well into the twenty-first century will wrestle with the difficult problem of explaining the odd pattern of quarks' masses.

Secaur

WORLD EVENTS	GELL-MANN'S LIFE
World War I 1914-18	
	1929 Murray Gell-Mann is born
World War II 1939-45	
Korean War 1950-53	
	1951 Gell-Mann earns doctoral degree from M.I.T.
	1955 Gell-Mann begins teaching at California Institute of Technology
	1963 Gell-Mann develops quark model of subatomic particles
	1969 Gell-Mann receives the Nobel Prize for Physics for work in particle physics
	1974-77 Research uncovers another two quarks in Gell-Mann's model
End of Vietnam War 1975	
	1984 Gell-Mann co-founds Santa Fe Institute
Dissolution of 1991 Soviet Union	
	1994 The top quark is added to Gell-Mann's model

For Further Reading:

Davies, P., ed. *The New Physics.* Cambridge, Mass.: Cambridge University Press, 1989.

Gell-Mann, M. *The Quark and the Jaguar: Adventures in the Simple and Complex.* New York: Freeman, 1994.

Lederman, Leon, and David Schramm. *From Quarks to the Cosmos: Tools of Discovery.* New York: Freeman, 1989.

Germain, Sophie

Pioneer in Number Theory and
Study of Elasticity
1776-1831

Life and Work

Sophie Germain's determined quest for understanding led her to new insights about number theory in mathematics and elasticity.

Germain was born on April 1, 1776, in Paris to a wealthy family. Her parents felt that women could be harmed by strenuous mental activities and tried to deny Germain an education. She persevered and read all she could from the family library. In doing so, she taught herself Latin and explored the writings of Greek mathematician ARCHIMEDES.

When the new school for sciences and math, École Polytechnique, was opened in Paris in 1794, Germain saw an educational opportunity. While women were not admitted to the institute, she found a way to obtain the lecture notes from the courses and, once again, taught herself. She became very interested in the work done by Joseph Lagrange at the school. She submitted a paper to him under the pseudonym M. LeBlanc as she believed that he would not read a manuscript written by a woman. Thus began her real education in mathematics.

Lagrange was impressed by M. LeBlanc's paper and they began a correspondence. She eventually confided her true gender and, much to his credit, Lagrange continued to support her completely. He met with her at her home and introduced her to the writings of many eminent mathematicians and scientists. Germain closely followed the work of CARL FRIEDRICH GAUSS and eventually wrote to him about her work in number theory. Their correspondence continued until her death.

Germain made significant progress on two puzzles that had confounded others for over a century. One was Fermat's last theorem, which states that no whole-number solutions exist for the equation $x^n + y^n = z^n$ when n is an integer greater than 2. Utilizing her work in abstract number theory, she developed what came to be called Germain's Theorem, which expanded the solution for Fermat's last theorem by creating a proof for primes less than 100.

The other puzzle she helped to solve was in the field of applied mathematics. The Institut de France established a contest seeking a theory about the elasticity of materials. Germain won the contest when it was offered the third

time in 1816. She reported her ideas to the institute in a paper called "Memoir on the Vibrations of Elastic Plates." Devised in response to the unexplained observations of Ernst Chladni, the German physicist who noted the symmetrical patterns sawdust would develop when lying on vibrating metal plates, the theorem described the predictable movement of particles due to vibration. The completed paper was published five years later in 1821.

Germain continued her work in the field of elasticity until her death in 1831. Her correspondence with others enabled her to gain the education denied her through traditional means. Gauss remained a staunch supporter of Germain. He had even convinced the University of Göttingen to present Germain with an honorary degree. Sadly, Germain died before it could be conferred.

Legacy

Germain's contributions to number theory allowed for new theories in both abstract and applied mathematics. Her unconventional approach earned her a citation by mathematics historian E. T. Bell as "the most creative algebraist in the world."

During her lifetime, Germain's work on elasticity enabled others to study the elastic properties of various materials. It has been applied since to studies of response patterns in solids that vibrate under stress, such as tall buildings swaying in the wind or responding to the stress of heavy traffic and the production of characteristic sounds in musical instruments. One specific example of this application is the Eiffel Tower in Paris, completed in 1889, which was designed using her ideas on elasticity of materials. Connections she made between mathematics and applied areas such as acoustics and material science continue to be of use today.

Germain's contributions to number theory remained important for many years. Her partial proof of Fermat's last theorem provided a major stepping stone toward its eventual solution. Mathematician Andrew Wiles of Princeton University finally derived a successful proof in 1995.

Wilson

For Further Reading:

Grinstein, Louise, and Paul Campbell. *Women of Mathematics.* New York: Greenwood Press, 1987.

Perl, Teri. *Math Equals.* Reading, Mass.: Addison-Wesley, 1978.

Singh, Simon. *Fermat's Enigma: The Quest to Solve the World's Greatest Mathematical Problem.* New York: Walker, 1997.

Goddard, Robert H.

Pioneer in Early Rocket Design
1882-1945

Life and Work

Robert H. Goddard conducted the first extensive rocket-design research in the United States. His numerous innovations contributed to the development of theoretical rocketry and rocket-powered spacecraft.

Goddard was born on October 5, 1882, in Worcester, Massachusetts, to a family that encouraged his early interest in tinkering. As a teenager, his imagination was fueled by H. G. Wells's science-fiction novel *The War of the Worlds,* and he began to dream of creating a machine capable of space flight. In 1899 he decided to devote himself to this pursuit.

Even while a student, Goddard considered ways of reaching the moon and investigated potential rocket fuels. Goddard received a doctorate in physics from Clark University, Worcester, Massachusetts, in 1911 and began teaching there in 1914. By 1915 he had obtained patents for the designs of a two-stage powder rocket, a cartridge-loading rocket, and a liquid-fuel-burning rocket. His experimental results indicated that use of liquid oxygen and liquid hydrogen would propel rockets with greater energy efficiency and higher exhaust velocity than previous models. He was the first to fire a rocket in a vacuum, which demonstrated that rocket engines could operate in the upper atmosphere.

Working for the United States military during World War I, Goddard devised single-charge and multiple-charge recoilless rockets. In 1919 he published a paper that outlined his research results and rocketry theory in general. The paper was widely misinterpreted because the media presented it as a fanciful discussion of a theoretical "moon rocket," and Goddard, embarrassed, shunned publicity thereafter.

In 1921, with funds from the Smithsonian Institution, Goddard focused his experimental work on rockets using liquid propellants and devised fuel pumps and special combustion chambers to employ this promising fuel source. After several failures, he successfully launched a rocket on his aunt's farm near Auburn, Massachusetts, in 1926, but it did not go far.

Goddard continued to refine his designs of liquid-propelled rockets, and in 1929 he launched the first rocket to carry camera-recorded instruments. This success attracted money from philanthropist David Guggenheim, so that the following year Goddard was able to develop an advanced rocket-testing program near Roswell, New Mexico. He invented systems of gyroscopic steering and propellant-cooled combustion, which in 1937 resulted in a rocket that reached an altitude of three kilometers (the greatest altitude his rockets had achieved).

Goddard joined the Naval Engineering Experimental Station at Annapolis, Maryland, in 1942 to develop defense-related rockets for use in World War II, but the war ended before any of his designs had completed testing procedures. He died in Annapolis on August 10, 1945.

Legacy

Goddard's research laid the foundations of rocket design and development, improving air travel technology and making space exploration possible. His 214 patents covered nearly every aspect of liquid-fuel rocketry.

Improvements to aircraft engines represented the first practical applications of Goddard's research. He helped develop jet-assisted take-off devices and variable-thrust motors; the latter provided the power for the Bell X-2 rocket plane, the first craft in the United States to have a throttled engine. His single-charge recoilless rocket served as the prototype for the bazooka used against tanks in World War II.

Much of Goddard's research was inaccessible to his contemporaries because he worked mostly alone and was reluctant to publish his findings in journals. Other rocket engineers of the time had to develop their own ideas, without the benefit of his experience and insight. After World War II, the United States government relied heavily on German expertise in the establishment of its rocketry program. However, by 1960 Goddard's contribution became more widely recognized, and the government awarded his widow one million dollars for the right to use his patents. She gave half the payment to the Guggenheim Foundation whose founder, David Guggenheim, had supported Goddard in his early work.

Rocketry revolutionized space flight. Liquid propellants, first introduced by Goddard, have been used in most subsequent space vehicles; rockets provide lift off and booster thrust in every modern craft that leaves Earth's atmosphere.

Schuyler

World Events	Goddard's Life
	1882 Robert Goddard is born
Spanish-American War 1898	
	1899 Goddard decides to pursue rocket research
	1911 Goddard receives doctorate in physics
World War I 1914-18	
	1917 Goddard develops military rocketry
	1926 Goddard launches first liquid-fueled rocket
	1930 Goddard establishes rocketry program near Roswell, New Mexico
	1937 Goddard sends rocket to three-kilometer altitude
World War II 1939-45	
	1942 Goddard moves to Annapolis, Maryland, to work for Navy
	1945 Goddard dies
Korean War 1950-53	
	1960 U.S. government pays $1 million for right to use Goddard's patents

For Further Reading:

Lehman, Milton. *Robert H. Goddard: Pioneer of Space Research.* New York: Da Capo, 1988.

Winter, Frank H. *Rockets into Space.* Cambridge, Mass.: Harvard University Press, 1990.

Goodall, Jane

Researcher and Ethologist of Chimpanzees

1934-

Life and Work

Jane Goodall pioneered research techniques and interpreted new information during her lengthy study of Gombe chimpanzees. Her work as an ethologist, a scientist who studies animal behavior patterns, has provided vital clues about the evolution of primate behavior.

Born in London on April 13, 1934, Goodall knew at an early age that she wanted to study animals. She was fascinated by the wildlife and barnyard creatures surrounding her country home. After reading *Dr. Doolittle,* Goodall became enamored with the idea of traveling to Africa. Once she had completed her secondary education, she was invited to visit the home of a friend in Kenya. Goodall worked to earn the necessary money and soon

WORLD EVENTS		GOODALL'S LIFE
	1934	Jane Goodall is born
World War II 1939-45		
Korean War 1950-53		
	1960	Louis Leakey asks Goodall to study chimps in Gombe
	1965	Goodall awarded doctorate from Cambridge University
End of Vietnam War	1975	Jane Goodall Institute created
	1986	*The Chimpanzees of Gombe: Patterns of Behavior* published
Dissolution of Soviet Union	1991	
	1993	*My Life with the Chimpanzees* published

was on her way. During her visit she met Dr. Louis Leakey, a famous anthropologist. Impressed by her desire to study in Africa, he hired her to help with his research on prehistoric humans. Goodall's patience, attention to detail, and insight seemed to Leakey the ideal qualities of a researcher. He asked her to undertake a special project for him, studying chimpanzees. He felt that by studying chimpanzees, important clues to early human behavior might be discovered. Goodall accepted the challenge. Thus began the research that would be her life's work.

Goodall studied the chimpanzee groups that lived in the Gombe Game Reserve in Tanzania. The study that began as a three-year project in 1960 continues today. It was the first field study of non-human primates.

Goodall pioneered techniques for observation and data collection. Unlike laboratory animals, the subjects were not always available or willing to be observed. Goodall spent months acclimating the chimpanzees to her presence. Her detailed, long-term study discovered many new facts about the primates. They were not strict vegetarians as previously had been thought: insects, small animals, and even young baboons were part of their diet. The chimpanzees also exhibited behaviors thought unique to humans, such as tool making. This forced a re-examination of the definition of human intelligence. They were like humans in other ways, too, having complex codes of behavior and social systems. Goodall's discoveries were groundbreaking. Cambridge University recognized her accomplishments and granted her a doctorate in 1965.

As the years went on, the research expanded. Goodall became more concerned with the preservation and treatment of chimpanzees. In 1975, the Jane Goodall Institute was created in the United States to encourage education about the endangered primates.

In *The Chimpanzees of Gombe: Patterns of Behavior,* published in 1986, Goodall outlined her observations, and in *My Life with the Chimpanzees,* published in 1993, she made her work accessible to a wide-ranging audience. She continues to work today, studying chimpanzees and teaching the rest of the world what she has learned.

Legacy

Goodall provided the scientific community and the world with powerful research techniques and valuable information about chimpanzee behavior.

Goodall showed how careful documentation and attention to detail could provide data not seen in the laboratory. Following Goodall's example, long-term study also became a hallmark of ethology. She has observed the chimpanzees for over 25 years and continues to discover new behaviors and to answer questions about their complex social system. Her methodology is now used by field researchers around the world in a wide range of habitats.

Through her writing and frequent interviews, Goodall has also become a spokesperson for conservation and the protection of endangered primates and for the ethical treatment of chimpanzees in particular. The Jane Goodall Institute furthers these objectives as well.

As a new generation of scientists continues her research, Goodall's findings about the behaviors of chimpanzees are offering clues to early human behavior. These insights into human evolution pose many questions that spark further research.

Her public role as a researcher and conservationist provides an important role model for future generations of young scientists who want to follow their dreams, as she did.

Wilson

For Further Reading:

Goodall, Jane. *My Life with the Chimpanzees.* New York: Silver Burdett, 1993.

———. *In the Shadow of Man.* Boston: Houghton Mifflin, 1971.

Nichols, Michael. *The Great Apes: Between Two Worlds.* Washington, D.C.: National Geographic Society, 1993.

Goodyear, Charles

Inventor of Vulcanized Rubber
1800-1860

Life and Work

Charles Goodyear discovered vulcanization, a process involving the treatment of natural rubber with chemicals and heat to make it more useful in manufacturing.

Goodyear was born on December 29, 1800, in New Haven, Connecticut, to a family of modest income. As a teenager he joined his father's hardware business, but in 1830 the business went bankrupt. During the next 10 years, both while in debtor's prison and in intervals when he was free, Goodyear turned to inventing and received several patents for improvements of basic items like buttons, air pumps, and faucets. Then Goodyear turned his attention to improving rubber products.

At that time rubber, tapped from a tree native to tropical America, had been introduced as a waterproofing material. Rubber is a polymer, a compound comprised of long, thin chains of repeating chemical sub-units. Its key characteristic is elasticity, which arises from its structure of tangled polymers that tend to spring back to their original positions after being stretched. Rubber's commercial future was uncertain because of its disadvantages: it becomes hard and brittle when cold and soft and sticky when hot, losing its elasticity in both cases. By 1830 attempts to solve these problems had met little success.

About 1837 Goodyear purchased the rights to a process of treating rubber with sulfur. Sulfur appeared to reduce rubber's stickiness. Experimenting further, Goodyear made his most significant discovery by accident. In 1839 he spilled some sulfur-treated rubber on a hot stove and left it there overnight as the oven cooled. Noting the next morning that the rubber on the stove was not hard or sticky but rather elastic and pliable, he tested it in cold and heat. It retained its desirable qualities.

In 1844 Goodyear patented the process of mixing rubber with sulfur and heating it, calling it vulcanization after Vulcan, the Roman god of fire. The invention was immediately exploited by patent infringers, and Goodyear fought numerous expensive court battles. He traveled to France and England in 1852 to apply for patents, but he was denied in both countries. In order to produce his award-winning exhibition of rubber products at the 1855 World's Fair in Paris, Goodyear accrued crushing debts; he was imprisoned for a short time for failure to repay his creditors.

Goodyear died on July 1, 1860, in New York City, leaving his family $200,000 in debt.

Legacy

Goodyear's introduction of vulcanization made rubber a durable, versatile material useful in a variety of applications, particularly in the automotive industry.

Goodyear and his contemporaries did not understand how vulcanization changes the properties of rubber. However, it is now known that the main effect of vulcanization is the creation of new chemical bonds that link the material's tangled polymers together at numerous locations. The resulting network of connections tightens the structure, making it snap back into place more firmly after being stretched.

Vulcanized rubber was first used to make machinery components such as belts and hoses, which needed to remain strong even when heated. However, few rubber products actually succeeded in the market until 1870, when American entrepreneur Benjamin Goodrich opened a rubber factory for the manufacture of heavy-duty fire hoses. His company also produced the first rubber gaskets, bottle stoppers, jar-sealing rings, and garden hoses.

Rubber's next major application was the bicycle tire. In 1887 British inventor John Boyd Dunlop covered the wheels of a tricycle with rubber tires to make the ride smoother and quieter.

The automotive industry took full advantage of vulcanized rubber. When cars and trucks emerged in the first decades of the twentieth century, their manufacturers became the dominant consumers of rubber. Vulcanized rubber was good for making various engine parts and ideal for making tires. Today, 60% of the world's rubber is used by the automotive industry.

In 1898 American industrialist Frank Seiberling founded the Goodyear Rubber and Tire Company in East Akron, Ohio. He named it in honor of the inventor of vulcanized rubber, but there are no other connections between Charles Goodyear and the company.

Natural rubber has been almost entirely replaced by synthetic rubbers (materials with similar chemical structure to rubber) and by mixtures of natural and synthetic rubbers. These materials are less expensive to make and more durable than natural rubber. Vulcanization is still a fundamental process in the manufacture of synthetic rubbers.

During the twentieth century the rubber industry employed hundreds of thousands of workers around the world. In the spirit of Charles Goodyear, its research scientists and engineers have synthesized innumerable new polymers that are at the base of many everyday products, including sweaters, playground equipment, and construction materials.

Schuyler

World Events		Goodyear's Life
French Revolution	1789	
	1800	Charles Goodyear is born
Napoleonic Wars in Europe	1803-15	
	1830	Goodyear's family's hardware business goes bankrupt
	1837	Goodyear buys rights to sulfur-treated rubber
	1839	Goodyear accidentally discovers vulcanization of rubber
	1844	Goodyear patents vulcanization
	1860	Goodyear dies
United States Civil War	1861-65	
	1870	Benjamin Goodrich opens rubber factory
Germany is united	1871	
	1887	First rubber tires are fitted to a tricycle
Spanish-American War	1898	

For Further Reading:

Barker, Preston Wallace. *Charles Goodyear: Connecticut Yankee and Rubber Pioneer.* Boston: G. L. Cabot, 1940.

Noonan, G. Jon. *Nineteenth Century Inventors.* New York: Facts On File, 1992.

O'Reilly, Maurice. *The Goodyear Story.* Elmsford, N.Y.: Benjamin Co., 1983.

Regli, Adolph C. *Rubber's Goodyear: The Story of a Man's Perseverance.* New York: J. Messner, 1941.

Gutenberg, Johannes

Inventor of Movable-Type Printing
c. 1400-1468

Life and Work

Johannes Gutenberg invented movable-type printing in Europe. His innovations revolutionized how people and societies communicate with each other.

Little is known about Gutenberg's life. He was born sometime between 1394 and 1400, most likely in Mainz, a small city in present-day southwestern Germany. Gutenberg trained as a goldsmith, which led him to experiment with the special metal alloys he used in typecasting. He may have also attended university, but that can only be inferred from his knowledge of Latin.

WORLD EVENTS	GUTENBERG'S LIFE
	c. 1400 Johannes Gutenberg is born
	c. 1452 Gutenberg joins with Johann Fust in partnership to explore new printing techniques
Ottoman Empire 1453 captures Byzantine capital, Constantinople	
	c. 1455 Gutenberg Bible is printed; it is first book to be printed with Gutenberg's movable-type press
	c. 1457 Fust prints 32-line Bible
	1468 Gutenberg dies
Columbus discovers 1492 Americas	
Reformation begins 1517	
	c. 1530 First Lutheran catechisms are printed

As with many European craftsmen of the period, Gutenberg's experiments in printing were encouraged by economic incentives as well as by the social and intellectual climate. Literacy was rising and universities were multiplying, leading to an increased demand for printed matter. In many university towns, large-scale duplication of literature was a thriving business, employing many scribes who hand-reproduced books one at a time. The need for a more efficient means of reproducing printed matter was mounting.

By 1450, Johann Fust, a businessman of Mainz, lent Gutenberg money for his experiments, and in 1452 they formed a partnership under which Gutenberg perfected his methods. He forged easily cast uniform typeface, formulated a special ink that would adhere to metal, and modified the standard wooden wine press to fit the task of printing on paper. While other craftsmen experimented with similar techniques, Gutenberg was the first to successfully integrate them all into a new printing process. In 1455 the partnership dissolved because of Fust's demand for repayment on his investment. By the end of the partnership, they produced the now famous 42-line Bible (often called the Gutenberg Bible).

Gutenberg produced few other books following the split with Fust. One is referred to as the 36-line Bible and another is the *Catholicon,* a popular encyclopedia, which Gutenberg printed in 1460. Gutenberg likely stopped printing after its production, perhaps because of blindness. In 1465 he received a pension from the Archbishop of Mainz Adolf II. Gutenberg died on February 3, 1468.

Legacy

Gutenberg's techniques, used without any substantial change until the end of the nineteenth century, profoundly changed society. Because information could be mass-produced, the wealthy no longer had exclusive access to it. Further, information could be disseminated quickly across wide areas, even across continents.

Much in demand, Gutenberg's new printing techniques spread quickly, blazing a trail of entrepreneurial success across Europe. Fust continued printing in Mainz with his son-in-law Peter Schöffer who had apprenticed under

Gutenberg; printing quickly spread from Mainz through the rest of Germany. By 1470 Venice was the center of a flourishing printing trade and by 1475 printing had spread to England and France. The desire for printed matter also spread to all levels of society. By the mid-1500s printers produced material targeted at the illiterate population, including illustrated calendars and broadsides.

The new printing methods allowed the ideas of the Renaissance to spread quickly. In fact, some scholars suggest that the Renaissance would have been stillborn without the invention of movable-type printing. Gutenberg's methods were instrumental in the spread of the Reformation as well. Luther printed massive quantities of broadsides and pamphlets explaining his reform efforts, which could be reprinted as soon as needed. One of Luther's tracts promoting "justification by faith alone" was printed seven times in one year to keep up with demand. By the 1530s, Lutheran catechisms were printed, which teachers used to educate children in the Lutheran program of Christian beliefs.

The extended results of Gutenberg's inventions were dramatic as well. Spelling and grammar were more easily standardized, literacy increased, mass education was made possible, and ideas spread rapidly throughout all levels of society. Of course, these effects progressed slowly across centuries and proceeded from concomitant cultural and social changes. Nevertheless, none of it could have happened without Gutenberg's invention.

Tomaselli

For Further Reading:

Kapr, Albert. *Johannes Gutenberg: The Man and His Invention.* Brookfield, Vt.: Ashgate Publishing, 1996.

Steinberg, S. H. *Five Hundred Years of Printing.* New Castle, Del.: Oak Knoll Press, 1996.

Haber, Fritz

Chemist; First to
Synthesize Ammonia
1868-1934

Life and Work

Fritz Haber developed a method for synthesizing ammonia from atmospheric nitrogen and hydrogen. This method was applied to the production of nitrogen fertilizer.

Fritz Haber was born December 9, 1868, in Breslau, Germany. His father was a wealthy businessman involved in industrial chemicals. As a boy Haber gained a solid education in German schools, and, after a few tentative starts in business and organic chemistry research, mainly associated with his father's firm, he was appointed to the Technical High School in Karlsruhe in 1893. There he refined his skills in mathematics, chemistry, physics, electricity, and thermodynamics, and developed a mastery of physical chemistry and chemical research. His extensive research efforts resulted in two books: *The Theoretical Basis of Technical Electrochemistry* (1898) and *The Thermodynamics of Technical Gas Reactions* (1905). The first book earned him the position of professor at the Hochschule.

Haber's next research project yielded his most significant achievement, a way to synthesize ammonia. The synthesized ammonia became the starting material that was oxidized into nitrates. By 1909 Haber had developed a high-pressure, high-temperature process for commercial production of ammonia. He identified catalysts to promote its synthesis, and his collaborator, Robert Le Rossignol, designed the ultra-strong vessel to contain the reactants.

The two textbooks and the ammonia synthesis catapulted Haber into the directorship of the new Kaiser Wilhelm Institute for Physical Chemistry of Berlin in 1911. In 1914 World War I began and Haber committed himself to the German war effort. He prepared chemicals for improving ammunition, cylinders of chlorine for trench warfare, and primitive gas masks. In 1916 he became Chief of Chemical Warfare Services in the War Ministry. Haber led Germany's effort in poison-gas warfare and directed the chlorine attacks during the Battle of Ypres.

Germany's defeat was a tragedy for Haber in many ways. His strong patriotism was frustrated. His wife, Dr. Clara Immerwahr, who was also a chemist, had pleaded with him to stop his work with poison gas. When he refused her requests, she committed suicide.

Haber was awarded the Nobel Prize for Chemistry in 1918 for his work in synthesizing ammonia but many disagreed with his selection. His promotion of poison-gas weapons, and his role in prolonging the war by providing nitrates for gunpowder, was not well-received by the scientific community, especially as he not only developed the poisons but also directed the attacks on the battlefields of Europe.

Haber attempted to extract gold from sea water to help pay for Germany's extensive war reparations. This project was declared a failure in 1926. The Kaiser Wilhelm Institute continued to be the center for superior training in physical chemistry; Haber expanded its influence by supporting the National Organization of Research and promoting relationships with scientists, especially the Japanese, worldwide.

The Kaiser Wilhelm Institute for Physical Chemistry began to crumble in 1933 as Hitler's power increased. Haber was a Jew and feared Hitler's anti-Semitism; he resigned the directorship of the institute and accepted a position in Palestine, now a part of Israel. Haber suffered a heart attack en route and died in Basel, Switzerland, on January 29, 1934.

Legacy

Haber's method of synthesizing ammonia had extensive effects in both weaponry and agricultural production.

At the beginning of the twentieth century, two-thirds of the world's fertilizer came from beds of nitrate mined in Chile. Germany alone was buying about one-third of Chile's output when the supply became nearly exhausted. This prompted a global crisis in food production and forced scientists to look for alternative sources of fertilizer.

Air is mainly nitrogen gas, but pure nitrogen is not readily absorbed by living tissue. It must be in soluble form, like ammonia. Haber's process synthesizes ammonia from hydrogen and atmospheric nitrogen. Haber made possible the large-scale production of ammonia. His method, in turn, led to increased food production based on industry-made fertilizers. NORMAN BORLAUG's "green revolution," which systematically helped to increase food production in poor countries, was made possible by Haber's work.

Yet there is a dark side to Haber's noteworthy achievement. Ammonia is the chemical feedstock for many other compounds of nitrogen that can be used to manufacture ammunition, explosives, and poison gases. As with many scientific innovations, destructive applications of ammonia synthesis were implemented alongside its beneficial uses.

Luoma

World Events	Haber's Life
United States 1861-65 Civil War	
	1868 Fritz Haber is born
Germany is united 1871	
	1893 Haber obtains junior position at Technical High School of Karlsruhe
Spanish-American 1898 War	*The Theoretical Basis of Technical Electrochemistry* is published
	1905 *The Thermodynamics of Technical Gas Reactions* is published
	1909 Haber develops method for synthesizing ammonia from nitrogen and hydrogen
	1911 Haber is appointed Director of Kaiser Wilhelm Institute for Physical Chemistry in Berlin
World War I 1914-18	
	1916 Haber becomes Chief of Chemical Warfare Services and plays leading role in development of poison gas
	1918 Haber receives Nobel Prize for Chemistry
	1933 Haber resigns directorship of Institute because of anti-Semitic government policies
	1934 Haber dies
World War II 1939-45	

For Further Reading:

Bowden, Mary Ellen. *Chemical Achievers.* Philadelphia: Chemical Heritage Society, 1997.

Partington, J. R. *A Short History of Chemistry,* 3rd ed. New York: Macmillan, 1957.

———*A History of Chemistry,* Vol. IV. New York: Macmillan, 1964.

Hahn, Otto

Co-discoverer of Nuclear Fission
1879-1968

Life and Work

Otto Hahn performed important research on radioactivity and, along with LISE MEITNER and Fritz Strassmann, discovered nuclear fission.

Hahn was born in Frankfurt am Main, Germany, on March 8, 1879. Although he exhibited some early interest in science, he was more taken by athletics, religious studies, and music. Even at the University of Marburg, where he pursued a degree in chemistry, he was inclined toward art and philosophy. He earned his doctorate in 1901, completed a year of military service, and returned to Marburg as a lecture assistant.

WORLD EVENTS		HAHN'S LIFE
Germany is united	1871	
	1879	Otto Hahn is born
Spanish-American War	1898	
	1906	Hahn and Lise Meitner first collaborate in studies of radioactivity
World War I 1914-18		
	1934	Hahn's work builds on Enrico Fermi's uranium research
	1939	Meitner conceives of nuclear fission
		News of nuclear fission reaches the United States
World War II 1939-45		
	1942	First nuclear reactor is built in Chicago
	1944	Hahn and Fritz Strassmann awarded Nobel Prize for Chemistry
	1945	Atom bombs dropped on Japan
Korean War 1950-53		
	1968	Hahn dies

From 1904 to 1906 Hahn worked in London and Montreal with expert radiochemists William Ramsay and ERNEST RUTHERFORD. Working with Ramsay and Rutherford, he gained proficiency at research techniques used in the study of radioactivity.

Returning to Germany in 1906 to take a post at the University of Berlin, he was joined by physicist Lise Meitner. Prohibited from working together in an official university laboratory (because Meitner was a woman), the two set up a makeshift laboratory in a basement. In 1912 Hahn and Meitner moved to the newly established Institute for Chemistry at Berlin-Dahlen, where the relatively radiation-free surroundings allowed the detection of several weak radioactive isotopes. They performed important experiments with the newly identified element protactinium, which they named.

Hahn was intrigued by the work of Italian physicist ENRICO FERMI, who in the early 1930s found that several mysterious radioactive products were formed when uranium was bombarded with neutrons. Fermi was unable to interpret these ambiguous results, and, in 1934, Hahn, Meitner, and newcomer Fritz Strassmann began investigating the question of what occurred when uranium was bombarded with neutrons. Four years later Meitner, who was Jewish, fled Germany to escape Nazi persecution, but Hahn and Strassmann continued their research. They remained in contact with Meitner, and it was she who, in 1939, interpreted their results correctly: she suggested that the uranium nucleus had split into two roughly equal parts (smaller radioactive nuclei). She termed this event "nuclear fission."

Hahn and Strassmann were awarded the 1944 Nobel Prize for Chemistry, and after World War II, Hahn helped rebuild the scientific community of West Germany. He was also active in the effort to curb misuses of nuclear energy.

Hahn died on July 28, 1968, in Göttingen, West Germany.

Legacy

Hahn's work was crucial to the development of the atomic age.

Within a few months of Hahn's experiments with uranium, the potential of nuclear fission had been recognized. Soon after Meitner published her interpretation of Hahn's experimental results, others showed that a great deal of energy is released during nuclear fission when the original nucleus splits. The news spread quickly, reaching the United States early in 1939 when Danish physicist NIELS BOHR met with Fermi and ALBERT EINSTEIN. Fermi proposed that neutrons, along with energy, might be emitted during fission, starting a chain reaction in which the newly emitted neutrons bombard and trigger fission in neighboring uranium atoms, releasing massive amounts of energy. By the spring of 1939, it was public knowledge that the enormous energy produced by nuclear fission could, theoretically, be used to create bombs or to generate controlled power.

In only a few years these theories were realized. A clandestine program, the Manhattan Project (directed by J. ROBERT OPPENHEIMER), began in the United States with the goal of creating an atom bomb. To fuel the project with plutonium, Fermi built the world's first nuclear reactor in Chicago in 1942. The first atom bomb was tested in 1945, and later that year two bombs were dropped on Japan, causing horrific devastation and ending World War II in the Pacific.

After the war, nuclear fission lent itself to the development of nuclear arsenals, nuclear power plants, and pure physics research.

Schuyler

For Further Reading:

Graetzer, Hans G., and David L. Anderson. *The Discovery of Nuclear Fission*. New York: Arno Press, 1981.

Shea, William R., ed. *Otto Hahn and Rise of Nuclear Physics*. Hingham, Mass: D. Reidel Publishing, 1983.

Hall, Charles; Hall, Julia

Developers of Aluminum
Manufacturing Process
1863-1914; 1859-1925

Life and Work

Charles and Julia Hall invented and marketed the process of using electrolysis to manufacture aluminum, the first method of producing aluminum cheaply.

Julia Hall was born on November 11, 1859, in the British West Indies to a missionary family. Her brother Charles was born on December 6, 1863, in Thompson, Ohio. Close companions in childhood, they remained confidantes throughout their lives. They both studied chemistry at Oberlin College, Ohio, Julia graduating in 1881 and Charles in 1885. Their mother died in 1885, and Julia undertook responsibility for the care of their two younger sisters and the general household tasks. Charles set up a chemistry laboratory in a shed next to the kitchen.

Having overheard Oberlin College professor Frank Jewett declare that whoever invented a way to produce aluminum cheaply would become wealthy and famous, Charles began investigating this challenge. Aluminum is present in the Earth's crust as aluminum oxide, or bauxite, but separating the metal from its ore was extremely expensive and difficult because of the aluminum atom's tight hold on the atoms of adjacent molecules.

In addition to providing scientific assistance, Julia kept careful notes on their experiments and recorded evidence substantiating the dates and results.

On February 23, 1886, the Halls discovered that passing an electric current through a mixture of aluminum oxide and cryolite (a mineral containing sodium, aluminum, and fluorine) yielded deposits of pure aluminum. Upon filing for a patent, they were challenged by French chemist Paul-Louis Héroult, who had devised a similar electrolytic method, also in 1886, and filed his patent request earlier. The

Halls were able to prove, due to Julia's expert testimony, that they had produced pure aluminum earlier than Héroult.

With financial assistance from the Mellon family, the Halls established the Pittsburgh Reduction Company, which later became the Aluminum Company of America (ALCOA). Charles served as vice-president from 1890 until his death. The production of aluminum was commercially successful, and upon his death in Daytona Beach, Florida, on December 27, 1914, Charles left a trust fund of several million dollars to Oberlin College. Among his other beneficiaries were 23 other universities and colleges around the world, from Bulgaria and Turkey to Japan and Korea. Julia died in obscurity in 1925.

Legacy

Charles and Julia Hall made possible the large-scale production of aluminum, a versatile metal that transformed industry and everyday life with its many applications. The electrolytic process they introduced remains the primary modern method of purifying aluminum.

Julia received little recognition for her contribution to the development of aluminum manufacture; it is likely that her work was obscured by an environment that generally overlooked the scientific endeavors of women. Even Charles failed to acknowledge publicly her role in his invention during a 1911 speech about the history of the Pittsburgh Reduction Company. Few modern sources mention her in association with her brother or aluminum production.

The electrolytic aluminum production method became known as the Hall-Héroult process, to credit both Charles and his French counterpart. During the last decades of the nineteenth century, several improvements were made to the process, and the aluminum market was launched. Cookware represented the first aluminum household products. The metal's lightness and strength also made it ideal for manufacture of early aircraft and internal combustion engines.

Throughout the twentieth century, aluminum's applications grew. Because of its weight, strength, and resistance to corrosion,

it is now commonly used in the production of buildings, aircraft, spacecraft, automobile and marine-vessel parts, food packaging, electrical wire and cable, sporting equipment, tools, kitchen appliances, paints, fuels, and explosives.

New technologies for extracting aluminum from raw materials other than bauxite are on the horizon. In Europe and Asia compounds such as clays, anorthosite, nepheline syenite, and alunite are sometimes used as sources of aluminum, although the extraction methods involved have not yet become financially competitive with the Hall-Héroult process.

Schuyler

WORLD EVENTS		HALLS'S LIVES
	1859	Julia Hall is born
United States Civil War	1861-65	
	1863	Charles Hall is born
Germany is united	1871	
	1881	Julia Hall graduates from Oberlin
	1885	Charles Hall graduates from Oberlin
	1886	The Halls discover electrolytic purification of aluminum
		Paul-Louis Héroult discovers electrolytic purification of aluminum
	1890	Charles Hall becomes vice-president of Pittsburgh Reduction Company
Spanish-American War	1898	1900
	1914	Charles Hall dies
World War I	1914-18	
	1925	Julia Hall dies
World War II	1939-45	

For Further Reading:

Craig, N. C. "Charles Martin Hall—the Young Man, His Mentor, and His Metal." *Journal of Chemical Education* 63 (1986).

———. "Julia Hall—Co-inventor." *Chemical Heritage* 15:1 (Fall 1997).

Kass-Simon, G., and Patricia Farnes, eds. *Women of Science: Righting the Record*. Indianapolis: Indiana University Press, 1990.

Trescott, M. M. *Dynamos and Virgins Revisited: Women and Technological Change in History.* Metuchen, N.J.: Scarecrow Press, 1979.

Halley, Edmond

Pioneering Investigator of Comets
1656-1742

Life and Work

Edmond Halley was the first to predict correctly the reappearance of a comet. He also charted the southern sky, discovered stellar motion, financed the publication of ISAAC NEWTON's *Principia,* and studied geophysical phenomena.

Halley was born on November 8, 1656, near London, England, into a wealthy family. He entered Oxford University in 1673 to study astronomy and published a paper on planetary motion. This paper earned him the opportunity to spend two years on St. Helena, a southern Atlantic island, charting the stars of the southern sky. This task yielded the first compilation of telescopically determined star positions and prompted Oxford to grant Halley a master's degree in 1678.

Halley befriended English physicist ISAAC NEWTON in 1684. Newton's theory of gravitational force extended to the celestial sphere

and helped to explain the elliptical shape of planetary orbits. Halley encouraged him to expand and publish this work. In 1687 Newton's *Principia (Mathematical Principles of Natural Philosophy),* edited and financed by Halley, was printed.

In 1686 Halley produced the first map of the world illustrating prevailing wind direction over the oceans. As ship's commander on the first sea voyage made expressly for scientific investigation (1698-1700), he did pioneering research on magnetism, monsoons, tides, evaporation, and salinity.

Upon viewing a comet that appeared in 1682, Halley began studying cometary motion. He gathered data on every comet that had been observed since 1337 and computed their orbits; his calculations suggested that the famous comets sighted in 1531, 1607, and 1682 were one and the same. In *A Synopsis of the Astronomy of Comets* (1705) he accurately predicted this comet would reappear in 1758 or 1759; this comet is now known as Halley's Comet.

In other astronomical research Halley devised a method for determining the distance of Earth from the Sun by using procedures he had already developed for observing the transit of Venus across the Sun—which would not occur until 1761. Finding extensive incongruities between historical and current star charts, he concluded, contrary to traditional doctrine, that stars move.

Halley died on January 14, 1742, in Greenwich, England.

Legacy

Halley's pioneering contributions to astronomy and geophysics laid groundwork for progressive research on comets, stars, astronomical distances, and navigation. His involvement in the publication of Newton's *Principia* made him important in the emergence of modern scientific thought.

In 1761 astronomers at 62 stations around the planet observed the transit of Venus mapped out by Halley earlier in the century. Based on measurements of the planet's passage, they calculated the distance from Earth to the Sun; their value was 153 million kilometers, surprisingly close to the modern value of 149.6 million kilometers.

Halley's support was crucial to the publication of Newton's *Principia,* considered a masterpiece of scientific thought. In it Newton codified the

laws of motion and universal gravitation, firmly uniting the heavens and Earth under a single system of mechanical motion.

Halley's announcement that the stars are not fixed in space was a revelation, and astronomers then strove to determine the nature of stellar motion. In 1904 part of the mystery was solved by Dutch astronomer Jacobus Kapteyn, who discovered that two streams of stars, moving in opposite directions, dominate the Milky Way.

In 1758 Halley's Comet (as it was thereafter known) appeared, confirming that comets are bodies that orbit the Sun and not, as some believed, meteorological phenomena. A frenzy of interest in comets began. In 1818 German astronomer Johann Encke calculated the orbit of a comet observed the previous year. Encke's comet has an orbital period of three and a half years, the shortest cometary period known to date; it is visible (with instruments) throughout its orbit.

Halley's Comet reappeared in 1835, 1910, and 1986, enabling astronomers to study its characteristics and those of comets in general. Comets consist of a nucleus containing rock, dirt, and ice, and a tail of dust and gas. They follow elliptical orbits, traveling in the opposite direction of the planets. Halley's Comet is the only comet to be visible to the unaided eye each time (every 76 to 79 years) it passes nearby. During its 1986 appearance, six unmanned spacecraft were launched to observe and make measurements of its chemical composition and to determine the nature and evolution of its nucleus.

Schuyler

WORLD EVENTS	HALLEY'S LIFE
Thirty Years' War 1618-48 in Europe	
	1656 Edmond Halley is born
	1673 Halley enters Oxford
	1682 Halley observes comet, begins studies of comets
	1687 Isaac Newton's *Principia* is published; Halley assists in its publication
England's Glorious 1688 Revolution	
	1705 Halley predicts return of comet in 1758-1759
Peace of Utrecht 1713-15 settles War of Spanish Succession	
	1742 Halley dies
	1758 Halley's Comet reappears
	1761 Astronomers use Halley's procedures for determining distance of Earth from the Sun

For Further Reading:

Cook, Alan. *Edmond Halley: Charting the Heavens and the Seas.* New York: Oxford University Press, 1998.

MacPike, Eugene Fairfield, ed. *Correspondence and Papers of Edmond Halley.* New York: Arno Press, 1975.

Ronan, Colin. *Edmond Halley: Genius in Eclipse.* London: Macdonald, 1970.

Thrower, Norman J., ed. *Standing on the Shoulders of Giants: A Longer View of Newton and Halley.* Berkeley: University of California Press, 1990

Hargreaves, James

Inventor of the Spinning Jenny
c. 1720-1778

Life and Work

James Hargreaves invented the spinning jenny, a machine that enables several threads or yarns to be spun simultaneously. The jenny increased the speed of thread production and helped transform the textile trade.

Little is known of Hargreaves's early life. He was born about 1720 in Blackburn, England, and worked as a carpenter and handloom weaver at nearby Standhill from 1740 to 1750. He is not known to have received any formal education. In 1760, he designed an improved carding machine that prepared fiber to be spun into thread.

About 1764 Hargreaves noticed that the spindle of an overturned spinning wheel, tipped over by his daughter Jenny, continued to revolve in a vertical position. The observation led him to posit that several spindles could be placed upright and side by side, and might be able to spin a number of threads at once. He designed and built a spinning machine that incorporated this innovation, naming it the "spinning jenny" after his daughter. At first, he and his family alone used spinning jennies, but soon he built and sold several machines to help support his many children.

In 1768 a number of local spinners became worried that the spinning jenny would put them out of work, and they angrily broke into Hargreaves's house and destroyed his machines. Hargreaves moved to Nottingham, where he and his new business partner, Thomas James, opened a mill and spun cotton with spinning jennies.

Hargreaves and James obtained a patent for the jenny in 1770 and soon brought legal action against local manufacturers that were using spinning jennies. Before the cases were settled, the partners' lawyer learned that Hargreaves had earlier sold several jennies in Blackburn, and he withdrew his services; all legal action ceased.

Hargreaves continued his moderately successful business until his death in Nottingham on April 22, 1778.

Legacy

Hargreaves's spinning jenny increased the output of spinners eightfold. It contributed to the transformation of the textile trade from a cottage industry to a factory industry.

The history of spinning is obscure, but it is known that in the thirteenth century, the invention of the first spinning wheel somewhat improved the ancient spindle-and-whorl spinning method. Two hundred years later, the Saxon Wheel (an improved wheel) appeared and made spinning more efficient. In the eighteenth century, weaving had received a boost with the invention of the flying shuttle, but prior to the invention of the jenny, spinning remained much the same as it had been for centuries. When the spinning jenny was introduced, its advantages were immediately recognized. It was valued particularly for its wool-spinning capability; by 1785, 20,000 jennies were in use in England.

The spinning jenny introduced several improvements to spinning. It enabled spinners to produce 16, rather than two, threads or yarns at once, and it was simple and could therefore be operated by children. This latter benefit was, of course, a mixed blessing. By the end of the 1700s, the exploitation of children in the textile trade and other industries was a social problem that would soon give rise to major reform efforts, such as Britain's Factory Act in 1833, which, among other measures, placed some restrictions on child labor.

The spinning jenny led to the invention of the spinning mule, which revolutionized the textile industry. In 1769, Richard Arkwright patented a water-powered spinning machine. Then Samuel Crompton combined Arkwright's invention with the spinning jenny and designed the spinning mule, an automatic machine that could easily spin any kind of yarn.

By 1785, steam power had been introduced to the textile industry, and the processes of carding, spinning, and weaving had moved into factories in order to centralize power sources. In Manchester, England, from 1782 to 1802, the number of textile factories increased from two to 52. And from 1780 to 1800, the amount of fabric produced in England multiplied tenfold.

Schuyler

World Events		Hargreaves's Life
Peace of Utrecht settles War of Spanish Succession	1713-15	
	c.1720	James Hargreaves is born
	1740	Hargreaves begins work in textiles
	1760	Hargreaves designs improved carding machine
	1764	Hargreaves invents spinning jenny
	1768	Spinners loot Hargreaves's house
	1770	Hargreaves and Thomas James obtain patent for spinning jenny
United States independence	1776	
	1778	Hargreaves dies
	c.1785	Steam power introduced into textile industry
French Revolution	1789	

For Further Reading:

Aspin, C. *James Hargreaves and the Spinning Jenny.* Helmshore, England: Helmshore Historical Society, 1964.

Cardwell, Donald. *The Norton History of Technology.* New York: Norton, 1995.

Ginsburg, Madeleine, ed. *The Illustrated History of Textiles.* London: Studio Editions, 1991.

Harrison, John

Inventor of Practical Chronometer
1693-1776

World Events	Harrison's Life
Thirty Years' War in 1618-48 Europe	
	1693 John Harrison is born
	1713 Harrison builds first clock
Peace of Utrecht 1713-15 settles War of Spanish Succession	
	1714 British Parliament passes Longitude Act
	1735 Harrison submits first chronometer to Board of Longitude
	1762 Harrison's fourth chronometer exceeds requirements for prize on Jamaica trip
	1763 Harrison is given 25% of longitude prize
	1773 King George III awards balance of prize to Harrison
United States independence	1776 Harrison dies
French Revolution 1789	

Life and Work

John Harrison was the first to make a practical chronometer, a portable, accurate timepiece that could perform aboard ocean-going ships. By doing so, he solved one of the biggest challenges of his time, determining longitude at sea.

Born in March 1693 in Yorkshire, England, Harrison was son of a carpenter. He became a craftsman and a local choir director. Harrison made his first pendulum clock before he was 20 years old; like others he built in 1715 and 1717, it was made almost entirely of wood.

Another clock, still working, was built by Harrison in a tower in Brockelsby Park where it runs without oiling, an advance unheard of before his design but that would prove to be an enormous advantage for a clock built to work at sea. With his brother James, Harrison also built clocks tuned to the positions of given stars. These timepieces were accurate within one second per month—the world's best watches of his time were off by 60 seconds per day. A prize of 20,000 pounds offered by the British Parliament encouraged him to try to build a different kind of clock that could perform at sea.

The problem of accurately calculating longitude had stymied the scientific community (including Isaac Newton and Edmond Halley) and loss of lives and cargo were high costs to maritime nations. Although latitude can be gauged by the height of the Sun or by specific guide stars, longitude, or meridian, is best determined by precise knowledge of the difference in time aboard ship compared to time at a place of known longitude (home port, for instance) at the same instant. Each hour's time difference represents 15 degrees longitude east or west of the starting point and, in conjunction with latitude, can be used to determine a ship's position relative to its destination.

Without an accurate method of determining longitude, sea captains often traveled longer routes than necessary and/or crashed on unanticipated landfalls. A single disaster from miscalculation of longitude in 1707 resulted in 2,000 British sailors and four warships lost about 20 miles from port. Parliament responded with the Longitude Act of 1714, which provided support for selected inventors plus a grand prize for a successful chronometer.

From 1730 to 1760 Harrison built five non-pendulum clocks with innovative counter-balanced, spring-driven works. The first one was submitted in 1735 for the Board of Longitude prize, as were subsequent smaller instruments. In 1762 Harrison's fourth version of a chronometer was in error by only five seconds during a six-week voyage to Jamaica, but he was given only 5,000 pounds in 1763. Only after King George III himself intervened in 1773 was Harrison paid in full.

Harrison died in London on March 24, 1776.

Legacy

Harrison's ingenious chronometers were built without using astronomical methods that most educated people assumed would underlie success in measuring longitude at sea. His knowledge of various woods and non-rusting metals and his superior craftsmanship contributed to successful designs of marine chronometers. His inventions saved time, lives, and cargo and thus helped Great Britain become the richest mercantile country in the world.

Captain Cook used a Harrison chronometer to make the first, and highly accurate, chart of the South Sea Islands during his second voyage, which ended in 1775.

Other watchmakers improved on Harrison's models. John Arnold finished several hundred chronometers including a pocket-sized one completed in 1779. He opened a factory to mass produce them in 1785. Thomas Earnshaw refined the chronometer further, reducing both its complexity and its size.

As naval vessels, merchant ships, and pleasure yachts began to rely on chronometers as navigational necessities, the world census of chronometers swelled from just one in 1737 to about 5,000 in 1815. When the *Beagle* set out in 1831 with a mission to determine longitude of foreign lands, she had Charles Darwin and 22 chronometers aboard.

Modern sailors have atomic clocks, global positioning satellites, and multiple communications systems to help them make quick and accurate determinations of longitude. John Harrison's inventions were the life-saving predecessors to these twentieth-century advances.

Simonis

For Further Reading:
Quill, Humphrey. *John Harrison: The Man Who Found Longitude.* London: Baker, 1966.
Sobel, Dana. *Longitude.* New York: Walker, 1995.

Harvey, William

Discoverer of Blood Circulation
1578-1657

Life and Work

William Harvey founded modern physiology with his conclusive demonstration that the heart pumps blood throughout the body in a closed loop of arteries and veins.

On April 1, 1578, Harvey was born in Folkestone, England, to a prosperous family. At age 16, he enrolled at Cambridge University and earned a bachelor's degree in 1597. He then attended the University of Padua, where he studied with the anatomist Fabricius, whose work involved detailed observations made during repeated and rigorous dissections. Before earning his medical degree in 1602, Harvey became fascinated by Fabricius's live demonstrations of the heart and circulatory systems of dissected animals; however, Harvey was puzzled by the heart's flap-like valves whose functions were not yet known.

Harvey returned to London to practice medicine and, in 1607, was elected to the Royal College of Physicians. Although he practiced medicine for 31 years at St. Bartholomew's, one of London's most prestigious hospitals, his true passion remained research. He conducted original investigations into the vascular systems of snails, frogs, chickens, and people, and he recognized similarities between these species. He

observed how arteries and veins were connected (conclusions made decades later by Italian biologist Marcello Malpighi) and reasoned that the blood must flow in a closed circle under the power of the beating heart. He proposed that the valves in the heart and veins are necessary to keep the blood flowing in one direction. In 1628, he published these revolutionary ideas in *An Anatomical Exercise on the Motion of the Heart and Blood in Animals.* Harvey's new theory of circulation challenged Galen's idea of an open system in which blood moved with an "ocean-like" ebb and flow.

Harvey's success as a medical practitioner earned him prestige. In 1618, he was appointed one of the physicians to King James I, and in 1631, Harvey became the main physician to King Charles I. The English Civil War forced him to leave London in 1642. Upon his return to London after the king's execution in 1649, he continued to study the embryology of deer and other animals, work he had begun before the civil war. His studies focused on investigation into the phases of mammalian development. His book, *Exercises on the Generation of Animals* (1651), emphasized the importance of observation and experimentation. Followers of his advice formulated modern physiological concepts of normal and abnormal functions within the human body.

Harvey died in London on June 3, 1657.

Legacy

Harvey's discovery of blood circulation represented a major advance in the understanding of animal biology and marked the beginning of modern physiology. His advocacy of rigorous scientific investigation raised the standards of science and scientific research.

Harvey's work was controversial during his lifetime. Because Harvey rejected the authority of Galen, especially his erroneous assumptions about the circulation of blood, he was ridiculed by many of his contemporaries. However, he was an inspiration to others, whose further research soon confirmed his claims.

René Descartes, the influential French philosopher who proposed an enduring mechanistic view of science and nature, incorporated Harvey's discovery into his work. Descartes's 1637 *Discourse on Method* includes a section on physiology and uses the circulation of the

blood to illustrate the author's approval of a quantitative, mechanistic approach to scientific methodology.

Harvey's methodology involved direct observation, dissection, record-keeping, and conceptualization based on first-hand knowledge. These rigorous methods advanced the practice of science as a discipline separate from history and philosophy.

Harvey's work was extended by other scientists during the second half of the seventeenth century. In 1661, Malpighi observed and described the function of the capillary blood vessels, completing the understanding of the circulatory system. English physiologist Richard Lower made the first successful direct blood transfusion in 1665, from the artery of one dog to the vein of another.

Harvey also contributed greatly to future medical practice by recognizing the concept of the pulse rate of the heart. He observed that the rate a heart beats can be measured by counting the number of pulses per minute in an artery. This technique is now a standard part of medical diagnostic procedure.

Harvey's work stands out as a significant milestone in the development of biological science, to which the understanding of the circulatory system has been immensely important.

Schuyler

World Events	Harvey's Life
Reformation begins 1517	
	1578 William Harvey is born
	1602 Harvey receives medical degree
	1607 Harvey is elected to Royal College of Physicians
Thirty Years' War 1618-48 in Europe	
	1628 Harvey's *An Anatomical Exercise on the Motion of the Heart and Blood in Animals* is published
	1657 Harvey dies
	1661 Marcello Malpighi describes capillaries
Peace of Utrecht 1713-15 settles War of Spanish Succession	

For Further Reading:

Cohen, I. Bernard. *Studies on William Harvey.* New York: Arno Press, 1981.

Pagel, Walter. *New Light on William Harvey.* New York: S. Karger, 1976.

Hawking, Stephen

Developer of Mathematical Proof
for Black Hole Theory
1942-

Life and Work

Stephen Hawking developed groundbreaking theories concerning black holes and the origin of the universe.

Hawking was born on January 8, 1942, in Oxford, England. He became interested in science and mathematics as a teenager and then attended Oxford University, where his peers and his professors noticed that he could handle very difficult problems with relative ease. In 1962 he graduated from Oxford University and in 1965 he received his doctorate in cosmology (the study of the universe

as a harmonious system) from Cambridge University, where he continued on as a research fellow and later as a professor.

In 1962, Hawking was diagnosed with amyotrophic lateral sclerosis (ALS), which causes a gradual loss of voluntary motor function and leads eventually to paralysis. However, ALS leaves thought and memory untouched, which has enabled Hawking to continue his theoretical work.

Hawking's doctoral work concerned the theory of black holes, originating with Karl Schwarzschild in 1916, which are thought to be cosmic bodies possessing such enormous gravity that nothing can escape from them. The center of a black hole is a space-time singularity, a place or moment at which the principles of physics break down. In 1968, Hawking began working on theoretical problems of space-time singularities, and in 1973 he co-authored with G. F. R. Ellis the technical book *The Large Scale Structure of Space-Time*. The book described the theory that the universe began with a space-time singularity, and that this singularity corresponds to the Big Bang theory of the origin of the universe (proposed by Aleksandr Friedmann and Abbé Georges Lemaître in the 1920s and updated by George Gamow in the 1940s). In 1974, Hawking issued the surprising theoretical proof that in some circumstances, black holes can emit thermal radiation.

In the late 1970s, Hawking began to attempt to link the theory of relativity with quantum mechanics in the hope of producing a unified theory of all physical principles. In the 1980s, this work led him to propose that the universe has no boundaries in either space or time and therefore no beginning and no ending.

In 1988, the publication of Hawking's *A Brief History of Time: From the Big Bang to Black Holes* catapulted him into international stardom. The book was Hawking's attempt to communicate his ideas to a general audience, and it remained on bestseller lists in the United States and Britain for several years. Hawking has received numerous honorary degrees and awards throughout his life, and the popular media have helped develop his reputation with articles and programs about his life and work.

Legacy

Hawking has reshaped the field of theoretical physics with his rare insight and new theories. He has also distinguished himself by successfully communicating his work to the general public.

Hawking's ideas have been crucial to the development of modern cosmology and the current understanding of black holes. The Big Bang theory of the origin of the universe was put forth in 1933, but Hawking's work in the 1970s concerning space-time singularities represented the first strong mathematical evidence supporting the theory. Prior to the introduction of Hawking's astounding black hole discoveries, many scientists thought it would be difficult, if not impossible, to learn anything concrete about the properties of black holes.

The success of Hawking's *A Brief History of Time* illustrated that the general public has great interest in the questions physicists attempt to address. Some scientists believed that most people are either too apathetic or uneducated to understand or engage in physical sciences. Hawking's book proved them wrong.

Hawking has attained a rare celebrity status among scientists. Despite suffering from an illness that confines him to a wheelchair and requires him to use a voice synthesizer to speak, he has pursued his work, taught, published, and spoken at many public functions. His life and work have been an inspiration to many.

Schuyler

WORLD EVENTS	HAWKING'S LIFE
World War II 1939-45	
	1942 Stephen Hawking is born
Korean War 1950-53	
	1962 Hawking graduates from Oxford University
	Hawking is diagnosed with ALS
	1965 Hawking receives doctorate in cosmology from Cambridge University
	1960s Hawking studies black holes, or space-time singularities
	1970s Hawking works toward a unified theory of physics
	1973 G. F. R. Ellis's and Hawking's *The Large Scale Structure of Space-Time* is published
End of Vietnam War 1975	
	1988 Hawking's *A Brief History of Time* is published
Dissolution of 1991 Soviet Union	

For Further Reading:

Hawking, Stephen. *A Brief History of Time: From the Big Bang to Black Holes.* New York: Bantam, 1988.

White, Michael, and John Gribbin. *Stephen Hawking: A Life in Science.* New York: Dutton, 1992.

Heisenberg, Werner

Developer of Matrix Mechanics
and the Uncertainty Principle
1901-1976

Life and Work

Werner Heisenberg was one of several physicists who developed quantum theory during the 1920s. His work led to the development of quantum mechanics and a new, mathematical conception of the universe.

Born on December 5, 1901, near Dusseldorf, Germany, Heisenberg attended school in Munich and studied physics at the university there, receiving his doctorate in 1923. He continued his studies with some of Europe's greatest physicists, including NIELS BOHR in Denmark. Bohr was already famous for his model of the atom, the miniature solar system that is often taught in middle school science classes. His model explained the properties of hydrogen, the simplest element, but it failed badly when applied to more complex atoms. In the early 1920s other scientists improved the model with versions of quantum theory, including Louis de Broglie, who imagined that waves guided the electrons around the nucleus.

In June 1925 Heisenberg took a radically different approach to quantum theory: he abandoned all pictures or models of atoms and considered instead only what could actually be seen or observed. No one can see electrons whirling around their nuclei, but the patterns of light emitted by hot atoms in a gas, such as in a neon sign, can be seen. He represented the brightness and frequency of the light waves as numbers in a chart and found that the arrays of numbers obeyed the laws of a branch of mathematics called matrix algebra. The new theory became known as matrix mechanics, and it was remarkably successful in predicting the properties of more complex structures. Heisenberg received the 1932 Nobel Prize for Physics for his application of matrix mechanics to molecules made of two hydrogen atoms.

Going deeper into the implications of quantum theory, Heisenberg realized in 1927 that the act of observing electrons would disturb them. To get information about the motion of electrons, observers would need to send in some light and record the reflection. But ALBERT EINSTEIN had shown that light, the most delicate probe we can imagine, travels in separate little droplets, called quanta (later known as photons), and even the smallest of them would affect the electrons when they hit. Heisenberg concluded that there is a limit to what can be known about electrons; the more we know of their position the less we know about their motion, and the better we measure their motion the more we cloud their position. He called this the Uncertainty Principle.

In 1942-1945 Heisenberg was in charge of Germany's project to make nuclear weapons during World War II. The program was poorly funded and many scientists involved were reluctant to make so much progress that Adolf Hitler would have access to nuclear bombs. After the war Heisenberg actively promoted the peaceful use of nuclear power and helped design Germany's first nuclear plant. He died in Munich of cancer on February 1, 1976.

Legacy

Heisenberg moved our understanding of the atom into the realm of quantified mathematical measurement, which refined the newly developing field of quantum mechanics. More importantly, his Uncertainty Principle and strict adherence to observation and mathematical measurement in scientific analysis of subatomic phenomena added a cautionary dimension to the field of nuclear physics.

Although Heisenberg's system of matrix mechanics was a powerful tool for predicting the behavior of atoms with multiple electrons and even of simple molecules, it was cumbersome to use. By 1926, only a year after Heisenberg developed matrix mechanics, the great Austrian physicist Erwin Schrödinger produced another mathematical representation of the atom, called wave mechanics, based on the probability of electrons appearing at various distances from the nucleus. Schrödinger's method was simpler, more elegant, and easier to use, and he showed that it was mathematically equivalent to Heisenberg's matrix mechanics. Scientists today generally use Schrödinger's version of quantum mechanics, but it is part of Heisenberg's legacy because he was the first to explain atoms successfully in terms of quantum theory.

Heisenberg's more lasting contribution was his Uncertainty Principle, setting a limit on what can be known about electrons and their subatomic relatives. Before Heisenberg scientists believed that any quantity could be measured as accurately as the equipment and the human mind would allow. When the objects under investigation are atoms or their parts, though, nature itself seems to prevent a total description. Einstein was never satisfied with the Uncertainty Principle, believing that the limits it recognizes are attributable to the design of the experiment. Repeated challenges by both theory and experiment throughout the twentieth century have always supported Heisenberg's position.

Perhaps Heisenberg's greatest legacy is his method of starting from what one can actually observe and measure, rather than from a mental picture or a mechanical model. It was a radical change in how scientific problems were addressed, and it was largely responsible for the success of quantum theory. Scientists have yet to develop an accurate picture or model of an atom, the basic unit of everything. Quantum theory transcends the human ability to see and imagine, giving researchers a powerful tool to explore nature while also making it difficult to discuss and understand the results.

Secaur

For Further Reading:

Cassidy, David. *Uncertainty: The Life and Science of Werner Heisenberg.* New York: Freeman, 1992.

MacPherson, Malcolm. *Fermi, Heisenberg, and the Race for the Atomic Bomb.* New York: Dutton, 1986.

Powers, Thomas. *Heisenberg's War: The Secret History of the German Bomb.* New York: Knopf, 1993.

WORLD EVENTS		HEISENBERG'S LIFE
Spanish-American War	1898	
	1901	Werner Heisenberg is born
World War I	1914-18	
	1923	Heisenberg earns doctorate at University in Munich
	1925	Heisenberg develops matrix mechanics
	1927	Heisenberg develops the Uncertainty Principle
	1932	Heisenberg receives Nobel Prize for Physics for work on matrix mechanics
World War II	1939-45	
	1942-45	Heisenberg leads German atomic bomb project
Korean War	1950-53	
End of Vietnam War	1975	
	1976	Heisenberg dies

Herschel, Caroline Lucretia

First Major Woman Astronomer
of Modern Times

1750-1848

Life and Work

Caroline Lucretia Herschel was the first important woman astronomer of modern times. She was the first woman credited with discovering a comet, and she catalogued and fixed the position of 2,500 nebulae.

In 1750 Caroline Herschel was born into a large family in Hanover, Germany. Her father, a musician in the Hanoverian guard, encouraged the development of her musical talents. However, because she was a girl, she received only a minimum of formal education and instead was expected to help to care for her family. She assisted with household management in Hanover until she was 22, when her brother WILLIAM HERSCHEL, who was a music instructor in Bath, England, invited her to come to

England to assist him. (She was allowed to go only after her brother promised to pay the family for a maid to replace her.) Once settled in Bath, Caroline Herschel trained and performed successfully as a singer. But her brother's interests also included astronomy, and she gave up a budding musical career to assist him in that field.

In 1781 her brother discovered the planet Uranus. The following year, both she and her brother gave their last musical performance and thereafter directed their attentions to astronomy. In 1782 her brother was appointed court astronomer to King George III. Five years later, she was appointed officially as William Herschel's paid assistant.

In 1782 her brother gave Herschel her own telescope. Her earliest independent discoveries included three new nebulae, and by the time a year had passed, she had discovered several new star clusters and 14 new nebulae. In 1786 she became the first woman recognized for the discovery of a comet. She also collaborated with and supported her brother in many studies of double stars.

She undertook extensive work in cataloging stars. She revised the star catalogues of the seventeenth-century astronomer John Flamstead, correcting their errors and facilitating their use. At the age of 77, she completed an immense work called *A Catalogue of the Nebulae* on the positions of some 2,500 nebulae, or faintly luminous patches among the stars (those within the Milky Way are now recognized as gas clouds; those beyond it are now known as galaxies). The Herschels' interest in and knowledge of nebulae led them to the inspired speculation that some of these mysterious cloudy-appearing objects might actually be independent galaxies far beyond the Milky Way.

With her brother, Herschel built new styles of improved reflecting telescopes. They not only made larger ones, they also invented methods of casting and mounting them. Caroline Herschel herself made the mirrors for a custom-built 12-meter telescope that they used for their work.

In her later years, Herschel received international recognition for her work as an astronomer: the Royal Astronomical Society elected her an honorary member in 1828, and the King of Prussia awarded her a gold medal in science in 1846. Caroline Herschel died in 1848 at the age of 97.

Legacy

Caroline Herschel's work in discovering and cataloguing stars and nebulae added significantly to the understanding of the night sky.

Herschel's revision of Flamstead's star catalogues helped to further the ongoing and accurate charting of astronomical objects. *A Catalogue of the Nebulae* became a benchmark for future observers, amateur and professional alike, who have searched for luminous bodies beyond our solar system.

She and her brother introduced pioneering ideas in cosmology, including the idea that some of the nebulae they studied might actually be extremely distant, independent galaxies rather than objects within the Milky Way. This was the first intimation of the true immensity of the universe, and it extended the scope of astronomy from interstellar to intergalactic space. The new telescopes that Herschel made with her brother also helped to extend the reach of astronomers.

Caroline Herschel has a unique place in the history of women in science: without any systematic education or training, she became a great astronomer and helped to open up astronomy to many other star gazers. Her willingness to learn a new discipline and work with her elder brother, coupled with personal initiative and independent thinking, helped make their partnership productive of both individual and shared accomplishments, and ultimately led to advances in the field of astronomy.

Hertzenberg

World Events		Herschel's Life
	1750	Caroline Lucretia Herschel is born
	1772	Herschel goes to England to join her brother William
United States independence	1776	
	1782-83	Herschel discovers new nebulae
	1786	Herschel becomes first woman recognized for discovering comet
French Revolution	1789	
	1797	Herschel announces discovery of seven new comets
Napoleonic Wars in Europe	1803-15	
	1822	Herschel completes *A Catalogue of the Nebulae*
	1846	Herschel awarded gold medal for science by King of Prussia
	1848	Herschel dies
Germany is united	1871	

For Further Reading:

Alic, Margaret. *Hypatia's Heritage: A History of Women in Science from Antiquity through the Nineteenth Century.* Boston: Beacon Press, 1986.

Ogilvie, Marilyn Bailey. *Women in Science: Antiquity through the Nineteenth Century.* Cambridge, Mass.: MIT Press, 1986.

Osen, Lynn W. *Women in Mathematics.* Cambridge, Mass.: MIT Press, 1975.

Herschel, William

Founder of Star-Related
Astronomical Research
1738-1822

Life and Work

William Herschel was the founder of star-related astronomical research. He constructed high-magnification telescopes, discovered Uranus, proposed that nebulae consist of stars, and identified infrared radiation.

Herschel was born Friedrich Wilhelm Herschel on November 15, 1738, in Hanover, Germany. Following in his father's footsteps, he became an army musician. Upon the French occupation of Hanover in 1757, he fled to Bath, England, where he developed a reputation as a music teacher, performer, and composer. He studied a book on harmonics by a professor of astronomy at Cambridge; he liked the book so much that he read the author's book on optics, which led to his interest in telescope construction and astronomical observation. He was dissatisfied with the quality of existing telescopes and was able to achieve better resolution by grinding larger mirrors and fabricating improved eyepieces.

In 1781 Herschel's first significant discovery, the planet Uranus, resulted in his appointment as court astronomer to King George III. This provided him and his sister, CAROLINE HERSCHEL, who had joined him in 1772, with sufficient funds to pursue astronomy exclusively. He soon spotted two unknown satellites of Uranus and two of Saturn.

The high magnification of Herschel's telescopes led to a series of astral discoveries. In 1784 he resolved nebulae (milky splotches in the sky), which had until then defied explanation, into individual stars. Herschel concluded that all nebulae are clusters of many stars, a theory he had to modify in 1790 when he observed a nebula that consisted of a single star surrounded by what appeared to be luminous fluid. Nebulae that contain many stars are now known as galaxies.

In 1789 Herschel and his sister completed construction of their largest telescope; it had a diameter of 122 centimeters and a focal length of 12 meters. He catalogued more than 2,000 nebulae and 800 two-star systems. In 1800 he was the first to detect infrared radiation. Tracking relative positions of stars across time, he produced calculations of the direction and velocity of the solar system through space.

Some invalid theories emerged from his research, such as the idea that all stars have the same brightness and that differences in appearance are attributable only to varying distance from Earth. He also incorrectly assumed that all star clusters inevitably condense into more tightly packed star clusters. But he was the first to find evidence that the Sun is moving in space and that stars are not distributed evenly but are arranged in a huge elongated disk (as with the Milky Way galaxy). These were novel conclusions that were upheld, with minor modifications, after his death.

Herschel died in Slough, England, on August 25, 1822.

Legacy

With his sister's assistance, Herschel added thousands of elements to the astronomical record, establishing the importance of high-magnification equipment and tireless observation of the night sky.

Herschel's conclusions concerning the movement of the Sun and the shape of the stellar system in which it lies formed the basis for further discoveries about the structure of stellar systems in general. By the 1830s instrumentation had become sophisticated enough to measure interstellar distances; in 1904 measuring distances with the use of statistical methodology helped Jacobus Kapteyn to conclude that stars in the Milky Way stream in opposite directions and that the shape of the stellar system is elliptical. In 1923 EDWIN HUBBLE showed that the "nebula" called Andromeda was actually a galaxy beyond our own Milky Way.

The telescopes the Herschels built, with their increased powers of magnification, helped succeeding generations of astronomers. Incremental improvements in telescopy gradually augmented astronomical observers' abilities to learn more about the solar system, our galaxy, and galaxies beyond.

Herschel's legacy is also reflected in the work of his son John, who explored a wide range of scientific pursuits. He experimented with polarized light, constructed a geomagnetic survey of the Antarctic, and performed important research in chemistry that contributed to the development of photography. From 1834 to 1839 he lived at the southern tip of South Africa and mapped most of the stars in the southern sky.

Schuyler

WORLD EVENTS		HERSCHEL'S LIFE
Peace of Utrecht settles War of Spanish Succession	1713-15	
	1738	William Herschel is born
	1757	Herschel relocates to England
	1772	Caroline Herschel joins her brother in England
United States independence	1776	
	1781	Herschel discovers Uranus
		Herschel becomes court astronomer
	1784	Herschel shows that some nebulae consist of many stars
French Revolution	1789	Herschel and sister build largest telescope
	1800	Herschel discovers infrared radiation
Napoleonic Wars in Europe	1803-15	
	1822	Herschel dies

For Further Reading:

Crowe, Michael J. *Modern Theories of the Universe: From Herschel to Hubble.* New York: Dover, 1994.

Hoskin, Michael, ed. *The Cambridge Illustrated History of Astronomy.* Cambridge: Cambridge University Press, 1997.

North, J. D. *The Measure of the Universe: A History of Modern Cosmology.* New York: Dover, 1990.

Hess, Harry Hammond

Originator of
Theory of Plate Tectonics
1906-1969

Life and Work

Harry Hammond Hess originated the theory of plate tectonics and revolutionized the study of geology in the twentieth century.

World Events	Hess's Life
	1906 Harry Hess is born
World War I 1914-18	
	1927 Hess earns degree in geology at Yale; starts field work in Africa
	1931 Hess begins research on oceans
	1932 Hess completes Ph.D. in geology at Princeton
	1934 Hess begins teaching at Princeton
World War II 1939-45	
	1941-45 Hess serves in U.S. Navy during World War II
Korean War 1950-53	
	1960 Revolutionary essay, "History of Ocean Basins," is shared with colleagues
	1962 "History of Ocean Basins" is published
	1969 Hess dies
End of Vietnam War 1975	

Hess was born in New York City on May 24, 1906. When he was five years old, his parents had a portrait made of him wearing a sailor suit; the likeness was titled "The Future Admiral." His parents seemed to anticipate his naval career and his life-long interest in ocean basins. Hess attended public schools in New York and New Jersey and earned a geology degree at Yale University in 1927.

After two years of fieldwork exploring for minerals in Rhodesia (now Zambia), he started graduate work at Princeton University. His research on oceans began in 1931 aboard a U.S. Navy submarine with Dutch geophysicist Felix Vening-Meinesz, who had developed a method for making gravity measurements at sea. Hess completed his Ph.D. in geology in 1932 and began teaching at Princeton in 1934 where he eventually became a professor in 1948. He maintained this post for nearly 20 years until 1966.

When World War II involved the United States in 1941, Hess was called from the Naval Reserve to active service. His first assignment led to the successful development of a system to detect German submarines in the North Atlantic. While captain of an assault transport, Hess did extensive echo-soundings for studies of flat-topped underwater seamounts (guyots). After the war, he returned to the Naval Reserve, attaining the rank of rear admiral in 1961.

As early as 1954, Hess declared that ocean floors are qualitatively different from continents. In 1959 he wrote that oceanic mantle ridges (undersea mountain ranges) were belts extruded over areas where hot rock material was pushing up through the ocean floor. Thus, unlike mountains on land, which are formed by folding and compression, seamounts resulted from mantle processes (spreading of the ocean floor and rising of hot fluid materials above cooler ones).

Hess clearly stated his revolutionary theory in geology in 1960 when he presented "geopoetry" in his essay "History of Ocean Basins" to colleagues in a mimeographed preprint; it was published in 1962.

Hess's dynamic synthesis considered the surface of Earth as an integrated system in which continents are passive passengers on moving ocean floors, not ships driving through them. He hypothesized that continents do not descend with sea floors into trenches but "ride over" them, colliding with each other, fusing, pushing up mountains, and/or breaking apart as they move.

Hess developed an X-ray method of analyzing composition of lunar rocks in 1968. He received numerous national and international awards for his contributions to science before his death on August 25, 1969, in Falmouth, Massachusetts.

Legacy

Harry Hess's theory on the origin of ocean basins provided the first comprehensive explanation of mountain building and continental drift. Spreading of the ocean floor was part of a dynamic solution, the plate tectonics model, to many geological puzzles.

Hess's model of Earth's history was championed and given the name of "sea floor spreading" by Robert Dietz, another American geologist, in 1961. Two years later, F. J. Vine and D. H. Matthews discussed magnetic anomalies along the North Atlantic ridge as evidence supporting sea-floor spreading.

At a London conference of the Royal Society in 1964, Edward Bullard of Cambridge presented a computer-guided map showing the close fit of the Atlantic continents at their true margins 1,000 meters underwater. This provided a convincing model to be tested by geochronological methods not available to ALFRED WEGENER when he proposed continental drift 50 years earlier. Confirming evidence of Hess's ideas came in less than a decade. Fossil records from underwater core samples, magnetic stripes in sea floor rocks, and oceanic rocks in Idaho were among many assorted facts that fit into the orderly framework Hess had provided.

Hess's revolutionary ideas about mobile crust (i.e., Earth's surface) contradicted popular views of a static Earth, but his systematic and comprehensive model became as central to geology as CHARLES DARWIN's ideas are to biology. His contributions and encouragement of colleagues also put North American scientists into the forefront of studies on plate tectonics.

Simonis

For Further Reading:

Anderson, A. H. *The Drifting Continents.* New York: Putnam, 1971.

Martin, Ursula. *Continental Drift: Evolution of a Concept.* Washington, D.C.: Smithsonian Institution Press, 1973.

Miller, Russell. *Continents in Collision.* Alexandria, Va.: Time-Life Books, 1983.

Hippocrates

Founder of Medicine

c. 460-c. 377 B.C.E.

Life and Work

Hippocrates founded clinical medical practice. He was an important figure in separating ancient medicine from superstition.

Hippocrates was born about 460 B.C.E. on the island of Cos, Greece. Little is known about his life except that he traveled widely in Greece and Asia Minor, taught medical students at Cos and abroad, and was renowned during his lifetime as an exceptional physician. Although a few biographies of Hippocrates were written within several hundred years of his death, modern scholars regard these as fictitious.

An anthology of 70 works concerning medical practice and philosophies makes up what is known as the *Hippocratic Collection*. Although the collection bears his name, Hippocrates probably wrote only six of the works. As the date of the earliest work in the collection precedes the latest (around 330 B.C.E.) by at least 100 years, the others were likely written by Hippocrates' students and their followers. The works therefore vary in style and reflect many beliefs. They include such topics as anatomy, clinical methods, diseases of women, epilepsy, prognosis, surgery, and medical ethics. Many authors

prescribe diet, rest, and exercise, rather than drugs, as disease treatments.

Unlike previous medical doctrine, the authors of the *Hippocratic Collection* doubted that divine intervention was the cause of most diseases; instead they blamed natural and environmental factors. They suggested that diet and climate can affect the health of a patient.

Hippocrates and his disciples advocated direct clinical observation and believed that a physician can predict the course of a disease after having witnessed numerous previous cases. They claimed that an effective healing method includes rational speculation.

The theory of the four bodily humors—blood, phlegm, yellow bile, and black bile—appears in the *Hippocratic Collection*. The authors believed that health relies on maintaining a proper balance of the bodily humors.

Hippocrates died about 377 B.C.E. in Larissa, Greece.

Legacy

Hippocrates' scientific legacy rests on the doctrines outlined in the *Hippocratic Collection*, many of which dominated medical theory for centuries.

Hippocrates transformed the basis of medical practice from religion to science. Prior to the introduction of his teachings, religious and magical beliefs dominated medicine in Europe. The Hippocratic traditions of observing symptoms and forming rational speculations moved medical methods in a scientific direction. Hippocrates' position that diseases can have environmental causes stood in opposition to the belief that diseases are mysterious, malicious, unnatural entities.

Modern scholars claim that the *Hippocratic Collection* probably belonged to the library at the Cos medical school, at which Hippocrates taught. The collection was likely passed to the library at Alexandria in the third or second century B.C.E. There the works were edited and then made available to medical instructors and students.

The Hippocratic doctrines were studied and developed by GALEN, a second-century Greek physician who practiced in Rome. Galen's medical approach, heavily informed by the Hippocratic tradition but with more emphasis on understanding anatomy and

physiology, was practiced for more than 1,400 years. Arab physicians systematized Galen's work in the eleventh and twelfth centuries, and medieval Western universities instituted Galen's views within their curricula.

Hippocrates' theory of the four bodily humors, accepted and promulgated by Galen, limited medical understanding throughout the 1,400 years of the dominance of Galen's ideas in medical practice. Not until the 1500s, with the work of such physicians and scientists as ANDREAS VESALIUS and PARACELSUS, did such misconceptions begin to be dislodged.

Many modern medical schools require graduates to take the Hippocratic oath, a promise embodying a code of ethical medical behavior. Although Hippocrates did not write the oath, his name lives on as a symbol of the ethics modern society values in its physicians.

Schuyler

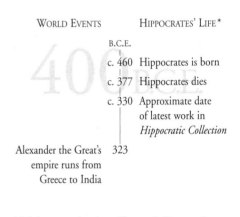

WORLD EVENTS		HIPPOCRATES' LIFE*
	B.C.E.	
	c. 460	Hippocrates is born
	c. 377	Hippocrates dies
	c. 330	Approximate date of latest work in *Hippocratic Collection*
Alexander the Great's empire runs from Greece to India	323	

* Scholars cannot date the specific events in Hippocrates' life with accuracy.

For Further Reading:

Pinault, Jody Rubin. *Hippocratic Lives and Legends*. Boston: Brill Academic Publishers, 1992.

Smith, Wesley D. *The Hippocratic Tradition*. Ithaca, N. Y.: Cornell University Press, 1979.

Hooke, Robert

Identifier of Cells; Formulator of
Laws of Elasticity
1635-1703

Life and Work

Robert Hooke was among the most versatile scientists in history. He devised and improved several scientific instruments, formulated fundamental ideas in physics, and was the first to describe cells.

Hooke was born on the Isle of Wight, England, on July 18, 1635. Confined to his home by illness, he kept busy by making a variety of mechanical toys. He was tutored by his father, a poor cleric. Orphaned at age 13, Hooke moved to London where he worked as an artist's apprentice until he saved enough for tuition to Westminster College and later to Oxford University.

In 1655 he helped chemist ROBERT BOYLE, one of his professors at Oxford, devise an improved air pump. Using the pump to create a vacuum in a jar, Hooke demonstrated that a feather and a coin released simultaneously from the top of the jar fall at the same speed in the absence of air. He also did original work on surface tension and capillary action.

Hooke proceeded to London in 1660 and joined Boyle, ISAAC NEWTON, and others in founding the Royal Society of London for the Improvement of Natural Knowledge. As Curator of Experiments, Hooke was responsible for presenting a demonstration at each weekly meeting.

Also in 1660, while investigating how materials respond to stress, Hooke formulated a law of elasticity stating that the amount of deformation of a material is directly proportional to the force (such as stretching, compressing, bending, and twisting) applied to it. Using this information, he designed a spring balance for controlling timepieces. However, CHRISTIAAN HUYGENS was the first to incorporate the device into a working model.

Hooke's *Micrographia* appeared in 1665; it contained observations made under his newly invented compound microscope, including detailed drawings of many living things, structures invisible to the naked eye. He described arrays of tiny compartments in a sliver of cork and named them *cells* from the Latin for "small chambers." That same year, he was appointed Professor of Geometry at Gresham College and remained there for the rest of his life.

He built a reflecting telescope, made numerous astronomical discoveries, and improved the pendulum clock and the barometer. He analyzed the nature of combustion and formulated theories of light diffraction and planetary motion. He devised several instruments to record weather conditions and methods to systemize that data.

By 1678 Hooke had developed a qualitative idea of gravitational force, but he failed to add a rigorous mathematical description; this task was left to Newton, who published his theory of gravity in 1687. Hooke thereafter engaged in bitter quarrels with Newton, claiming credit for his theory of gravity.

Hooke died in London on March 3, 1703.

Legacy

Hooke's innovative experiments and various theories were influential across a wide range of disciplines. His ideas often preceded those of others who developed his ideas more thoroughly and obtained full credit.

Hooke's first scientific endeavor, the construction of an air pump, led to Boyle's investigations of the properties of air. Boyle used the pump to demonstrate the role of air in combustion, respiration, and sound transmission. His examination of other gases resulted in the formulation of Boyle's Law, which states the mathematical relationship between the pressure and volume of gases.

Hooke's work with springs and pendulums contributed to the technology of timepieces. His efforts, combined with those of Huygens, produced clocks that could tell time to the nearest minute.

In 1722 astronomer EDMOND HALLEY acquired an advanced telescope constructed by Hooke and with it made daily observations of the moon for 18 and a half years. The data enabled him to create accurate lunar tables, useful for calculating longitudes on the open sea.

Hooke's observation of plant cells proved to be far ahead of the understanding of microscopic biology. In 1839, nearly two centuries later, the pioneering work of Matthias Schleiden and Theodor Schwann showed cells to be the building blocks of all life.

Hooke founded the study of the effects of stresses and strains on elastic materials, which has been fundamental to the development of architectural design and various engineering fields. He also played a key role in making a professional body of the Royal Society, which remains the primary forum for the exchange and encouragement of scientific ideas and experimentation in England.

Schuyler

For Further Reading:
Drake, Ellen. *Restless Genius: Robert Hooke and His Earthly Thoughts.* New York: Oxford University Press, 1996.
Nichols, Richard. *The Diaries of Robert Hooke, the Leonardo of London, 1635 to 1703.* Lewes, England: Book Guild, 1994.

Hopper, Grace Murray

Inventor of the Computer
Language Compiler
1906-1992

Life and Work

Grace Murray Hopper made groundbreaking contributions to the early development of computer technology; she invented the first computer language compilers and spearheaded the development of the first standardized computer language, COBOL.

Hopper was born in New York City on December 9, 1906. She did her undergraduate work in mathematics and physics at Vassar College, earning her degree in 1928. She earned her Ph.D in mathematics from Yale University in 1934. Hopper returned to Vassar as a faculty member in mathematics. She remained on the faculty until 1943, when she joined the Naval Reserve. The following year, HOWARD AIKEN's Computation Project at Harvard hired Hopper to work on the Mark I, the first full-scale programmable computer in the United States, which debuted that year.

Hopper joined the Eckert-Mauchly Corporation in 1949. Eckert-Mauchly completed UNIVAC I, the first large-scale commercial computer, and Hopper headed the UNIVAC programming group. The company

merged with Sperry Rand; Hopper continued to work as a senior programmer in the UNIVAC division. In 1952, while at Sperry Rand, she developed A-0, the first compiler. Compilers translate higher-level programmed instructions to machine-level code. She also developed B-0, an English-language-based compiler, later called FLOW-MATIC; it allowed UNIVAC I and II to understand over 20 programmed English-based statements.

Hopper then headed the team that developed the first standardized computer language, COBOL (Common Business-Oriented Language). Previously, each new computer that was developed had used its own specific coding for programmed instructions. COBOL was introduced in 1960.

Hopper was forced to retire from the Naval Reserves in 1967 due to her age; however, the Navy recalled her to active duty seven months later to help standardize the Navy's computer languages. She lectured on the topic of standardization frequently, and covered other topics as well, including her criticism of the industry's bias toward large computers.

Hopper retired from Sperry Rand in 1971. On her retirement, the UNIVAC division established the Association of Computing Machinery's (ACM's) annual Grace Murray Hopper Award; the award goes to an impressive young computer professional active in the field.

She retired from the Navy as a rear admiral in 1986 but continued as a senior consultant with Digital Equipment Corporation. She died on January 1, 1992, in Arlington, Virginia.

Legacy

Hopper came to computer science as the field was being invented. She helped to steer its course through her groundbreaking innovations in computer languages, including the development of the first compilers and standardized computer language (COBOL), and through her insistence on establishing industry standards.

Hopper's work to create standards in computer languages was important to the early development of the field. Lack of standards was not only costly to government and private companies, it limited the development of computers, as incompatibilities between languages used by different machines, even within single

companies, caused duplication of effort and great waste. Frequently, hardware and software had to be discarded as new generations of machines and languages developed that were not backwards-compatible to older ones.

Hopper's FLOW-MATIC, the first English-language data-processing compiler, represented a huge step forward in computing capabilities. Early computer languages were composed of mathematical symbols; Hopper's insistence that such languages could be written in English made the development of new languages and programs an easier task. The common business-oriented language COBOL, which Hopper's team developed, was the first English-based programming language to have wide-scale commercial applications. While it is not used today to write new programs, its legacy is still felt; many older programs still in service today were written in COBOL.

Steinberg

WORLD EVENTS	HOPPER'S LIFE
1906	Grace Hopper is born
World War I 1914-18	
1928	Hopper graduates from Vassar
1934	Hopper receives Ph.D. from Yale
World War II 1939-45	
1944	Hopper works programming Mark I
1949	Hopper begins work on UNIVAC
Korean War 1950-53	
1952	Hopper develops first compiler, called A-0
1960	Hopper heads team that develops COBOL
End of Vietnam War 1975	
Dissolution of 1991 Soviet Union	
1992	Hopper dies

For Further Reading:
Goldstine, Herman. *The Computer from Pascal to von Neumann.* Princeton, N.J.: Princeton University Press, 1972.
Palfreman, Jon, and Doron Swade. *The Dream Machine: Exploring the Computer Age.* London: BBC Books, 1991.

Howe, Elias

Inventor of the Sewing Machine
1819-1867

World Events	Howe's Life
Napoleonic Wars 1803-15 in Europe	
	1819 Elias Howe is born
	1836 Howe is apprenticed to machinist
	1846 Howe patents his new sewing machine and leaves for England soon after
	1849 Howe returns to U.S. and initiates lawsuits against patent infringement
	1854 Howe is granted royalties from sewing-machine sales after settlement of lawsuits
United States 1861-65 Civil War	
	1867 Howe dies
Germany is united 1871	

Life and Work

Elias Howe invented the first sewing machine practical for home and industry; the basic elements of his model are used in the design of modern sewing machines.

Howe was born on July 9, 1819, in Spencer, Massachusetts. He grew up on a farm and was fascinated by the machines used in his father's gristmills and sawmills. Apprenticed to a manufacturer of textile machinery at age 17, he subsequently worked as a machinist in the cotton-machinery industry. In the late 1830s, he overheard his superior tell someone that whoever could devise a small automatic sewing machine would make a fortune; he began to work diligently on the design of such a machine.

In 1846 Howe patented his sewing machine, which incorporated two threads, the first lock-stitch mechanism, a curved needle with its eye near the point, and an automatic shuttle that moved one thread beneath the cloth. There was almost no local interest in his invention, so he soon sailed for England, where a corset manufacturer bought his patent and paid him a low wage to customize the machines for use in the sewing of corsets. Howe's financial situation declined in England, so he returned to the United States in 1849.

Upon his return the United States, he found that others (in particular Isaac Singer) were successfully selling sewing machines using designs protected under Howe's patent, and he brought lawsuits against them for patent infringement. A long court battle ensued, ending in 1854 with the establishment of the seniority of Howe's patent. He collected royalties from all sewing-machine sales until his patent expired in 1867; he died in the same year on October 3, in Brooklyn, New York, leaving an estate worth $2 million.

Legacy

Howe's introduction of the sewing machine revolutionized the garment-manufacturing industry and made sewing less laborious for those who sewed at home.

Like many technological inventions, the earliest sewing machines were resisted by tradespeople whose jobs were threatened by the prospect of mechanization. Indeed, in the 1840s the shop of French machinist Barthélemy Thimmonier, who had invented a crude sewing machine that was not to have a commercial future, was destroyed by tailors and seamstresses fearing loss of their livelihood.

The Howe-Singer court case brought public attention to the sewing machine. As people recognized the potential for speed and efficiency offered by the machine, they were willing to pay the relatively high cost. Improvements in the machine and manufacturing efficiencies eventually made the machines more affordable.

The sewing machine is credited by some historians as the first item of the Industrial Revolution to significantly lighten the workload of the housewife. While an experienced seamstress could complete as many as 30 stitches per minute, Howe's earliest machines could reel out 250 stitches per minute. Machines could also sew tough fabrics that were difficult, dangerous, or impossible to sew by hand.

Technological refinements made since Howe's time have produced many versions of the sewing machine, from the common household appliance to the sophisticated, powerful machines used in various industries. Many machines are designed to sew particular items, such as hats, shoes, embroidered material, denim, leather goods, silk, and hardy outdoor clothing. Howe's machines were powered by a hand-driven wheel; the advent of electricity in the late nineteenth century made sewing machines faster and easier to operate. The newest machines take advantage of microelectronics and incorporate push-button controls.

However, Howe's sewing machine included basic elements that remain unchanged in modern sewing machines. Needles still have their eyes located near their points. Modern machines use two threads, one below and one above the cloth, to create a lock stitch.

Schuyler

For Further Reading:
Cooper, Grace Rogers. *The Sewing Machine: Its Invention and Development.* Washington, D.C.: Smithsonian Institution Press, 1976.
Eastley, Charles M. *The Singer Saga.* Devon, England: Merlin Books, 1983.
Godfrey, Frank P. *An International History of the Sewing Machine.* London: R. Hale, 1982.

Hubble, Edwin

Discoverer of
Galaxies Beyond Milky Way
1889-1953

Life and Work

Edwin Hubble not only verified the existence of millions of galaxies like the Milky Way, he also discovered that most galaxies are moving away from each other, expanding the universe.

Edwin Hubble was born on November 20, 1889, in Marshfield, Missouri. Excelling at school and interested in many subjects, he received a scholarship to the University of Chicago. Upon graduating in 1910 with a degree in mathematics and astronomy, he decided to study law at Oxford University and then worked briefly as a lawyer in Kentucky. Quickly bored with law he returned to astronomy in 1914 when he joined the staff of Yerkes Observatory at the University of Chicago.

In 1917 Hubble received a Ph.D. in astronomy from the University of Chicago for research on nebulae, a word then used to describe distant celestial objects not recognizable as stars (most nebulae were thought to be clouds of gas or clumps of stars). After serving in World War I Hubble joined the Mount Wilson Observatory in Pasadena, California, in 1919. The observatory housed a 100-inch telescope, then the largest in the world.

In 1923 Hubble identified a Cepheid star (a pulsating star with a characteristic luminosity) in the Andromeda nebula. Using equations formulated by American astronomer Henrietta Leavitt, Hubble calculated the distance from Earth to the Andromeda Cepheid to be about 900,000 light years. It was known that our galaxy—which was thought to be the whole universe—was only about 100,000 light years across. Thus Hubble's discovery, announced in 1924, showed that the universe is much larger than had been thought.

Hubble's extensive observations of nebulae also established that these mysterious objects were actually entire galaxies. In 1925 he introduced a classification plan for galaxies based on their shapes.

Hubble then began to consider the work of American astronomers Vesto Slipher and Milton Humason, who had calculated the radial velocities of numerous galaxies (radial velocity is the rate of an object's movement away from or toward Earth). Analyzing their data Hubble noted a correlation between a galaxy's distance from Earth and its radial velocity. In 1929 he proposed a mathematical relationship, now known as Hubble's Law, stating that galaxies' radial velocities increase the further away they get. This supported the theory that the universe is expanding. It also allowed a calculation of the age of the universe, which Hubble estimated to be about two billion years.

In his later years Hubble wrote, lectured, served in World War II, and supported the development of the Mount Palomar Observatory in California. He died in San Marino, California, on September 28, 1953.

Legacy

Hubble's discoveries overturned traditional astronomical ideas and provided foundations for the modern understanding of the universe's size and age.

Hubble's calculation of the distance to the Andromeda galaxy dislodged the long-held assumption that the universe extends only to ends of Earth's galaxy (the Milky Way). This required a major perceptual shift as it suggested that the universe was billions of times larger than was previously assumed.

The new understanding of the universe's size launched an explosion of research aimed at discovering what lies beyond the Milky Way. It also led to speculations that if the universe is so immensely huge, it is possible that appropriate conditions for life exist in other galaxies.

Hubble's Law represented the first coherent confirmation of the expanding-universe theory. In 1917 Dutch astronomer Willem de Sitter and Russian astronomer Aleksandr Friedmann devised a mathematical model of universe expansion. Their model attracted little support, but Hubble's Law made it a basic element of the most widely accepted theories in cosmology (the study of the origin and structure of the universe). The Big Bang theory (as it came to be known), first proposed in the late 1920s, grew out of Hubble's Law. It suggests that the universe began as a highly condensed point of matter that exploded, expanding outward and continuing to expand today.

Hubble's Law has also allowed continuously refined calculations of the age of the universe. The calculations rely on a number called the Hubble constant (part of Hubble's Law), which is the ratio of the radial velocity of galaxies to their distance from Earth. Hubble's calculation of this constant was higher than current research suggests, and his estimate of the universe's age was consequently two billion years, much smaller than the 12 to 20 billion years currently estimated.

The National Aeronautics and Space Administration (NASA) launched a highly advanced telescope into space in 1990. In recognition of Hubble's legacy, this outer space observatory was named the Hubble Space Telescope. While plagued with technical problems, which were fixed in December of 1993 on a mission of NASA's space shuttle *Endeavor,* the telescope now sends images to NASA's scientists, and continuously compiles further records of celestial phenomena, a mission Edwin Hubble would be proud of.

Schuyler

For Further Reading:

Christianson, Gale E. *Edwin Hubble: Mariner of the Nebulae.* New York: Farrar, Straus & Giroux, 1995.

Clark, Stuart. *Stars and Atoms: From the Big Bang to the Solar System.* New York: Oxford University Press, 1995.

Sharov, Aleksandre S. *Edwin Hubble, The Discoverer of the Big Bang Universe.* Translated by Igor D. Novikov. New York: Cambridge University Press, 1993.

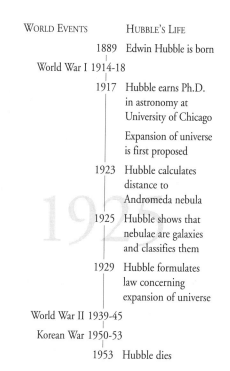

World Events	Hubble's Life
1889	Edwin Hubble is born
World War I 1914-18	
1917	Hubble earns Ph.D. in astronomy at University of Chicago
	Expansion of universe is first proposed
1923	Hubble calculates distance to Andromeda nebula
1925	Hubble shows that nebulae are galaxies and classifies them
1929	Hubble formulates law concerning expansion of universe
World War II 1939-45	
Korean War 1950-53	
1953	Hubble dies

Huygens, Christiaan

Originator of Wave Theory of Light;
Inventor of Pendulum Clock

1629-1695

Life and Work

Christiaan Huygens introduced the wave theory of light, accurately described the rings of Saturn, invented the pendulum clock, and established some of the basic principles of the physics of colliding and rotating bodies.

Huygens was born on April 14, 1629, in The Hague, Holland (the Netherlands), to a prominent, wealthy family. At an early age he exhibited a talent for mathematics and art, impressing his father's scientist friends with his geometrical drawings. His father was a friend of the French philosopher-scientist RENÉ DESCARTES, who established the dominance of mechanistic explanations in post-seventeenth-century science. Huygens's exposure to Descartes's ideas influenced his later scientific methodology. He studied law

and mathematics at the University of Leiden and the College of Breda in Holland.

In 1655 Huygens discovered Titan, Saturn's largest satellite, by utilizing a telescope equipped with homemade, sharply defining lenses. Observing the rings of Saturn in 1659, he was the first to describe them accurately. In 1656, motivated by the need for precise time measurements associated with astronomical observations, he followed GALILEO's suggestion and constructed the first pendulum-regulated clock. It was an immediate success; its accuracy was better than that of any other clock available at that time. Huygens himself improved on it later by inventing a balance spring regulator that made pocket watches possible.

Huygens's publications in astronomy and mathematics earned him wide recognition. At the invitation of King Louis XIV, who asked him to supervise current French scientific research, he moved to France in 1666. There he became friendly with noted scientists BLAISE PASCAL and GOTTFRIED WILHELM LEIBNIZ and helped found the French Academy of Sciences.

In 1669 Huygens published a groundbreaking paper on the dynamics of colliding elastic bodies. Four years later, his influential work *Horologium Oscillatorium* appeared. It encompassed theories and derivations concerning the mathematics of curvatures, pendulums, dynamics, and centripetal force (a force that seeks a central point, causing objects to travel in a circle).

Huygens visited The Hague in 1681, and the conflict between Holland and France prevented him from ever returning to France. In 1689 he traveled to London, where he met ISAAC NEWTON and presented to the Royal Society his own (incorrect) theory of gravitation. His *Treatise on Light* was published in 1690 (although he had written it in 1673); it introduced the principle that light is an advancing wave and that every point on this wave front is a source of secondary spherical wavelets. His wave theory successfully explained most light phenomena in terms of wavelength and frequency. He also spent years constructing lenses of focal length as long as 210 feet to improve telescopes available at the time.

Huygens died on July 8, 1695, in The Hague, after years of chronic illness.

Legacy

Huygens is sometimes considered, after Newton, to be the most influential physical scientist of the late seventeenth century.

Many of Huygens's mathematical proofs, representing solutions to longstanding problems, advanced several areas of physics. Dynamics, the study of collisions, was essentially established by his seminal formulations on colliding bodies, and his work on the physics of centripetal force provided a basis for subsequent research on rotating bodies. His derivations regarding the physics of pendulums contributed to Newton's development of the theory of gravity.

Huygens's wave theory of light was perhaps his most significant achievement. It provided the best-known explanation for the laws of reflection and refraction of light, and it allowed him to predict, correctly, that light travels more slowly through denser media. He differed with Newton, who advocated a particle theory of light, and it was only in the twentieth century that the two theories were reconciled by the understanding that they are both correct but applicable to different experimental situations.

Huygens also introduced new methods of grinding and polishing telescope lenses and invented a telescope eyepiece that bears his name. Such technical improvements contributed to the progress of astronomy in following generations.

Schuyler

WORLD EVENTS	HUYGENS'S LIFE
Thirty Years' War 1618-48 in Europe	
1629	Christiaan Huygens is born
1655	Huygens discovers Saturn's moon, Titan
1656	Huygens invents pendulum clock
1659	Huygens describes Saturn's rings
1669	Huygens publishes work on colliding elastic bodies
1673	Huygens's *Horologium Oscillatorium* is published
England's Glorious Revolution 1688	
1689	Huygens meets Isaac Newton
1690	Huygens's *Treatise on Light* is published
1695	Huygens dies
Peace of Utrecht 1713-15 settles War of Spanish Succession	

For Further Reading:

Chappell, Vere, ed. *Seventeenth-century Natural Scientists*. New York: Garland Publishing, 1992.

Yoder, Joella G. *Unrolling Time: Christiaan Huygens and the Mathematization of Nature*. Cambridge: Cambridge University Press, 1988.

Hypatia of Alexandria

First Recognized Woman
Mathematician and Scientist
c. 370-415

Life and Work

The first woman mathematician in recorded history, Hypatia was also an astronomer, inventor, and philosopher. She is the earliest documented woman scientist.

Hypatia was born in Alexandria, Egypt, around the year 370. We know nothing of her mother, but we do know that she was educated by her father, Theon, a distinguished mathematician and astronomer at the Museum, the Alexandrian institution of higher learning, which contained the famous Alexandrian library. After her formal education, which evidently included mathematics and astronomy, she traveled to Italy and Athens, where she was the guest of Plutarch the Younger, the biographical writer and philosopher, and his daughter Ascelepigenia.

After travelling abroad to institutions of higher learning around the Mediterranean, she returned to teach mathematics and philosophy in Alexandria. In the year 400, she became the leader of the Neoplatonic school of Alexandria. The Neoplatonists were the last important school of Greek philosophy, active during the third century C.E., and were dedicated to the ideas of Plato. She became a widely recognized authority, and her fame attracted numerous students; she was visited by outstanding thinkers of her time.

Hypatia is best known as a philosopher and mathematician. She is the first woman known to have written on mathematical subjects, including conic sections, which she studied extensively, and she suggested refinements of the algebra of Diophantus, an earlier Egyptian mathematician.

Hypatia wrote several treatises in philosophy, mathematics, and astronomy, which are referred to in other contemporary writings; unfortunately, none have survived except for fragments of a treatise that some scholars think may have been co-authored with her father.

She was also an inventor, and has been credited with invention of the astrolabe, used to measure star positions relative to Earth, and a still for purifying water.

Hypatia died in Alexandria in 415 in a brutal assassination when she was tortured to death by religious zealots devoted to Cyril, the new Christian patriarch (religious leader) of Alexandria. Some accounts link her assassination to her association with Orestes, the non-Christian prefect of Alexandria, whom Cyril opposed.

Legacy

For 15 centuries, Hypatia was often considered the only female scientist in history. Indeed, she was the most famous of all women scientists until MARIE CURIE.

During her lifetime Hypatia became a widely recognized scholar and an authority in her fields. She led the Neoplatonist school until its downfall following her death. The Neoplatonists were responsible for keeping the thinking of Plato alive in the waning years of Roman Empire, particularly his belief in an absolute world of ideas distinct from physical reality.

Hypatia attracted widespread attention not only in her own time, but throughout subsequent history. Charles Kingsley (1815-1875) wrote a historical novel about her called *Hypatia,* which was reprinted in 1968. Her fame and influence was assured by the atrocity of her martyrdom—her violent death caused other scholars to leave Alexandria. Together with the complete destruction of the great libraries of Alexandria by 391, her death is regarded by many as marking the end of the influence of Greek philosophers and scientists of antiquity.

Hertzenberg

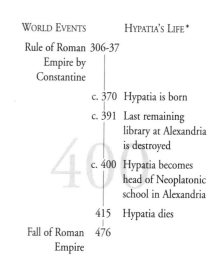

WORLD EVENTS		HYPATIA'S LIFE*
Rule of Roman Empire by Constantine	306-37	
	c. 370	Hypatia is born
	c. 391	Last remaining library at Alexandria is destroyed
	c. 400	Hypatia becomes head of Neoplatonic school in Alexandria
	415	Hypatia dies
Fall of Roman Empire	476	

* Scholars cannot date the specific events in Hypatia's life with accuracy.

For Further Reading:

Dzielska, Maria. *Hypatia of Alexandria.* Cambridge, Mass.: Harvard University Press, 1995.

Kapsis, Eda C. "Hypatia (c. 370-c. 415)." In *Women in Chemistry and Physics: A Biobibliographic Source Book,* edited by Louise S. Grinstein, Rose K. Rose, and Miriam H. Rafailovich. Westport, Conn.: Greenwood Press, 1993.

Ogilvie, Marilyn Bailey. *Women in Science: Antiquity through the Nineteenth Century.* Boston: MIT Press, 1986.

Ibn Sina (Avicenna)

Physician;
Author of Canon of Medicine
980-1037

WORLD EVENTS		IBN SINA'S LIFE
Northern Sung Dynasty begins	960	
	980	Ibn Sina is born near Bukhara (in present-day Uzbekistan)
	990	Ibn Sina has memorized the Koran
	995	Ibn Sina heals Samanid ruler
	c. 1000	Saminid Dynasty falls; Ibn Sina is forced to leave
		Ibn Sina flees to Isfahan; he eventually settles in Hamaden and begins the *Canon of Medicine* and *the Book of Healing*
	1037	Ibn Sina dies
First Crusade	1095	
	1100s	*Canon of Medicine* is translated into Latin

Life and Work

Ibn Sina produced a vast amount of written work including the influential *Canon of Medicine,* an encyclopedia covering all Western medical knowledge up to and including that of his contemporaries and himself. This volume was used as a textbook in the Middle East and Europe for centuries. (The Latinized form of Ibn Sina's name is Avicenna.)

Ibn Sina was born in 980 near Bukhara, in present-day Uzbekistan, then the capital of the Persian Samanid dynasty. He demonstrated exceptional talent at an early age, memorizing the *Koran* (the collection of divine revelations forming the basis of Islamic belief) and numerous poems by age 10. He taught himself medicine and at age 15 attended to the sick Samanid ruler, Nuh ibn Mansur. His successful treatment of the ruler earned him access to the library of the Samanid princes, where he mastered mathematics, law, Arabic literature, and ancient Greek philosophy; he became particularly proficient in Aristotelian logic and metaphysics.

After a period of travel forced by the overthrow of the Samanid dynasty (c. 1000), Ibn Sina settled in Hamadan, in west-central Iran, as court physician to the Buyid prince Shams ad-Dawlah. There he began his two great works, the *Book of Healing* and the *Canon of Medicine.* The *Book of Healing,* an immense achievement, covers nearly all of ancient theoretical and practical knowledge, including logic, natural sciences, psychology, metaphysics, and mathematics. His ideas show the influence of Aristotle and the Neoplatonists (philosophers from the third century onward who developed and incorporated Plato's metaphysical theories, particularly his idea of dualistic reality).

Ibn Sina's *Canon of Medicine* is a grand exposition of medical knowledge consisting of more than one million words. It clearly and systematically discusses all the medical and pharmaceutical material he was able to collect. It follows the basic principles of GALEN, a second-century physician who revitalized the teachings of HIPPOCRATES, Greek founder of Western medicine. Ibn Sina embraced the Hippocratic-Galenic theory of the four bodily humors (blood, phlegm, black bile, and yellow bile). The *Canon* also describes Ibn Sina's varied clinical experience. For example, he distinguished two kinds of jaundice and provided the first good description of meningitis.

In 1022 Shams ad-Dawlah was killed and his court overthrown; Ibn Sina and his followers fled to Isfahan, in modern-day Iran, and found refuge in the court of another prince, 'Ala al-Dawlah.

He spent his last years finishing the *Book of Healing,* the *Canon of Medicine,* and nearly 200 other treatises. He died in 1037 while visiting Hamadan.

Legacy

Ibn Sina's work has influenced philosophical and medical ideas and practices throughout Europe and the Islamic world, his medical treatises enjoying as much popularity as those of Hippocrates and Galen.

Ibn Sina's *Canon of Medicine* quickly became a leading medical authority and was used as a textbook in Europe until the late seventeenth century. It was translated into Latin in the twelfth century, into Hebrew in the fifteenth century, and into Arabic in the sixteenth century. Its dogmatic views supported the dominance of Galenic and Hippocratic medicine throughout the Middle Ages and into the Renaissance.

The philosophy of Ibn Sina, combined with that of fifth-century Christian theologian Saint Augustine, helped shape the thought of the medieval Franciscan monks. Their scholarship, in turn, contributed to European university curricula—thus Ibn Sina's ideas broadly infiltrated Western thought.

Ibn Sina's works are still a focus of study within modern Islam. According to some historians, he was the most influential medieval Islamic physician, philosopher, and writer, and one of the most illustrious figures of Islamic history.

Schuyler

For Further Reading:

Afnan, Soheil Muhsin. *Avicenna: His Life and Works.* Westport, Conn.: Greenwood Press, 1980.

Arberry, Arthur. *The Legacy of Persia.* Oxford: Clarendon Press, 1953.

Elgood, Cyril. *A Medical History of Persia.* New York: Cambridge University Press, 1951.

Jenner, Edward

Developer of
Modern Smallpox Vaccine
1749-1823

Life and Work

Edward Jenner developed and actively promoted the practice of injecting samples of cowpox-infected matter to inoculate people against smallpox.

Jenner was born the son of a clergyman on May 17, 1749, in Berkeley, England. As a youth he developed a love of nature and an inquiring attitude. From the ages of 12 to 21, he was apprenticed to a surgeon, and he subsequently studied anatomy and surgery in London with the prominent physician John Hunter. Under Hunter's tutelage, Jenner developed a firm belief in the importance of experimental investigation. In 1773, Jenner returned to Berkeley and began a medical practice that he would maintain for the rest of his life.

Smallpox was a major cause of death in the eighteenth century. When smallpox did not kill its victim, scars or blindness often followed the disease's extremely painful duration. In Jenner's time, physicians attempted to inoculate patients against smallpox by injecting them with pustule fluids from mildly infected people. This practice had been introduced into England by Lady Mary Wortley Montagu (1689-1762), 30 years before Jenner was born. She had observed the practice in 1717 while living in Turkey with her ambassador husband. Although the inoculation was generally successful, it was sometimes fatal, causing full-blown smallpox itself.

Jenner's unique innovation was based on reports by rural patients that previous infection with cowpox, a milder disease transferred by cattle, made them immune to smallpox.

In 1796 Jenner injected matter from a cowpox victim's sores into eight-year-old James Phipps and six weeks later inoculated him again, but with smallpox, not cowpox, matter. Young Phipps remained healthy. Jenner repeated this experiment successfully several times, and then wrote in 1798 *An Inquiry into the Causes and Effects of the Variolae Vaccinae,* a description of his experimental results.

The tract's publication met with resistance from physicians who did not accept his findings, and Jenner had initial difficulties obtaining and preserving cowpox inoculate. However, the procedure was soon accepted, and it spread quickly throughout Europe and North America. The rate of smallpox deaths plummeted.

Jenner received worldwide fame, honorary degrees from Oxford and Harvard, and money from the British Parliament. He died in his hometown on January 26, 1823.

Legacy

Jenner's smallpox vaccine relieved the suffering and saved the lives of millions of people and led eventually to the complete eradication of the disease in the twentieth century.

Jenner's inoculation method represented the second step toward ending the spread of smallpox. Once his discovery was accepted as safe, governments in Europe and North America began requiring smallpox inoculations of their citizens. The practice slowly spread to other places, and the worldwide smallpox incidence dropped.

In the nineteenth century, LOUIS PASTEUR drew upon Jenner's discovery and created vaccines against other infectious diseases, such as cholera and rabies. The study of infectious diseases and the development of vaccines progressed rapidly, and in the twentieth century, scientists investigated the physiology behind the body's immune response.

In 1977 the last known case of smallpox occurred in Somalia; in 1980, the World Health Organization announced that the disease had been completely eradicated.

Schuyler

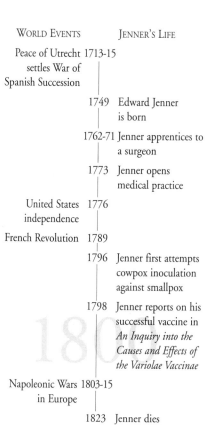

World Events	Jenner's Life
Peace of Utrecht 1713-15 settles War of Spanish Succession	
	1749 Edward Jenner is born
	1762-71 Jenner apprentices to a surgeon
	1773 Jenner opens medical practice
United States 1776 independence	
French Revolution 1789	
	1796 Jenner first attempts cowpox inoculation against smallpox
	1798 Jenner reports on his successful vaccine in *An Inquiry into the Causes and Effects of the Variolae Vaccinae*
Napoleonic Wars 1803-15 in Europe	
	1823 Jenner dies

For Further Reading:

Baxby, Derrick. *Jenner's Smallpox Vaccine: The Riddle of Vaccinia Virus and Its Origin.* London: Heinemann Educational Books, 1981.

Jenner, Edward. *Vaccination against Smallpox.* Amherst, New York: Prometheus Books, 1996.

Vare, Ethlic, and Greg Ptacek. *Mothers of Invention.* New York: William Morrow, 1988.

Jones, Amanda

Inventor of Vacuum Canning

1835-1914

Life and Work

Amanda Jones, an American inventor, patented a process that made it possible to preserve food without cooking it. Her vacuum canning process provided a safe way to store fruits, vegetables, and a variety of other foods.

Not much is known about the details of Jones's early life. Born in upstate New York in 1835, she grew up in a small, rural town. After her graduation from the local school, Jones took up a career as a teacher at age 15. While working as an educator, she began writing poetry, and many of her poems were published in magazines or in collections.

Jones gave up teaching and focused on her writing when tuberculosis weakened her health. In search of cures for her illness, Jones moved often and tried her hand at various occupations throughout her life.

One question that preoccupied Jones in 1872 was how to safely preserve food. She went on to devise a process in which it was not necessary to first cook the food. This had been a major drawback in conventional methods. Certain foods became tasteless, and things like fruit were better fresh.

Her method avoided heating the food itself. The process involved placing the food in a glass jar and using a system of valves that pulled the air out of the container. Once the air had been evacuated, hot water was injected and the jar sealed. Jones applied for and received nine patents for this process.

To promote her techniques, she formed the U.S. Woman's Canning and Preserving Company in 1890. Beginning with puddings, they expanded production to include canned fruits and lunchmeats. Nearly all of the company officers, stockholders, and employees were women. Although she was forced out of management after several years, the company continued until 1923.

Jones is credited with another, unrelated patent during the 1880s. At the time she was living near the Pennsylvania oil fields. The workers could not safely use the liquid fuel source they were extracting from the ground. She developed a safety valve that allowed workers to control the amount of oil being released into a liquid fuel burner. This solution caught the attention of the Navy, which decided to use it on their coal-powered vessels. Jones received an additional three patents in the field of steam engineering.

Never very successful in any of her business ventures and receiving little recompense for her inventions, Jones was dependent for much of her life on her family. She continued to write throughout her life and published an autobiography a few years before her death in 1914.

Legacy

With unreliable refrigeration and few methods of safely storing foods, Jones's method of vacuum canning provided a process by which companies and individuals could preserve many types of food. Her innovation helped to spawn the development of the canning industry and changed the way people ate.

Jones did not receive much credit for her inventions during her lifetime. While her canning process was eventually used by several different companies, she had already sold her interest in the patents to a meat-packing firm.

A supporter of the feminist movement, her female-run business was unique in its time. In many areas of her life, Jones sought to help others but her lack of money, few business skills, and poor health interfered with the development of her ideas.

Through her vacuum canning process Jones did find a way to make an enduring contribution. As with many other inventions, the process lasts much longer than the name of the inventor. The availability of fresh-tasting food that required no refrigeration impacted everyone. While her process was eventually refined decades later, Jones provided the nation with a simple, safe source of preserved food.

Wilson

For Further Reading:

Macdonald, Anne. *Feminine Ingenuity.* New York: Ballantine Books, 1992.

Vare, Ethlie, and Greg Ptacek. *Mothers of Invention.* New York: William Morrow, 1988.

Joule, James Prescott

Discoverer of Mechanical
Equivalent of Heat
1818-1889

Life and Work

James Joule was the first scientist to understand the connection between heat and other forms of energy, and he discovered the mechanical equivalent of heat.

Joule was born on December 24, 1818, at Salford, near Manchester, England. His father was a wealthy brewer who hired private tutors for James and his brother. By about age 18, he was attempting to build an electrical perpetual-motion machine, a device that could generate its own power to keep running. He soon decided that the task was impossible, although some people are trying even today. He studied at the University of Manchester in 1835 where he worked with English chemist JOHN DALTON.

Soon he turned his attention to studying the energy lost to heat in the resistance of wires carrying the electricity. By 1840 he had demonstrated that the power lost in a circuit is proportional to the resistance in the circuit times the current squared. This basic relationship is called Joule's Law.

His greatest discovery was derived from ongoing experiments on the nature of heat and how it can be produced by moving or falling objects. To aid in this investigation, Joule produced an experimental device featuring an insulated container with paddle wheels inside, connected by pulleys to large falling weights. As the weights descended, the paddles spun, churning up the liquid in the container and raising its temperature. Joule found that the actual amount of heat produced did not depend on what fluid he used, but only on how much weight was falling and how far it fell. In his best trials, he determined that 4.15 units of mechanical energy (the sum of the energy of movement and energy stored in parts of a system) were equivalent to one calorie of heat: Joule's measurement was very close to our current calculation of 4.184. Thus he established the mechanical equivalent of heat and published his findings in 1843. The unit of energy has been named the joule, in his honor, and is abbreviated as J.

In Joule's time, energy was called *vis viva*, or "living force." In a public lecture in 1847 Joule explained how this living force could be converted to heat, or the other way around. He stated that nothing is lost in such conversions: a given amount of heat will be converted into the same amount of "living force." He was the first to clearly understand and explain what we now call the Law of Conservation of Energy, which is the first law of thermodynamics.

His last major scientific contribution came in 1852, when he and physicist William Thomson (later known as LORD KELVIN) worked together on the thermal properties of gases. They discovered that if a gas is allowed to expand into an empty chamber, the temperature of the gas decreases. The principle is now known as the Joule-Thomson effect.

Joule lived as a wealthy amateur scientist until 1875, when he lost most of his fortune. His health also began to fail and his scientific work slowed to a stop. He lived until 1889, dying in Sale, England.

Legacy

Joule's exploration of the connection between heat and other forms of energy laid a crucial foundation for future theoretical work. Joule's ideas also had extensive practical

applications, including innovations in refrigeration and electrical systems.

Two of Joule's contemporaries, Hermann von Helmholtz and RUDOLF CLAUSIUS, expanded on his work by establishing the mathematical principles that underpinned Joule's experiments. Their formulas are still in use by contemporary scientists and engineers.

The practical applications of Joule's work are significant. Refrigerators, freezers, air conditioners, and dehumidifiers all make use of the Joule-Thomson effect, which states that a gas cools as it expands. The Joule-Thomson effect is also at work in electric heat pumps, which are used to collect diffuse, low-temperature heat from outdoor air or from underground and concentrate it to warm homes.

Joule's Law, relating the heat loss in a circuit to resistance and current, has an important application today in distributing electricity. Calculations of how much electricity is lost because of resistance as it is carried across wires helps electric utility companies supply power at the right voltage to homes. The efficiency of our electrical systems is a consequence of Joule's Law and a tribute to his work.

Secaur

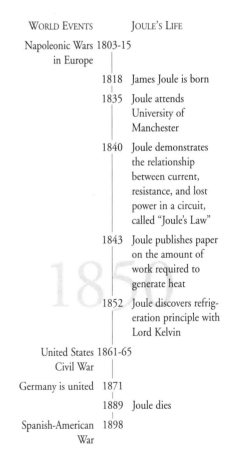

WORLD EVENTS	JOULE'S LIFE
Napoleonic Wars 1803-15 in Europe	
	1818 James Joule is born
	1835 Joule attends University of Manchester
	1840 Joule demonstrates the relationship between current, resistance, and lost power in a circuit, called "Joule's Law"
	1843 Joule publishes paper on the amount of work required to generate heat
	1852 Joule discovers refrigeration principle with Lord Kelvin
United States 1861-65 Civil War	
Germany is united 1871	
	1889 Joule dies
Spanish-American 1898 War	

For Further Reading:

Cardwell, D. S. L. *James Joule: A Biography.* New York: St. Martin's Press, 1989.

Segrè, Emilio. *From Falling Bodies to Radio Waves.* New York: Freeman, 1984.

Steffens, Henry John. *James Prescott Joule and the Concept of Energy.* New York: Science History Publications, 1979.

Julian, Percy Lavon

Organic Chemist; Developer of Treatment for Glaucoma
1899-1975

Life and Work

Percy Lavon Julian devised a method of synthesizing the chemical physostigmine, used to treat the eye disease glaucoma. He also developed several pharmaceutical and industrial applications of soybean extracts.

Julian was born on April 11, 1899, in Montgomery, Alabama, to an African-American family. He was interested in science as a boy, but had few chances to study it. He could not even attend the local high school, because virtually all high schools in Alabama were for whites only. He left home to attend the one public high school for blacks (in Birmingham) and did so well that he won a

scholarship to DePauw University in Indiana. He majored in chemistry at DePauw, graduating at the top of his class in 1920. As his father had warned him, he had limited job prospects in chemical industries because of widespread discrimination. After teaching for two years at Fisk University, he received a fellowship to Harvard University, where he earned his master's degree in one year.

In 1928 Julian became head of the chemistry department at Howard University in Washington, D.C. He traveled to Austria in 1929 to investigate the chemical properties of soybean extracts. The University of Vienna awarded him a Ph.D. in 1931. He then returned to Howard to study the structure of the chemical physostigmine, which was used to treat glaucoma, an eye disease leading to blindness.

Julian transferred to a post at DePauw in 1932. Three years later, having identified the chemical steps in the formation of physostigmine, he performed the first synthesis of the drug. Despite this accomplishment, he was denied two academic positions because of his race, but he secured the post of chief chemist at the Glidden Company, a Chicago industrial laboratory. Within a few years, he had found a way to use the soybean protein in the manufacture of textiles, paints, paper coatings, and Aero Foam fire extinguishers.

Using soy extracts, Julian achieved the synthesis of the sex hormones progesterone and testosterone and subsequently synthesized, also from soy extracts, the chemical cortisone, recently found to be an effective treatment of rheumatoid arthritis.

In 1954 Julian established Julian Laboratories, Inc., in Chicago, and a sister plant in Mexico City, where he used extracts from wild yams to synthesize cortisone. He sold the business when he retired in 1964.

He received the 1947 Spingarn Medal from the National Association for the Advancement of Colored People. He continued to experiment with synthetic chemicals until his death in 1975.

Legacy

Julian's pharmaceutical successes reduced the cost and increased the availability of several drugs, leading to treatment for many people suffering from debilitating diseases.

Julian's synthetic production of physostigmine, which was cheaper than extracting it from beans, proved invaluable to people with glaucoma. The disease slowly damages the retina, eventually blinding its victim. As a treatment for glaucoma, physostigmine was eventually replaced by other drugs and surgical interventions, but it remains a valuable pharmaceutical. In 1998 preliminary findings of a clinical experiment showed that physostigmine increases short-term memory in humans.

The soy-based drug syntheses that Julian perfected reduced the cost of manufacturing sex hormones and cortisone, making their production more viable. Progesterone and testosterone had previously been made using chemicals taken from cows' spinal cords and brains, an expensive and restrictive procedure. These hormones have numerous uses in modern medicine, but in the early twentieth century, progesterone was used primarily to prevent miscarriages, and testosterone was used to treat diminishing sex drive in older men.

The traditional method of producing cortisone utilized cholesterol from the bile of oxen. This slow and costly process required the bile of more than 14,000 animals to produce cortisone for one arthritis patient for one year. Successful synthesis of cortisone reduced its cost from hundreds of dollars to pennies per dose.

Julian's accomplishments in organic chemistry and drug synthesis resulted in more than 100 patents and paved the way for future progress in those areas.

Schuyler

World Events		Julian's Life
Spanish-American War	1898	
	1899	Percy Julian is born
World War I	1914-18	
	1920	Julian graduates from DePauw University with degree in chemistry
	1928	Julian becomes head of chemistry at Howard University
	1931	Julian earns doctorate in chemistry from University of Vienna
	1935	Julian synthesizes physostigmine
World War II	1939-45	
	1947	Julian awarded Spingarn Medal
Korean War	1950-53	
	1954	Julian forms his own company
End of Vietnam War	1975	Julian dies

For Further Reading:

Cobb, Montague W. "Percy Lavon Julian." *Journal of the National Medical Association* (March 1971).

Jenkins, Edward S., ed. *American Black Scientists and Inventors.* Washington, D.C.: National Science Teacher Association, 1975.

Just, Ernest Everett

Pioneer in Developmental Biology
1883-1941

Life and Work

Ernest Everett Just made groundbreaking discoveries in the newly emerging field of developmental biology, the study of the development of organisms from cellular reproduction to embryo stage.

Just was born on August 14, 1883, in Charleston, South Carolina. His father died when he was young, and his mother, a schoolteacher, supervised his education. She enrolled him at a college preparatory school in New Hampshire, highly unusual for an African American at that time; he excelled at academics, edited the school newspaper, and served as president of the debate team. He entered Dartmouth College, the only African American in a freshman class of 288 students. After taking every biology class offered, he graduated in 1907 and was elected to the academic honor society Phi Beta Kappa.

Upon graduation, Just gained a teaching position at Howard University in Washington, D.C. He taught zoology and physiology, and his busy schedule left him no time for research. However, he began graduate work at the Marine Biological Laboratory in Woods Hole, Massachusetts, during the summer of 1909 and continued

to spend summers there conducting research for the next 20 years.

At Woods Hole, Just studied the fertilization process and early embryological development of marine organisms. In 1915 his first results, which showed how the developmental orientation (the positioning of the head and tail ends) of the fertilized egg is established, earned him the first Spingarn Medal awarded by the National Association for the Advancement of Colored People. Just received a doctorate from the University of Chicago in 1916.

Just subsequently studied parthenogenesis, the development of egg cells without fertilization, a process possible in sea urchin and frog eggs. Exposing various invertebrate eggs to altered concentrations of salt-water solutions and acids, Just formulated innovative theoretical explanations for parthenogenesis.

In 1939 Just published *The Biology of the Cell Surface,* which outlined the results of his research on cell biology. At the time, it was believed that the nucleus controls all the activities of the cell. Just hypothesized that the cytoplasm, which refers to all the material inside a cell except the nucleus, plays an important role in directing a cell's behavior with respect to fertilization, development, and heredity.

Just's achievements earned him membership in many scientific societies, including the American Society of Zoologists, of which he became vice president. He believed that racial discrimination prevented him from being elected to the National Academy of Sciences, and he is said to have become increasingly bitter about such injustices during his career.

Seeking opportunities beyond the limited ones available to African-American researchers in the United States, Just spent time working in European laboratories after 1929. He died in Washington, D.C., on October 27, 1941.

Legacy

Just's work contributed to the progress of embryology and developmental genetics, fields that are crucial to the modern understanding of biological systems. The study of developmental processes, which occur in both embryos and adult animals, offers a framework for addressing problems in such diverse areas

as physiology, immunology, fertility, evolution, ecology, and cancer research.

Just's discoveries led to the recognition that immediately after fertilization, the egg loses its ability to fuse with any other sperm. It was later shown that this blockade is achieved by a change in the electrical potential across the egg-cell membrane. This proved crucial to producing the appropriate conditions for in vitro fertilization.

In Just's later years, other scientists were beginning to investigate genetic mutations in order to understand developmental events. Just criticized them for emphasizing the development of specific characteristics, such as eye color and wing shape (in fruit flies); he advocated instead a search for the mechanisms involved in how an embryo determines the overall body plan. Just was ahead of his time, for it is only in the 1990s that the tools of molecular technology have revealed the answers to such questions, including evidence supporting his insistence on the active roles of cytoplasm.

Schuyler

WORLD EVENTS		JUST'S LIFE
Germany is united	1871	
	1883	Ernest Just is born
Spanish-American War	1898	
	1907	Just earns B.A. and begins teaching at Howard University
	1909	Just begins research at Woods Hole, Mass.
World War I	1914-18	
	1915	Just is awarded Spingarn Medal for studies of fertilization and embryos of marine organisms
	1916	Just earns Ph.D.
	1920s	Just conducts developmental biology research
	1939	Just's *Biology of the Cell Surface* is published
World War II	1939-45	
	1941	Just dies

For Further Reading:

Haber, Louis. *Black Pioneers of Science and Invention.* San Diego: Harcourt, Brace, 1970.

Manning, Kenneth R. *Black Apollo of Science: The Life of Ernest Everett Just.* New York: Oxford University Press, 1983.

Kekulé von Stradonitz, August

Discoverer of
Carbon Benzene Rings
1829-1896

WORLD EVENTS	KEKULÉ'S LIFE
Napoleonic Wars 1803-15 in Europe	
	1829 August Kekulé is born
	1852 Kekulé awarded doctorate from University of Giessen
	1856 Kekulé begins teaching chemistry at University of Heidelberg
	1857 Kekulé publishes theory of carbon tetravalence
United States 1861-65 Civil War	
	1865 Kekulé publishes theory of ring structure of the hydrocarbon benzene
	1867 Kekulé becomes chair of chemistry at the University Bonn
Germany is united 1871	
	1895 Kekulé is awarded title of nobility
	1896 Kekulé dies
Spanish-American 1898 War	

Life and Work

August Kekulé, later von Stradonitz, pioneered the field of structural organic chemistry. His discovery that carbon formed rings transformed the study of chemistry.

Born in 1829 in Darmstadt, Germany, Kekulé was talented at drawing and mathematics. He followed the wishes of his affluent family and attended school with the intention of becoming an architect. While at the University of Giessen, however, he was most intrigued by Justus von Liebig's lectures in chemistry. He switched his academic focus to chemistry and studied in France, Switzerland, and England.

By 1852 Kekulé had worked in various labs across the continent and earned a Ph.D. from the University of Giessen. Taking advantage of numerous opportunities for advanced study, Kekulé worked in hospitals, universities, and private labs until he began to teach organic chemistry in 1856 at the University of Heidelberg.

During the next few years Kekulé turned his attention to the behavior of carbon atoms. He formulated the idea that carbon could be tetravalent; that is, carbon could have a valence of four, which would enable it to bond with both metals and nonmetals. He published this theory in 1857. Soon after Kekulé expanded on this to propose that in compounds containing more than one carbon atom, the carbon atoms could be joined in long chains. There still remained molecules that did not fit this framework. Kekulé worked on this problem for many years. He used his mathematical abilities and spatial skills to help determine the structure of compounds.

One night Kekulé sat dozing in front of his fire. As he gazed into the flames, he dreamt that the fingers of fire seemed to become snakes that chased themselves in circles. One even caught its tail to form a ring. When Kekulé awoke, he knew he had the key: certain carbon compounds do not form long, open chains; rather they arrange themselves in closed rings. He worked through the rest of the night and eventually worked out a six-carbon ring. Published in 1865 this ring structure of the hydrocarbon benzene heralded a new era in organic chemistry.

Kekulé continued to study the structure of organic compounds (compounds containing carbon) for the remainder of his life. In 1867 he accepted the chair of chemistry at the University of Bonn, where he established a chemical institute.

The year before he died, Kekulé was granted a title of nobility, "von Stradonitz," by the King of Prussia. He died on July 13, 1896, in Bonn, Germany.

Legacy

Kekulé's discovery of the structure of various carbon compounds enabled scientists to better understand how atoms combine. He can be credited with establishing the field of structural organic chemistry.

Prior to Kekulé's discoveries scientists had been unable to reconcile the behavior of carbon with the current theories of chemical bonding. The idea that an atom could be tetravalent was inconceivable. Kekulé's discovery changed how scientists viewed the valence of carbon, that is, the property of carbon that indicates the number of other atoms it can bond with.

He instigated another shift in thinking when he proposed that molecules could have structures that were nonlinear; all organic compounds up to that time were seen as long chains. His ring structure was a radical idea. It literally changed the shape of chemistry and became the foundation of modern organic chemistry. Although Kekulé was not able to directly test his theory, current experimental techniques support his findings.

Today the field of organic chemistry has expanded on the ring theory of Kekulé. It is used to examine the structure of other non-linear molecules because it provides an accurate model of how molecules are spatially arranged. This is integral to today's research with complex carbon molecules. As scientists continue to explore coal- and petroleum-based compounds and how to synthesize them, Kekulé's insight into the structure of organic compounds has proved invaluable.

Wilson

For Further Reading:

Bowden, Mary Ellen. *Chemical Achievers*. Philadelphia, Penn.: Chemical Heritage Society, 1997.

Brock, William H. *The Norton History of Chemistry*. New York: Norton, 1993.

Crosland, Maurice. *The Science of Matter: A Historical Survey*. Philadelphia, Penn.: Gordon & Breach Science, 1992.

Wotiz, John H. *The Kekulé Riddle: A Challenge for Chemists and Psychologists*. Clearwater, Fla.: Cach River Press, 1993.

Kelvin, Lord

(William Thomson)

Pioneer in Electromagnetism,
Heat, and Mechanics
1824-1907

Life and Work

Lord Kelvin accomplished groundbreaking theoretical and technical work that advanced and synthesized the studies of heat, electricity, magnetism, and mechanics.

Kelvin was born William Thomson on June 26, 1824, in Belfast, Ireland. (He adopted the name Kelvin when he became a baron in 1892.) From his father, a university mathematics professor, he learned recent, advanced mathematics that was not yet taught at British schools. He entered the University of Glasgow at age 10, where he was introduced to Joseph Fourier's application of abstract mathematics to investigate heat flow, which was controversial at the time. After transferring to Cambridge University in 1841, Thomson published two papers defending Fourier's approach and became the first to suggest that the approach be used in other fields, including the study of fluids and electricity.

Graduating from Cambridge in 1845, he was elected the following year to become head of the natural philosophy (later physics) department at the University of Glasgow. He held the position for 53 years.

Kelvin's scientific approach was shaped by his belief that physical theories of matter and energy were converging toward one unified theory, an idea based on evidence hinting that all forms of energy are related. Studying others' experimental results, he extracted generalizations about various physical phenomena and developed theories on hydrodynamics, elasticity, and the electrodynamic properties of metals. Applying mathematics formulated by Irish physicist George Stokes to models of rotating elastic solids, he was able to discuss some of the forces acting between electrical current and magnetism.

In 1848 Kelvin introduced the scale of temperature that starts at absolute zero, the theoretically lowest possible temperature. The scale's unit, the kelvin (K), corresponds in caloric value to one degree Celsius (C); no negative numbers exist on the Kelvin scale, and the temperature of melting ice at atmospheric pressure is 273.15°K, which corresponds to 0°C.

Kelvin was intrigued by JAMES PRESCOTT JOULE's proposition stating that heat is a form of mechanical motion, which opposed accepted doctrine that heat is a special substance that has no fixed relationship to the amount of mechanical work generated by it. In 1851 Kelvin offered mathematical support for Joule's theory with the influential paper "On the Dynamical Theory of Heat." The following year Kelvin and Joule collaborated and developed ideas on work and heat, including that when a gas expands without doing work (not moving a piston or a turbine, for example), the gas cools slightly. Their ideas opened the area of low-temperature physics (cryogenics).

Kelvin was also an inventor who used his successful patents to finance his research. In 1854 Kelvin became involved in Cyrus Field's plan to lay submarine telegraph cable across the Atlantic Ocean. He formulated equations describing heat flow in a solid that gave the velocity of electrical current flowing through cable wire. His inventions, the mirror galvanometer and siphon recorder, served as the telegraph-receiving mechanism in most undersea telegraph cables. Queen Victoria granted him knighthood, primarily based on the success of the transatlantic cable, which stretched between Ireland and Newfoundland, Canada, the first cable to connect both sides of the Atlantic.

In his later years Kelvin owned a yacht, and his sea-going adventures resulted in several navigational patents: an improved compass, which compensated for the effects of iron and steel components of ships, a tide-measuring device, and fathometers and sounding equipment, which measured water depth. He died at his estate near Largs, Scotland, on December 17, 1907.

Legacy

Kelvin's achievements represented significant progress in nineteenth-century physics, resulting in the turn-of-the-century concept that all physical change is related to energy.

Kelvin's contributions helped lay the groundwork for the First and Second Laws of Thermodynamics, which are fundamental to the modern understanding of nature. The First Law of Thermodynamics states that energy cannot be created or destroyed; its quantity is constant. The Second Law can be expressed in a number of ways. It states that a system's entropy, a measure of its disorder, can never decrease; in other words, all systems tend toward a more disordered state. The law similarly posits that heat will flow spontaneously from a hotter system to a colder system, but will not flow in the opposite direction.

Kelvin's work on the relationship between electricity and magnetism led directly to JAMES CLERK MAXWELL's 1865 theory that light is a form of electromagnetic radiation. This theory unified the understanding of light, electricity, and magnetism.

The laying of transatlantic telegraph cables, to which Kelvin's electrical engineering skills contributed, resulted in an international submarine communication network by the end of the nineteenth century.

Kelvin's nautical inventions had a direct impact on maritime safety. His improved compass, tidal gauges and predictors, fathometers, and sounders helped make navigation more accurate, thus saving lives by reducing shipwrecks.

Schuyler

For Further Reading:

Sharlin, Harold I. *Lord Kelvin: The Dynamic Victorian.* University Park: Pennsylvania State University Press, 1979.

Smith, Crosbie. *Energy and Empire: A Biographical Study of Lord Kelvin.* Cambridge: Cambridge University Press, 1989.

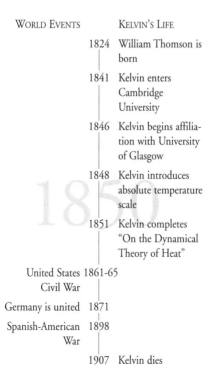

WORLD EVENTS		KELVIN'S LIFE
	1824	William Thomson is born
	1841	Kelvin enters Cambridge University
	1846	Kelvin begins affiliation with University of Glasgow
	1848	Kelvin introduces absolute temperature scale
	1851	Kelvin completes "On the Dynamical Theory of Heat"
United States Civil War	1861-65	
Germany is united	1871	
Spanish-American War	1898	
	1907	Kelvin dies

Kepler, Johannes

Developer of
Laws of Planetary Motion
1571-1630

Life and Work

Johannes Kepler formulated the three laws of planetary motion, which constitute the basis of the modern understanding of the solar system.

Kepler was born to a poor family in Weil der Stadt, Germany, on December 27, 1571. He was a sickly child and nearly died of smallpox when he was four years old. He worked intensely throughout his life as if to justify his

own survival. His interest in astronomy may date from age six, when he was greatly impressed by the appearance of a comet. In 1587 he obtained a scholarship to attend the University of Tübingen. There he studied astronomy under Michael Mästlin, an adherent of COPERNICUS's view that the planets orbit the Sun. Kepler immediately accepted the Copernican theory.

In 1600 Kepler joined Danish astronomer TYCHO BRAHE at his observatory near Prague. Upon Brahe's death in 1601, Kepler replaced him as court astronomer to the Holy Roman Emperor. Kepler attempted to calculate the orbit of Mars, and this eventually led him to develop two theories, which he introduced in *New Astronomy* in 1609. The first theory stated that the planets follow elliptical orbits, with the Sun at one focus. The second stated that an imaginary line from the center of the Sun to the center of a planet always sweeps over an equal area of its ellipse in equal time, meaning that planets travel faster when they are closer to the Sun.

After moving to Linz, Austria, in 1611, Kepler wrote the treatise that contains his third theory of planetary motion. In *Harmonics of the World*, published in 1619, he showed that the square of the time a planet takes to orbit the Sun (its period) is proportional to the cube of the planet's mean distance from the Sun. All of Kepler's astronomical discoveries were brought together in *Epitome of Copernican Astronomy*, which appeared between 1618 and 1621. The work became the first textbook of astronomy grounded on Copernican principles.

In 1627 Kepler finished the Rudolfine Tables, improved tables of planetary motion based on Brahe's painstaking observations and research. He died in Regensburg on November 15, 1630.

Legacy

Kepler's three theories of planetary motion, now known as Kepler's laws, are essential elements of the modern understanding of astronomy.

Kepler's revolutionary ideas emerged just as the views of Copernicus became the focus of a violent intellectual controversy. The Ptolemaic system, in which Earth was viewed as the cen-

ter of the universe, had been in vogue for nearly 1,400 years. Copernicus voiced his heliocentric theory in 1543, and the Roman Catholic church was so offended at the suggestion that it banned Copernicus's major work in 1616. Tycho Brahe attempted to verify Copernicus's claim by vastly improving measurements of stellar and planetary positions, and his data influenced Kepler's ideas.

Kepler's *Epitome of Copernican Astronomy* became the most influential book in astronomy for three decades. It converted many astronomers from Ptolemaic to Copernican views. GALILEO, already strongly supportive of the Copernican system, was encouraged by Kepler's calculations. Galileo published his views in 1632 and was subsequently persecuted by the Roman Catholic Church during the Inquisition.

ISAAC NEWTON also relied on Kepler's work, particularly his third law, to formulate the theory of gravitational force. In 1687, Newton published his *Principia*, in which he described the gravitational force that governs objects on Earth's surface as well as the orbits of the moon and the planets. He used the gravitational force to measure the mass of the Sun and to calculate the orbit of a comet. Gravitational theory was both derived from and helped to confirm Kepler's three laws.

Kepler's Rudolfine Tables were the first modern tables of planetary motion and enabled future astronomers to calculate the positions of the planets at any past, present, or future time.

Schuyler

WORLD EVENTS		KEPLER'S LIFE
Reformation begins	1517	
	1571	Johannes Kepler is born
	1587	Kepler obtains scholarship to attend university
	1600	Kepler joins Tycho Brahe at Brahe's observatory
	1601	Brahe dies
	1609	*New Astronomy* is published and explains first two laws of planetary motion
	1618-21	*Epitome of Copernican Astronomy* is published
Thirty Years' War in Europe	1618-48	
	1619	*Harmonics of the World* is published; it explains third law of planetary motion
	1627	Kepler completes the Rudolfine Tables
	1630	Kepler dies
	1632	Galileo publishes his views
	1687	Newton's *Principia* is published
England's Glorious Revolution	1688	

For Further Reading:

Casper, Max. *Kepler*. New York: Dover, 1990.

Koestler, Arthur. *The Watershed: A Biography of Johannes Kepler*. Lanham, Md.: University Press of America, 1985.

Kozhamthadam, Job. *The Discovery of Kepler's Laws: The Interaction of Science, Philosophy, and Religion*. Notre Dame, Ind.: University of Notre Dame Press, 1994.

Al-Khwarizmi, Mohammed ibn Musa

Developer of Arabic Numerals;
Pioneer in Algebra
c. 780-850

Life and Work

Mohammed ibn Musa al-Khwarizmi helped popularize the use of Arabic numerals. His text on solving equations helped to systematize modern algebra.

Not much is known about al-Khwarizmi's life. He was born around 780 in Khwarizm, a town in present-day Uzbekistan, and he lived much of his life in Baghdad, in what is now Iraq.

Al-Khwarizmi was one of the most important Arabic mathematicians to adopt the Hindu number system and make improvements on it. Hindu numerals, with Islamic improvements, were the basis for the Arabic number system. Al-Khwarizmi refined that system and did much to popularize it. His innovations included adopting the use of zero, which was not widely used by Arabic mathematicians.

His text *Al-jabr wa-al-muqabilah* (c. 830) helped to systemize the subject of algebra, and the word "algebra" comes from the title (*Al-*

jabr). The text describes methods of *Al-jabr,* or restoring equations. By restoring, he meant that if you alter one side of the equation, you have to alter the other to restore the balance. For example, if you wanted to solve $x + 5 = 11$ for x, you would have to take 5 away from the left side and, then, to restore the balance you would take 5 away from the right side.

He also wrote about astronomy. His tables of future planetary and stellar positions are the earliest complete astronomical work to survive from Islamic astronomers. In addition, he described adaptations of the Greek astrolabe, which his colleagues improved to measure celestial positions using both altitude and azimuth (a word of Arabic origin designating a horizontal arc along the horizon).

Al-Khwarizmi died around 850.

Legacy

Al-Khwarizmi is thought by some scholars to be the founder of modern algebra, and his popularization of and innovations with Arabic numerals provided a convenient number system for future generations.

Al-Khwarizmi's text on *"Al-jabr,"* or restoring equations, presented and described methods of manipulating numbers that we still use today. This text introduced algebra to European scientists and mathematicians. Girard of Cremona translated it into Latin in the twelfth century, and this translation became the principal mathematical textbook used in universities well into the Renaissance. Algebra provided foundations for other fields in mathematics including the calculus.

His adoption of zero into the Arabic number system helped to solidify the concept of place value and provided a foundation for the decimal system. Counting and calculating in Roman numerals was cumbersome and inadequate. The Arabic system soon replaced the Roman. It allowed future mathematicians and lay people alike an ease of calculation not previously possible, and it has remained with us to this day.

His work on the Hindu-Arabic numeral system has been preserved only in a Latin translation *Algoritmi de numero Indorum*. The English title is known as *Al-Khwarizmi on the Hindu Art of Reckoning.* The title, in particular the use of his name Al-Khwarizmi, gave rise to

the word algorithm, which refers to a finite series of detailed instructions that aim to accomplish a specific task. Algorithms are a fundamental approach in many areas of mathematics and, in the twentieth century, have been widely used in computer science.

Steinberg

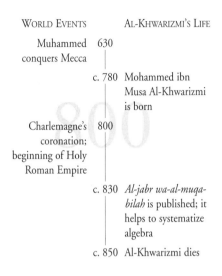

WORLD EVENTS		AL-KHWARIZMI'S LIFE
Muhammed conquers Mecca	630	
	c. 780	Mohammed ibn Musa Al-Khwarizmi is born
Charlemagne's coronation; beginning of Holy Roman Empire	800	
	c. 830	*Al-jabr wa-al-muqabilah* is published; it helps to systematize algebra
	c. 850	Al-Khwarizmi dies

* *Scholars can not date the specific events in Al-Khwarizmi's life with accuracy.*

For Further Reading:

Dunham, William. *Journey Through Genius.* New York: John Wiley, 1990.

Kline, Morris. *Mathematical Thought from Ancient to Modern Times.* New York: Oxford University Press, 1972.

Ronan, Colin. *Science: Its History and Development Among the World Cultures.* New York: Facts On File, 1982.

Koch, Robert

Founder of Medical Bacteriology
1843-1910

Life and Work

Robert Koch identified and isolated several disease-causing bacteria, including those of tuberculosis and cholera, and

WORLD EVENTS		KOCH'S LIFE
	1843	Robert Koch is born
United States 1861-65 Civil War		
	1866	Koch graduates from the University of Göttingen
	1870	Koch serves as surgeon in the Franco-Prussian War
Germany is united	1871	
	1876	Koch demonstrates that anthrax is caused by bacteria
	1882	Koch identifies the tuberculosis bacterium
	1883	Koch identifies the cholera bacterium
Spanish-American War	1898	
	1905	Koch wins Nobel Prize for Physiology or Medicine
	1910	Koch dies
World War I 1914-18		

developed methods of cultivating and studying bacteria. He is considered to be the founder of the modern discipline of medical bacteriology.

Koch was born in Klausthal-Zellerfeld, Germany, on December 11, 1843. He studied botany, physics, and mathematics at the University of Göttingen, graduating in 1866. After briefly running a private medical practice, he served from 1870 to 1871 as a field surgeon in the Franco-Prussian War. He then became a district surgeon at Wollstein, Germany, and led the life of a small-town doctor. After his wife gave him a microscope for a birthday gift, however, he neglected his practice. He built a laboratory in his home, where he began to investigate pathogenic (disease-causing) organisms.

Friedrich Henle, who in 1840 had been among the first to suggest that many diseases are caused by microscopic organisms, was one of Koch's instructors at Göttingen. Koch had embraced Henle's idea. In 1876, Koch announced the results of studies on anthrax: he had isolated the anthrax bacterium and shown that it multiplies via spores that can remain dormant yet infectious for years. It was the first time that a bacterium had been demonstrated definitively to cause a specific disease. Koch theorized that each infectious disease has its own particular bacterium as its cause—a new idea in medicine.

Koch introduced and improved methods for the isolation and cultivation of microorganisms in the laboratory. Using microscopes, microtomes (tools for cutting microscope specimens), incubators, and nutrient solutions, he observed microorganisms' complete life cycles. He applied his laboratory techniques to the study of wounds, isolating the microorganisms responsible for several types of infection. He experimented with stains, which make the smallest microbes visible, and then took photographs of the microbes.

In 1881 Koch began to study tuberculosis, a disease that was responsible for one in seven deaths in Europe at that time. He successfully isolated the tuberculosis bacterium in 1882. In 1883 he traveled to India (where cholera was epidemic), identified the cholera bacterium, and showed that it is transmitted primarily through drinking water.

In the 1890s, Koch broadened his research to include leprosy, rinderpest, bubonic plague, surra, Texas fever, and malaria. In 1905 he won the Nobel Prize for Physiology or Medicine for his work on tuberculosis. He died in Baden-Baden, Germany, on May 27, 1910.

Legacy

Koch's investigations of bacteria revolutionized the understanding of infectious disease and led to improvements in public health.

Koch's work crystallized the theory that infectious diseases are caused by microorganisms. This theory was first put forth in the 1830s and 1840s, but scientists were reluctant to accept it and held fast to the predominant idea that bad air was responsible for disease. In the 1860s, LOUIS PASTEUR, JOSEPH LISTER, and others made strides toward proving the microorganism theory. Koch's demonstration that a bacterium causes anthrax convinced non-believers of the theory's accuracy.

Within 15 years of Koch's discovery of the anthrax bacterium, the bacterial agents of numerous diseases were identified. A school of adept students trained in Koch's effective lab techniques flourished around him and inaugurated the field of medical bacteriology. By 1892 the organisms responsible for typhoid, pneumonia, diphtheria, colic, meningitis, and salmonella had been identified.

A groundbreaking advance in public health evolved from Koch's studies of the tuberculosis and cholera bacteria. An effective method of diagnosing the presence of tuberculosis developed immediately, and, as cases could be identified at earlier stages, the spread of the infection slowed. Koch's identification of drinking water as the primary mediator of cholera led to a drop in cholera infections.

Koch's work established the modern criteria (now known as Koch's postulates) for determining that a specific organism causes a particular disease. The criteria include: the organism's presence in every examined case of the disease; the preparation of a laboratory culture of the organism; and the ability of the culture to remain infectious through several generations. The set of criteria remains a helpful tool in battling new, unknown, and uninvestigated diseases.

Schuyler

For Further Reading:

Brock, Thomas D. *Robert Koch: A Life in Medicine and Bacteriology.* New York: Springer-Verlag, 1988.

De Kruif, Paul. *Microbe Hunters.* New York: Harcourt Brace, 1954.

Kovalevsky, Sonya

Developer of
Partial Differential Equations
1850-1891

Life and Work

Sonya Kovalevsky (born Krukovsky) made valuable contributions to the field of differential equations and became a celebrity in the European scientific community. She also wrote several published novels that reflected her experiences as a child in Russia.

Krukovsky was born in Moscow on January 15, 1850. When she was a young child, the family moved to a large country home. The family miscalculated the amount of wallpaper needed for Sonya's bedroom and temporarily used pages from a mathematics lecture. These abstract diagrams and formulae fascinated the child. When she later studied the calculus at age 15, many concepts seemed familiar to her, to the astonishment of her tutor.

Because Russian universities did not accept female students at that time, Krukovsky sought ways to study abroad and promote her ideas about the emancipation of women. Her solution was a marriage of convenience to Vladimir Kovalevsky, who also wished to attend a university outside Russia. They were married in 1868 and, once established in Germany, the two went their separate ways. In Heidelberg, Kovalevsky quickly came to the attention of her professors

because of her mathematical abilities. There she learned of the famous mathematician Karl Weierstrass who taught at the University of Berlin. Kovalevsky moved to Berlin, and, as that university did not enroll women, she became a private student of Weierstrass for the next four years. During this time she greatly expanded her understanding of differential equations. She also wrote three doctoral dissertations because she and Weierstrass believed that, as a woman, she would not be awarded a degree without extraordinary proof of her abilities. Her work on partial differential equations resulted in a doctorate in absentia (because women were denied residential study) from the University of Göttingen, summa cum laude, in 1874.

Neither her degree nor her outstanding recommendations were any help to her back in Russia. After several disheartening years in her homeland, Kovalevsky returned to Europe in 1884 to become a lecturer at the University of Stockholm. This was a time of intense creativity and work. She eventually was appointed a professor at the university, became a spokesperson for women's rights groups in Sweden, and was an integral part of the intellectual community of Europe. In 1888 the French Academy of Sciences awarded Kovalevsky the Prix Bordin (an equivalent of the Nobel Prize) for her paper on how a rigid body revolves around a fixed point, its center of mass. Although others had studied this phenomenon in symmetric solids, Kovalevsky constructed an elegant explanation of the motion of asymmetrical crystals. Her work was deemed so strong that the prize amount was nearly doubled.

During the next few years Kovalevsky exhausted herself physically and emotionally. One of the things she accomplished was the publication of *The Rajavski Sisters* (1889), a novel based on her childhood. She eventually contracted pneumonia while traveling between Moscow and Stockholm. Her death in 1891 came at the peak of her career.

Legacy

Kovalevsky not only broke new ground in mathematics and had an impact in nuclear physics, she also was influential in the women's rights movement, particularly in education.

Kovalevsky's credibility enabled other women to enter scientific and mathematical fields. As

she gained the respect of the mathematics community, many academics acknowledged that a woman was capable of rigorous, intellectual work. During her time in Germany she helped several other young women gain entrance into previously male-only research labs. Her nickname, "Princess of Science," reflected her popularity and renown in Europe.

Kovalevsky's work on infinite series had direct impact in nuclear physics; other implications of infinite series are still being explored. The foremost successor of Kovalevsky, LISE MEITNER, based some of her calculations of radioactive chain reactions on Kovalevsky's work on infinite series. This made controlled nuclear fission possible—both for bombs and for electrical generation. A geometric series is one where the ratio of each term to its predecessor is as constant as the number of vibrations per second (the frequency) in the musical scale of a well-tuned piano. But in an infinite series, the ratio gets repeatedly larger or smaller as when computing compound interest or declining balances in home mortgages, or in carrying out the value of pi to many decimal places. The limits and uses of infinite series have continued to interest mathematicians, scientists, and engineers since Kovalevsky's time, as practical applications of this method of analysis continue to expand.

Wilson

WORLD EVENTS		KOVALEVSKY'S LIFE
Napoleonic Wars in Europe	1803-15	
	1850	Sonya Krukovsky is born
United States Civil War	1861-65	
	1868	Krukovsky marries Vladimir Kovalevsky
Germany is united	1871	
	1874	Kovalevsky receives doctorate from University of Göttingen for her work on partial differential equations
	1884	Kovalevsky becomes lecturer at University of Stockholm
	1888	Kovalevsky is awarded Prix Bordin
	1891	Kovalevsky dies
Spanish-American War	1898	

For Further Reading:
Kennedy, Don H. *Little Sparrow: A Portrait of Sonya Kovalevsky*. Athens: Ohio State University, 1983.
Koblitz, Ann H. "Sonya Kovalevsky." In *Women in Mathematics: A Bibliographic Sourcebook*. Edited by Louise S.
Grinstein and Paul J. Campbell. New York: Greenwood Press, 1987.
Osen, Lynn. *Women in Mathematics*. Cambridge, Mass.: MIT Press, 1974.

Landsteiner, Karl

Immunologist; Identifier of
Human Blood Types
1868-1943

Life and Work

Karl Landsteiner's accomplishments spanned the fields of immunology (the study of the body's defenses against foreign organisms), pathology (the branch of medicine concerned with causes and changes associated with diseases), and serology (the study of blood). He is best known for the identification and characterization of human blood types.

Landsteiner was born on June 14, 1868, in Vienna, Austria. He received a medical degree from the University of Vienna in 1891, and for the next five years he learned to apply chemical principles to medical problems while working with prominent biochemists such as Emil Fischer and Max von Gruber. By 1897 he had begun independent serological investigations at Vienna's Institute of Pathology.

In 1900 Landsteiner published a paper describing the agglutination (clumping) that occurs when the blood of two or more people is combined. He attributed the phenomenon to incompatibility among three distinct types of blood, dubbing them A, B, and C (later called O). Two colleagues added the group AB the following year. Landsteiner subsequently demonstrated the biochemical difference among blood types: blood of each type contains unique substances (called antigens) that trigger an immune response, such as agglutination, when they come in contact with blood of a different type.

From 1908 to 1920 Landsteiner served as director of Wilhemina Hospital in Vienna, where he continued to study serological immunity (immunity relating to blood fluids). He was among the first to dissociate antigens from antibodies (the substances that are produced in response to antigens), a distinction important in the study of immunological mechanisms. He developed novel experimental techniques that advanced the understanding of syphilis and poliomyelitis and discovered that the latter's causative agent is a virus rather than a bacterium.

Landsteiner moved to the United States in 1922 to take a position at Rockefeller Institute in New York. There he identified the M, N, P, and Rh factors present in human blood, which further characterized the differences among the properties and immune reactions of people's blood. In 1930 his achievements in serology earned him the Nobel Prize for Physiology or Medicine. He completed a summary of his life's work, titled *The Specificity of Serological Reactions,* in 1936. He died in New York City on June 26, 1943.

Legacy

Landsteiner's achievements revolutionized the understanding of blood-related medical conditions and helped establish the biochemical basis of immunity.

The identification of blood types made transfusions possible. Prior to 1900, blood transfusions were only occasionally successful; Landsteiner's discovery of incompatible blood types explained the many failures. By 1907 the laboratory techniques for determining blood type had been refined and compatible blood types could be matched for transfusions. During World War I, blood transfusions saved the lives of scores of injured soldiers and civilians.

Blood transfusions are used today to overcome numerous potentially fatal situations. Transfusions replace blood lost during hemorrhaging, restore blood plasma depleted during the healing of severe burns, maintain healthy levels of red blood cells and hemoglobin in cases of chronic anemia, and provide the blood coagulation factors that are missing in certain blood disorder cases.

The discovery of the M, N, and P blood factors provided a valuable forensic tool. Scientists could describe and identify people's blood using the combination of their blood type and the presence or absence of blood factors. They could therefore match blood samples found at crime scenes to some suspects and rule out those suspects whose blood did not match.

In the 1920s blood factors were shown to be inherited, which allowed their use in paternity disputes, genetics research, and studies of human evolution.

Understanding the Rh factor allowed the detection and prevention of a devastating pregnancy-related condition. When a pregnancy involves a woman and fetus with mismatched Rh factors, erythroblastosis fetalis can result, leading to brain damage and sometimes death of the fetus. Tests of Rh factor can identify incompatible blood, and precautions can be taken.

Schuyler

For Further Reading:

Duin, Nancy. *A History of Medicine: From Pre-history to the Year 2000.* New York: Simon & Schuster, 1992.

Speiser, Paul P. *Karl Landsteiner: The Discoverer of the Blood-Groups and a Pioneer in the Field of Immunology.* Wein, Austria: Hollinek, 1975.

Lavoisier, Antoine

Originator of
Law of Conservation of Mass
1743-1794

Life and Work

Antoine Lavoisier's experimental approach to chemistry led him to explore the properties of oxygen and its role in combustion, and to discover a pivotal law of science, the conservation of mass.

Born in Paris on August 26, 1743, Antoine Lavoisier enjoyed the support of a wealthy extended family. Educated to become a lawyer, Lavoisier also used his time at the College Mazarin to attend science lectures. He quickly decided that he would rather pursue a career in chemistry.

Lavoisier felt that science should serve society and to that end he helped design better street lighting, water hydrants, and sewer systems. In 1766 his essay on how to generate light for a big town earned him first place in a contest sponsored by the Academy of Sciences. In 1775 he was appointed to the post of director of the gunpowder administration and set about improving the production of gunpowder and practices within the industry. During this time Lavoisier decided to supplement his income by joining a group of government tax collectors. He joined the tax administration in 1768 and soon after, in 1771, he met and married Marie Paulze, the daughter of another tax collector.

Lavoisier made several key discoveries concerning chemical compounds, oxygen, and combustion. In 1774 he experimented with oxygen and recognized that air was not an element but a mixture of gases. Lavoisier also defined combustion as a combination of a burnable material with oxygen. This conclusion overturned the common assumption that all burnable material contains "phlogiston," which escapes as fire, leaving only ash behind. Lavoisier became dissatisfied with this explanation when his careful weighing proved that some ashes weighed more than the original samples he burned. He published his theory on combustion in 1783. Related to his work with oxygen, Lavoisier determined that water was composed of hydrogen and oxygen.

Marie Lavoisier helped her husband in countless ways. She translated scientific articles and correspondence from Latin and English, which Antoine did not speak, kept lab records, illustrated his writings, and assisted in laboratory work.

Perhaps his most innovative experimentation resulted in his theory of conservation of mass. It is said that Marie encouraged him to think about the gases that escape during reactions and capture them, procedures which enabled Lavoisier to quantitatively measure the products of chemical reactions. This work led him to conclude that, while matter may change its form during chemical reactions, it may not be created or destroyed—that is, it is conserved.

Lavoisier produced two major works that gave chemistry its modern form: *Nomenclature,* published by Lavoisier and a group of other scientists in 1787, described a concise system of naming chemical compounds; his *Elementary Treatise of Chemistry,* published in 1789, summarized his new discoveries and view of chemistry, including his conclusions on the conservation of mass.

Rising tensions during the French Revolution (1787-1789) eventually led to Lavoisier's death. As resentment grew against the government, tax collectors were singled out. Although he had never abused the position, Lavoisier was sentenced to die by guillotine in 1794.

Legacy

Antoine Lavoisier offered generations of scientists new ways of thinking about and investigating the chemical foundations of the physical world.

Although Lavoisier was very involved in the scientific community of his time, his discoveries and ideas were not all immediately accepted by his peers. His practical applications were most valued during the time he was living. For instance, Lavoisier's improvements in gunpowder proved critical to the success of the American Revolution.

His challenge to the phlogiston theory was controversial, contributing to intense debate about fire and flammability. Lavoisier's experimental method supported his findings that phlogiston was a myth and that oxygen was necessary for combustion. His *Elementary Treatise on Chemistry* promoted his revolutionary ideas and spread them quickly during his lifetime and after. Marie helped to ensure the legacy of their work. Following his death, Marie Lavoisier compiled his research and expanded his unfinished papers into an eight-volume work, *Memoirs of Chemistry,* which she published under her husband's name in 1803, nine years after he died. It included an explanation of his Law of Conservation of Mass.

The chemical names and terms that Lavoisier created are still used today. This system allows compounds to be described in a uniform manner, enabling researchers everywhere to communicate clearly.

Finally, Lavoisier's Law of Conservation of Mass became a foundation of chemistry and all physical sciences. In the twentieth century, it was amended to describe the conservation of matter and energy. ALBERT EINSTEIN used this law as part of his famous formula, $E = mc^2$ (energy equals mass times the speed of light squared). Scientists are still exploring the implications of that finding.

Wilson

WORLD EVENTS		LAVOISIER'S LIFE
	1743	Antoine Lavoisier is born
	1771	Antoine Lavoisier and Marie Paulze marry
	1774	A. Lavoisier concludes that air is composed of a mixture of gases
United States independence	1776	
	1783	A. Lavoisier publishes theory of combustion
	1787	*Nomenclature* is published
French Revolution	1789	*Elementary Treatise of Chemistry* is published
	1794	A. Lavoisier is executed
	1803	M. Lavoisier publishes *Memoirs of Chemistry,* including Law of Conservation of Mass

For Further Reading:

Donovan, Arthur. *Antoine Lavoisier: Science, Administration, and Revolution.* Cambridge, Mass.: Blackwell, 1993.

Fuerlac, Henry. *Antoine Lavoisier, Chemist and Revolutionary.* New York: Scribner's, 1975.

Poirier, Jean P. *Antoine Laurent de Lavoisier.* Philadelphia: University of Pennsylvania Press, 1996.

Leakey, Mary

Paleoanthropologist; Researcher of
Early Human Ancestors
1913-1996

World Events	Leakey's Life
	1913 Mary Douglas Nicol is born
World War I 1914-18	
	1936 Nicol marries Louis S. B. Leakey
World War II 1939-45	
	1948 Leakey finds *Proconsul africanus* skull
Korean War 1950-53	
	1959 Leakey uncovers *Australopithecus* fossils
	1960s Leakey discovers *Homo habilis* fossils
End of Vietnam War 1975	
	1978 Leakey's team finds upright-walking human ancestor footprints
Dissolution of 1991 Soviet Union	
	1995 Leakey discovers *Australopithecus anamensis* fossils
	1996 Leakey dies

Life and Work

Mary Leakey made groundbreaking discoveries of fossils, tools, and footprints in Africa that helped shape the modern understanding of human evolution.

Mary Leakey was born Mary Douglas Nicol in London on February 6, 1913. She developed a passion for archaeology as a child and participated in several excavations in England during the early 1930s. In 1935, she traveled to Africa as part of a team headed by Louis S. B. Leakey, a prominent paleoanthropologist. She married Leakey in 1936, and they worked together on much of their subsequent work—Mary often unearthing and identifying clues, and Louis often helping to interpret and publicize them. Although she received little higher education beyond courses in anthropology and geology at University College in London, Leakey's revolutionary finds earned her numerous honorary degrees.

In 1948, Leakey found the skull of the 18-million-year-old human ancestor *Proconsul africanus* on an island in Africa's Lake Victoria. This find overturned many anthropologists' belief that humans evolved in Asia. In 1959, Leakey uncovered more than 400 fragments of the skull of a 1.75-million-year-old hominid at the Olduvai Gorge in Tanzania. The fragments were the oldest fossils of the genus *Australopithecus* (a group that includes human ancestors and their relatives) that had been found at that time and the first ones located in eastern, rather than southern, Africa.

In the early 1960s, Leakey unearthed the first known fossils of *Homo habilis,* named by her husband, which was a direct ancestor of modern humans who lived about two million years ago. In 1978, Leakey participated in what she considered to have been her most important contribution to the study of human evolution. Members of her team found the footprints of two (or possibly three) hominids in volcanic ash at Laetoli on the Serengeti Plain in northern Tanzania. Leakey recognized that the footprints, dated 3.65 million years ago, showed the toes, heel and arch characteristic of hominid feet, indicating upright walking. When the discovery become public, most paleoanthropologists agreed that the footprints belonged to *Australopithecus afarensis,* a direct ancestor of *H. habilis* and modern humans.

Leakey remained in East Africa for the rest

of her life. She died on December 9, 1996, in Nairobi, Kenya.

Legacy

Leakey was among the most prominent paleoanthropologists in her field, and her discoveries, as part of her husband's team, had a major impact on the study of human evolution during her lifetime.

Her work moved the birthplace of modern humans from southern to eastern Africa. The footprints at Laetoli revealed how early in evolutionary history human ancestors walked upright, an ability that is considered crucial to the evolution of modern humans.

Leakey also left an archaeological legacy within her family. Her son, Richard, remained in Africa after his upbringing there and became a prominent paleontologist. Richard Leakey and his wife, Meave Leakey, have made numerous important fossil finds in eastern Africa. In 1995, Meave Leakey found the fossils of a new species, *Australopithecus anamensis,* near Lake Turkana, Kenya. *A. anamensis,* who lived more than four million years ago, is the oldest confirmed upright-walking human ancestor.

The vast number of fossils, tools, and other artifacts unearthed by Leakey and her family continue to reveal hints of how human ancestors lived. These discoveries also have helped to define the questions challenging contemporary paleoanthropologists, including how long ago and where the oldest upright-walking human ancestors inhabited the earth.

Schuyler

For Further Reading:

Leakey, Mary. *Disclosing the Past: An Autobiography.* New York: Doubleday, 1984.

Morell, Virginia. *Ancestral Passions: The Leakey Family and the Quest for Humankind's Beginnings.* New York: Simon & Schuster, 1995.

Le Châtelier, Henri-Louis

Originator of Le Châtelier's
Principle of Equilibrium
1850-1936

Life and Work

Henri-Louis Le Châtelier's interest in thermodynamics and how chemical systems react to stress resulted in his discovery of a principle of equilibrium for chemistry. This law predicts how chemical reactions will respond to changes in their environment.

Le Châtelier was born in Paris on October 8, 1850, into a family of scientists and engineers. His father encouraged him to pursue a scientific career; Le Châtelier attended the prestigious École Polytechnique after spending time in the military.

In 1877 Le Châtelier became a professor of chemistry at École des Mines and remained an educator throughout his career, obtaining posts at various schools including the College de France and the Sorbonne. He also became an advocate for educational reform and championed a more hands-on teaching methodology. His initial studies focused on applied science dealing with cements and ceramics.

This research required subjecting materials to high temperatures and pressures. From this research he delineated his principle of equilibrium in 1884. It states that a system in equilibrium will react characteristically to changes: if one of the conditions of the system is altered, a shift will occur to minimize the effects of that change. An application of this principle is that if a reaction is difficult to initiate, then the reactants can be subjected to certain stresses, which will accelerate the chemical reaction process.

During this research into high-temperature reactions, Le Châtelier required more precise instruments to measure accurately the changes in temperature. In 1887 he invented a thermocouple that used the combination of a platinum wire and a platinum alloy wire. The difference in temperature of the two conductive metals produced an electrical current. Changes in the current strength could be measured and used to determine the temperature.

Le Châtelier refined many other devices, including miners' lamps and metallurgical microscopes. His work in thermodynamics also involved studying gas explosions and blast furnaces. Le Châtelier conducted experiments on many gases to determine under what conditions and temperatures they would explode. One result of these studies was the invention of the acetylene torch.

Le Châtelier continued to teach and conduct research for many years. In 1916 he was awarded the Davy Medal by the British Royal Society. He retired from his position at École des Mines in 1919. Le Châtelier died in France on September 17, 1936.

Legacy

Le Châtelier's legacy rests on his law of chemical equilibrium, which enabled scientists to predict how a system will react to changes in pressure and temperature. In addition, the many connections he made between theoretical science and industrial applications resulted in technological advances that directly influenced industry and daily life.

During his lifetime Le Châtelier gladly shared his techniques for studying thermodynamics and what he learned. He taught hundreds of students and published over 500 books and articles. He felt that students needed to learn the general principles of chemistry and how to expand their knowledge rather than memorize exhaustive lists. He argued for educational reform in France.

His refinements and inventions saved the lives of miners in Europe, preventing deadly mine explosions. Blast furnaces incorporated his research to increase productivity. Another scientist, FRITZ HABER, using Le Châtelier's law of equilibrium, mastered the process of producing ammonia on a commercial scale, something Le Châtelier also attempted.

Several of Le Châtelier's inventions remain in use today. Thermocouples continue to be used for accurate, high-temperature measurements. Metal workers and welders use the acetylene torch in construction and other industries. Blast furnaces still produce steel using his more efficient technique of preheating gases before they enter the main chamber by using the hot exhaust. Finally, his principle of equilibrium led to applications for the phase rule, which predicts how temperature, pressure, and concentration of components affect phase changes of matter.

Throughout his career, Le Châtelier showed how science could be used to solve practical problems. He firmly established the study of thermodynamics as both a theoretical and applied science.

Wilson

WORLD EVENTS		LE CHÂTELIER'S LIFE
	1850	Henri-Louis Le Châtelier is born
United States Civil War	1861-65	
Germany is united	1871	
	1877	Le Châtelier is appointed professor of chemistry at École des Mines
	1884	Le Châtelier publishes principle of equilibrium
	1887	Le Châtelier invents platinum thermocouple
Spanish-American War	1898	
World War I	1914-18	
	1916	Le Châtelier is awarded Davy Medal by Royal Society
	1936	Le Châtelier dies
World War II	1939-45	

For Further Reading:
Farber, Eduard. *Milestones of Modern Chemistry.* New York: Basic Books, 1966.
Travers, B. *World of Scientific Discovery.* Detroit, Mich.: Gale Research, 1994.

Leeuwenhoek, Antoni Van

Pioneer Microscopist and
Microbiologist
1632-1723

Life and Work

Antoni van Leeuwenhoek was a microscopist who was a pioneer in the field of microbiology. He built microscopes fitted with hand-crafted, high-quality lenses through which he made the first observations of numerous microscopic living things such as bacteria, protozoa, and red blood cells.

Leeuwenhoek was born Antoni Thoniszoon on October 24, 1632, in Delft, Holland. He later took the surname Leeuwenhoek ("Lion's Corner") from his family's house located at a street corner near the Lion's Gate in Delft.

He received little formal schooling, but was apprenticed to a draper in Amsterdam, who taught him to examine textiles closely with a hand lens. In 1652, Leeuwenhoek opened a drapery shop in Delft. In 1660 he gained a position within the Delft city government, a situation that left him relatively free to pursue his consuming interest in microscopy.

In 1671, Leeuwenhoek built his first microscope, which bore one doubly convex glass lens he ground himself. Throughout his life he crafted more than 400 lenses that magnified objects up to almost 300 times their original size.

In 1674, Leeuwenhoek was the first to observe living cells, which he called "animalcules." The cells were protozoa, members of the kingdom now known as Protista. Two years later, he communicated this discovery to the Royal Society of London, causing a sensation across Europe.

In 1677, he became the first person to describe spermatozoa. He discovered yeast cells in 1680 and bacteria in 1683. In 1684, he discovered red blood cells and confirmed WILLIAM HARVEY's theory of blood circulation with observations of capillaries carrying blood between arteries and veins. Under his lenses, he studied plant anatomy and physiology, the structure of the lens of the eye, and life histories of hydra and insects. He identified the complete metamorphosis of fleas, which eradicated popular belief in their spontaneous generation.

He sent drawings of his observations to the Royal Society in more than 350 letters over his lifetime, and his findings were published in the Society's Philosophical Transactions. *Arcana Naturae*, a description of Leeuwenhoek's discoveries, was published in 1696.

Leeuwenhoek continued his studies until he died at age 90 in Delft on August 26, 1723.

Legacy

Leeuwenhoek was a pioneering and enthusiastic microscopist whose record of observations founded the field of microbiology.

Prior to Leeuwenhoek's investigations, the abundant single-celled creatures that populate the planet were unknown. The significance of Leeuwenhoek's discoveries was recognized immediately, and the sensation surrounding the industrious lens crafter from Delft brought visits from European royalty.

Leeuwenhoek's microscopes were of higher quality than those of his contemporaries because of his superior lenses, but he kept his techniques confidential. Therefore, his influence on the development of the microscope cannot be easily traced.

His observations of biological processes that take place at the microscopic level opened up new ways of thinking to scientific investigators. His observations on the life cycle of the flea, for instance, began to break down beliefs in spontaneous generation, which was an accepted explanation of reproduction in lower levels of animal life.

Such observations encouraged scientists to investigate and reformulate popularly held beliefs through microscopic inquiry, but it was not until the nineteenth century that the understanding of living cells progressed greatly. In 1838, Theodor Schwann and Matthias Jakob Schleiden proposed that all living things are made up of cells, which is a basic tenet of biology now known as the cell theory. In the late 1800s, scientists first recognized that some microscopic organisms cause diseases; such infectious diseases are still among the leading causes of human death in most countries. Studies of microscopic organisms escalated dramatically in the twentieth century, when advances in the compound microscope increased magnification to about 2,000 times.

Schuyler

For Further Reading:

Dobell, Clifford. *Antoni van Leeuwenhoek and His "Little Animals."* New York: Dover, 1960.

Ford, Brian J. *The Leeuwenhoek Legacy.* Champaign, Ill: Balogh Scientific Books, 1991.

Ruestow, E. G. *The Microscope in the Dutch Republic.* Cambridge: Cambridge University Press, 1996.

Leibniz, Gottfried Wilhelm

Inventor of Differential
and Integral Calculus
1646-1716

Life and Work

Largely self-educated, Gottfried Leibniz made significant contributions to mathematics, including the invention of the calculus as well as results in combinatorics, a mathematical field closely related to logic and focused on selection, arrangement, and operations within closed systems. He was active and influential in other fields as well, including philosophy, law, and religion.

Leibniz was born in Leipzig, Germany, on July 1, 1646. Six years later his wealthy father, a book collector, died. Most of Leibniz's early education, including the classics, history, and philosophy, came from reading in his father's library. When he was 15, Leibniz entered the University of Leipzig to study law. He received a bachelor of philosophy degree in 1663 after defending his thesis on logic, "On the Principle of the Individual." In 1666 he wrote two important papers and received a doctorate of philosophy from the University of Altdorf. His papers, "On the Art of Combination," laid groundwork for symbolic logic, which Leibniz proposed as a "universal language of ideas."

Leibniz refused a professorship in law at the University of Altdorf and instead began his career as a statesman (one of his goals was to reunify the Roman Catholic and Protestant churches). While on a diplomatic mission to Paris in 1672 he met the mathematician CHRISTIAAN HUYGENS and began to explore mathematics. In 1673 another mission took him to London where he was elected a foreign member of the Royal Society for his calculating machine. BLAISE PASCAL had made an adding machine in 1642 that made carrying of tens automatic. Leibniz's invention was able to add, subtract, multiply, and divide based on a binary instead of a decimal system.

In 1675 Leibniz worked out much of the detail for both differential and integral calculus, developing systematic rules and formulas, and he connected the two with the fundamental theorem of calculus. In England, ISAAC NEWTON claimed that he had already invented calculus. Each criticized the other's work and there was a nationalistic disagreement on who had invented the calculus first. Since Leibniz published first, in 1684, custom would credit him, but the remarkable fact is that Newton and Leibniz each independently created the calculus at roughly the same time.

Although Leibniz continued to work in mathematics, most of the remaining 40 years of his life was devoted to philosophy. From 1685 on he was the historian and genealogist for the House of Brunswick in Germany, making the case supporting his employer's claim to the throne of England in 1714 as King George I. Leibniz died in Hanover, Germany, on November 14, 1716, at the age of 70.

Legacy

Although he contributed to many fields, Leibniz's primary legacy is his work in mathematics, particularly his invention of differential and integral calculus and promotion of symbolic logic.

Leibniz was a prominent advocate of the binary system to speed calculations. Because of his recognition of the simplicity of using only one and zero as symbols to express all numbers, and given his philosophical concept of reducing all ideas to an artificial but universal language of symbols, like algebraic symbols, Leibniz has been called the Patron Saint of

Cybernetics. Modern computers are tangible derivatives of his ideas.

The controversy over who invented the calculus arose because of the time elapsed between Leibniz's and Newton's claims and the dates of publication. Leibniz published first; Newton had shown his work to selected mathematicians as early as 1665 but didn't publish his ideas until 1704 in *Optiks*. Both Leibniz's notation style and his analytical methods were easier to understand and develop. Newton's technique of fluxions (derivatives) was described in symbols like x with a dot over it to indicate rate of change of some variable. Leibniz described the same relationship as dx/dt where dx is the tiny part of x that changes during a tiny increment of time, dt. In complex equations the descriptive notation by Leibniz suggested relationships that were hidden in Newton's cryptic symbols. Consequently Leibniz's calculus influenced a wider range of mathematicians and scientists, including the BERNOULLI family, LEONHARD EULER, Luigi Lagrange, and Pierre Laplace.

Steinberg

WORLD EVENTS	LEIBNIZ'S LIFE
Thirty Years' War 1618-48 in Europe	
1646	Gottfried Wilhelm Leibniz is born
1666	Leibniz earns Ph.D.
1672	Leibniz meets Christiaan Huygens in Paris
1673	Leibniz demonstrates calculating machine to Royal Society
1675	Leibniz works out details of the calculus
1684	Leibniz publishes work on calculus
England's Glorious 1688 Revolution	
1704	Isaac Newton discusses his calculus in *Optiks*
Peace of Utrecht 1713-15 settles War of Spanish Succession	
1716	Leibniz dies

For Further Reading

Bell, E. T. *Men of Mathematics*. New York: Simon & Schuster, 1937.

Calinger, Ronald, ed. *Classics of Mathematics*. Englewood Cliffs, N.J.: Prentice-Hall, 1995.

Kline, Morris. *Mathematical Thought from Ancient to Modern Times*. New York: Oxford University Press, 1972.

Linnaeus, Carolus

Botanist; Originator of Taxonomic
Classification System

1707-1778

Life and Work

Carolus Linnaeus introduced fundamental principles of botany and developed binomial nomenclature for the classification of plants and animals in the eighteenth century.

Linnaeus was born on May 23, 1707, into a religious family in South Råshult, Sweden. His father's passion for gardening and natural his-

tory instilled in Linnaeus an enthusiasm for botany, which remained the primary focus of his life's work. At university, Linnaeus studied medicine, a discipline that included botanical training, and, by 1734, he had identified more than 100 new plant species. By 1730, he became dissatisfied with the prevailing system of classifying plants by seed comparisons and began to develop a new method based on the number and arrangement of flowering plants' sexual organs (pistils and stamens). This idea emerged from his belief, unlike that of many of his contemporaries, that plants reproduce sexually. In 1735, Linnaeus traveled to Holland where he published his first book, *Systema Naturae,* in which he presented his new botanical classification system.

In 1738 Linnaeus received a medical degree from the University of Harderwijk, Holland, and returned to Sweden to practice medicine. He joined the University of Uppsala in 1741 as a professor of medicine and was soon appointed head of the botany department. Linnaeus collected and classified a vast number of plants during his life, procedures that required extensive notes. Plant species were designated by their genus name plus a lengthy description of their appearance. Linnaeus began to record a shortened designation, which included the organism's genus name and a single species name to replace the lengthy version. Linnaeus outlined this binomial classification system in *Philosophia Botanica,* published in 1751, and applied it to the classification of 8,000 plant species in *Species Plantarum,* published two years later.

Linnaeus continued to teach botany and classify plants until his death on January 10, 1778, in Uppsala, Sweden.

Legacy

Because of Linnaeus's innovations, the cataloguing of living things progressed greatly during his lifetime. As an enthusiastic teacher, Linnaeus elevated the perceived importance of taxonomy within the field of biology and inspired students to travel widely in search of unknown plants. In the eighteenth century, previously unknown organisms were identified at an increasing rate all over the planet, and Linnaeus's classification method allowed a systematic recording of

those discoveries. The Linnaean Society was established in 1788, a testimony to the immediate significance of his work.

Linnaeus's binomial nomenclature created an international language for biological studies. Humans have long had names for different kinds of organisms, but popular names vary with language and location, hindering accurate cross-cultural communication. Scientists began to contemplate the production of biology books in the late fifteenth century, after the invention of moveable-type printing. To write books with universal scientific appeal, they needed a universal biological language, and so they developed the multiple-word naming system that preceded that of Linnaeus. Linnaeus's binomial system dismissed the problems associated with such an inconvenient system, gave structure to the task of naming organisms, and simplified biological discourse. Binomial nomenclature was accepted as an international standard during Linnaeus's lifetime, and scientists still use it to name organisms today.

The theory of evolution has altered some of the classifications mapped by Linnaeus and his successors. Linnaeus believed that his classification system was simply a recognition of a natural arrangement of immutable species created by God. However, evolutionary theory posits that species evolve over time. Taxonomists now use genetics and biochemistry to investigate evolutionary history, to distinguish species, and to determine taxonomic relationships.

Schuyler

World Events		Linnaeus's Life
	1707	Carolus Linnaeus is born
Peace of Utrecht settles War of Spanish Succession	1713-15	
	1730	Linnaeus begins to develop new plant classification system
	1730s	Linnaeus collects unknown plant specimens
	1735	*Systema Naturae* is published
	1741	Linnaeus begins teaching botany at University of Uppsala
	1751	*Philosophia Botanica* is published; it outlines Linnaeus's binomial classification system
	1753	*Species Plantarum* is published
United States independence	1776	
	1778	Linnaeus dies
French Revolution	1789	

For Further Reading:

Frängsmyr, Tore, ed. *Linnaeus: The Man and His Work.* Canton, Mass.: Science History Publications, 1994.

Larson, James L. *Interpreting Nature: the Science of Living Form from Linnaeus to Kant.* Baltimore: Johns Hopkins University Press, 1994.

Lister, Joseph

Founder of Antiseptic Surgery
1827-1912

Life and Work

Joseph Lister introduced antisepsis (procedures to hinder the growth of microorganisms) to surgery and strengthened the microorganism theory of disease in the nineteenth century.

Born in Upton, England, on April 5, 1827, Lister was educated at various Quaker schools, where science was emphasized. As a teenager, he became interested in anatomy and decided on a career in surgery.

He received a medical degree from University College, London, in 1852, and the following year became an assistant to surgeon James Syme in Edinburgh. In 1856 Lister was appointed physician at the Edinburgh Royal Infirmary; in 1861 he became head of surgical wards at the Royal Infirmary, Glasgow.

In Lister's time, nearly half of surgical patients died of sepsis (blood poisoning). Believing that sepsis was caused by bad air, he had attempted to decrease the death rate by keeping wards exceedingly clean and wounds tightly covered with bandages. These attempts were futile.

Thanks to his wife Agnes, who could read French and translated LOUIS PASTEUR's works for him, Lister learned of Pasteur's theory that organic decay is caused by living matter. Lister theorized that sepsis also resulted from such living particles, and in 1865 he began to use a carbolic-acid solution to clean wounds and surgical instruments. In 1867 he announced that carbolic-acid treatment had kept his wards free of sepsis for nine months, but his news was met with hostility from traditional surgeons.

It took several critical successes before Lister's theories on antiseptics became accepted. He successfully treated Queen Victoria for an abscess, using carbolic acid as a disinfectant. During the Franco-Prussian war in 1871, his disciples reported that antiseptic surgery resulted in virtual disappearance of gangrene and blood poisoning among their patients. Lister performed a public demonstration of a very dangerous knee surgery in 1877, and again his methods proved successful,

finally leading to widespread acceptance of antiseptic surgical procedures.

In 1891 Lister helped found the British Institute of Preventative Medicine, later renamed the Lister Institute. In his last years, Lister was honored with numerous awards recognizing his achievement. He died at Walmer, England, on February 10, 1912.

Legacy

Lister's introduction of antisepsis made surgery much safer, and it provided strong support for the theory that living matter causes infection, one of the basic tenets of modern medicine.

Lister's innovation hit the medical community at a time when physicians were reluctant to accept that microscopic life could significantly affect human health. In the early 1800s, many people believed that changes in the air caused most diseases. Then in the 1840s, Justus von Liebig and Jacob Henle performed the first experiments that hinted at the existence of living disease-causing agents. Pasteur, in the 1850s and 1860s, showed that living organisms caused fermentation and putrefaction. Lister drew upon Pasteur's theories and attempted to clean living infectious agents from wounds and incisions. Pasteur's work was not yet widely accepted, and Lister's idea was at first rejected.

However, Lister was fortunate to see his work fully embraced during his lifetime, and this acceptance helped strengthen the microorganism theory of disease. Lister's successful public surgery in 1877 forced physicians to accept that carbolic acid worked as an antiseptic agent, and soon all surgeries were performed with antiseptic precautions.

The work of ROBERT KOCH in the 1870s then led to aseptic surgeries (surgeries free from disease-causing agents). He advised the use of steam to sterilize surgical materials; steam treatment left the equipment completely aseptic. Aseptic procedures drastically reduced the number of surgical deaths, and physicians in Europe and North America quickly began to see that a sterile environment prevents infection.

Lister's legacy remains the basis of surgical methodology in modern medicine. Wounds, the sites of surgical incision, and the instruments and general surroundings of a surgery must be kept free of bacteria.

Schuyler

World Events	Lister's Life
Napoleonic Wars 1803-15 in Europe	
	1827 Joseph Lister is born
	1852 Lister earns medical degree from University College
	1853 Lister becomes assistant to surgeon James Syme
	1856 Lister becomes practicing surgeon
	1861 Lister takes charge of surgical wards at Glasgow infirmary
United States 1861-65 Civil War	
	1865 Lister learns of Louis Pasteur's work and germ theory
	1867 Lister announces success of antiseptic methods
Germany is united 1871	
	1877 Lister performs public surgery under antiseptic conditions
	1891 Lister founds British Institute of Preventative Medicine
Spanish-American 1898 War	
	1912 Lister dies
World War I 1914-18	

For Further Reading:

Fisher, Richard. *Joseph Lister, 1827-1912.* New York: Stein & Day, 1977.

Pasteur, Louis. *Germ Theory and Its Applications to Medicine and on the Antiseptic Principle of Surgery.* Amherst, N.Y.: Prometheus Books, 1996.

Lovelace (Countess of), Augusta Ada

First Computer Programmer
1815-1852

Life and Work

Augusta Ada Byron, later called Countess of or Lady Lovelace, was a mathematician who worked on CHARLES BABBAGE's nineteenth-century "Analytical Engine," a prototype of twentieth-century computers. She was first to describe a method of programming such a machine.

Augusta Ada Byron, daughter of the poet Lord Byron, was born on December 10, 1815, in London. Her parents separated when she was a month old and Lord Byron left England. Ada, as she was known, was educated privately. Her abilities in mathematics were so great that at 15 she began working with the mathematician and logician Augustus De Morgan.

In 1833 Ada first saw plans for Charles Babbage's "Difference Engine," a machine intended to automate the calculation of astronomical tables. Impressed by the machine and its possibilities, she began an association with

Babbage. She also began corresponding with other scientists and mathematicians such as MICHAEL FARADAY and Babbage's friend from Cambridge, the astronomer Sir John Herschel. Ada married Lord King in 1835. Three years later he was elevated to an earl, and Ada was afterwards known as Lady Lovelace.

While Babbage was in Turin, Italy, he gave a series of talks on the successor to his difference machine, the Analytical Engine, considered to be the first automatic calculator in the world. General Luigi Menabrea reported on Babbage's new invention in Italian in 1842. Lady Lovelace decided to translate this into English, and Babbage suggested that she supplement the translation with her own notes. Her notes more than tripled the length of the original text and added a great deal of insight, information, and clarity.

Babbage adapted the operation of the Jacquard loom to his design for the Analytical Engine; the loom used punch cards to guide the weaving of specific patterns. Lady Lovelace made this connection explicit, writing that "the Analytical Engine weaves algebraic patterns, just as the Jacquard loom weaves flowers and leaves." Although the Analytical Engine was never built, the specifications were complete and it was constructed twice within the following 150 years. The Analytical Engine separated operators from the objects on which they operated and from those that were moved in and out of storage. With Babbage's guidance, Lovelace constructed a computer program, a set of instructions for the Analytical Engine, that would carry out operations on the objects. She also understood and explained the machine's limitations, stating that it "has no pretensions to originate anything. It can do whatever we know how to order it to perform."

Lady Lovelace had never been healthy. She died on November 27, 1852, in London at age 36, the same age at which her father had died.

Legacy

Lady Lovelace is considered to be the first computer programmer; her explanation and elaboration of the Analytical Engine left a legacy for computer designers in the twentieth century.

While Lovelace did not design the Analytical Engine, her understanding of it and her

collaboration with Babbage on programming it allowed her to explain what it was capable of and how it accomplished its work. In preserving and explicating this crucial step in computer history, she offered engineers in the twentieth century a grasp of the possibilities seen by Babbage. Babbage's vision for his computing machine far outdistanced the technological capacity of his times. In the middle of the twentieth century, when technology caught up with the Analytical Engine, Lovelace's documentation—as well as Babbage's own notes—allowed engineers to incorporate his ideas into their new designs. HOWARD AIKEN, for example, referred to the Analytical Engine in developing the Mark I, the first full-scale programmable computer in the United States, in 1944. His machine adapted the punch card idea developed and explained by Babbage and Lovelace.

Lovelace originated the concept of modern computer programming. She explained that you could use a sequence of expressions on punch cards repeatedly, just as a subroutine in computer programming does today. Instructions fed into computer via punch cards, punched paper tape, and punched movie film were used in the earliest computers. In recognition of Lovelace as the first computer programmer, the Pentagon named a computer language, called Ada, after her in the late 1970s. Much of the work contracted by the United States Navy in the 1980s and 1990s was written in Ada.

Steinberg

WORLD EVENTS	LOVELACE'S LIFE
Napoleonic Wars 1803-15 in Europe	
	1815 Augusta Ada Byron is born
	1830 Byron begins work with mathematician and logician Augustus De Morgan
	1833 Byron begins work with Charles Babbage
	1838 Byron becomes Lady Lovelace
	1842 Lovelace translates and supplements Luigi Menabrea's work on the Analytical Engine
	1852 Lovelace dies
United States 1861-65 Civil War	

For Further Reading:

Goldstine, Herman. *The Computer from Pascal to von Neumann.* Princeton, N. J.: Princeton University Press, 1972.

Newman, James R. *The World of Mathematics.* New York: Simon & Schuster, 1956.

Palfreman, Jon, and Doron Swade. *The Dream Machine: Exploring the Computer Age.* London: BBC Books, 1991.

Stein, Dorothy. *Ada: A Life and a Legacy.* Cambridge, Mass.: MIT Press, 1985.

Lumière, Auguste; Lumière, Louis

Inventors of Cinema
1862-1954; 1864-1948

Life and Work

The brothers Auguste and Louis Lumière launched the film industry by inventing and manufacturing the first motion picture camera and projector, dubbed the cinématographe.

Auguste Lumière was born on October 19, 1862, and Louis Lumière was born on April 10, 1864, both in Besançon, France. At school in Lyons, the boys excelled in science and upon completion of secondary education joined the family's photography business. Their father encouraged their attempts to improve upon existing photographic technology. By 1880 they had devised a dry plate that improved the quality of commercially developed film, and two years later Louis opened a factory for manufacturing the plates.

In 1894 the boys' father witnessed a presentation of American inventor THOMAS EDISON's kinetoscope, an early motion-picture machine introduced in 1887. Informed of this device, the Lumière brothers were inspired to design a machine that would combine animation with projection (Edison's kinetoscope lacked a built-in projection mechanism). They patented their resulting cinématographe, from which the term "cinema" is derived, in 1895.

The machine involved celluloid film run through a projector at 16 frames per second and a semicircular shutter that dampened the light between the lens and the film. It was considerably smaller and lighter than Edison's kinetoscope.

On December 28, 1895, the brothers held the first demonstration of their invention at a Parisian café packed with a full audience, showing a short film called *Quitting Time at the Lumière Factory* and several other clips. By 1898 they and their staff had produced more than 1,000 short motion pictures, including newsreels, comedy shorts, and documentaries; however, they regarded their invention as a novelty and did not believe it to have much of a future.

Louis continued with photography, devising an apparatus for producing panoramic shots and a process of using dyed starch grains for color printing. In his later years he experimented with stereoscopy and three-dimensional films.

He died on June 6, 1948, in Bandol, France. His brother Auguste died on April 10, 1954, in Lyons, France.

Legacy

The Lumière brothers' cinématographe marked the beginning of cinema, a revolutionary form of twentieth-century entertainment and information.

Although the Lumières failed to foresee the potential of their invention, cinema culture spread quickly. During the first decade of the twentieth century, most films were short, lasting 15 to 20 minutes, and were geared toward a public that could not afford to attend live theater. People became movie fans in the thousands around the world. In 1908 the United States had an estimated 10,000 movie houses, and as many as 250,000 people went to the cinema every day in New York City alone. Russia had 1,200 movie houses by 1910.

The technology of cinema progressed rapidly as well. Animated cartoons, series of slightly altered drawings projected in rapid sequence, appeared in 1907. A major advance was related to sound; attempts at talking projections began in 1902 but were not fully successful until 1927, when the first sound movie *The Jazz Singer* was produced. Its sound source was a synchronized phonograph record.

During World War I, newsreels took on a new importance that they were to retain after the war. People became accustomed to receiving real-life moving images, and the demand for filmed news was established.

From its beginning as a curious attraction, cinema has become a huge industry in the late twentieth century, involving hundreds of professional careers and billions of dollars per year.

Schuyler

WORLD EVENTS	THE LUMIÈRES' LIVES
United States 1861-65 Civil War	
1862	Auguste Lumière is born
1864	Louis Lumière is born
Germany is united 1871	
1880	The Lumière brothers invent an improved photographic dry plate
1887	Thomas Edison devises kinetoscope
1894	The Lumières' father learns of Edison's kinetoscope
1895	The Lumières obtain patent for the cinématographe
	The first moving picture is shown to an audience
Spanish-American 1898 War	The Lumières have produced over 1,000 short motion pictures
World War I 1914-18	
World War II 1939-45	
1948	Louis Lumière dies
Korean War 1950-53	
1954	Auguste Lumière dies

For Further Reading:

Gianetti, Louis D. *Flashback: A Brief History of Film.* Englewood Cliffs, N.J.: Prentice-Hall, 1996.
Letters: Auguste and Louis Lumière. Boston: Faber & Faber, 1995.

Lyell, Charles

Originator of Fundamental
Principles of Geology
1797-1875

Life and Work

Charles Lyell established the backbone of modern geology by gathering and expounding support for the principle of uniformitarianism, which states that gradual physical processes shape Earth's surface features over long periods of time.

Lyell was born in Kinnordy, Scotland, on November 14, 1797, to a family that encouraged his early interest in natural history. Entering Oxford University at age 19 to study mathematics and geology, he spent vacations indulging a growing enthusiasm for fieldwork. His parents, however, pushed him to pursue a law career, and he was admitted to the bar in 1825. Given his passion for outdoor work, Lyell finally persuaded his parents that he should give up legal service and become a full-time geologist.

In 1825 Lyell published his first scientific papers, including evidence for vertical movements of Earth's crust and observations of sediment layers in lake beds. Drawing on his acute observational and reasoning skills, he embraced the theory of uniformitarianism, first put forth by eighteenth-century geologist James Hutton. Lyell traveled throughout Europe and gathered evidence for the idea that all geological formations result from the cumulative effects of ordinary physical processes still at work, such as erosion and deposition. This idea opposed popular belief in catastrophism, which posits that only individual global cataclysms reshape Earth periodically, changing both land and life forms.

The three volumes of Lyell's influential *Principles of Geology* were published in 1830, 1831, and 1833. The book, appearing in 11 more editions during Lyell's lifetime, offered numerous global examples supporting uniformitarianism. Its popularity and superior descriptions of rapid erosion and deposition in selected areas of the world earned him wide recognition and a professorship at King's College, London.

He completed a second treatise in 1838: *Elements of Geology* discussed the sedimentation of rocks and fossils. He introduced a method of classifying the layers of Earth's surface based on how many fossils of modern species they contain, proposing the names Eocene, Miocene, and Pliocene for three distinct epochs.

In the 1840s and 1850s Lyell traveled and lectured in North America and wrote two more books, describing unique aspects of that continent, both natural wonders and social progress.

In his later years, he participated in efforts to prevent mine disasters and to reform education, and continued to study geological phenomena. At first reluctant to accept CHARLES DARWIN's *On the Origin of Species,* which appeared in 1859, he was quickly won over; Lyell's 1863 work, *The Geological Evidences of the Antiquity of Man,* supported Darwin's ideas.

After his death on February 22, 1875, he was honored as a geologist and humanitarian with burial in Westminster Abbey in London.

Legacy

Lyell established fundamental principles of all subsequent geology, contributing to the development of evolutionary biology and the modern understanding of Earth's history.

The eloquence and persuasiveness of Lyell's *Principles of Geology* popularized uniformitarianism. The theory challenged the long-held belief that Earth was only a few thousand years old. Lyell noted that Earth must be at least several millennia older for slow processes such as erosion and deposition to have produced its current features.

Lyell was among the founders of the geological sub-discipline stratigraphy, the study of the layers, or strata, of Earth's crust. The names Eocene, Miocene, and Pliocene are still used today to designate three epochs of the Tertiary period, which lasted from 65 million to two million years ago.

Darwin was strongly influenced by Lyell's work. On his famed voyage aboard the Beagle, he read the first volume of *Principles of Geology* and witnessed many phenomena convincing him of Lyell's arguments. He later claimed that the book opened his mind to ideas that led him to question the immutability of species and to construct his theory of evolution. Lyell's acceptance of and supporting evidence for evolution added credence to the controversial theory.

Lyell's reputation waned in the next generation, as Darwin's fame overshadowed him. Scientists began to attribute some of his ideas to Darwin, but in fact much modern geological methodology originated with Lyell.

Schuyler

WORLD EVENTS		LYELL'S LIFE
French Revolution	1789	
	1797	Charles Lyell is born
	1816	Lyell enters Oxford
	1825	Lyell publishes first scientific papers, which include important geological concepts
	1830-33	Lyell's *Principles of Geology* is published
	1838	Lyell's *Elements of Geology* is published
	1841	Lyell first travels to North America
	1859	Charles Darwin's *On the Origin of Species* is published
United States Civil War	1861-65	
	1863	Lyell's *The Geological Evidences of the Antiquity of Man* is published
Germany is united	1871	
	1875	Lyell dies
Spanish-American War	1898	

For Further Reading:
Brush, Stephen G. *Transmuted Past: The Age of the Earth and the Evolution of the Elements from Lyell to Patterson.*
Cambridge: Cambridge University Press, 1996.
Oldroyd, D. R. *Thinking About the Earth: A History of Ideas in Geology.* Boston: Harvard University Press, 1996.

Malpighi, Marcello

Early Pioneer in Medical Research
1628-1694

Life and Work

Marcello Malpighi discovered capillary blood vessels, the structure of the lungs, and numerous other microscopic elements of anatomy and physiology.

Malpighi was born on March 10, 1628, in Crevalcore, Italy. Little is known about his youth. He attended the University of Bologna and earned doctorates in both medicine and philosophy in 1653. Upon graduation, he began teaching at Bologna and subsequently devoted himself to anatomical investigations, to describing what he observed under his microscope, and to inferring relationships between form and function.

In 1656 Malpighi was invited to teach theoretical medicine at the University of Pisa, where he began to question prevailing medical teachings including the theory that blood was changed into flesh at the outer edge of the body. He also advocated experimental methods to investigate problems in anatomy, physiology, and medicine.

Three years later, Malpighi returned to Bologna and focused on microscopic research. In a letter published in 1661, he described the network of capillary blood vessels that links the veins and arteries. He also discovered the

alveoli (air sacs) of the lungs. These significant achievements provoked jealousy among his colleagues, and in 1662 Malpighi left the hostile atmosphere for a post at the University of Messina, in Sicily. There he identified the taste buds and described for the first time many microscopic structures, including those of the brain, optic nerve, blood, and fat reservoirs.

In 1667 Malpighi returned again to Bologna, where he practiced medicine and continued to scrutinize body tissues under the microscope. He observed minute substructures within the liver, brain, spleen, kidneys, bone, and skin. He incorrectly concluded that all organs are organized into glandular subcomponents, and that the glandular mixing of body fluids was the main function of organs.

Malpighi later investigated the microscopic structures and properties of insect larvae, chick embryos, and plants. He recognized some correspondence between plant and animal anatomy. His thesis on plant anatomy included a remarkable inference that green plants manufacture their own food.

In 1669 Malpighi was named an honorary member of the Royal Society in London, and all of his work was thereafter published in the society's *Philosophical Transactions*.

In Malpighi's later years, opposition to his unorthodox views heightened. In 1684, opponents burned his villa and ruined his equipment, books, and papers. Despite this opposition, Pope Innocent XII invited him to serve as his personal physician in 1691 in recognition of his accomplishments. Malpighi died in Rome on November 30, 1694.

Legacy

Malpighi's extensive work revolutionized the understanding of microscopic anatomy, which advanced the study of human physiology. He is considered to be the founder of histology, the study of tissue structure and organization, and the father of descriptive embryology and plant anatomy.

Malpighi's discoveries of lung structure and of the capillary system completed the understanding of how blood circulates through the body. In 1628, English physiologist WILLIAM HARVEY proposed his hypothesis that blood is pumped by the heart and flows in a circular path through the arteries and veins. Malpighi's

findings both confirmed Harvey's claim and extended the theory of blood circulation.

Malpighi's work toppled the already crumbling dominance of Greek and Roman teachings in Renaissance medical theory. Malpighi's predecessors—Harvey, ANDREAS VESALIUS, and PARACELSUS—began the attack on the authority of Roman physician GALEN. Malpighi's rejection of traditional medical teachings encouraged his contemporaries and followers to question established beliefs.

Malpighi's masterful technique with the microscope advanced its role in the biological sciences. After Malpighi, microscopic investigation was a prerequisite to progress in physiology, embryology, and medicine.

Schuyler

For Further Reading:

Adelman, Howard. *Marcello Malpighi and the Evolution of Embryology.* Ithaca, N.Y.: Cornell University Press, 1966.

Meli, Domenico Bertoloni, ed. *Marcello Malpighi: Anatomist and Physician.* Florence, Italy: L.S. Olschki, 1997.

Wilke, J. S. "Marcello Malpighi." In *Late Seventeenth Century Scientists.* New York: Pergamon Press, 1969.

WORLD EVENTS	MALPIGHI'S LIFE
Thirty Years' War 1618-48 in Europe	
	1628 Marcello Malpighi is born
	William Harvey's blood-circulation theory is published
	1653 Malpighi receives doctorates in medicine and philosophy from University of Bologna
	1656 Malpighi takes post at University of Pisa
	1659 Malpighi moves to University of Bologna
	1661 Malpighi discovers capillaries and lung structure
	1662 Malpighi moves to University of Messina, where he identifies taste buds
	1667 Malpighi returns to Bologna
	1669 Malpighi elected to Royal Society
	1684 Malpighi's home is vandalized
England's Glorious Revolution 1688	
	1691 Malpighi becomes papal physician
	1694 Malpighi dies
Peace of Utrecht 1713-15 settles War of Spanish Succession	

Marconi, Guglielmo

Developer of Radio Communication

1874-1937

Life and Work

Guglielmo Marconi devised the first practical systems of radio and short-wave wireless communication.

Marconi was born in Bologna, Italy, on April 25, 1874, to an Italian father and an Irish mother. His early education occurred mostly at home but was supplemented by schoolwork in Florence and Leghorn, and by his own practical experiments with electricity. In 1894 he learned of Heinrich Hertz's 1888 discovery of radio waves. Inspired, he audited the classes of Augusto Righi, a radio-wave expert, at the University of Bologna. Where the professor saw only an interesting scientific phenomenon, Marconi had a vision of mass communication—an ability to signal across long distances.

In 1895 Marconi built a radio transmitter (an oscillator in which an electrical circuit produced radio waves) and receiver (a glass tube of metal filings, known as a coherer). Improving the apparatus, he added insulated wire antennae and grounded the transmitter and receiver. In fall 1895 his equipment sent and received radio signals across a distance of about two and a half kilometers.

Marconi obtained an English patent for his invention in 1897 and formed the Wireless Telegraph and Signal Company (the name was changed to Marconi's Wireless Telegraph Company in 1900). Continually engineering improvements, in 1899 he accomplished radio communication across a distance of 50 kilometers and soon more than doubled that range. In 1900 he was granted a patent for a system that enabled several radio stations to operate simultaneously on different wavelengths.

Convinced that radio could eventually connect people all around the globe, he set up a transmitter in England and a receiver in Newfoundland, Canada. A signal was successfully sent across this 2,800-kilometer distance on December 12, 1901, causing a worldwide sensation. Marconi thought that radio waves followed the Earth's curvature, but it was later shown that they are reflected by the atmosphere and can thus travel great distances from point to point on the planet's surface.

Marconi continued to stretch the limit of radio signaling, improving transmission and reception, and establishing several long-distance radio stations. His achievements won him the Nobel Prize for Physics in 1909. He served as a technical consultant to the Italian military during World War I and was instrumental in England's 1918 radio transmission to Australia.

The radio waves Marconi generated had wavelengths of about 8,000 meters; in the 1920s, he began investigating short-wave wireless transmission. Short waves are only a few meters in wavelength. Unlike longer-range radio waves their range is limited if there are no relay stations along their path. Marconi found that short wavelengths produced louder signals, and he instituted a short-wave wireless system connecting England with its colonies around the globe.

In his last years Marconi experimented with microwaves, those with wavelengths shorter than one centimeter. He died in Rome, Italy, on July 20, 1937.

Legacy

Marconi's introduction of radio and short-wave devices created the foundation of

modern wireless technology, which has a broad range of applications in radio, television, telephones, and remote controls.

When first introduced Marconi's radio system transmitted only simple impulses; it was a wireless telegraph using Morse code. In 1906 Canadian Reginald Aubrey Fessenden devised the modulation system that made radio a practical source of communication, a carrier of news, and a popular form of entertainment. Modulation translates speech and music into electrical impulses for transmission, and then demodulates them at the receiving end.

The first widespread use of Marconi's radio systems occurred aboard ships. Within a few years of the first wireless transatlantic communication, passenger and cargo ships traveling between Europe and North America used radio regularly to communicate with the shore and other ships. Such communication greatly improved navigation and enabled ships to steer clear of foul weather and collisions. The 1912 sinking of the Titanic fixed the significance of radio in the public view: radio communication to a nearby ship was crucial in saving the lives of many passengers.

The first military use of radio took place during the Russo-Japanese War, 1904 to 1905, and it was universally used aboard ships and airplanes during World Wars I and II. Regular radio broadcasts began in England in 1922 and in 1927 the British Broadcasting Corporation (BBC) was founded as a public company.

Schuyler

World Events		Marconi's Life
Germany is united	1871	
	1874	Guglielmo Marconi is born
	1894	Marconi learns of radio waves
	1895	Marconi sends radio signal
	1897	Marconi patents radio apparatus
Spanish-American War	1898	
	1901	Marconi achieves transatlantic radio transmission
	1906	Reginald Fessenden introduces modulation
	1909	Marconi is awarded Nobel Prize for Physics
World War I	1914-18	
	1918	England transmits radio signal to Australia
	1937	Marconi dies
World War II	1939-45	

For Further Reading:

Dunlap, Orrin. *Marconi: The Man and His Wireless*. New York: Arno Press, 1965.

Garratt, G. R. M. *The Early History of Radio: From Faraday to Marconi*. London: Institution of Electrical Engineers, 1994.

Gunston, David. *Marconi: Father of Radio*. New York: Crowell, Collier Press, 1965.

Maxwell, James Clerk

Founder of Modern Electrical Theory
1831-1879

Life and Work

One of the greatest theoreticians of the nineteenth century, James Maxwell did for the difficult subjects of electricity and magnetism what ISAAC NEWTON had done for force and motion: he developed a set of simple laws that became the basis of modern electrical theory.

Clerk (pronounced "Clark") was his family's original name, but his father added Maxwell as a second surname after the family inherited an estate from ancestors named Maxwell. James was an only child, born on June 13, 1831, in Edinburgh, Scotland. His mother died of abdominal cancer when he was only eight, and his father, with whom he developed a close relationship, raised him. By age 14 he had published his first scientific paper, describing how to draw geometric curves with the aid of pins and thread. His teacher at Cambridge University said that Maxwell was the most extraordinary student he had ever known, and that it seemed impossible for him to think wrongly on any physical subject. He taught in Scotland and England from 1860 until 1865, when he retired to his Scottish family estate.

Maxwell tended to work on several projects at a time, often setting one aside and taking up another, to come back later to the first. In 1861 his interest in color and color vision culminated in the first color photograph, taken and projected through three separate color filters. By 1866 he had completed his important work on the physics of gases, proving for the first time that the velocities of molecules of a gas, previously assumed to be equal, had to be distributed in a particular statistical way. In 1871 Maxwell was elected the Cavendish Professor of Physics at Cambridge University. He designed and oversaw the construction of the famous laboratory there.

His greatest work was in the theory of electromagnetism, which he laid out in a book published in 1873 called *Treatise on Electricity and Magnetism*. Many other scientists— MICHAEL FARADAY, CARL FRIEDRICH GAUSS, Hans Christian Oersted, and ANDRÉ AMPÈRE, to name a few—had made discoveries about electricity and magnetism. But no one could see how to put them all together, because one key piece was missing. Faraday had shown, in the year that Maxwell was born, that a changing magnetic field generates an electric field. Maxwell was able to demonstrate the converse, that a changing electric field generates a magnetic field. If an electric field can generate a magnetic field and a magnetic field can generate an electric field, then the sequence of changes can go on forever, generating electromagnetic waves. Maxwell showed that these waves traveled at exactly the speed of light; he was the first person to understand that light is a continual series of changes in electric and magnetic fields.

During the spring of 1879 he became ill with abdominal cancer. He died of the disease on November 5 of that year, at age 48. He was buried in the small village of Parton, Scotland.

Legacy

Maxwell's far-reaching influence has had both theoretical and practical effects: it helped lay the groundwork for ALBERT EINSTEIN's theory of relativity and provided a foundation for the whole array of contemporary communications technologies.

Eight years after Maxwell's death, Heinrich Hertz demonstrated that electromagnetic waves of the kind we now call radio waves could be created in a laboratory. Before long they were being transmitted across a lab, then through lab walls, and soon across oceans. Today, of course, the entire planet is brought closer together by radio and television waves, and on a more personal level, pagers and cordless and cellular phones extend the reach of these waves to anyone with the proper equipment. Maxwell's discovery led directly to these innovations.

Albert Einstein, born the year that Maxwell died, would use Maxwell's work on electromagnetism in his work in the twentieth century. Maxwell's theory about the constancy of the speed of light became a foundation for Einstein's relativity theory.

Maxwell's ideas had applications in many disparate fields; his work touched on what has become information theory, cybernetics, fluid theory, and optics. The fish-eye lens, for example, came from his research.

One concrete influence of Maxwell's work is from the lab at Cambridge that he designed and supervised. It became one of the most fruitful laboratories in Europe, where countless discoveries in physics have since been made.

Secaur

WORLD EVENTS		MAXWELL'S LIFE
Napoleonic Wars 1803-15 in Europe		
	1831	James Maxwell is born
	1860	Maxwell becomes professor at King's College, London
	1861	Maxwell produces first color photograph
United States 1861-65 Civil War		
	1866	Maxwell lectures on the physics of gases
Germany is united	1871	Maxwell supervises building of the Cavendish Laboratory at Cambridge
	1873	Maxwell publishes *Treatise on Electricity and Magnetism.*
	1879	Maxwell dies
Spanish-American War	1898	

For Further Reading:

Segrè, Emilio. *From Falling Bodies to Radio Waves.* New York: Freeman, 1984.

Tolstoy, I. *James Clerk Maxwell.* Chicago: University of Chicago Press, 1981.

McClintock, Barbara

Pioneer in Genetic Analysis
1902-1992

Life and Work

Barbara McClintock conducted pioneering work on genetic analyses of corn plants. Her cell studies led to innovative conclusions about how chromosomes transfer genetic information.

Born on June 16, 1902, in Hartford, Connecticut, McClintock was an independent child who loved reading, music, sports, and "thinking about things." She developed a passion for science while in high school in Brooklyn, New York, and insisted on her right to know more. In 1919 she began her university education at Cornell where she completed her Ph.D. in botany in 1927. By then she was a professional scientist fully absorbed in her work.

Using new staining techniques that allowed the viewing and describing of individual corn chromosomes by their length, shape, and pattern, in 1929 McClintock showed that all 10 chromosomes of corn are structurally distinguishable. Next she matched microscopic investigations with outcomes of agricultural experiments; she collected evidence that chromosomes cross-over (mutually exchange genetic material during division), a fundamental concept that she established in 1931 with publication of her work with Harriet Creighton. That same year Curt Stern presented support for this idea based on studies with fruit flies.

After years of temporary assignments at Cornell, which hired women faculty only in home economics, McClintock was hired by the University of Missouri in 1936. Working on chromosomes damaged by exposure to X-rays, she detected evidence of non-random rejoining of chromosomes that she interpreted as another mechanism for generating hereditary changes. After she left Missouri in 1941, McClintock moved to New York to work at Cold Spring Harbor Laboratory where she spent the rest of her life.

In 1944 she was elected to membership in the National Academy of Sciences, only the third woman in its history to be so honored.

McClintock's experiments revealed complex and dynamic relationships between the cell and the genome (total chromosomal material in a germ cell). Her work described the importance of the organizer region of the nucleolus (spherical body inside a cell nucleus) and recognized transpositions in genomes as responses to internal and external stresses that affect both form and function of an organism. Her ideas about "jumping genes" gained recognition and verification in the late 1970s. In 1978 a paper describing her work on mechanisms that reorganize the gene garnered international attention, and in 1981 she received the first MacArthur Laureate award as well as other honors. In 1983 she was awarded the Nobel Prize for Physiology or Medicine.

McClintock died on September 2, 1992, in Long Island, New York.

Legacy

McClintock's work opened up new paths in the field of genetic analysis, including the possibility of a new, dynamic model of genetic behavior.

McClintock's work encouraged the use of corn, in particular maize, in genetic analyses. Corn proved to have advantages that scientists could exploit for genetic testing: each of the many kernels on an ear of corn represented a distinct fertilization episode and the variegation in color offered a chance to study many instances of genetic mutations.

More controversial were McClintock's experiments and observations suggesting complex "nature-nurture" effects in hybrid corn plants. Physicists and chemists in the 1950s and 1960s relied on molecular structure alone to describe the simplest possible "genetic code." McClintock, however, started claiming as early as 1944 that chromosomes could rearrange themselves to respond to "activators" within the cell. This possibility of functional transposition or movement of genetic material did not fit with the Watson-Crick model of a limited number of static configurations of DNA.

In the late 1970s and beyond the patterns McClintock saw in corn cells began to be noticed by others studying bacteria, yeast, and fruit flies. The genome was seen to be a very complex and dynamic structure that does not behave only in response to known physical laws. What these discoveries imply for the development of individual organisms and for the evolution of species remains a topic for ongoing research.

Simonis

For Further Reading:

Fedoroff, Nina, and David Botstein, eds. *The Dynamic Genome: Barbara McClintock's Ideas in the Century of Genetics.* Cold Harbor, N.Y.: Cold Harbor Press, 1992.

Keller, Evelyn Fox. *A Feeling for the Organism: The Life and Work of Barbara McClintock.* San Francisco: Freeman, 1983.

McGrayne, Sharon B. *Nobel Prize Women in Science: Their Lives, Struggles, and Momentous Discoveries.* Secaucus, N.J.: Carol Publishing Group, 1993.

McCormick, Cyrus Hall

Inventor of Mechanical Reaper
1809-1884

Life and Work

Cyrus Hall McCormick designed and built the first successful mechanical reaper (an apparatus for harvesting grain), improving the machine throughout the nineteenth century. He also introduced innovative business practices, such as advertising and the offer of guarantees and deferred-payment plans.

McCormick was born on February 15, 1809, in Rockbridge County, Virginia, to a blacksmith father who had invented several useful farm implements. Young McCormick received little formal education but acquired mechanical skills in his father's workshop.

In the early nineteenth century, harvesting grain was a labor-intensive activity. Inventors in England and the United States (including McCormick's father) had made a number of failed attempts to construct a machine that would collect the grain more efficiently. In 1831 McCormick approached the problem, devising a two-wheeled, horse-drawn vehicle with an apparatus for reeling in the stalks and a vibrating blade for removing the grain. He obtained a patent for the reaper in 1834.

At that time McCormick concentrated on his father's iron foundry, which was in financial straits, and his invention remained unexploited. He returned to it when the foundry failed in 1837, improved it, and by 1844 had sold 88 machines.

McCormick moved to Chicago, Illinois, in 1847, convinced that the fertile soil of the Midwestern states would become the hub of the world's wheat production. With financial help from Chicago's mayor, William Ogden, he opened a reaper factory and sold 800 in the first year. When his patent expired in 1848, rival manufacturers attempted to prevent renewal, and he was involved in a series of costly lawsuits. He was not granted a new patent.

McCormick decided to beat his competitors by outselling them. He initiated an advertising campaign, and, traveling through the surrounding farming communities, he performed public demonstrations, gave out warranties, and extended credit to customers. By 1850 the McCormick reaper was the best-selling harvesting machine in the United States.

In 1851 McCormick presented his reaper at the Great Exhibition of London and won the grand prize. Numerous international awards followed. By 1856 the machine was world-renowned, and McCormick's wealth was secured.

McCormick's factory was destroyed by the Chicago fire of 1871, but upon rebuilding, the business continued to grow. He died on May 13, 1884, in Chicago.

Legacy

McCormick's invention contributed significantly to the development of modern mechanized agriculture and embodied elements incorporated into all subsequent harvesting machines. Later improvements included mechanized equipment for raking, twine-binding, and swathing.

The efficiency of McCormick's reaper reduced the number of workers required to harvest grain. Grain production thus became less costly, grain farmers prospered, and bread became one of the most affordable foods in the United States. Grain output in the United States was concentrated in the prairie states, perhaps as a direct result of the introduction of mechanical harvesting.

After the London exhibition of 1851, McCormick's reaper enjoyed success in European grain fields. The machine was a vital part of an international agricultural revolution kindled by the industrialization of the late nineteenth century.

In 1902 McCormick's son became president of the International Harvester Company, a large prosperous corporation formed from the unification of his father's company and other harvesting-machinery businesses. The company no longer exists. J. I. Case bought its farm equipment division in 1986 and the truck business, a substantial part of International Harvester at that time, was reorganized as Navistar. However, Cyrus McCormick's invention still benefits every consumer of flour products; they are inexpensive and widely available thanks to the agricultural revolution he helped start.

Schuyler

WORLD EVENTS	McCORMICK'S LIFE
Napoleonic Wars 1803-15 in Europe	
	1809 Cyrus McCormick is born
	1831 McCormick builds mechanical reaper
	1834 McCormick obtains patent for mechanical reaper
	1847 McCormick opens factory in Chicago
	1848 McCormick's patent-renewal request is denied
	1850 Innovative practices make McCormick's business successful
	1851 McCormick's reaper wins at London exhibition
United States 1861-65 Civil War	
Germany is united 1871	
	1884 McCormick dies
Spanish-American 1898 War	

For Further Reading:

Casson, Herbert N. *Cyrus Hall McCormick: His Life and Work.* Freeport, N.Y.: Books for Libraries Press, 1971.

Dies, Edward J. *Titans of the Soil: Great Builders of Agriculture.* Chapel Hill: University of North Carolina Press, 1976.

Hutchinson, William Thomas. *Cyrus Hall McCormick.* New York: Da Capo Press, 1968.

Mead, Margaret

Pioneer in Cultural Anthropology
1901-1978

WORLD EVENTS	MEAD'S LIFE
Spanish-American War 1898	
	1901 Margaret Mead is born
World War I 1914-18	
	1923 Mead graduates from Barnard
	1925 Mead travels to Samoa
	1926 Mead appointed assistant curator at American Museum of Natural History
	1928 *Coming of Age in Samoa* is published
	1929 Mead earns Ph.D. in anthropology from Columbia University
	1935 *Sex and Temperament in Three Primitive Societies* is published
World War II 1939-45	
Korean War 1950-53	
	1972 Mead writes an autobiography
End of Vietnam War 1975	
	1978 Mead dies
Dissolution of 1991 Soviet Union	

Life and Work

Margaret Mead was one of the most prominent cultural anthropologists of the twentieth century. She observed indigenous cultures and wrote numerous books and articles describing her research and theories regarding child development, sexuality, and cultural determinism.

Mead was born on December 16, 1901, in Philadelphia, Pennsylvania, to a family that encouraged her to pursue an education and to retain her independence and identity after marriage. From her paternal grandmother, a pioneer child psychologist, she learned to observe children's behavior when she herself was still a child. She graduated from Barnard College in 1923 and earned her Ph.D. in anthropology from Columbia University in 1929.

In 1928 she published *Coming of Age in Samoa*, which was based on observations of Samoan culture conducted during a 1925 research visit. In the book, she set forth the theory of cultural determinism, which states that people's personalities and values are formed largely by the ideologies of their culture. The book was a controversial best seller: critics attacked her data-gathering techniques and her conclusions. Most anthropologists of the time believed that biology represented the basis of cultural differences among groups of people.

In 1926 Mead was appointed assistant curator at the American Museum of Natural History, in New York City, beginning a lifelong association with that institution. She became associate curator in 1942, curator in 1964, and emeritus curator in 1969. She also lectured at Columbia University, Vassar College, New York University, and Fordham University.

Mead studied firsthand the indigenous cultures of North America and islands of the western Pacific including New Guinea, Bali, and the Admiralty Islands. She focused on women, children, adolescence, and sexuality, used photography to document her visits to these areas, and pioneered innovative methods of gathering anthropological data. She wrote 23 widely read books, most of which, like her first book, provoked controversy. Her works covered analyses of child development, psychoanalytic theory, and ethnography (a branch of anthropology concerned with the observation and description of traditional cultures); one of her more popular books was *Sex and Temperament in Three Primitive Societies,* published in 1935.

Mead was involved, often in an executive capacity, in professional and scientific organizations throughout her life. She also participated in public debates surrounding social issues such as child rearing, female sexuality, race relations, population control, environmental degradation, and nuclear proliferation. She died on November 15, 1978, in New York City.

Legacy

Mead was an influential force in academic circles, contributing to the evolution of cultural theories that occurred during the second half of the twentieth century.

The aftermath of Mead's first book set the tone for her whole career; her critics continued to question her non-traditional approach and revolutionary ideas.

Although Mead remains a controversial figure in academic circles, many historians concede that she founded the modern field of cultural anthropology. Her work turned others toward the comparative study of native societies and was responsible for the spread of cultural determinism to other academic disciplines, such as literature, history, psychology, and sociology. She initiated anthropological study of women and children in non-literate societies, topics her predecessors had largely ignored.

Many of Mead's books were successfully aimed at the general reader, and she thus opened the field of cultural anthropology to non-scholars. Her writings contributed to a wider understanding of other contemporary cultures, which led people in Western societies to take a critical look at their own assumptions and culturally manifested values. Her involvement in controversial social issues, coupled with her authoritative status, shaped popular opinion and helped initiate progressive social movements such as environmentalism and feminism.

Schuyler

For Further Reading:

Grosskurth, Phyllis. *Margaret Mead.* New York: Penguin Books, 1988.

Howard, Jane. *Margaret Mead: A Life.* New York: Ballantine, 1990.

Mark, Joan. *Margaret Mead.* New York: Oxford University Press, 1998.

Mead, Margaret. *Blackberry Winter: My Earlier Years.* New York: Morrow, 1972.

Meitner, Lise

Co-discoverer of Nuclear Fission
1878-1968

Life and Work

Lise Meitner is known as the co-discoverer of nuclear fission, one of the most important scientific discoveries of the twentieth century.

Born in Vienna on November 7, 1878, Meitner studied under LUDWIG BOLTZMANN, who described her as his best, but also his shyest, student. In 1907 she moved to Berlin to study with MAX PLANCK. Because of discrimination against women in science, Meitner had difficulty finding work, and Plank had to personally intervene to get her a job at the Chemical Institute in Berlin. There she was teamed with OTTO HAHN, beginning 30 years of friendship and productive collaboration. Together they pioneered the field of radiochemistry, separating and identifying new radioactive elements and isotopes, including the element protactinium in 1918.

In the 1930s Italian physicist ENRICO FERMI showed that heavy elements like uranium can capture incoming neutrons, making new heavy atoms. When these atoms decayed, they emitted patterns of radiation that did not correspond to any known heavy atom. Hahn and Meitner began investigating the phenomenon.

In 1938 Meitner, who was Jewish, was forced to flee to Sweden to avoid Nazi persecution.

Nonetheless Hahn and Meitner continued their research, and in December 1938 they reached the startling conclusion that the uranium atoms were not just swallowing neutrons and growing heavier, but were splitting in two—the radiation did not match the pattern of any heavy atoms because it was from two lighter ones. Hahn and Fritz Strassmann, a young chemist at the Institute, published their result, the discovery of nuclear fission, in January of 1939. Hahn did not include Meitner's name in the article for fear that it would not be published, or the work recognized, if the Jewish woman's name appeared in it. Meanwhile, Meitner and her nephew, Otto Frisch, published the results independently a few weeks later. The word "fission" was first used in Meitner's publication in 1939.

Hahn was awarded the Nobel Prize for Physics 1944 for the discovery that he and Meitner had made together. Despite their 30-year friendship, after the war Hahn never mentioned Meitner's contribution or the leadership she provided to their team. She moved from Stockholm to Cambridge, England, in 1960 and died there on October 27, 1968.

Legacy

Meitner's legacy rests on her contribution to the discovery of fission, which changed the course of nuclear physics and its application to the real world, in particular to the development of nuclear energy and weapons.

While not fully understanding the process of fission she had helped to uncover, Meitner understood its importance. Once in Sweden she sent a message to Danish scientist NIELS BOHR just as his ship was leaving for America and told him about fission. He understood the results immediately and carried the news to American scientists, who in turn urged President Roosevelt to start a crash program to build a nuclear bomb before the Germans could complete one. President Roosevelt launched the Manhattan Project to undertake the development of an atomic bomb.

Bohr, Fermi, and other scientists including John Wheeler of the United States continued investigations into the fission process. Soon fission was understood more fully—particularly the nature of the products of uranium fission including the uranium isotope (uranium

235) and the potential energy released from chain reactions launched during fission events.

In 1942 Fermi and a team of physicists under the auspices of the Manhattan Project completed a nuclear reactor to produce the first self-sustained and controlled fission chain reaction. Three years later, the atomic bombs that the United States dropped on Japan utilized this fission reaction.

Today fission has both peaceful and military applications. Nuclear reactors used to generate electrical power are employed worldwide, although new production of these power plants has slowed because of concern about the safety of their operation. Nuclear weapons, most typically hydrogen bombs, use fission along with the process of fusion to generate their huge destructive explosions. In the fusion reaction, atoms of lighter elements "fuse" to form heavier elements, and these reactions are ignited by the heat generated in a fission reaction.

As co-discoverer of nuclear fission, Meitner's contribution to both the peaceful and military uses of these nuclear reactions are undeniable.

Secaur

WORLD EVENTS		MEITNER'S LIFE
Germany is united	1871	
	1878	Lise Meitner is born
Spanish-American War	1898	
	1907	Meitner meets Otto Hahn, begins work in Berlin
World War I	1914-18	
	1918	Meitner separates and identifies the new element protactinium
	1938	Meitner flees to Sweden
	1939	Meitner announces discovery of fission at the same time Hahn does
World War II	1939-45	
	1944	Hahn is awarded Nobel Prize for Physics
Korean War	1950-53	
	1960	Meitner moves to England
	1968	Meitner dies
End of Vietnam War	1975	

For Further Reading:
Crawford, E., R. L. Sime, and M. Walker. "A Nobel Tale of Postwar Injustice." *Physics Today* (September 1997).
Meitner, Lise. "Looking back." *Bulletin of the Atomic Scientists* (November 1964).
Sime, R. L. "Lise Meitner and the Discovery of Nuclear Fission." *Scientific American.* (January 1998).

Mendel, Gregor Johann

Founder of Modern Genetics
1822-1884

Life and Work

Gregor Johann Mendel performed extensive plant breeding experiments from which he developed revolutionary theories of inheritance in the nineteenth century. His investigations and results laid the foundation of modern genetics.

Mendel was born on July 22, 1822, in Heinzendorf, Austria (now Hyncice in the Czech Republic), a rural area where all schoolchildren received agricultural training. In 1842, he entered the Augustinian monastery

in Brünn, Moravia (now Brno in the Czech Republic), where dedicated scholars and researchers sustained Mendel's passion for science. After completing his theological education, Mendel entered the University of Vienna in 1851 to study the natural and physical sciences. There he learned to apply mathematics to experimental results and developed an interest in variation among plants, two factors that greatly influenced his subsequent work.

In 1854, Mendel began to grow peas in the monastery garden to investigate how characteristics are inherited from parent to offspring. The pea plant was ideal for inheritance experiments because of its availability in distinguishable varieties, its short generation time, its production of many offspring, and its ability either to cross- or self-pollinate. Mendel carefully controlled the pollination of each plant so that he knew which parents gave rise to which offspring. From 1854 to 1863, he dedicated himself to the cultivation of more than 28,000 pea plants and to the analysis of the patterns of heredity for characteristics such as flower color and seed shape.

Many scientists at the time regarded heredity as a blending of the characteristics of the parents within the offspring, but this theory fell short of explaining many observable phenomena in plants and animals. An alternative idea suggested that parents pass on discrete heritable elements that retain separate intact identities within the offspring. Mendel's experimental results illustrated the latter theory; today we call those hereditary elements genes. Mendel reported his conclusions to the local science society in 1865 and published them in *Experiments with Plant Hybrids* in 1866. He was elected abbot of his monastery in 1868. Mendel died in Brünn on January 6, 1884.

Legacy

Mendel is considered to be the founder of genetics, a field that has illuminated and is crucial to the study of all biological disciplines.

The importance of Mendel's remarkable achievements remained unrecognized until 16 years after his death. In 1900 three scientists, Hugo de Vries, Carl Correns, and Erich Tschermak von Seysenegg, independently rediscovered Mendel's papers as they searched the existing literature for material to support the results of their own hybridization experiments. When they published their work, they cited Mendel as the first to have completed similar research.

Mendelian genetics underlie the modern understanding of how organisms inherit characteristics. Modern geneticists credit Mendel with having discovered the two basic laws of inheritance. In its modern version, Mendel's law of segregation states that alleles (different forms of the same gene) are equally distributed among an organism's gametes, or sex cells. This accounts for the even distribution of different alleles among offspring. Mendel's law of independent assortment states that the alleles for one character are distributed among gametes independently of the alleles for other characters. This accounts for the abundance of variation in the characteristics of individual offspring from the same parents. Mendel's laws apply to the basic genetics of all organisms.

Mendel's work introduced the experimental and theoretical approach to heredity that is used in modern genetic analysis. Selective breeding programs, such as those employed by Mendel, increase the quality and productivity of the organisms from which humans obtain food, clothing, and pharmaceuticals. Mendelian analysis also enables genetic counselors to construct disease pedigrees of human families to determine the genetic makeup of potential parents and to predict the chance of a heritable disease appearing in their progeny.

Schuyler

WORLD EVENTS	MENDEL'S LIFE
Napoleonic Wars 1803-15 in Europe	
	1822 Gregor Mendel is born
	1842 Mendel enters the Augustinian monastery in Brünn
	1854 Mendel begins pea plant experiments
United States 1861-65 Civil War	
	1865 Mendel communicates his results to the scientific community
	1866 Mendel publishes *Experiments with Plant Hybrids*
	1868 Mendel is elected abbot of his monastery
Germany is united 1871	
	1884 Mendel dies
Spanish-American 1898 War	
	1900 Mendel's work is rediscovered

For Further Reading:

Lerner, I. Michael, and William J. Libby. *Heredity, Evolution, and Society.* San Francisco: Freeman, 1976.

Orel, Vitezslav. *Mendel.* New York: Oxford University Press, 1984.

Mendeleyev, Dmitry

Creator of Periodic Table of Elements
1834-1907

Life and Work

Dmitry Mendeleyev created a logical system of categorizing the elements that is called the Periodic Table of Elements.

Born in the isolated town of Tobolsk, in Siberia, Russia, in 1834, Mendeleyev was the last of 14 children. His widowed mother, who operated a glassworks factory, sought a better education for him than was available in Siberia and took him to St. Petersburg. There he studied chemistry, physics, and mathematics. Mendeleyev traveled to Europe to study with great chemists in France and Germany. Eventually, he returned to Russia. During his studies Mendeleyev investigated the characteristics of the 63 known elements.

Mendeleyev wrote an organic chemistry text, *Principles of Chemistry,* that was published from 1868 to 1870. This earned him a full professorship at the University of St. Petersburg, where he had earned a doctorate with a dissertation on "The Union of Alcohol with Water." While working on the text, he began to search for a way to systematically describe the elements. Certain groups acted in similar ways. Mendeleyev created cards for each element that included its physical properties and all known characteristics of how it behaved chemically.

He knew the atomic weights of each element and so used that to begin arranging them. In addition to the linear progression Mendeleyev put certain elements in groups by shared characteristics. By 1869 he published a table that organized the elements horizontally by weight and vertically by families of properties; he produced a revised table that was published in 1871. He had noticed that these traits repeated every seventh element, which led him to describe the table as "periodic." Mendeleyev soon revised his original table to incorporate blank spaces. These voids, he predicted, were the locations of undiscovered elements. This was a revolutionary idea. He described the relative weights and properties of the unknown substances. Within 15 years, three of the missing elements had been discovered.

Mendeleyev worked to apply science to problems of his country and was lauded in Russia for his work with petroleum and his suggestions on how to improve the efficiency of the petroleum industry. His other civic work included written recommendations on standardization of weights and measures; shipbuilding and trade routes; and manufacture of smokeless powder.

Mendeleyev was also a champion of human rights; he advocated progressive reform of Russia's tsarist regime, an opinion that caused him to lose his academic post at the University of St. Petersburg. He continued to be involved with chemistry and how it could help society for the remainder of his life. He was nominated for a Nobel Prize in 1906 but lost by one vote. Mendeleyev died on February 2, 1907.

Legacy

Mendeleyev's legacy rests on his insightful organization of the elements in the Periodic Table. This system of organizing the elements, based upon atomic weight and shared characteristics, helped future scientists understand the nature of matter and paved the way for future discoveries of unknown elements.

Mendeleyev spurred other researchers across Europe to seek out the missing elements. The table gave them clues and inspiration. The first discovery of a new element, gallinium, came in 1875. Soon after, in 1879, Lars Fredrick Nilson discovered scandium in its oxide form, scandia, and German chemist Glemens

Winkler discovered germanium in 1886. Mendeleyev had predicted the existence of all three of these new elements.

Mendeleyev's recognition of the patterns of properties among the elements helped others to better understand matter. He was also willing to question commonly accepted knowledge. When several elements fit the pattern by characteristic but not weight, Mendeleyev suggested that the weights were incorrect. He was eventually proven correct.

As scientists accepted Mendeleyev's Periodic Table, they sought to add to it. Researchers also strove to understand why certain groups of elements acted in similar ways. Changes were made in Mendeleyev's original scheme when six new gases were discovered late in the nineteenth century. These chemically inactive elements, the inert or noble gases (helium, neon, argon, krypton, xenon, and radon), were assigned to a column of their own, expanding the original table. In 1912 Henry Moseley discovered that the atomic number or number of protons was the key to the order of elements. In 1955 element 101 was named mendelevium in Mendeleyev's honor.

Mendeleyev's work and reports on practical matters facing Russian society, including the petroleum industry and the standardization of weights and measures, contributed to the modernization of Russia.

Wilson

WORLD EVENTS	MENDELEYEV'S LIFE
Napoleonic Wars 1803-15 in Europe	
	1834 Dmitry Mendeleyev is born
United States 1861-65 Civil War	
	1869 First version of Periodic Table is published
	1870 *Principles of Chemistry* is published
Germany is united 1871	Revised table is published
	1875 First missing element, gallinium, is identified
Spanish-American War 1898	
	1906 Mendeleyev is nominated for a Nobel Prize
	1907 Mendeleyev dies
World War I 1914-18	

For Further Reading:

Brock, William. *The Norton History of Chemistry.* New York: Norton, 1992.

Kelman, Peter. *Mendeleyev: Prophet of Chemical Elements.* Englewood Cliffs, N.J.: Prentice-Hall, 1970.

Serres, Michel. *A History of Scientific Thought.* Cambridge, Mass.: Blackwell Reference, 1995.

Mercator, Gerardus

Cartographer; Creator of
Mercator Map Projection
1512-1594

Life and Work

Cartographer Gerardus Mercator (born Gerhard Kremer) introduced the groundbreaking mapping technique now known as the Mercator projection, which enabled navigators to plot and follow a straight-line course without changing compass direction.

Mercator was born on March 5, 1512, in Rupelmonde, Flanders (now Belgium). After earning a master's degree in philosophy and theology from the University of Louvain in 1532, he studied under Gemma Frisius, a noted mathematician and astronomer. At this time he also improved his artistic and mechanical skills, mastering calligraphy, engraving, and instrument making.

In collaboration with Frisius and engraver Gaspar à Myrica, Mercator constructed a terrestrial globe in 1536. During the next few years, he produced a booklet of italic lettering and beautifully printed maps of Palestine, Flanders, and the world, establishing a reputation as a master cartographer.

Because he was a Protestant in a Catholic-dominated region, Mercator was arrested for heresy in 1544 and imprisoned for several months. He moved permanently to Duisberg (now in Germany) in 1552, where he was free to practice his religion. He taught grammar school and became court cosmographer to Duke Wilhelm of Cleve.

Between 1554 and 1564 Mercator completed maps of Lorraine, the British Isles, and Europe, and perfected his projection method, which he applied to the 1569 world map that became his most famous accomplishment. The projection method entailed straightening the meridians (longitude lines) into parallel lines and lengthening the degrees of latitude, which distorted distance and land/sea areas but kept direction relatively intact. With this method, Mercator achieved his goal of devising a way for sailors to plot a straight line (crossing all meridians on a map at the same angle) and to follow the projected line without continual adjustment of compass direction.

In his later years, Mercator printed an updated edition of Ptolemy's cartographic works and compiled a collection of accurate, detailed maps of western and southern Europe that remained unfinished at his death. A completed version of this collection, which was the first map collection to be titled *Atlas*, was published in 1595 (after Mercator's death) by his son.

Mercator died in Duisberg on December 2, 1594.

Legacy

Mercator changed the nature of cartography with his artistic enhancements and

forever altered ocean navigation with his projection method, which was the most influential invention of early cartography.

Mercator's talents inspired following generations of cartographers to hone their artistic skills in calligraphy and engraving. Thus, maps of Renaissance Europe were generally elegant pieces of art. In addition, the work of Mercator, Frisius, and Myrica made Louvain a center of cartography and navigational-instrument making during the sixteenth and seventeenth centuries, times of expanding world exploration.

The value and importance of Mercator's projection method was recognized only at the end of the sixteenth century, after his death. Navigators at first regarded the projection with suspicion because it greatly distorted distances; they were accustomed to their own methods and were reluctant to change. Material promoting Mercator's projection was published in 1597 by William Barlowe and in 1599 by Edward Wright. Wright provided a table of "meridional parts" that explained how to compute a distance correctly, and sailors were eventually persuaded of the projection's merit. It vastly facilitated navigation, sped ocean passages, and reduced sailing disasters from the seventeenth century to the present.

The Mercator projection is still used for most flat maps in modern cartography despite its gross distortion of high-latitude land masses, such as Greenland and Antarctica, on many world maps.

Schuyler

World Events		Mercator's Life
Columbus discovers Americas	1492	
	1512	Gerardus Mercator is born
Reformation begins	1517	
	1532	Mercator earns master's degree from University of Louvain
	1536	Mercator constructs globe
	1544	Mercator accused of heresy
	1552	Mercator moves to Duisberg
	1569	World map with Mercator projection is printed
	1594	Mercator dies
	1595	Mercator's *Atlas* is published
	1599	Edward Wright publishes key to Mercator projection
Thirty Years' War in Europe	1618-48	

For Further Reading:

Crone, G.R. *Maps and Their Makers: An Introduction to the History of Cartography.* Hamden, Conn.: Archon Books, 1978.
Wilford, John Noble. *The Mapmakers.* New York: Knopf, 1981.

Metchnikoff, Élie

Immunologist;
Discoverer of Phagocytes
1845-1916

Life and Work

Élie Metchnikoff discovered phagocytes, specialized cells that destroy microbes and other foreign substances, thereby uncovering one of the fundamental elements of the human immune system.

Metchnikoff was born in the village of Ivanovka, near Kharkov, Ukraine, on May 16, 1845. In high school he developed a keen interest in biology; in 1862 he entered the University of Kharkov, where he became fascinated with living forms seen under a professor's rare microscope. After completing a four-year program in two years, he traveled to Germany to study at the University of Würzburg, where he read CHARLES DARWIN's *On the Origin of Species* before beginning classes. He was a passionate convert to the theory of organic evolution, which influenced his work on the evolution of nematodes, worms that alternate between sexual and asexual reproduction. Continuing his studies in Naples, Italy, he helped demonstrate that different animals have similar cell-layer structures in their embryonic stage.

For his thesis on the early development of fish and crustaceans, Metchnikoff was awarded a doctorate from the University of St. Petersburg, Russia, in 1867. That same year he received the Karl Ernst von Baer Prize for his work in embryology. He subsequently taught at St. Petersburg, at Odessa, Ukraine, and at Messina, Italy, where he made his most significant discovery.

Biologists at that time knew that white blood cells were part of the human immune system, but they incorrectly thought that such cells simply transported material throughout the body, aiding in the spread of infectious organisms. Metchnikoff showed that white blood cells actually have a protective function.

Studying transparent starfish larvae, he observed that certain cells surround, engulf, and destroy foreign substances. He noted similarities between the action of these cells and that of the white blood cells present in inflamed areas of human infections. He theorized that the starfish cells and the white blood cells, both of which he named phagocytes ("cell-eaters"), are derived during development from the same embryonic cell layer and perform the same function. He concluded that phagocytes clear the body of disease-causing organisms.

It is now known that phagocytosis, the ingestion of foreign particles by phagocytes, is the body's second line of defense against infections. Phagocytes attack unwelcome organisms and other substances that bypass the body's external defenses, such as skin and secretions from mucous membranes.

Metchnikoff continued to study phagocytes in animals as the director of the Bacteriological Institute at Odessa. Attacked by the press for his lack of medical training, he moved to Paris in 1888 to accept a post in a new laboratory at the Pasteur Institute, where he worked until his death on July 15, 1916.

In 1908, Metchnikoff received the Nobel Prize for Physiology or Medicine (jointly with PAUL EHRLICH, who conducted independent immunological research) for his contributions to immunology. In his later years he became interested in the effects of diet on health and aging and advocated the consumption of lactic-acid-producing bacterial cultures, such as those present in yogurt.

Legacy

Metchnikoff's assumptions about phagocytes formed the basis of further immunological research, which led to the modern understanding of the body's immune system.

Metchnikoff joined the scientific community just as the study of immunity was emerging as a distinct discipline within biology. Immunology developed through a convergence of microbiology, pathology, and embryology. Scientists had shown that pathogenic organisms cause disease and that inflammation occurs at sites of infection. Metchnikoff's identification of phagocytes revealed the basic mechanism of the immune response and allowed directed investigations of the immune system to continue.

The theory of phagocytosis was unique in suggesting that the body mounts an active response to invaders and is capable of internal reparative processes. The body had previously been viewed as relatively passive, easily succumbing to external impacts.

Metchnikoff inspired other researchers to explore immunology. One of his students at the Pasteur Institute, Jules Bordet, went on to discover the first of the nine existing "complements," enzyme systems that are responsible for the destruction of a distinct group of pathogens.

Human immunity, including action of T-cells, is much more complex than Metchnikoff and his students imagined. The search for comprehensive understanding of the body's natural defenses has continued throughout the twentieth century, revealing the physiological, chemical, and genetic factors involved; however, details are still clouded enough to maintain active debates concerning the interpretation of immunological studies. Metchnikoff's version, correct in its basic concepts of immune function, still inspires researchers and medical practitioners.

Schuyler

WORLD EVENTS		METCHNIKOFF'S LIFE
	1845	Élie Metchnikoff is born
United States Civil War	1861-65	
	1862	Metchnikoff enters University of Kharkov
	1867	Metchnikoff earns doctorate from University of St. Petersburg
Germany is united	1871	
	1880s	Metchnikoff discovers phagocytosis
	1888	Metchnikoff takes post at Pasteur Institute
Spanish-American War	1898	
	1908	Metchnikoff receives Nobel Prize for Physiology or Medicine
World War I	1914-18	
	1916	Metchnikoff dies

For Further Reading:

De Kruif, Paul. *Microbe Hunters.* New York: Harcourt Brace, 1954.

Metchnikoff, Olga. *Life of Élie Metchnikoff, 1845-1916.* Translated by E. Ray Lankester. Boston: Houghton Mifflin, 1921.

Tauber, Alfred I., and Leon Chernyak. *Metchnikoff and the Origins of Immunology: From Metaphor to Theory.* New York: Oxford University Press, 1991.

Mitchell, Maria

First American Woman
Astronomer
1818-1889

Life and Work

Maria Mitchell was the first woman astronomer in the United States; she became world famous for her discovery of the first comet visible only with the aid of a telescope.

Mitchell was born in 1818 and grew up on Nantucket Island, off the coast of Massachusetts. Her father, William Mitchell, bank teller and private school teacher, maintained a small observatory; he guided her studies in astronomy and enlisted her to time a solar eclipse when she was 12 years old. Her formal education ended when she reached age 16, and she did not go to college, as there was only one college in the United States that accepted women at that time. She worked as a

librarian at the Nantucket Library for 20 years and read the most challenging treatises available in mathematics and science.

On October 1, 1847, Mitchell discovered a new comet using a two-inch telescope from the roof of her family home. This discovery, which was later referred to as "Miss Mitchell's comet," earned her a gold medal from the King of Denmark. In 1848 she became the first woman elected to the Academy of Arts and Sciences in Boston. (For nearly 100 years—until 1943—no other woman was elected to the Academy.) She was hired by the U.S. Coast Survey to compute tables on daily planetary positions for its nautical almanac, work she continued for 19 years. In 1853 she was awarded an honorary degree from Hanover College, Indiana, the first given to a woman by an American college.

Mitchell went on to become one of the most noted American astronomers of her time. When Vassar College opened in 1865, Mitchell was asked to join the faculty, and became America's first woman astronomy professor. She was also appointed director of the observatory at Vassar College, which had a 12-inch telescope, the third-largest telescope in America at that time.

Her research at Vassar included solar astronomy and observations of the planetary surfaces of Jupiter and Saturn. In 1873 she began taking daily photographs of the Sun, the first such series made in the United States. Unlike other observers who described sunspots as clouds in the solar atmosphere, she saw them as cooler vortices of gas on the surface of the Sun. She studied the four satellites of Jupiter visible with Vassar's telescope and hypothesized that their covering was ice or frozen gas; she noticed that the rings of Saturn were of a different texture and composition than the planet itself. She observed that nebulae vary in luminosity over time and guessed that they must rotate as well as revolve around each other.

Mitchell worked as professor of astronomy at Vassar College until she retired on December 25, 1888. Throughout her academic career, she worked for the advancement of women: she lobbied for women's employment rights, worked on boards of trustees of women's colleges, and fought against paying women in academia less than men.

Mitchell died in Lynn, Massachusetts, on June 28, 1889.

Legacy

As a pioneer in telescopic astronomy and a popularizer of findings that expanded people's ideas of the size and scope of the universe, Maria Mitchell was one of the best-known American scientists of the nineteenth century. After her death a crater on the moon was named for her, as was a public school in Denver.

Her encouragement of women as scientists is a lesser known but lasting legacy. She helped advance the status of women at a time when basic women's rights were a source of controversy and even ridicule. In her lobbying efforts and work on college boards of trustees, she drew attention to issues that, unfortunately, continued to hobble women for another century.

Mitchell was a teacher who insisted on challenging women to do rigorous work, to acquire the tools of mathematics and the critical spirit of science, and to plan for marriage and careers simultaneously. "A sphere is not made up of one, but of an infinite number of circles; women have diverse gifts and to say that women's sphere is the family circle is a mathematical absurdity," she wrote.

Her legacy can be measured in the accomplishments of her students; a half century after her death, 25 of Mitchell's students were listed in *Who's Who in America* for accomplishments in science.

Hertzenberg

WORLD EVENTS		MITCHELL'S LIFE
Napoleonic Wars in Europe	1803-15	
	1818	Maria Mitchell is born
	1847	Mitchell discovers comet
	1848	Mitchell becomes first woman elected to American Academy of Arts and Sciences
	1853	Mitchell awarded honorary degree by Hanover College
United States Civil War	1861-65	
	1865	Mitchell joins faculty at Vassar College upon its founding
Germany is united	1871	
	1873	Mitchell begins photographing Sun
	1889	Mitchell dies
Spanish-American War	1898	

For Further Reading:

Gormley, Beatrice. *Maria Mitchell: The Soul of an Astronomer.* Grand Rapids, Mich.: William B. Eerdmans Publishing, 1995.

Shearer, Benjamin F., and Barbara S. "Maria Mitchell." In *Notable Women in the Physical Sciences.* Westport, Conn.: Greenwood Press, 1997.

Wright, Helen. *Sweeper in the Sky: The Life of Maria Mitchell, First Woman Astronomer in America.* New York: Macmillan, 1949.

Morgan, Garrett

Inventor of Gas Mask
and Traffic Light
1877-1963

Life and Work

Garrett Augustus Morgan, African-American entrepreneur and inventor, invented the gas mask and the standard three-color traffic light.

Morgan was born on March 4, 1877, in Paris, Kentucky; he was the seventh of 11 children. He attended school only until the fifth grade, at which time he left home and moved to Cincinnati, Ohio.

In Cincinnati he worked as a handyman until he moved to Cleveland and found work repairing sewing machines. Morgan was an excellent businessman, and he soon left his job and started his own sewing machine repair shop. He subsequently started his own tailor shop.

When Morgan was not running his businesses, he was inventing. In 1901 he patented his first invention, a tonic for straightening hair. His keen business sense and experience allowed Morgan to market his own invention. His new business, Morgan Hair Refinishing Company, brought him great financial security, allowing him to devote more of his time to inventing new products. The most famous of those products were the gas mask and traffic light.

Morgan filed to patent his "Safety Hood" in 1912. Despite endorsements and use by the Akron, Ohio, fire chief, this invention did not receive very wide recognition until 1916 when a tunnel that was being constructed under Lake Erie collapsed. As rescue workers attempted to save the men from the collapsed tunnel, they were overcome by a deadly gas. Morgan was called and asked to bring his masks. He wore his own invention as he descended 228 feet below the lake and rescued many workers. For this, Morgan was awarded a gold medal at the Second International Exposition of Safety and Sanitation. Soon the

orders for the mask poured in. The gas mask was used by firefighters and soldiers.

Morgan's business and invention successes brought him great wealth. He is said to have been the first black man in Cleveland to own a car. This opportunity led to Morgan's second famous invention, the traffic signal. He filed for a patent for this invention in 1922.

Morgan remained in Cleveland until his death in 1963.

Legacy

Garrett Morgan's inventions have increased public safety and saved many lives.

His gas mask has saved the lives of many people, allowing firefighters to enter buildings full of deadly gases. His masks were also widely used to protect World War I soldiers from chlorine gas fumes used as weapons on the battlefield. In addition, the traffic light made the roadways safer as the number of automobiles grew.

However, Garrett Morgan's legacy is more far-reaching than the benefits society has gained from his inventions. Morgan has become a role model for many people, especially African Americans. Despite an impoverished background, meager educational opportunities, and racial prejudice, Morgan became a successful businessman and inventor.

Success did not come easily: for example, many orders for his gas mask were cancelled once business owners found out he was an African American. It is said that he even had to pretend he was an Indian assistant while a white man demonstrated how the gas mask worked to potential buyers. His creative mind and diligence, however, did not allow the prejudices to stop him from making his contributions.

Once he became successful, Morgan continued to work to improve society for others. In 1916 he started a weekly newspaper, the *Cleveland Call* (now the *Cleveland Call and Post*), that was devoted to news about and for African Americans. Cincinnati and Columbus editions were added in 1950 and 1960, respectively.

Weaver

WORLD EVENTS		MORGAN'S LIFE
Germany is united	1871	
	1877	Garrett Morgan is born
Spanish-American War	1898	
	1901	Morgan patents tonic for straightening hair
	1912	Morgan patents gas mask called "Safety Hood"
World War I	1914-18	
	1916	Morgan uses his "Safety Hood" to rescue men from collapsed tunnel
		Morgan begins publishing *Cleveland Call*
	1922	Morgan patents the traffic signal
World War II	1939-45	
	1950	Cincinnati edition of the *Cleveland Call* launched
Korean War	1950-53	
	1960	Columbus edition of the *Cleveland Call* launched
	1963	Morgan dies
End of Vietnam War	1975	

For Further Reading:
Jenkins, E. S., ed. *American Black Scientists and Inventors*. Washington, D.C.: National Science Teachers Association, 1975.

Morse, Samuel

Inventor of First Single-Wire
Telegraph
1791-1872

Life and Work

Samuel Morse devised the first single-wire electric telegraph, built the first telegraph line in the United States, and invented the Morse code. He was also an accomplished portrait painter.

Morse was born on April 27, 1791, in Charlestown, Massachusetts. He attended Yale College (now Yale University), graduated in

World Events		Morse's Life
French Revolution	1789	
	1791	Samuel Morse is born
	1811-15	Morse studies art in England
Napoleonic Wars in Europe	1803-15	
	1825	Morse settles in New York City
	1837	Morse engineers single-wire electric telegraph
	1838	Morse formulates Morse code
	1844	Morse sends message over first U.S. telegraph line
	1854	Morse acquires patent rights for telegraph
United States Civil War	1861-65	
Germany is united	1871	
	1872	Morse dies
Spanish-American War	1898	

1810, and became a clerk at a book publisher in Boston. From an early age he harbored a passionate desire to paint, and his parents finally agreed to finance an art education and sent him to London in 1811. There he became enamored of the English style of painting—bold, brilliant portrayals of romantic legends and historical events. Meanwhile the War of 1812, between the United States and England, created an American hostility toward English tastes, and Morse found upon his return in 1815 that his adopted style was out of favor.

He settled in 1825 in New York City and took up portraiture to earn a living, acquiring some distinction as an artist. In 1826 he helped found the National Academy of Design, organized to promote United States painters, and served as its president until 1845.

In 1832 Morse became interested in the principles of electromagnetism, recently discovered and developed by Hans Christian Oersted, Joseph Henry, and MICHAEL FARADAY. He formulated the design of an electric telegraph, apparently unaware of similar technological progress simultaneously occurring in Europe. In 1837, with financial and mechanical help from friends, he constructed the first single-wire electric telegraph, which was able to relay signals more quickly than the multiple-wire telegraphs used in Europe. But an economic depression discouraged further investments.

Morse worked for years to persuade Congress to subsidize the building of a telegraph line from Washington, D.C., to Baltimore, Maryland; he finally won a grant to link the two cities in 1843. By 1844 he sent the first message over this line: "What hath God wrought." After costly legal suits brought by rival inventors, the Supreme Court granted patent rights to Morse in 1854.

Perhaps his best-known contribution came in 1838, when Morse introduced a telegraphic code consisting of varying combinations of dots and dashes. It was soon known across the world as the Morse code and was used for telegraphic and navigational communication.

He died on April 2, 1872, in New York City.

Legacy

Morse's work, although later overshadowed by the introduction of telephone and radio, ushered in the era of mass communication.

Like the railroad, the telegraph helped unite the geographically vast United States. By 1852 telegraph lines spread to every town east of the Mississippi, revolutionizing newspapers and commerce.

Morse's electronic telegraph, along with the technological modifications and additions brought by rapid progress in the field of electromagnetism, increased the speed of communication, at first locally and then internationally. By 1865 the first Trans-Atlantic submarine cables had been installed. Telegraphs sent messages more than 100 million times faster than a horse or ship. The resulting improvement in communications made travel and trade more convenient and efficient. Used by Union forces during the American Civil War, telegraphy was credited with assisting their victory by providing immediate information about distant situations.

By the early twentieth century, telegraphy had been surpassed by telephone and radio communication, and Morse's fame had waned. He is remembered mostly for the invention of the Morse code. The longest-lasting legacy of the code is SOS, the internationally recognized signal for help. The abbreviation SOS is not an acronym for anything, but was chosen because of the three-unit Morse codes for the letters S and O, which were efficient and easy to memorize. Three dots represent S and three dashes represent O.

Schuyler

For Further Reading:

Hawke, David. "One More Song and Dance Man." In *Nuts and Bolts of the Past: A History of American Technology 1776-1860.* New York: Harper & Row, 1989.

Kloss, William. *Samuel F. B. Morse.* Washington, D.C.: Smithsonian Institution, 1988.

Quackenbush, Robert M. *Quick Annie, Give Me a Catchy Line! A Story of Samuel F. B. Morse.* New York: Prentice-Hall, 1983.

Morton, William

Developer of General Anesthesia
1819-1868

Life and Work

William Morton introduced pain-free dental and medical surgery.

Morton was born on August 19, 1819, in Charlton, Massachusetts. He studied dentistry at the Baltimore College of Dental Surgery and Harvard Medical School. In 1844, after a brief partnership with dentist Horace Wells, he opened his own dental practice in Boston, Massachusetts, and began to investigate ways to deaden patients' pain during dental procedures.

Morton consulted the physician Charles Jackson, his former teacher, who suggested administering sulfuric ether as a general anesthetic. Morton experimented with this method and in 1846 successfully extracted the tooth of a patient anesthetized with sulfuric ether. Later in the same year, he removed a tumor from a patient's face in the first public demonstration of surgery under ether anesthetic.

The successful demonstration, performed at Massachusetts General Hospital, convinced spectators of ether's efficacy. In 1847 Morton published directions for the use of ether and its potential dangers in *Remarks on the Proper Mode of Administering Sulphuric Ether by Inhalation;* this new method of anesthesia quickly spread through the United States and Europe.

Morton and Jackson obtained a patent for the use of ether as a general anesthetic in 1846, but Morton subsequently attempted to claim sole credit for the discovery. Jackson contested Morton's false claim, and the two spent years in costly litigation. Another physician, Crawford Long, also claimed credit for pioneering ether anesthesia; Long had successfully removed a neck tumor from a patient anesthetized with ether in 1842, but he had not published his results until 1849.

The British government offered Morton a monetary reward for the introduction of ether anesthesia, but Jackson opposed the gesture, and the offer was withdrawn. The French Academy of Medicine proposed to pay a sum of money to both Morton and Jackson, but Morton refused. In the 1850s the United States Congress failed to pass three bills that would have awarded Morton $100,000 in recognition of his achievement. Morton died in poverty in New York City on July 15, 1868.

Legacy

Morton's pioneering work led to the widespread adoption of sulfuric ether anesthesia by dentists and surgeons, which improved surgical conditions and broadened the range of possible surgical procedures.

Sulfuric ether anesthesia was firmly established as part of the surgeon's repertoire by the end of 1847. In that year U.S. physician and writer Oliver Wendell Holmes coined the term "anesthesia."

The introduction of sulfuric ether anesthesia led to the discovery that chloroform behaves as an anesthetic agent as well. In 1847 Marie Flourens of France presented a paper announcing the anesthetic effects of chloroform and comparing them to those of sulfuric ether. Her paper went unnoticed until James Young Simpson performed a public demonstration of chloroform anesthesia later that same year. Chloroform is less irritating to the eyes and nose than sulfuric ether.

Chloroform became prevalent as an anesthetic in childbirth. In the 1850s John Snow delivered the children of Queen Victoria using chloroform anesthesia. Chloroform was soon being used more often than ether in childbirth. Snow specialized in anesthesiology (the first physician to do so) and developed the chloroform inhaler, a device that controlled the amount of vapor the patient inhaled.

In modern times, anesthesiology is a specialized branch of medicine in which anesthesiologists use the pain-killing properties of numerous substances to control pain and sensation during and immediately following surgical procedures.

Schuyler

WORLD EVENTS	MORTON'S LIFE
Napoleonic Wars 1803-15 in Europe	
	1819 William Morton is born
	1842 Crawford Long uses ether anesthesia but doesn't publish findings until 1849
	1844 Morton opens dental practice
	1846 Morton publicly demonstrates ether anesthesia
	Morton and Charles Jackson obtain patent for ether anesthesia
	1847 Morton publishes *Remarks on the Proper Mode of Administering Sulphuric Ether by Inhalation*
United States 1861-65 Civil War	
	1868 Morton dies
Germany is united 1871	

For Further Reading:

Davison, Meredith. *The Evolution of Anesthesia.* Baltimore, Md.: Williams & Wilkins, 1965.

MacQuitty, Betty. *The Battle for Oblivion: The Discovery of Anesthesia.* Edinburgh: Harrap, 1969.

Rutkow, Ira M. *Surgery: An Illustrated History.* St. Louis, Mo.: Mosby-Year Book, 1993.

Newcomen, Thomas

Inventor of Steam Engine
1663-1729

Life and Work

Thomas Newcomen designed and built the first practical steam engine, which served as the primary means of pumping water out of mines until it was replaced by JAMES WATT's improved model in the late eighteenth century.

Newcomen was born in 1663 in Dartmouth, England. He came from a land-holding family, and, while not much is known about his youth, his letters give the impression that he was well educated.

By the end of the seventeenth century, Newcomen was working as a metal worker and toolsmith in his hometown. Little else is known of this part of his life.

Dartmouth was near an extensive system of coal mines from which an increasing amount of coal, largely for iron smelting, was demanded in the late seventeenth century. The mines required large pumps for water removal. Thomas Savery had designed a steam-driven pump for that purpose. His pump was severely limited by inefficiency and technical problems; nevertheless, Newcomen based his invention on Savery's model.

Newcomen introduced his new steam engine in 1712. During the following years, Newcomen appears to have experimented (helped by assistant John Calley) with various versions of the completed, working engine.

Unable to obtain a patent because Savery's earlier patent included any machine that raises water by use of heat, he formed a partnership with Savery. Newcomen's engine used both atmospheric pressure and low-pressure steam to operate pistons, and jets of cool water to condense the steam. It also included automatic valves and was more efficient and less dangerous than Savery's pump.

Newcomen's engine was adopted throughout Europe for draining mines and raising water to power water wheels. Newcomen added technical improvements to the design in 1725. He died on August 5, 1729, in London.

Legacy

The Newcomen engine, which played a significant role in the Industrial Revolution, was widely used in coal mines for most of the eighteenth century and served as the model for many subsequent engine designs.

Coal was crucial to industrialization. At first it was used mostly to smelt iron and other metals that were used to produce various alloys with a wide range of applications. Metal had rapidly replaced wood as a stronger and longer-lasting material for machine parts and large structures. Newcomen's steam engine increased the amount of coal available and was partly responsible for England's leading role in industrialization.

The heaviest use of Newcomen engines was in England's northeast coalfields; records indicate that by 1769, 98 Newcomen engines had been erected there.

The steam engine took hold in continental Europe as well. It was used throughout the 1720s, with varying degrees of success, in Austria, France, and Sweden, and it was introduced to North America in 1755.

In 1769 Scottish engineer James Watt was granted a patent for a steam engine that was vastly more efficient than Newcomen's because of the addition of a separate cooling condenser. Watt's innovations widened the range of the steam engine's industrial applications, and his engine gradually replaced Newcomen's (although a few Newcomen engines operated into the twentieth century). Later improvements on Watt's model led to high-pressure steam engines, which led to the invention of the steam locomotive in the early nineteenth century.

Schuyler

For Further Reading:
Hills, Richard Leslie. *Power from Steam: A History of the Stationary Steam Engine.* New York: Cambridge University Press, 1989.
Rolt, L. T. C. *Thomas Newcomen: The Prehistory of the Steam Engine.* Dawlish, England: David & Charles, 1963.

Newton, Isaac

Discoverer of Laws of Motion
and Gravitation
1642-1727

Life and Work

Isaac Newton is considered to be one of the greatest scientists of all time. He established the universal laws of motion and gravitation, and made significant contributions in optics, mathematics, and telescopy.

Newton was born on December 25, 1642, in Woolsthorpe, England. His father died shortly before his birth and left the family with few resources. His mother remarried when Newton was three, and he was sent to live with his grandparents and later with a family named Clark. In the Clarks' tool shop his genius for mechanical devices and toys began to emerge.

At age 18 Newton enrolled in Trinity College at Cambridge, where he excelled in mathematics. Four years later, in 1665, an outbreak of bubonic plague closed the university, and he was sent home. During this time, he formulated the binomial theorem in algebra and developed the calculus (also established independently by GOTTFRIED LEIBNIZ in 1675). He also discovered that white light is a mixture of pure colors and explored ways to project and reflect images in telescopes.

Newton later told a story that he was sitting in an apple orchard, pondering what keeps planets in their orbits, when an apple fell near him. In a flash he saw that the same force that pulls the apple to the ground also pulls the moon to Earth (but the moon's sideways speed is great enough to keep it from hitting Earth). He told no one about his discovery for nearly 20 years.

Newton returned to Trinity for a fellowship in 1667. He developed a reflecting telescope in 1668 and was appointed to a professorship at Trinity College in 1669. His first series of lectures explained his work on the nature of light and optics.

Newton cautiously made his work public through the Royal Society of London starting in 1672. The papers he delivered there soon led to controversy between Newton and scientist ROBERT HOOKE, who contended that Newton used his ideas without credit. Newton bitterly withdrew and suffered a nearly complete breakdown in 1679, following the death of his mother.

In 1684 EDMUND HALLEY convinced Newton to resume his work. Halley was the King's official astronomer and one of many scientists close to discovering what Newton had known for years, that a force of gravitation controlled all planetary motion. Halley urged Newton into writing his theory, and in 1687, Newton produced what is still considered to be one of the greatest scientific books of all time, *Mathematical Principles of Natural Philosophy*, also known as the *Principia*. In it he laid out his three laws of motion, which were: 1. Every object remains at rest or in motion until it is compelled to change by a force acting on it (the law of inertia). 2. The motion of an object changes in proportion to and in the direction of the force acting on it (the law of force). 3. Every action has an equal opposing reaction (the law of action and reaction).

Newton became interested in politics in the 1690s and was put in charge of the Royal Mint in 1696. In 1704 he finally published *Optiks*, which compiled all of his work in the field. In 1705 he was knighted by Queen Anne. Newton died on March 20, 1727.

Legacy

While Newton's contributions in optics, telescopy, and calculus were significant, his laws of motion were truly Earth-shaking. With one great stride, Newton codified all of planetary motion, which laid the groundwork for much of modern physics and changed the way the universe was viewed.

Newton's influence was immediate. He solved outstanding questions about planetary movement that other contemporary scientists, including Halley and Hooke, struggled with. Following the publication of the *Principia*, Newton's followers immediately adopted his theories, which spread across Europe.

The great influence of Newton's laws of motion lies in the fact that they had universal application. They were used to explain the paths and speeds of planets and moons, eclipses and tides.

The laws also enabled future astronomers to analyze all observations and calculations against a systematic understanding of how bodies in space move and operate. In the early years of the twentieth century, deviations in the predicted paths of Uranus and Neptune led astronomers at the Lowell Observatory in Flagstaff, Arizona, to look for and find the planet Pluto. Engineers have used Newton's laws to correctly lay out the flight paths of probes sent to explore the far reaches of the solar system.

Later in the twentieth century work in subatomic physics led to the realization that the movements of the smallest particles were controlled by a slightly different set of laws, called quantum mechanics. Together, classical (Newtonian) and quantum mechanics define on a macro and micro scale the movement and behavior of the largest objects and tiniest particles in the most vast and smallest of spaces.

Secaur

WORLD EVENTS	NEWTON'S LIFE
1642	Isaac Newton is born
1661	Newton enters Trinity College
1665-68	Newton develops laws of motion and gravitation; experiments with nature of light and color; develops reflecting telescope; develops the calculus
1669	Newton appointed to a professorship at Trinity College
1687	*Mathematical Principles of Natural Philosophy*, or the *Principia*, published
England's Glorious Revolution 1688	
	1704 *Optiks*, Newton's theory of light and color, is published
Peace of Utrecht 1713-15 settles War of Spanish Succession	
1727	Newton dies

For Further Reading:

Christianson, Gale E. *Isaac Newton and the Scientific Revolution.* New York: Oxford University Press, 1996.

Segrè, Emilio. *From Falling Bodies to Radio Waves.* New York: Freeman, 1984.

Strathern, Paul. *Newton and Gravity.* New York: Doubleday, 1998.

Nightingale, Florence

Founder of Nursing Profession
1820-1910

Life and Work

Florence Nightingale improved the conditions of soldiers in the British Army, reformed health-care administration, and founded the modern profession of nursing.

Nightingale was born in Florence, Italy, on May 12, 1820, to a wealthy English family. She spent her youth in England and received from her father a thorough education in the classical disciplines: Greek, Latin, literature, history, philosophy, and mathematics. Her family expected her to marry, raise children, and entertain, but she had no interest in such activities. In 1837 she believed that she heard the voice of God telling her she had a mission, and for the next decade she pursued interests related to health, hoping to identify exactly what that mission should be. She trained to care for the sick at the Institution of Protestant Deaconesses in Karsersuerth, Germany, in 1850.

Nightingale was asked to take charge of the Harley Street Nursing Home in London in 1853. She revolutionized the institution by improving nursing care; the experience convinced her that the major problem confronting the medical community was insufficient training of nurses.

The Crimean War, fought between Russia and British-French-Turkish forces, broke out in 1854 and diverted Nightingale from her concerns about nurse training. She traveled to Scutari, Turkey, to care for wounded British soldiers, who were living in a filthy, rat-infested barrack hospital with grossly inadequate supplies. With funds she brought from England, Nightingale stocked the hospital with cleaning and medical equipment. She proved to be both a strong administrator and devoted to patients. Eventually she took charge of the whole hospital, implementing sanitary practices that reduced the death rate from 42% in 1854 to 2% in 1855.

Nightingale become famous in London, and a grand reception was organized for her return at the end of the war in 1856. However, she arrived clandestinely and never again appeared in public. She continued to work for health care reform from her home. She completed a lengthy report in 1857, including statistics and diagrams to support her claims, describing the horrible living conditions of soldiers in times of both war and peace. She also wrote *Notes on Matters Affecting the Health, Efficiency, and Hospital Administration of the British Army* (1857) and *Notes on Nursing* (1859), giving basic instruction in modern nursing. In 1860 she established a nurse-training school at St. Thomas Hospital in London.

For the last 40 years of her life, Nightingale lived as an invalid, conducting all her business through correspondence and visitors. She died in London on August 13, 1910.

Legacy

Nightingale's nursing and humanitarian work improved the lives of soldiers, and her innovations in medical administration established nursing as a profession.

Nightingale's influential report of 1857 prompted major reforms in military health care. Her statistical analyses were pioneering intellectual studies that complemented her practical strategies in sanitation and hospital care of military personnel. Nightingale's achievements earned her celebrity status that enabled her to secure funds for continuing improvements of medical services. One outcome was her establishment in 1860 of the Army Medical School where physicians were trained to work under conditions of war.

Nightingale's work served as both the practical and inspirational origin for the field of nursing. Women all over the world wanted to emulate her and aspired to become nurses. During the Civil War in the United States, larger numbers of women entered nursing than ever before. As a result of Nightingale's influence and authority, many medical institutions in Europe and North America opened nurse-training branches, using the nursing school Nightingale established in London as a model.

Nightingale firmly established nursing as a woman's activity, offering a professional opportunity to many women who otherwise might not have worked outside the home.

Schuyler

World Events	Nightingale's Life
Napoleonic Wars 1803-15 in Europe	
	1820 Florence Nightingale is born
	1837 Nightingale believes she hears the voice of God
	1853 Nightingale runs nursing home
	1854 Crimean War begins; Nightingale travels to Turkey to care for wounded soldiers
	1856 Crimean War ends; Nightingale returns to England
	1857 Nightingale completes *Notes on Matters Affecting the Health, Efficiency, and Hospital Administration of the British Army*; army medical school opens
	1859 *Notes on Nursing* is published
	1860 Nightingale opens a nursing school
United States 1861-65 Civil War	
Germany is united 1871	
	1910 Nightingale dies
Spanish-American 1898 War	

For Further Reading:

Cohen, I. B. "Florence Nightingale." *Scientific American* 250: 128, (March 1984).

Huxley, Elspeth. *Florence Nightingale.* New York: Putnam, 1975.

Nightingale, Florence, and M. Calabria. *Florence Nightingale in Egypt and Greece: Her Diaries and "Visions."* Albany, N.Y.: SUNY Press, 1997.

Woodham-Smith, Cecil. *Florence Nightingale, 1820-1910.* New York: McGraw-Hill, 1951.

Noether, Emmy

Pioneer in Modern Algebra
1882-1935

Life and Work

Emmy Noether's work in abstract algebra paved the way for group theory, an entirely new branch of mathematics. Her theorems promoted modern algebra as "the foundation tool of all mathematics," as she claimed in 1931.

Noether was born in Erlangen, Germany, into a family of tradesmen on March 23, 1882. Crippled by polio, her father was excused from family business concerns and became a mathematics professor at University of Erlangen. Noether herself followed the usual educational path for women of her time, which culminated in language studies, piano practice, and dancing lessons. But all along she aspired to a more active career.

One of the few professional options for women was teaching children. After passing exams for certification as a teacher of English and French, Noether petitioned to audit math classes at the University of Erlangen. Women were not accepted as students in German universities at that time and could attend lectures only with the permission of individual professors. After several years of visiting classes, Noether and several other women were permitted to enroll at Erlangen in 1904. Noether earned her doctorate in mathematics, summa cum laude, in 1907.

Over the next 25 years, Noether worked at various universities as a privatdozent, a person who had earned the right to teach but was not paid. In 1915 she went to the University of Göttingen where she assisted David Hilbert and Felix Klein in work on general relativity theory. From 1923 to 1933 she finally got a small stipend as a lecturer and official permission to guide doctoral students.

During these lean years Noether developed her theorems in the areas of abstract algebra and physics. Her most famous theorem relates the physical laws of conservation to mathematical properties of symmetry. She generated foundations of the general theory of ideals, beginning in 1921, and vastly expanded the theory of noncommutative systems. In the mathematical theory of ideals Noether developed a process of normalization in which all integers of a field are sorted so that any two elements of a class—and no two elements of different classes—are congruent. Her years of collaboration and editorial work with the journal *Mathematische Annalen* extended understanding of modern mathematics.

When the Nazis seized power in 1933, Noether's university position was lost as were those of many of her friends and students. She represented all of the things Hitler despised; not only was she a woman but also she was a Jew and a pacifist. In this climate Noether accepted an invitation to go to Bryn Mawr College in Pennsylvania as a guest lecturer. While there she taught young women interested in mathematics and did some research at the Institute for Advanced Study in Princeton, New Jersey, meeting with ALBERT EINSTEIN on a regular basis.

After less than two years in the U.S., Noether died unexpectedly after routine surgery on April 14, 1935.

Legacy

Noether's investigations in abstract algebra proved to be crucial in changing what is known about numbers and their relationship to physical reality. She also helped to provide the necessary mathematical groundwork for the atomic age. Her contributions helped break down biases against women working in mathematics.

During her time in Germany Noether's presence on university campuses caused many to question whether women could have a place in academic life. Her repeated attempts to enter the University of Erlangen eventually resulted in acceptance of women as students there, but coeducation was not generally accepted in Germany until 1908. Noether's ideas became known and respected internationally, but she never did secure a position with benefits and pension rights that professors receive.

After World War II a new generation of German scholars recognized the scope and originality of Noether's work. In 1958 the University of Erlangen sponsored a conference of her distinguished former students and their students to discuss her work, its applications, and its influence on research. In 1960 the city of Erlangen named one of its streets Noetherstrasse and in 1982 dedicated the Emmy Noether Gymnasium, a coeducational high school emphasizing mathematics and science.

In the decades since Noether's death her technique of normalization evolved as a fundamental tool in commutative algebra. Topology, the study of properties of shapes and figures that remain unchanged as an object is distorted, is another area that advanced by use of Noether's algebra.

Noether's contributions to Einstein's work and the development of nuclear energy are perhaps her most significant legacy. Her theory of symmetry and conservation was crucial to Einstein's theory of relativity. She helped develop the mathematical bases that support Einstein's elegant equation $E = mc^2$.

Wilson

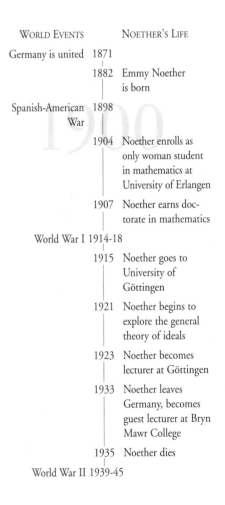

WORLD EVENTS	NOETHER'S LIFE
Germany is united 1871	
	1882 Emmy Noether is born
Spanish-American 1898 War	
	1904 Noether enrolls as only woman student in mathematics at University of Erlangen
	1907 Noether earns doctorate in mathematics
World War I 1914-18	
	1915 Noether goes to University of Göttingen
	1921 Noether begins to explore the general theory of ideals
	1923 Noether becomes lecturer at Göttingen
	1933 Noether leaves Germany, becomes guest lecturer at Bryn Mawr College
	1935 Noether dies
World War II 1939-45	

For Further Reading:
Dick, Auguste. *Emmy Noether*. Boston: Birkhauser, 1981.
Grinstein, L., and P. Campbell. *Women of Mathematics*. New York: Greenwood Press, 1987.
Srinivasan, B., and J. D. Sally, eds. *Emmy Noether in Bryn Mawr*. New York: Springer-Verlag, 1983.

Noguchi, Hideyo

Isolator of Cause of Paralysis in
Syphilis Patients
1876-1928

Life and Work

Hideyo Noguchi, a pioneer in twentieth-century bacteriology, attained important experimental results pertaining to the diseases syphilis, trachoma, and Bartonellosis.

WORLD EVENTS		NOGUCHI'S LIFE
Germany is united	1871	
	1876	Hideyo Noguchi is born
	1897	Noguchi earns medical degree from Tokyo Medical College
Spanish-American War	1898	
	1899	Noguchi goes to United States
	1905	Noguchi and Simon Flexner confirm syphilis-causing bacterium
	1913	Noguchi proves link between syphilis and paresis
World War I	1914-18	
	1918	Noguchi begins traveling to study rare diseases
	1920s	Noguchi's yellow fever research is discredited
	1928	Noguchi dies
World War II	1939-45	

Noguchi was born on November 24, 1876, in the village of Inawashiro, Japan. His father deserted the family soon after Noguchi's birth, and his mother worked in the rice fields to support her two children. In his youth Noguchi fell into an open hearth fire, an accident that left him severely burned and without the use of the fingers on his left hand.

Noguchi's superior performance in grade school brought him to the attention of the superintendent, who eventually arranged for Noguchi to receive reconstructive surgery. Thus developing an interest in medicine, Noguchi became apprenticed to his surgeon, entered Tokyo Medical College in 1894, and earned a medical degree in 1897.

He obtained a position as research assistant to bacteriologist Shibasaburo Kitasato at the Institute for Infectious Diseases in Tokyo. In 1899 a medical commission from the United States visited the institute, and Noguchi became interested in studying in that country. Later that year he traveled to Philadelphia and obtained an assistantship to snake-venom researcher Silas Weir Mitchell. Noguchi investigated the venomous substances hemolysin and agglutinin, and devised sera to protect against them. This work earned him a one-year fellowship to study in Copenhagen, Denmark.

In 1904 bacteriologist Simon Flexner (who had been Noguchi's first contact in the United States) became director of the newly founded Rockefeller Institute in New York, and he invited Noguchi to join his laboratory. In 1905 Flexner and Noguchi were among the first to confirm German biologist Fritz Schaudinn's identification of the bacterium *Treponema pallidum* as the causative agent of syphilis. In 1913 Noguchi proved the long-suspected connection between syphilis and paresis, a type of slow paralysis, by isolating *T. palladium* from the brains of people who had died of paresis.

At about this time, his credibility began to diminish. In 1911 he asserted that he had isolated a pure culture of *T. palladium*, but no other scientist was able to repeat this finding; he also introduced a diagnostic test for syphilis that proved unreliable. He and Flexner were then discredited in their claim of cultivating the virus that causes poliomyelitis in monkeys.

In 1918, Noguchi began traveling to investigate the causes of rare infectious diseases.

Visiting the American West, Central and South America, and the Peruvian Andes, he attained important results concerning Bartonellosis (then called Carrion's disease) and trachoma (a disease of the eyes).

Controversy erupted around Noguchi's work on yellow fever. In the 1920s, he announced that he had isolated the disease's bacterial agent, but he was later shown to have obtained the organism from misdiagnosed patients who had actually been suffering from hemorrhagic jaundice. Noguchi was criticized for drawing conclusions from insufficient research.

In 1927 Adrian Stokes determined that yellow fever is caused by a virus, and Noguchi immediately traveled to Africa to conduct further research. He died of yellow fever on May 21, 1928, as he was about to return to New York.

Legacy

Noguchi's research led to several disease treatments and contributed to the growing understanding of the organisms responsible for infectious diseases.

By 1910, prior to Noguchi's work on syphilis, Polish scientist PAUL EHRLICH developed a syphilis treatment, Salvarsan, which contained arsenic. While it was effective in curing syphilis, Salvarsan had to be administered carefully, because of the presence of arsenic. After Noguchi's work in isolating the bacterial cause of syphilis, work continued unabated on finding safer and more effective drugs. By 1943, an effective one-week treatment of antibiotics proved successful in treating syphilis in its early stages.

Noguchi's discovery that paresis is associated with syphilis represented a breakthrough in understanding the organic aspect of some psychiatric disorders. It was among the first demonstrations that pathogenic organisms can invade and damage tissues of the nervous system.

Noguchi's studies of Bartonellosis and trachoma aided the development of treatments for these obscure diseases.

Despite his unsubstantiated claims, Noguchi received numerous honors and is remembered as a tireless and accomplished researcher.

Schuyler

For Further Reading:

Flexner, Simon. "Hideyo Noguchi, A Biographical Sketch." *Science* 69 (1929).

Plesset, Isabel R. *Noguchi and His Patrons.* Rutherford, N.J.: Fairleigh Dickinson University Press, 1981.

Ohm, Georg

Investigator of Electricity;
Originator of Ohm's Law
1789-1854

Life and Work

Georg Ohm was a physicist whose work in electricity led to groundbreaking discoveries including Ohm's law, which describes the relationship of electrical circuit, voltage, and resistance.

Ohm was born in Erlangen, Germany, on February 16, 1789, to a mechanic father from whom he acquired an early interest in science. He studied physics at the University of Erlangen and held an academic position at Cologne from 1817 to 1826.

In 1827 Ohm joined the faculty at the Military Academy in Berlin. That year he announced his most famous discovery, now known as Ohm's law. His research was inspired by the work of French mathematician Jean Baptiste Joseph Fourier, who studied the physics of heat in the early 1820s. Fourier demonstrated that the flow of heat between two points depends on the temperatures of the two points and on the heat-conducting capacity of the material between the two points. Ohm predicted that the flow of electrical current between two points would similarly depend on the points' electrical potential and electrical conductivity.

To test this hypothesis, Ohm performed experiments with electricity flowing through wires of various thicknesses and lengths and found mathematical relationships between the amount of current transmitted and these two variables. The relationships led him to establish that electrical resistance within a circuit is determined both by the voltage (the difference in electrical potential) and the current. But the importance of Ohm's law, which states that current is directly proportional to voltage and inversely proportional to resistance, was not recognized by the scientific community for many years.

Ohm took a position at The Polytechnic School of Nuremberg, Germany, in 1833. In 1841 his work began to be publicly appreciated; that year he received the Copley Medal of the Royal Society of London.

In 1843 he conducted research on sound and found that the human ear is capable of analyzing the numerous component frequencies present in complex musical sounds. His last teaching post, at the University of Munich, began in 1849. He died on July 7, 1854, in Munich, Germany.

Legacy

Ohm's work played a significant role in the progress of the field of electricity and in the development of a standard international system of units for energy-related concepts.

Ohm's work topped off the advances in the understanding of electricity and magnetism that were made in the first quarter of the nineteenth century. In 1800 ALESSANDRO VOLTA devised the first battery, and in 1820, Hans Oersted discovered that an electric current creates a magnetic field. Through the 1820s, MICHAEL FARADAY and Joseph Henry independently conducted experiments with electricity and formulated the principles of the electric motor and electromagnetic induction. In 1827 the accomplishments of ANDRÉ AMPÈRE may have overshadowed those of Ohm. Ampère discovered the mathematical relationship between magnetic force and electric current and distinguished current from voltage.

Ohm's discovery was extended by Gustav Kirchhoff in 1854. Kirchhoff showed that the total voltage driving a current through a given set of connected circuits must equal the total of the voltages opposing it. From this he formulated laws that enabled the mathematical analysis of complex networks of circuits, which made possible all subsequent development of electronic systems.

In 1881 the members of the International Congress of Electrical Engineers adopted standard, interrelated international units for measurable electrical properties. The ohm (named for Ohm) was defined as the unit of resistance; volt (named for Alessandro Volta) represented electrical potential; and ampere (named for André Ampère) measured the strength of electrical current. Ohm's law can be stated as "amperes equal volts over ohms," thus honoring three pioneers in electricity in one formula.

Schuyler

WORLD EVENTS		OHM'S LIFE
French Revolution	1789	Georg Ohm is born
Napoleonic Wars in Europe	1803-15	
	1827	Ohm discovers relationships among current, voltage, and resistance, known as Ohm's law
	1833	Ohm moves to Nuremberg
	1841	Ohm receives Copley Medal
	1843	Ohm conducts research on sound
	1849	Ohm moves to Munich
	1854	Ohm dies
		Gustav Kirchhoff extends Ohm's law
	1861	Unit of resistance is named the "ohm"
United States Civil War	1861-65	
Germany is united	1871	

For Further Reading:

Bordeau, Sanford P. *Volts to Hertz: The Rise of Electricity*. Minneapolis, Minn.: Burgess Publishing, 1982.

Nye, Mary Jo. *Before Big Science: The Pursuit of Modern Chemistry and Physics, 1800-1940*. London: Prentice-Hall, 1996.

Purrington, Robert D. *Physics in the 19th Century*. New Brunswick, N.J.: Rutgers University Press, 1997.

Oppenheimer, J. Robert

**Head Scientist on
First Atomic Bomb Project
1904-1967**

Life and Work

J. Robert Oppenheimer led the international team of scientists who worked in the United States to develop and detonate the first atomic bomb.

Born in New York City to a wealthy German immigrant family on April 22, 1904, Oppenheimer studied at a boarding school in the desert in Los Alamos, New Mexico, and then attended Harvard University, where he excelled in both the sciences and languages. Graduating in 1925, he went to England to study atomic physics with the great physicist ERNEST RUTHERFORD at the Cavendish Laboratory in Cambridge. He was soon invited to the University of Göttingen, one of the great German schools where the radically new science of quantum mechanics was being developed. He

received his doctorate from Göttingen in 1927 and returned to the United States to teach and conduct research on the energy of subatomic particles at the University of California-Berkeley.

In 1939, at the beginning of World War II, NIELS BOHR brought word from Europe that German scientists OTTO HAHN and LISE MEITNER had discovered nuclear fission: that uranium atoms could be split into smaller particles, releasing huge amounts of energy. Germany had leading scientists in nuclear physics, plus access to all the raw materials to build an atomic bomb. Leo Szilard and ALBERT EINSTEIN warned United States President Franklin D. Roosevelt of the danger of the German leader Adolf Hitler developing such a weapon. After much delay the American government launched the secret Manhattan Project to make a bomb to counter the German threat. Oppenheimer was chosen as the leader in 1942 and, recalling the New Mexico desert where he had attended school, he built a secret laboratory there.

The task was daunting, the pressure tremendous, and the risk terrifying, as scientists knew the enormous destruction that splitting uranium atoms could cause. Within three years the group tested the first nuclear weapon on July 16, 1945. Two bombs, one built from uranium-235 and the other from plutonium-239, were ready to use within three more weeks. The scientists who remained in Germany had never succeeded in making a bomb and that nation had already surrendered, but the war in Japan was raging on. After a great controversy the bombs were dropped on Hiroshima and Nagasaki, Japan, on August 6 and 9, 1945. Japan surrendered soon after and World War II was over.

In 1947, because of his unique blend of scientific and leadership skills, Oppenheimer became the director of the Institute for Advanced Study at Princeton University, where Einstein spent the last years of his career.

Because of his opposition to the hydrogen bomb, nuclear devices with 50 times the explosive power of the bombs of the Manhattan Project, and his acquaintance with members of the Communist Party in his youth, he was stripped of his security clearance and role as adviser to the Atomic Energy Commission in 1953. In 1963 President Lyndon Johnson rehabilitated Oppenheimer by awarding him the Enrico Fermi Award, given to scientists who advance the understanding of nuclear physics.

Oppenheimer retired in 1966 and died of throat cancer on February 18, 1967.

Legacy

The Manhattan Project, under the direction of Oppenheimer, changed the course of history in the twentieth century. It ushered in the atomic age, with the threat of nuclear annihilation hanging over all countries of the world.

Oppenheimer realized, on the morning of the first successful bomb test, the long-term impact of what his group had accomplished. He wrote, "We knew the world would not be the same. A few people laughed, a few people cried. Most people were silent. I remembered the line from the Hindu scripture . . . 'Now I am become Death, the destroyer of worlds.'" Oppenheimer's realization of the impact of the development of nuclear weapons and his opposition to the building of hydrogen bombs helped to galvanize anti-nuclear protests early in the atomic age. While thousands of hydrogen bombs are stockpiled around the world today, the ongoing protests against nuclear arms build-up gathers strength from the fact that the originators of the first nuclear bombs disapproved of their continued development and deployment.

Soon after the Manhattan Project's development of the atomic bomb, scientists began developing nuclear reactors for commercial energy production. In 1956 the first nuclear power plant was completed in England, and the next year the United States followed with its first nuclear plant. Until the 1970s nuclear energy production steadily increased. Since then, however, few new plants have been built and production has slowed on many plants, as concern for safety increased. Some countries today, such as France, which lack adequate sources of other forms of energy, still rely on nuclear power.

Secaur

WORLD EVENTS	OPPENHEIMER'S LIFE
	1904 J. Robert Oppenheimer is born
World War I 1914-18	
	1925 Oppenheimer graduates from Harvard
	1927 Oppenheimer earns doctorate from University of Göttingen and joins U. of California–Berkeley
	1939 Niels Bohr informs United States of discovery of nuclear fission
World War II 1939-45	
	1942 Oppenheimer becomes leader of Manhattan Project to build atom bomb
	1945 First nuclear weapons tested and used
Korean War 1950-53	
	1953 Oppenheimer is discredited because of opposition to hydrogen bomb
	1963 Oppenheimer gains Enrico Fermi Award
	1967 Oppenheimer dies

For Further Reading:
Driemen, John Evans. *Atomic Dawn: A Biography of Robert Oppenheimer.* Minneapolis, Minn.: Dillon Press, 1989.
Rhodes, Richard. *The Making of the Atomic Bomb.* New York: Simon & Schuster, 1988.
York, Herbert F. *The Advisors: Oppenheimer, Teller, and the Superbomb.* Stanford, Calif.: Stanford University Press, 1989.

Otto, Nikolaus August

Inventor of Four-Stroke Internal Combustion Engine

1832-1891

Life and Work

Nikolaus August Otto designed, constructed, and marketed the first successful four-stroke internal combustion engine.

Otto was born on June 10, 1832, in Holzhauzen, Germany, the son of a farmer. The details of his early life are unknown. At age 16 he left school and worked at a merchant's office, never obtaining any formal higher education.

In 1861 Otto became interested in the first coal-gas engine, developed the previous year by the Belgian steam-engine manufacturer Jean Lenoir. The engine, a horizontal, double-acting single-cylinder machine, was inefficient and tended to overheat, but nevertheless achieved a modest commercial success. In 1864 Otto formed a company with industrialist Eugen Langen and began marketing gas-powered engines that represented modifications of Lenoir's design.

In a competition with 14 other gas-powered engines at the 1867 Paris Exhibition, Otto's engine won first prize. It won because of its efficiency, despite being large, heavy, and noisy. Otto opened a new factory near Cologne, Germany, in 1869 to accommodate the flood of orders. Joined by talented engineers Gottlieb Daimler and Wilhelm Maybach in 1872, the business flourished.

In 1876 Otto introduced a radically different engine. This quiet, efficient machine had a horizontal cylinder, conventional piston, connecting rod, and flywheel; it was single-acting and utilized a four-stroke cycle. The engine's cycle consisted of four steps: as the piston moved outward the fuel (coal gas) and air were drawn into the cylinder; as the piston moved inward the fuel-air mixture was compressed and ignited by a spark; the ignition drove the piston outward for the third stroke; and as the piston moved inward again, the waste gases were forced out an exhaust valve. Then the cycle repeated.

Otto's four-stroke engine sold extremely well. However, in 1886 his competitors unearthed an obscure pamphlet written in 1862 by Frenchman Alphonse Beau de Rochas describing the four-stroke cycle. This finding invalidated Otto's patent, he lost exclusive rights to sell four-stroke engines, and sales dropped because of competition.

Otto died on January 26, 1891, in Cologne.

Legacy

Otto's engine created gas-based industrial power and made gas-powered vehicles possible. It is the basis of four-stroke internal-combustion engines of modern technology.

Otto's earliest engines were used immediately in numerous industries, powering pumps, sewing machines, printing presses, lathes, looms, and saws. These engines were limited to towns because they required a gas supply and to stationary settings because they were loud and heavy.

In 1882 Otto's engineers, Daimler and Maybach, left his company and developed the first engine that ran on gasoline, or petrol. They added a carburetor to vaporize the gasoline, which ignited upon insertion of a red-hot metal bar. Their engines were lighter and more efficient than even Otto's improved engines of 1876, and they were the first to be used in automobiles.

Daimler and Maybach built a motorcycle in 1885 and an automobile in 1886; both ran on four-stroke internal combustion engines powered by gasoline. Karl Benz, a German automobile pioneer, replaced two-stroke engines with engines built by Daimler and Maybach, and then introduced electric ignition, which became the standard.

In the 1890s RUDOLF DIESEL developed an engine that used cheaper fuel and ignited with the heat produced by compression rather than with electrical assistance. The resultant high pressures required the use of heavy metal to construct the engines, which were (and are) used predominantly in large vehicles such as trucks, buses, and ships.

As the early twentieth century progressed, continually improved internal combustion engines were built to power automobiles, tractors, trucks, ships, dirigibles, and airplanes, as well as various large industrial machines and relatively small appliances, such as lawnmowers.

Schuyler

WORLD EVENTS		OTTO'S LIFE
Napoleonic Wars in Europe	1803-15	
	1832	Nikolaus Otto is born
	1848	Otto leaves school at age 16; he begins working at merchant's office
	1860	Jean Lenoir introduces first coal-gas-powered engine
United States Civil War	1861-65	
	1864	Otto forms company to market modified gas-powered engines
	1867	Otto's engine wins first prize at Paris Exhibition
	1869	Otto opens new factory to accommodate orders
Germany is united	1871	
	1872	Otto recruits Gottlieb Daimler and Wilhelm Maybach
	1876	Otto introduces four-stroke engine
	1886	Otto's patent is revoked
	1891	Otto dies
Spanish-American War	1898	

For Further Reading:

Ganesan, V. *Internal Combustion Engines.* New York: McGraw-Hill, 1996.

Heywood, John B. *Internal Combustion Engine Fundamentals.* New York: McGraw-Hill, 1988.

Stone, Richard. *Introduction to Internal Combustion Engines.* New York: Macmillan, 1992.

Paracelsus

Revolutionary Physician
and Alchemist
1493-1541

Life and Work

Paracelsus accelerated the application of chemistry to the practice of medicine, moved alchemy closer to the realm of scientific investigation, and weakened the dominance of traditional Greek and Roman medical theory in the sixteenth century.

Paracelsus was born Theophrastus Bombastus von Hohenheim in November 1493 in Einsiedeln, Switzerland, the son of an impoverished physician. In his youth, he was introduced to alchemy, a medieval art aimed at changing base metals to gold or silver, at a school for mining and metallurgy. From 1507 through 1512, he traveled across Europe and is thought to have attended seven universities and been disappointed with them all. Historians believe he received a degree in

medicine from the University of Vienna in 1510.

Around 1516, Paracelsus began to publicly attack the prevailing medical views of HIPPOCRATES, GALEN, and IBN SINA. He criticized the theory of the four bodily humors, originating with Hippocrates, and adopted the name Paracelsus, which means "surpassing Celsus," referring to a first century Roman physician and implying his superiority over the ancients.

For several years, Paracelsus worked as an army physician in the Netherlands and Italy. He sought out the most experienced alchemists across Europe and gained new practical knowledge, which he used to great success while treating diseases and wounds on the battlefield. His reputation for healing powers spread quickly during his service.

On the heels of his newly acquired reputation as a miracle healer, he received an appointment to lecture at the University of Basel in 1527. Despite his popularity among students, he was forced to flee in 1528 after offending university authorities by burning the works of Galen and Ibn Sina. For the next eight years, he remained in exile and wrote medical treatises outlining his theories. He proposed, for instance, that disease was caused by foreign parasites that disrupted normal body chemistry and that mental health was dependent on physical conditions, not on evil spirits.

His innovative medical treatments were rooted in an advanced understanding of chemistry and mineralogy, which allowed him to make great improvements over traditional alchemical treatments. In 1530, Paracelsus introduced a clinical description of syphilis and prescribed mercury as treatment. He correctly claimed that silicosis (a miners' disease) was caused by inhalation of metal vapors. He was the first to state that small doses of the agent that causes an illness can be used as a treatment for that illness. He passionately advocated the use of minerals and chemical compounds to cure illnesses.

Paracelsus died under mysterious circumstances in Salzburg, Austria, on September 24, 1541.

Legacy

Paracelsus contributed to the evolution of medicine by challenging traditional dogma

and by insisting on the importance of chemistry in human health.

Paracelsus inspired his contemporaries to question the authority of Greek and Roman medicine. The teachings of Hippocrates and Galen, translated and preserved by Arab scholars during the Middle Ages, were followed blindly during the European Renaissance. Physicians stood by the theory that health requires a balanced state of bodily humors, and bloodletting was a common treatment for a variety of ills. Paracelsus opposed bloodletting and believed that a balance of chemicals—not humors—within the body maintains health. His prescription of minerals and chemicals shaped the development of his followers' methodologies.

Paracelsus's disciples developed new medicines and discovered methods of preparation and purification. Paracelsus stressed the importance of separating out the active ingredients of medicines and consequently purified a great many substances. His followers honed his experimental methods, which aided the development of pharmaceutical and experimental chemistry, both of which blossomed during the seventeenth century.

Among the best-known followers of Paracelsus are: Oswald Croll, who isolated succinic acid from amber; Leonhard Thurneysser, a famous alchemist; and J. Duchesne and Theodore Tourquet de Mayerne, who discovered benzoic acid.

Schuyler

World Events		Paracelsus's Life
Columbus discovers Americas	1492	
	1493	Theophrastus Bombastus von Hohenheim is born
	1507	Theophrastus travels across Europe
	1510	Theophrastus earns medical degree from University of Vienna
	c.1516	Theophrastus takes the name Paracelsus
Reformation begins	1517	
	1520s	Paracelsus works as army physician
		Paracelsus gains fame for healing powers
	1527	Paracelsus teaches at University of Basel
	1528	Paracelsus flees for his life after challenging university authority and begins to write medical treatises
	1530	Paracelsus develops clinical description of syphilis
	1541	Paracelsus dies

For Further Reading:
Connell, Evan S. *The Alchymist's Journal.* New York: Random House, 1995.
Pagel, Walter. *Paracelsus: An Introduction to Philosophical Medicine in the Era of the Renaissance.* New York: S. Karger, 1982.

Paré, Ambroise

Founder of Modern Surgery
1510-1590

Life and Work

Ambroise Paré was a devotedly humane sixteenth-century physician who introduced improvements in surgical techniques and whose work made surgery a distinct field of medicine.

Paré was born the son of an artisan in 1510 in Bourg-Hersent, France. About 1533, he went to Paris, where he apprenticed to a barber-surgeon at the hospital Hotel-Dieu.

In 1537, he began working as an army surgeon. For much of the rest of his life, he alternated between practicing surgery in Paris and living on the front lines ministering to French troops.

The treatment in Paré's time for gunshot wounds, which were thought to be poisonous, was an application of boiling oil. Paré did away with this method when he ran out of oil and was forced to improvise. He created a mixture of rose oil, egg yolk, and turpentine, and found that wounds treated with the mixture healed better than those treated with boiling oil. He published these findings in 1545 in *The Method of Treating Wounds Made by Harquebuses and Other Firearms*.

Paré contributed other innovations to surgical practice. He advocated the ligature (or tying), rather than the cauterization, of blood vessels to prevent hemorrhaging during amputation. He built artificial limbs, invented various surgical devices, and abandoned the practice of castration for patients receiving hernia surgery.

Paré became swiftly renowned for his great practical skill in treating patients. In 1552, he became surgeon to the French king, Henry II, and subsequently served three others: King Francis II, King Charles IX, and King Henry III.

Paré is said to have used surgery only as a last resort, unlike his contemporaries. He was honored for his humility and dedication, and is reported to have claimed that, although he provided the treatment for an illness or injury, God brought about the cure.

Paré died in Paris on December 20, 1590.

Legacy

Paré's surgical innovations influenced the practice of medicine in his lifetime and resulted in the dismissal of many outdated methodologies and ideologies.

Paré's improved surgical techniques and medical inventions saved the lives of numerous French soldiers and civilians and decreased the suffering of many others.

Unlike many of his contemporaries, Paré wrote in French, because he did not know Latin; he was therefore able to reach the general public. Because the French were able to read his scientific treatises, Paré became quite popular. Many of his innovations were adopted by other physicians, possibly because of patient demand.

Paré's focus on surgical procedures led to the development of surgery as a distinct field within medicine. During Paré's time, surgery was primarily performed by barbers whose services were often requested by physicians who needed some type of invasive operation performed. By the eighteenth century, as medicine and associated understandings in the fields of anatomy and physiology advanced, surgery advanced also. Since physicians understood the internal structure and function of the body more fully, they could increase the speed with which operations were performed. And, in 1846, surgery made a great leap forward with the introduction of ether anesthesia by American dentist WILLIAM MORTON.

Today surgical techniques have greatly advanced including highly refined and exact means of making incisions, including laser surgery.

Schuyler

WORLD EVENTS		PARÉ'S LIFE
Columbus discovers Americas	1492	
	1510	Ambroise Paré is born
Reformation begins	1517	
	c. 1533	Paré travels to Paris and apprentices at Hotel-Dieu, a hospital in Paris
	1537	Paré first works as army surgeon
	1545	Paré's influential book on gunshot wound treatments is published
	1552	Paré becomes surgeon to King Henry II and subsequently to three other French kings
	1590	Paré dies
Thirty Years' War in Europe	1618-48	

For Further Reading:

Doe, Janet. *A Bibliography of the Works of Ambroise Paré*. Chicago: University of Chicago Press, 1937.

Hamby, Wallace B. *Ambroise Paré: Surgeon of the Renaissance*. St. Louis, Mo.: W.H. Green, 1967.

Packard, Francis R., ed. *Life and Time of Ambroise Paré, 1510-1590*. North Stratford, N.H.: Ayer Company, 1972.

Pascal, Blaise

Pioneer in Geometry, Probability, and Hydrostatics;
Inventor of Calculating Machine
1623-1662

World Events	Pascal's Life	
Thirty Years' War 1618-48 in Europe		
	1623	Blaise Pascal is born
	1639	Pascal proves important geometry theorem on a hexagon in a conic
	1642	Pascal devises first digital calculating machine
	1648	Pascal tests barometric pressure
		Pascal formulates law of fluids and pressure (called Pascal's Law)
	c. 1654	Pascal and Pierre de Fermat establish mathematics of probability
	1655	Pascal enters Jansenist retreat
	1662	Pascal dies
	1663	Pascal's *Pensées* is published
England's Glorious Revolution	1688	

Life and Work

Blaise Pascal, a mathematical genius of the seventeenth century, conducted theoretical work in projective geometry and formulated the mathematics of probability. He also performed experiments in the fields of hydrostatics and barometry.

Pascal was born on June 19, 1623, in Clermont-Ferrand, France, the son of noted mathematician Étienne Pascal, who provided most of the boy's education. His mother died when he was four, and his sisters, who were devoutly religious, had a strong influence on his upbringing. The family adhered to Jansenism, a strict reformist movement of the Roman Catholic Church (which condemned it), and religion played a significant role throughout Pascal's life.

Pascal exhibited mathematical brilliance as a youth; at age 16 he proved one of the most important theorems of projective geometry, now sometimes called "Pascal's mystic hexagram." The theorem states that, for a hexagon inscribed in a conic, the intersections of the three pairs of opposite sides are collinear. He later completed more advanced work in geometry.

At age 19 he devised the first digital calculating machine to help his father, who had been appointed tax collector of Rouen in 1640. The instrument consisted of a series of connected wheels inscribed with numbers one through 10; the first wheel's numbers represented units, the second wheel's numbers represented tens, and so on. He built about 50 machines, and a standard model was in production by 1652. It made Pascal famous, but it was not a commercial success.

In 1648 Pascal tested the theory behind Evangelista Torricelli's mercury-column barometer, invented five years earlier. He enlisted his brother-in-law to take a mercury column up a mountain to an altitude of 1,200 meters. The air pressure is lower at the top of a mountain than at sea level, and the mercury height dropped. This confirmed Torricelli's proposition that atmospheric pressure regulates the height of the mercury column. The same year Pascal completed experiments that yielded a universal physical principle of fluid (now Pascal's Law), stating that when pressure is exerted on a confined liquid, the pressure is constant in all directions.

About 1654 Pascal and mathematician Pierre de Fermat developed the mathematics of probabilities and combinatorial analysis. Pascal was the first to use the "Pascal triangle" (the triangular arrangement of the coefficients of successive integral powers of a binomial) for calculating probabilities.

In 1655, on the heels of two profound religious experiences, Pascal entered a Jansenist retreat where one of his sisters was a nun. His health had been weak since his youth; he died at age 39 on August 19, 1662, in Paris. His philosophical treatise, *Pensées* ("Thoughts") appeared in 1663.

Legacy

Pascal's scientific legacy consists of his accomplishments in geometry and probability, his explanation of barometric pressure changes, and his successful experiments with pressure and fluids.

Portions of Pascal's work in geometry were influential in the formulation of differential and integral calculus, which was consolidated in the eighteenth century and which served as the basis for much subsequent progress in mathematics and physics.

Combinatorial analysis, co-founded by Pascal, was crucial to the development of the mathematics involved in computers in the twentieth century. The modern computer language Pascal is a testament to this seventeenth-century contribution.

When Pascal's brother-in-law took a mercury column up a mountain, he not only demonstrated that air pressure is responsible for the height of the mercury column, but also confirmed that air pressure decreases with increasing altitude. This showed that the atmosphere has a finite height, and when it was understood that air becomes less dense with increasing altitude, the concept that "outer space" is a vacuum took hold.

Pascal's Law established the field of hydrostatics, a branch of physics dealing with the properties of fluids, especially in relation to pressure dynamics. The invention of hydraulic devices, which use liquid under pressure to apply force, emerged directly from the development of this field, yielding the hydraulic press, the hydraulic jack, and automobile brakes.

Schuyler

For Further Reading:

Adamson, Donald. *Blaise Pascal: Mathematician, Physicist, and Thinker About God.* New York: St. Martin's Press, 1995.

Davidson, Hugh. *Pascal and the Art of the Mind.* New York: Cambridge University Press, 1993.

Krailsheimer, A. J. *Pascal.* New York: Oxford University Press, 1980.

Pasteur, Louis

Pioneer Microbiologist;
Developer of Pasteurization
1822-1895

Life and Work

Louis Pasteur played a central role in the fields of chemistry and microbiology in the nineteenth century. He studied the arrangement of atoms within compounds, examined the process of fermentation, and developed vaccines against infectious diseases.

Louis Pasteur was born on December 27, 1822, in Dôle, eastern France. He was a mediocre student in primary school but showed promise at painting, and his early ambition was to teach art. However, by 1843 his university work was good enough to place him at the École Normale Supérieure, a teacher-training school in Paris, from which he received his doctorate in chemistry in 1847.

Pasteur's early research concerned tartaric acid, a by-product of wine making. He analyzed a solution of tartaric acid that was optically inactive (did not rotate polarized light). He found that the solution contained two kinds of tartaric acid molecules that were mirror images of each other. The molecules were stereoisomers, alternative forms of the same compound in which the atoms are arranged differently. Pasteur separated the two kinds and

determined that they had distinct properties; he thus founded stereochemistry, the study of the arrangement of atoms within compounds.

In the early 1860s, Pasteur studied fermentation in beer and wine making. In 1864 he found that souring of a fermented product can be prevented if it is heated to kill harmful microscopic agents. This began the practice of heating foods to kill harmful microorganisms, particularly tuberculosis bacilli in dairy products, a process now called pasteurization after its inventor.

In the 1880s Pasteur focused on microbial virulence, borrowing ideas from the success of EDWARD JENNER, who in the 1790s had developed a vaccine against smallpox by injecting people with weakened cowpox bacteria. This gave Pasteur the idea that a "safe attack" on disease could be developed using weak cultures of specific germs to vaccinate healthy animals. Serums he developed against chicken cholera, anthrax, and rabies brought him international renown.

In 1888, Pasteur established the Pasteur Institute, an agency devoted to biological research, in Paris. He headed the institute until he died in Paris on September 28, 1895.

Legacy

Pasteur's scientific insight and innovative experimentation contributed to improvements in human health and paved the way for further progress in the battle against infectious diseases.

Prior to pasteurization, health problems related to food spoilage were rampant. As pasteurization became accepted, the storage time of milk increased and the incidents of food poisoning decreased.

Pasteur's fermentation research and the development of pasteurization influenced surgical procedures during his lifetime. British surgeon JOSEPH LISTER read Pasteur's publications in 1865 and hypothesized that microorganisms cause infection in wounds, just as microorganisms cause souring in liquids. Lister introduced antiseptic procedures to combat infection and sepsis. His innovations spread through Europe quickly, and the incidence of fatal post-surgical infection decreased drastically.

Pasteur founded the field of stereochemistry, the study of the three-dimensional structure of molecules and compounds. The concepts

underlying stereochemistry are important in the pharmaceutical industry, as exemplified by the drug thalidomide. Thalidomide occurs in two mirror-image stereoisomers, only one of which is harmful to a developing fetus. The other is completely benign. Stereochemical analysis is applied in fields beyond medicine, such as material technology, to help understand how molecules and compounds will interact with each other.

Pasteur's final legacy was left with those who have fought the war against infectious diseases. Pasteur provided a general theory to guide physicians' research: the concept that infectious diseases are caused by microbial agents. Pasteur also established standard experimental techniques subsequently used by others to identify treatments for infectious diseases. His revolutionary work with anthrax, cholera, and rabies laid the groundwork for the development of numerous other vaccines against debilitating diseases.

Schuyler

WORLD EVENTS	PASTEUR'S LIFE
Napoleonic Wars 1803-15 in Europe	
	1822 Louis Pasteur is born
	1847 Pasteur receives doctorate from the École Normale Supérieure
	1850s Pasteur studies stereoisomers
	1860s Pasteur studies fermentation
	1864 Pasteur introduces pasteurization
United States 1861-65 Civil War	
	1865 Joseph Lister applies Pasteur's ideas to antiseptic surgery
Germany is united 1871	
	1880s Pasteur studies infectious diseases and develops vaccines
	1888 The Pasteur Institute is founded
	1895 Pasteur dies
Spanish-American 1898 War	

For Further Reading:

Dubos, René. *Louis Pasteur: Free Lance of Science.* New York: DaCapo, 1986.

Geison, Gerald L. *The Private Science of Louis Pasteur.* Princeton, N.J.: Princeton University Press, 1995.

Pauling, Linus

Pioneer Researcher in Nuclear
and Molecular Structure
1901-1994

Life and Work

Linus Pauling was a prolific theoretical chemist who greatly contributed to the modern understanding of chemical bonding, nuclear structure, and protein structure. He was also active in efforts to halt nuclear weapons testing.

Pauling was born on February 28, 1901, in Portland, Oregon. He developed an interest in chemistry as a child, and in high school he conducted simple experiments with materials obtained from an abandoned metallurgy plant near his home. He received his doctorate from the California Institute of Technology (CIT) in 1925. During the following two years, he visited the laboratories of scientists working in the new field of quantum mechanics.

In 1927 Pauling began teaching and conducting research at CIT. He used X-ray diffraction and electron diffraction techniques to determine the molecular structure of scores of organic and inorganic compounds. In the 1930s, Pauling used the principles of quantum mechanics to develop three aspects of theoretical chemistry: hybridization, bond character, and resonance theory. These theories concern the behavior of electrons and the energy involved in chemical bonding. This work led to Pauling's most famous book, *The Nature of the Chemical Bond* (1939), and also to the 1954 Nobel Prize for Chemistry.

Pauling's work in late 1930s revolved around the structure of biological molecules. He found that the hemoglobin molecule changes shape when it acquires or unloads an oxygen atom. He further determined that hemoglobin's molecular structure involves a helix. In the 1940s, he discovered that a change in a single amino acid within the hemoglobin molecule is the cause of the debilitating disease sickle cell anemia.

In the 1950s Pauling became actively involved in efforts to promote peace and end the atmospheric testing of nuclear weapons. He spoke and wrote unflaggingly about the threat of nuclear armament to world peace and the dangers of nuclear fallout, authoring the 1958 book, *No More War!* Pauling's activism earned him the 1963 Nobel Peace Prize.

In the late 1960s Pauling began a crusade promoting the use of vitamin C for treating ailments ranging from the common cold to cancer. His claims of the efficacy of vitamin C in combating illness caused debates and controversy, which continue to the present. From 1964 through 1973, he held positions at several research and teaching institutions. Pauling died on August 19, 1994, near Big Sur, California.

Legacy

Pauling's extensive contributions to chemistry and his concern for the welfare of human society made him one of the most prominent and influential scientists of the twentieth century.

Pauling solved pressing problems in theoretical structural chemistry. His work formed the basis for further research into chemical bonding, electron orbitals, and atomic nuclear structure. His book, *The Nature of the Chemical Bond*, has been and still is used in college coursework and has directed many paths of chemistry research.

Pauling's achievements in biochemistry represented significant progress in the fields of molecular biology and medicine. His discovery that hemoglobin changes shape when interacting with oxygen molecules was a key step in understanding how hemoglobin functions in the body. Similar structural changes in other biological compounds have provided clues to unknown physiological processes. When Pauling identified a helix in the hemoglobin molecule, he revolutionized protein research. The helix was subsequently found to be one of the basic structures of proteins, and it surfaced again in 1953 as the long-sought structure of deoxyribonucleic acid (DNA).

With his vitamin C postulate, Pauling helped found orthomolecular medicine, a field of study based on the premise that extraordinarily large doses of minerals and vitamins can be used to prevent and treat diseases. In 1973 Pauling co-founded the Institute of Orthomolecular Medicine; it was later renamed the Linus Pauling Institute of Science and Medicine and is located in Palo Alto, California. The efficacy of orthomolecular medicine is debatable; research studies have not provided support for Pauling's assumptions.

Schuyler

WORLD EVENTS	PAULING'S LIFE
	1901 Linus Pauling is born
World War I 1914-18	
	1925 Pauling receives doctorate from California Institute of Technology
	1930s Pauling studies chemical structural theory, hemoglobin, and other biological molecules
	1939 *The Nature of the Chemical Bond* is published
World War II 1939-45	
	1950s Pauling becomes active in global peace efforts
Korean War 1950-53	
	1954 Pauling receives Nobel Prize for Chemistry
	1960s Pauling founds orthomolecular medicine
	1963 Pauling receives Nobel Peace Prize
	1973 Pauling co-founds the Institute of Orthomolecular Medicine
End of Vietnam War 1975	
Dissolution of 1991 Soviet Union	
	1994 Pauling dies

For Further Reading:

Goertzel, Ben and Ted. *Linus Pauling: A Life in Science and Politics.* New York: Basic Books, 1995.

Marinacci, Barbara, ed. *Linus Pauling in His Own Words.* New York: Touchstone, 1995.

Pavlov, Ivan

Originator of Theory of
Conditioned Reflexes
1849-1936

Life and Work

Ivan Pavlov conducted research on the physiology of the digestive system. He is most famous for his experiments on conditioned reflexes, in which he trained dogs' salivary glands to respond to the sound of a bell.

Pavlov was born on September 26, 1849, in Ryazan, in central Russia. His father, a priest and scholar, taught the boy to read and encouraged him to read all books at least twice. Pavlov's family expected him to enter the clergy, but while studying at divinity school, he developed an interest in physiology and CHARLES DARWIN's recently published theory of evolution.

Pavlov entered St. Petersburg University to study science and graduated in 1875. He earned a medical degree in 1879 and a doctoral degree in 1883 from the Military Medical Academy, St. Petersburg.

In 1890 Pavlov became director of physiology at a new Institute for Experimental Medicine in St. Petersburg. He performed experiments showing that the secretion of digestive juices does not require the presence of food in the stomach. He allowed dogs to see, smell, and swallow food, but then removed the food from their throats through surgery, and digestive juices were secreted just as if the food had reached the dogs' stomachs. He also studied the connections between nerves and gastric glands and the stimulation of the salivary glands. For his work on digestion, he received the 1904 Nobel Prize for Physiology or Medicine.

Pavlov published his classic work *Conditioned Reflexes* in 1907, and in 1910 he performed a set of experiments that led to the term "Pavlov's dog." He rang a bell each time certain dogs were given food and demonstrated that those dogs would later salivate at the sound of the bell, even without the presence of food. They had been trained (or "conditioned") to respond to a specific stimulus.

After the 1917 Bolshevik Revolution in Russia, Pavlov was an outspoken opponent of the new Communist regime, but his fame as an accomplished scientist protected him from the government suppression that silenced many of his colleagues. He died on February 27, 1936, in Leningrad, Russia.

Legacy

Pavlov advanced the understanding of animal physiology and influenced the development of theoretical and practical psychology.

Pavlov's work launched a field of psychological research exploring the influence of conditioned reflexes on learning, behavior, and neurosis. Pavlov's followers believed that conditioned reflexes were responsible for much human behavior and neurological disorder. They treated some mental patients by isolating them in quiet surroundings where no physiological or psychological stimuli could trigger conditioned responses. Such investigations led to a greater understanding of some psychopathological disorders, such as hysteria, obsessive neurosis, and paranoia.

A debate quickly arose between followers of Pavlov and SIGMUND FREUD. Freud taught that the individual's thought processes (particularly the unconscious) drive human behavior. While Freudian psychology has somewhat overshadowed Pavlovian ideas and shaped treatment of psychological distress in Western societies, Pavlov's ideas have still had a profound effect in many areas of psychological treatment.

Pavlov's influence led directly to the development of a school of psychological study and treatment called behavioral psychology, which was championed by American psychologist B. F. Skinner in the 1940s. Behavioral psychology analyzes human behavior as a set of responses to the individual's environment and behavior therapy focuses on reconditioning patterns of behavior by adjusting the rewards for certain types of actions.

Variations of such behavior-modification techniques using these principles have developed; these include everything from the benign assertiveness-training workshops developed in the 1970s in which outwardly expressive behavior is rewarded with positive feedback from a group, to the more controversial aversion-therapy techniques in which a painful chemical or electrical stimulus is delivered to a subject following undesirable behavior.

Pavlov pioneered the humane treatment of laboratory animals. He was among the first to maintaining healthy living conditions for them and to understand the problems associated with pharmaceutical use and with allowing disease to spread among a laboratory population. He instituted aseptic surgical procedures and believed in minimizing pain.

Schuyler

WORLD EVENTS	PAVLOV'S LIFE
1849	Ivan Pavlov is born
1879	Pavlov earns medical degree from Military Medical Academy in St. Petersburg
1890	Pavlov becomes director at new Institute for Experimental Medicine in St. Petersburg
Spanish-American War 1898	
1904	Pavlov receives Nobel Prize for Physiology or Medicine for work on digestion
1907	*Conditioned Reflexes* published
1910	Pavlov performs conditioned-response experiments on dogs
World War I 1914-18	
1917	Pavlov opposes post-revolutionary government
1936	Pavlov dies
World War II 1939-45	

For Further Reading:

Gray, Jeffrey Alan. *Ivan Pavlov.* New York: Viking Press, 1980.

Stephen Everson, ed. *Psychology.* New York: Cambridge University Press, 1991.

Pincus, Gregory

Developer of Oral Contraceptives
1903-1967

World Events	Pincus's Life
	1903 Gregory Pincus is born
World War I 1914-18	
	1927 Pincus receives Ph.D. in genetics and reproductive physiology from Harvard
	1930 Pincus starts teaching at Harvard
World War II 1939-45	
Korean War 1950-53	
	1951 Pincus is exposed to the ideas of feminist Margaret Sanger, who supports birth control
	1953 Pincus begins collaboration with Min-Chueh Chang
	1954 Pincus and Chang develop and test oral contraceptive
	1960 F.D.A. approves the Pill
	1967 Pincus dies
	1969 Barbara Seaman's *A Doctor's Case Against the Pill* is published
End of Vietnam War 1975	

Life and Work

Applying his research on the physiology of reproduction to the demand for better birth-control methods, Gregory Pincus developed the first oral contraceptives.

Pincus was born on April 9, 1903, in Woodbine, New Jersey, to a farming family. Following his parents' example, he studied agriculture as an undergraduate but moved to genetics and reproductive physiology at Harvard University, where he earned his doctorate in 1927. His graduate research concerned the links between hormone secretion and egg development, and the stages of fertilization and embryonic growth.

After studying in Europe with pioneers of endocrinology (the study of the hormonal system), Pincus returned to teach at Harvard in 1930. In 1938 he moved to Clark University, in Massachusetts, where he conducted research for the United States Air Force on the link between hormone secretion and the stresses of flying.

In 1951 Pincus was encouraged to explore the possibility of hormone-controlled contraception by feminist Margaret Sanger, a vocal advocate of controlling human population growth with the introduction of a new, effective birth-control method. In 1953 Pincus began collaborating with Min-Chueh Chang to investigate the effects on fertility of steroids, a family of substances including the reproductive hormones.

With the help of human-reproduction specialists, Pincus and Chang soon produced a pill containing synthesized progesterone and estrogen, the two main female reproductive hormones. This pill prevented (in laboratory animals) ovulation, or the release of a mature egg (ovum) from the ovary, by disrupting the cascade of hormone secretions that regulates the female reproductive cycle. In 1954 the researchers conducted clinical trials of their product in Haiti and Puerto Rico, showing it to be effective. They thus created a nearly 100% effective form of birth control.

Pincus worked and traveled during his later years, despite encroaching illness. He died on August 22, 1967, in Boston, Massachusetts.

Legacy

"The Pill," as Pincus's oral contraceptive came to be known, eventually became a popular, effective, and relatively safe contraceptive used around the world. It helped to liberate many women from the fear of unwanted pregnancy, allowing them more social and economic freedom than ever before.

The Pill was approved by the U. S. Food and Drug Administration in 1960; such an effective and convenient birth-control method created great excitement. Some observers argued that too few tests had been done to warrant the immediate wide-scale prescriptive use that ensued. By 1962 an estimated 1.2 million women in the United States were taking the Pill; by 1973 the number was 10 million.

By the late 1960s research began to reveal the potential dangers of Pincus's original formula for the Pill. High doses of synthetic hormones caused numerous negative effects, such as weight gain, skin problems, hypertension, headaches, depression, decreased libido (sexual drive), nausea, diabetes, urinary tract infections, and other complications. For women with certain pre-existing conditions, such as diseases associated with excess blood clotting, liver diseases, and breast cancer, the Pill was shown to be dangerous and, in some cases, lethal.

In 1969 physician Barbara Seaman published *A Doctor's Case Against the Pill.* The book brought to the public's attention the Pill's potential dangers and the lack of medical knowledge about the long-term effects of subjecting the body to daily doses of synthetic hormones.

Popular concern led to modifications of the Pill, including incorporating greatly reduced levels of hormones in its formula. Today, millions of women choose the Pill as a convenient, effective way to prevent unwanted pregnancies, although some dangerous side effects may still exist.

Pincus's work led to further development of hormone-based birth control, including several progesterone-only pills, implantation and injection of slow-release hormones, and single-dose abortion pills.

Hormone research also led to the 1965 discovery that estrogen therapy for post-menopausal women (whose estrogen levels decline with age) decreases the risk of osteoporosis, a debilitating disease that causes bone disintegration.

Schuyler

For Further Reading:

Asbell, Bernard. *The Pill: A Biography of the Drug That Changed the World.* New York: Random House, 1995.

Gunn, Alexander. *Oral Contraception in Perspective: Thirty Years of Clinical Experience With the Pill.* New York: Parthenon, 1987.

Planck, Max

Developer of Foundation for
Quantum Theory
1858-1947

Life and Work

Max Planck was a theoretical physicist who established the basis for quantum theory, one of the most important and far-reaching ideas of the twentieth century.

Plank was born on April 23, 1858, in Kiel, Germany. His family taught him the principles of loyalty, devotion to excellence, and dedication to education. Educated at one of the premier schools in Munich, where his father was appointed to the university, Planck became a well-rounded pupil who excelled in all subjects. Deciding finally on pursuing physics, Planck was only 21 when he received his doctoral degree from the University of Munich, where his work had centered on the laws of thermodynamics.

Planck was appointed to the University of Berlin in 1889, where he remained for the rest of his career. In the 1890s Plank began analyzing "black bodies," theoretical objects that emit amounts of energy equivalent to the energy they absorb. Attracted by the seeming perfect symmetry of emission and absorption in black bodies, Plank explored theoretical formulas for how the electromagnetic energy emitted by black bodies related to varying frequencies and temperatures. The best theoretical models did not fit the actual experimental results, and Planck wanted to find a way to bring the theory and results into a closer fit. Planck made a simple assumption that the energy of the atoms in the black body could only be transferred in or out in certain-sized pieces, depending on the frequency of the light given off or absorbed. He regretted the step because it was so arbitrary; there was no known physical reason why energy should "clump up" in that way.

However, his new theoretical model worked perfectly and matched the results of experiments very well. In 1900, at age 42, Planck published his results, which described how electromagnetic radiation was emitted and absorbed in little bundles, or quanta. For his new model, he developed what later became known as Planck's constant, which established the relationship between the energy content of the radiation and its frequency. He received the Nobel Prize for Physics for this work in 1918.

Planck's age, demeanor, and character, coupled with his successful theoretical work, elevated him to a revered status among his fellow scientists. In 1930 he became president of the Kaiser Wilhelm Society, the most prestigious scientific organization in Germany, which continues today as the Max Planck Society.

While other scientists fled Germany before World War II or took up the absurd scientific policies of the new Nazi government, Planck chose to remain in the country and do what he could to preserve German science. He died in Göttingen on October 4, 1947.

Legacy

Planck created the basis for quantum theory, one of the most important propositions in contemporary physics. His influence was both immediate and extensive, stimulating other scientists to confirm his results and build upon them.

ALBERT EINSTEIN, working independently, extended Planck's ideas, especially in relationship to light. In 1905 Einstein proposed the idea of wave-particle duality, stating that some entities have properties of both waves and particles: if light waves can act like particles, then particles of matter, like electrons, might have wave properties. Mathematician HENRI POINCARÉ established a mathematical proof for quantum theory in 1911, and in 1913 NIELS BOHR applied quantum theory to the hydrogen atom, thereby further establishing its validity in the scientific community.

Advances came quickly in quantum theory largely because Planck had gathered an unprecedented pool of brilliant scientists in Berlin. He brought Einstein there in 1914, along with other scientists. His support of other young scientists was often direct and of vital importance. LISE MEITNER secured her position at the Chemical Institute in Berlin because Planck spoke up for her to the Institute's director, and Einstein's first paper on relativity would not have been published without Planck's intercession with a publisher.

By the mid-1920s physicists had developed quantum mechanics, offering a full system of mechanics to explain the behavior of atomic and subatomic particles based on quantum theory. In several formulations, including MAX BORN's and WERNER HEISENBERG's wave matrix theory and Louis V. de Broglie's and Erwin Schrödinger's wave mechanics, quantum mechanics completely reshaped the understanding of the rules at play on the atomic and molecular level. And the discoveries keep coming; even today there is serious disagreement about what quantum theory means, and it continues to stretch our ability to grasp the real nature of the universe.

Secaur

WORLD EVENTS		PLANCK'S LIFE
	1858	Max Planck is born
United States Civil War	1861-65	
Germany is united	1871	
	1879	Planck receives doctoral degree from University of Munich
Spanish-American War	1898	
	1900	Planck publishes his paper on black body radiation, introducing idea of quanta
World War I	1914-18	
	1918	Planck earns Nobel Prize for Physics
	c. 1925	Quantum mechanics is established
	1930	Planck becomes president of Kaiser Wilhelm Society
World War II	1939-45	Planck chooses to remain in Germany during World War II
	1947	Planck dies

For Further Reading:

Hermann, A. *The New Physics: The Route into the Atomic Age.* New York: Verlag, 1979.

Keller, A. *The Infancy of Atomic Physics: Hercules in His Cradle.* New York: Oxford University Press, 1983.

Segrè, Emilio. *From X-rays to Quarks: Modern Physicists and the Discoveries.* San Francisco: Freeman, 1983.

Poincaré, Henri

Influential Modern Mathematician
1854-1912

Life and Work

Not only did Henri Poincaré understand a wide range of disciplines in mathematics in an era when mathematicians were becoming overspecialized, but like the great mathematicians before him, he made significant contributions in many areas. He also wrote on science and mathematics for the general public.

Poincaré was born on April 29, 1854, in Nancy, France. He was physically awkward with poor eyesight and a selective memory. By the time he was studying mathematics in school, he was unable to see the blackboard but could remember all that he had heard (but not whether he had eaten that day or not). Although much of the mathematics he later created was very visual, he learned it aurally.

Poincaré entered the École Polytechnique in Paris in 1873. Three years later he moved on to study engineering at the School of Mines. In 1879 he received his doctorate for work on differential equations. He spent two years on the faculty at the University of Caen and then in 1881 moved to the University of Paris. His commitment to teaching and to research continued until his death.

His early work was on differential equations, following up on his doctoral thesis. He also constructed and explored the properties of what he called a Fuchsian function, a transformation of a complex variable that sends a curve into itself. His papers on probability were influential and far reaching; he wrote a pioneering book on topology in 1895 called *Analysis Situ,* which looked at the continuity of properties in a universe that is stretched or distorted.

The relative motion of the planets has long fascinated mathematicians. Poincaré used differential equations to solve the *n*-body problem, an extension of the three-body problem of describing the gravitational effects of the Earth, Sun, and moon on each other. He concluded in 1898 that there is no place in the universe that is not moving, and in 1905 and 1906 Poincaré published a special theory of relativity independent of Albert Einstein.

Poincaré's mathematical ability had been recognized since he was young, and the honors and recognition continued. He was put in charge of the programs in mechanics and experimental physics at the University of Paris. In 1887 he was elected to the Academy of Sciences and would later serve as its president.

Toward the end of his life Poincaré began to write books for the general public such as *Science and Hypothesis* (1903), *The Value of Science* (1905), and *Science and Method* (1908). Because of the literary quality of these writings, Poincaré was elected to the Academie Francais in 1908. Poincaré died on July 17, 1912, in Paris.

Legacy

Poincaré's impact was extensive and immediate, since his talents were recognized early and his writing was so prolific. Beyond his immediate impact, however, were his contributions to the public's understanding of science arising from his writings on the processes of science and mathematics.

Poincaré influenced each field in which he worked. For example, his work in probability anticipated concepts used in statistical mechanics, a specialty in physics that uses statistical methodology and classical and quantum mechanics to explain the behavior of a system based on its constituent elementary particles; his explorations of Fuschian functions in the field of differential equations were influential in non-Euclidean geometry. One of his most direct influences arose from his book on topology, which offered specialists an early systematized treatment of the subject. Indeed, the title of the book—*Analysis Situ*—was the name given that specialty for a while.

Poincaré was the last of the global mathematicians in an age of specialization. His ability to consider mathematics as a whole is considered a key component of his genius. Poincaré's life and works are often studied today by psychologists, educators, and historians of science for clues to the psychology of discovery and induction that seem to be independent of logic. Poincaré himself described the stages of his problem-finding, problem-solving thinking in "Mathematical Creation" and other essays.

Poincaré wrote elegant prose that introduced the public to the importance of science and mathematics. His books, which were translated into many languages, demonstrate his desire to share with the public not just the results, but the processes of science.

Steinberg

World Events	Poincaré's Life
Napoleonic Wars 1803-15 in Europe	
	1854 Henri Poincaré is born
United States 1861-65 Civil War	
Germany is united 1871	
	1879 Poincaré receives doctorate for work on differential equations
	1881 Poincaré joins faculty of University of Paris
	1887 Poincaré elected to Academy of Sciences
	1895 Poincaré publishes work on topology, *Analysis Situ*
Spanish-American War 1898	
	1903 Poincaré publishes *Science and Hypothesis*
	1906 Poincaré completes paper on special relativity
	1908 Poincaré elected to Academie Francais
	1912 Poincaré dies
World War I 1914-18	

For Further Reading:

Bell, E. T. *Men of Mathematics.* New York: Simon & Schuster, 1937.

Callinger, Ronald. *Classics of Mathematics.* Englewood Cliffs, N.J.: Prentice-Hall, 1995.

Kline, Morris. *Mathematical Thought from Ancient to Modern Times.* New York: Oxford University Press, 1972.

West, Thomas G. "Genius with Mixed Abilities: Henri Poincaré." In *In the Mind's Eye.* Amherst, N.Y.: Prometheus Books, 1977.

Priestley, Joseph

Discoverer of Oxygen
1733-1804

Life and Work

Although trained as a minister, Joseph Priestley made a lasting contribution to the field of chemistry. His isolation of gases, most importantly oxygen, laid the groundwork for new theories and discoveries.

Priestley was born on March 13, 1733, near Leeds, England. Orphaned while young, he lived with his aunt who enrolled him in a rigorous school. He excelled in languages and went on to divinity school at the Dissenting Academy in Daventry. In 1761 Priestley taught at the Warrington Academy in Warrington, England. He instituted a curriculum incorporating science and modern literature that replaced a traditional syllabus based on classical readings.

Priestley was deeply religious and his religious beliefs played a prominent role throughout his life; his ideas evolved from his family's Calvinist beliefs to unorthodox interpretations of traditional Christian doctrine. In 1762 he was ordained as a minister in the Dissenting church, a group of denominations (including Presbyterian) that opposed the Church of England.

Priestley met American scientist, inventor, and statesman BENJAMIN FRANKLIN in 1766. It was this meeting that catalyzed Priestley's interest in electricity, one of Franklin's main areas of study. The discussion with Franklin resulted in Priestley's first published science book, *History of Electricity*, in 1767. During this time he became a member of the Royal Society because of his findings on electricity.

Priestley then focused his research on gases and their properties. At a nearby brewery, he noticed that certain gases were given off during fermentation, which led to his work with carbon dioxide. He identified its fire-extinguishing capabilities and found that it could be dissolved under pressure in water. Thus soda water was invented. He also studied the air quality around factories. Drawing on his previous work in electricity, Priestley found he could initiate chemical reactions that released gases. By sending electric charges through various compounds or by heating them, he isolated many new gases. Previously only three gases, hydrogen, carbon dioxide, and air (thought to be an element at that time), had been described, but during his lifetime Priestley identified nitrous oxide, sulfur dioxide, hydrogen chloride, and ammonia.

His use of a pneumatic trough in the early 1770s enabled him to collect gases released during reactions. He substituted mercury in the trough to capture gases that were soluble in water.

In 1774 Priestley began experiments that led to his discovery of oxygen. He used a magnifying lens to heat a mercury compound. It gave off a gas that bubbled through the liquid mercury into a glass tube. This new gas had several fascinating properties: it made a glowing splint burst into flame and it was given off by plants, a fact he recognized by doing experiments in closed containers of normal air. Other research demonstrated that this gas was beneficial to animals since they lived twice as long in closed vessels of this new gas than in ordinary air. He calculated that his new gas made up one fifth of the atmosphere. He called this new gas dephlogisticated air since it seemed to lack phlogiston, a particle scientists at the time believed was an essential component for combustion. His new gas was later renamed oxygen.

Priestley's political and religious views were as inflammatory as his newly discovered gas. He was forced to leave England and moved to America in 1794. He continued his research and found two other gases, nitrous oxide and carbon monoxide (1797). Priestley died on February 6, 1804, in Pennsylvania.

Legacy

Priestley's isolation and identification of oxygen and other gases provided scientists with clues about our atmosphere and combustion.

During his lifetime Priestley's influence was widespread. His discovery of soda water initiated a craze throughout Europe for the carbonated beverage. The refinements of the pneumatic trough enabled other researchers to capture water-soluble gases for nominal equipment costs, helping to demonstrate that sound experimental techniques did not require enormous amounts of expensive equipment. His addition of 10 more gases to the catalog of known substances dramatically furthered the study of chemistry by demonstrating that substances like ammonia can exist in both liquid and gas form. Priestley also connected chemistry and electricity, spawning a new branch of science, electrochemistry.

But it was the discovery of oxygen that made the biggest impact. It provided other scientists, namely French chemist ANTOINE LAVOISIER, with the key to understanding combustion, which requires oxygen. Lavoisier's careful experiments found that combustion occurred only when oxygen was present. The foundation Priestley laid for understanding the cycles of oxygen and carbon dioxide in living things eventually led to the contemporary knowledge about vital processes in the environment.

Priestley's legacy can be seen in many areas today. The discovery of hydrogen chloride gas paved the way for the development of products and applications based on chlorine, including bleach and chlorination. Chlorination currently provides safe drinking water for millions of people around the world. His air-quality testing is critical to our management of air pollution. Also, huge business conglomerates are based on his technology of carbonating water.

Wilson

WORLD EVENTS	PRIESTLEY'S LIFE
Peace of Utrecht 1713-15 settles War of Spanish Succession	
	1733 Joseph Priestley is born
	1766 Priestley meets Benjamin Franklin and becomes interested in electricity
	1767 Priestley publishes *History of Electricity*
	1774 Priestley discovers oxygen
United States 1776 independence	
French Revolution 1789	
	1794 Priestley leaves England for U.S.
	1797 Priestley discovers carbon monoxide
Napoleonic Wars 1803-15 in Europe	
	1804 Priestley dies

For Further Reading:

Asimov, Isaac. *A Short History of Chemistry.* New York: Doubleday, 1965.

Schofield, Robert E. *The Enlightenment of Joseph Priestley: A Study of His Life and Work from 1733 to 1773.* University Park: Pennsylvania State University Press, 1997.

Pythagoras

Early Pioneer in Mathematics
c. 580-c. 500 B.C.E.

Life and Work

Pythagoras was a Greek mathematician and philosopher whose ideas influenced the way we think about numbers. In addition to the Pythagorean Theorem, he introduced concepts in geometry and number theory.

Pythagoras was born around 580 B.C.E. on the Greek island of Samos. (There are no reliable contemporary records of his life, so dates cited are approximate.) He was probably a student of Thales (c. 640-546 B.C.E.), a merchant-philosopher who predicted the solar eclipse of 585 B.C.E. and who first recognized some mathematical patterns in the physical world. For example, Thales is credited with the proposals that any diameter bisects a circle and that angles of base/sides of an isosceles triangle are identical. Those generalizations were groundbreaking, going beyond mere measurement, which was the only way numbers were used before Thales.

Around 530 B.C.E. Pythagoras fled Samos, which was controlled by despotic rulers, and moved to Crotona, a Greek colony in southern Italy, where he began to teach mathematics and philosophy. He was a popular lecturer and his audiences included the upper class and, notably, women. His most loyal followers organized the secret Order of Pythagoras, symbolized by a five-pointed star. Because none of Pythagoras' original writings have survived, it is difficult to know which works were his and which were developed by his followers. It is certain that his influence on his students was profound, both in practice of mathematics and in attribution of mystical significance to numbers.

The Pythagoreans would represent numbers as dots in a symmetrical grid. Those that could be arranged as two rows of two dots, three rows of three dots, etc., were called square numbers because their visual representation was a square array and each could be calculated from its predecessor by adding an L-shaped border like a carpenter's square. They could see relationships in large arrays, such as the fact that to build onto a square with n^2 points, $2n+1$ points must be added to complete another square, and that there is always an odd number difference (5, 7, 9, 11 . . .) between successive squares (4, 9, 16, 25, 36 . . .).

Pythagoreans explored properties of odd vs. even numbers, primes (whose only divisors are one and itself), and perfect numbers (those that are both sums and products of their divisors). They noted that pleasant musical harmonies were possible when strings of different lengths were vibrated in particular whole number ratios. They assumed that the planets (including Earth) were moving while separated by lengths of similarly pleasing ratios producing the "harmony of the spheres."

The Pythagorean Theorem is probably older than Pythagoras but he and his followers are credited with its proof. The theorem states that the diagonal c of a rectangle with sides of lengths a and b is the square root of the sum of the squares of its sides: $c^2 = a^2 + b^2$. Applied to a square with side length of 1, the diagonal is the square root of 2, which is not a rational number (that is, it is not a ratio of two whole numbers). This insight challenged the Pythagorean assumption that all mathematical relationships would be whole numbers. Hippasus, one of the Pythagoreans, reputedly was thrown overboard while on a sea voyage as punishment for this discovery.

Pythagoras died around 500 B.C.E. The Pythagorean brotherhood continued to work for at least 50 years afterward.

Legacy

Pythagoras and his followers established the basis for much of the mathematics, particularly in number theory and geometry, used today.

Pythagoras and his followers developed mathematics in two ways. They made functional generalizations about whole numbers that we call laws and theorems today, and they developed them by seeking logical proofs. This marriage of mathematics and reason was a new way to find truth, one that contributed immeasurably to science and to civilization in general.

The Pythagorean assumption that each number has its own inner meaning and secret vibrations that are the secret to all knowledge, however, has led to some practices, based on superstition rather than science, that continue to the present day. Numerology, gematria (converting letters and words into numbers), the Elliot Wave Theory of stock market prediction, fear of the numbers 13 and 666, and Bible-numerists can trace their heritage to Pythagoras.

Not all of the ancient intuitions were superstition, however. Pythagoreans were among the first to assume that Earth itself moves around a central fire, a concept that NICOLAUS COPERNICUS acknowledged was helpful to him in formulating his Sun-centered model of the local planetary system. DMITRY MENDELEYEV assumed whole number atomic weights in assembling his successful periodic table in the nineteenth century, attributing fractional results to weighing errors when isotopes were unknown. Modern atomic numbers, products of twentieth-century science, do support Pythagoras' hunch that all matter can be described with whole numbers.

Steinberg

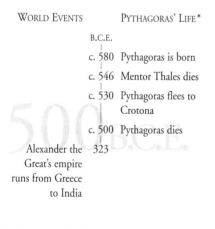

WORLD EVENTS	PYTHAGORAS' LIFE*
B.C.E.	
	c. 580 Pythagoras is born
	c. 546 Mentor Thales dies
	c. 530 Pythagoras flees to Crotona
	c. 500 Pythagoras dies
Alexander the 323 Great's empire runs from Greece to India	

* Scholars cannot date the specific events of Pythagoras' life with accuracy.

For Further Reading:

Dudley, Underwood. *Numerology: What Pythagoras Wrought.* Washington, D.C.: Mathematical Association of America, 1997.

Dunham, William. *Journey through Genius.* New York: John Wiley, 1990.

Kline, Morris. *Mathematical Thought from Ancient to Modern Times.* New York: Oxford University Press, 1972.

Meschkowski, Herbert. *Ways of Thought of Great Mathematicians.* San Francisco: Holden-Day, 1964.

Raman, Chandrasekhara Venkata

Pioneer Researcher in
Modern Optics
1888-1970

Life and Work

Chandrasekhara Venkata Raman, a pioneering researcher in light, optics, and sound, is best known for his discovery that the particles of a beam of light are scattered by the individual molecules of the substance through which the light is passing.

Raman was born on November 7, 1888, in Trichinopoly (now Tiruchirapalli), India. His father, a professor of physics, mathematics, and geography, collected books and played the violin, hobbies that Raman adopted also. Raman excelled at school and at age 16 won a scholarship to attend the Presidency College, from which he graduated first in his class with a degree in physics. He continued there as a graduate student, conducting independent experiments on the diffraction of light passing through rectangular slits. He sent a summary of his findings to a British scientific magazine and achieved publication at age 18. He earned a master's degree in 1907.

Because few opportunities for research were open to Indians in British colonial India, he worked as an accountant for the Indian government before becoming head of the physics department at Calcutta University in 1917.

At Calcutta Raman began investigating the scattering of light by matter. In 1928, he discovered the phenomenon that became known as the Raman effect: the particulate nature of light causes it to be scattered (or diffracted) by the molecules of a gas, liquid, or solid, and this scattering results in changes in the light's frequency (and color). This finding won him the 1930 Nobel Prize for Physics.

Raman also explored the physics of musical instruments; he studied the behavior and effects of the components of the violin, piano, tambla, and tambura.

Combining his interest in sound and light, he outlined the properties and effects of light as it passes through a liquid in which a sound wave is propagating. The explanation is known as the Raman-Nath theory.

Indulging his passionate interest in diamonds, Raman collected approximately 300 of them and used them in the study of optical phenomena such as light absorption, light scattering, and X-ray diffraction.

He established the Indian Academy of Sciences and founded the academy's journal in 1934. In 1948 he opened the Raman Research Institute, a privately funded institution for scientific research.

Raman died in Bangalore, India, on November 20, 1970.

Legacy

Raman's accomplishments represented several steps toward the modern understanding of the physics of light and sound, and his efforts gave rise to India's modern scientific community.

Raman lived at an opportune moment to influence India's scientific development. Until the twentieth century, most progressive endeavors in India were associated with literature, art, and architecture. The British colonialists encouraged the pursuit of science that would aid in furthering commerce, but they did not allow Indians much participation in that task. When the British introduced western education to India in the twentieth century, Indians became exposed to European science for the first time. Raman's founding of the Indian Academy of Sciences and its journal, and his training of hundreds of young Indian scientists were instrumental in establishing India's strong position in modern science.

The Raman-Nath theory became important in experimental physics, to studies of the passage of starlight through the atmosphere and the properties of laser beams sent through plasma.

The Raman effect explained the diffraction of light at the molecular level and led to advanced studies of the quantum theory of scattering light, investigations of various types of spectroscopy, and research into the physical and chemical properties of condensed matter.

Schuyler

WORLD EVENTS		RAMAN'S LIFE
	1888	Chandrasekhara Raman is born
Spanish-American War	1898	
	1906	Raman's first published research appears; it focuses on light defraction
World War I 1914-18		
	1917	Raman becomes head of Calcutta University's physics department
	1928	Raman discovers the Raman effect of scattering light
	1930	Raman wins Nobel Prize for Physics
	1934	Raman founds the Indian Academy of Sciences
World War II 1939-45		
	1948	Raman opens the Raman Research Institute
Korean War 1950-53		
	1970	Raman dies
End of Vietnam War	1975	

For Further Reading:

Jayaraman, A. *Chandrasekhara Venkata Raman*. New Delhi, India: Affiliated East-West Press, 1989.

Venkataraman, G. *Journey into Light: Life and Science of C. V. Raman*. New York: Oxford University Press, 1988.

Ar-Razi, Abu Bakr Muhammad

Pioneering Islamic Physician
of Middle Ages
c. 865-c. 935

WORLD EVENTS		AR-RAZI'S LIFE*
Charlemagne's coronation; beginning of Holy Roman Empire	800	
	c. 865	Abu Bakr Muhammad ar-Razi is born
	c. 895	Ar-Razi begins study of medicine
	c. 935	Ar-Razi dies
First Crusade	1095	
	1279	*Continens* is translated into Latin

* Scholars cannot date the specific events of Ar-Razi's life with accuracy.

Life and Work

Abu Bakr Muhammad ar-Razi, Persian physician, philosopher, and alchemist, wrote hundreds of papers demonstrating his extensive medical knowledge and astute clinical observations. His experimental practices helped advance medical science.

Ar-Razi was born about 865 in Rai, Persia (near modern Tehran, Iran). As a young student, he was first interested in music and he wrote a book called *On the Beauty of Music.* He is reputed to have had a career as a moneychanger.

When he was about 30 years old, he decided to study medicine after his first visit to a hospital in Baghdad. After an apprenticeship with a mentor in Rai, he became a hospital administrator in his hometown. Later he returned to Baghdad, took charge of the Muqtadiri Hospital as chief physician, and began his habit of writing detailed treatises on medical, alchemical, and philosophical topics.

Among Ar-Razi's numerous treatises are a monograph on diseases in children, which is the basis for calling him the father of pediatrics; an enormous encyclopedia of Greek, Syrian, and Arabic medical doctrines and his own conclusions called *System of Medicine;* and the *Continens,* a basic medical text that was first translated into Latin in 1279 and became a standard reference in Middle Eastern and European medical colleges.

He was the first to recognize smallpox and measles as separate diseases; to go beyond use of plants as medicinals (for example, he was first to use mercury as a purgative, white lead ointment for eye infections, and animal gut for surgical thread); and to describe cataract extraction and the reaction of the pupil of the eye to light.

Despite his excellence in clinical medicine, Ar-Razi became controversial because of his alchemical and philosophical ideas. Notable were his views that all men are equal, miracles are impossible, religion is unnecessary, and science is subject to change as experimental evidence accumulates.

Unlike other Arabic physicians of his time, he questioned reliance on Greek and Roman authorities, in particular the views of Aristotle and GALEN that had become dogma.

Ar-Razi performed original investigations on specific gravity, and he supported the atomic ideas of ancient Greek philosopher Democritus (c. 500 B.C.E.) that the universe is composed of minute basic particles. He introduced into modern thinking the idea of classifying matter as animal, vegetable, or mineral, and he further recognized that the last category has both volatile and non-volatile components.

Ar-Razi died around 935 in Rai after blindness slowed his extraordinary literary production.

Legacy

Ar-Razi's legacy rests in the portion of his works that were preserved and translated after his death, especially those on medical and scientific topics. For centuries his medical texts were required reading for student physicians, and his philosophical writings have been reviewed by scholars from the Middle Ages to modern times.

Ar-Razi's insistence on empirical evidence as the basis of scientific inquiry predates by 800 years the similar views of FRANCIS BACON, seventeenth-century thinker whose work influenced the development of scientific methodology. Ar-Razi is credited with transforming Islamic alchemy into the science of chemistry. He described many chemical techniques and processes clearly, and he developed atomic explanations for differences among the basic elements recognized during his time (earth, air, fire, and water).

Willing to experiment and to use inorganic materials in medical treatments, he was more science-oriented than a later, more celebrated philosopher and physician, IBN SINA (Avicenna). Ar-Razi still ranks among the most influential physicians and thinkers of the Islamic world.

Simonis

For Further Reading:

Arberry, Arthur. *The Legacy of Persia.* Oxford, England: Clarendon Press, 1953.

Elgood, Cyril. *A Medical History of Persia.* Cambridge: Cambridge University Press, 1951.

Roentgen, Wilhelm

Discoverer of X-rays
1845-1923

Life and Work

Wilhelm Roentgen is most famous for discovering X-rays; he also conducted research in diverse fields of physics.

Roentgen was born on March 27, 1845, in Lennep, Prussia (now Remscheid, Germany). In 1864 his college education was interrupted by expulsion from the Utrecht Technical School when he confessed to having drawn a caricature of an unpopular instructor. After a brief enrollment at the University of Utrecht, Holland, he continued at the Swiss Federal Institute of Technology in Zurich, Switzerland, and received a mechanical engineering degree in 1868 and a physics doctorate in 1869.

In 1875 Roentgen obtained his first academic appointment, at Hohenheim, subsequently filling teaching posts at Strasbourg, Giessen, and Würzburg. He studied the specific heats of gases, the heat conductivity of crystals, the compressibility of solids and liquids, and the production of electricity in crystals by heat and pressure.

In 1894, while at Würzburg, Roentgen began to investigate the nature of cathode rays. In the 1870s English physicist William Crookes had shown that when an electrical current is discharged within a vacuum tube, negatively charged radiation (cathode rays, later shown to be streams of electrons) is produced and causes the tube's glass walls to glow, or fluoresce.

Attempting to observe this fluorescence more clearly, Roentgen dimmed the lights and wrapped his vacuum tube in black paper. When the electrical current was turned on, a chemical-covered screen across the room began to glow. As cathode rays travel only a few centimeters in air, Roentgen knew that another type of radiation was responsible. Exploring further, he found that the mysterious radiation was strong, that it passed easily through wood and glass (but not through thick metal), that it was not deflected by electric fields, and that it could expose photographic plates. He named the radiation "X-rays."

In December 1895 Roentgen announced his discovery, described physical properties of X-rays, and publicized an X-ray image of his wife's hand. After 1897 Roentgen paid little attention to X-rays, choosing not to patent his discovery because of his belief that scientific inventions belong to all people. He was awarded the Nobel Prize for Physics in 1901 for the discovery and donated the proceeds of the award to the University of Würzburg.

Roentgen taught at the University of Munich, Germany, from 1900 to 1920. He died on February 10, 1923, at his home near Munich.

Legacy

Roentgen's identification of X-rays had a momentous impact on the progress of physics research, the development of structural chemistry, and the initiation of important medical practices.

The advent of X-rays inspired extensive research. In 1896 French physicist HENRI BECQUEREL discovered radioactivity while performing experiments with fluorescence and photographic plates. MARIE CURIE and her husband Pierre continued Bequerel's work in 1898 with the discovery of new radioactive substances. In 1912 German physicist Max von Laue showed that X-rays consist of electromagnetic radiation similar to visible light but with a much shorter wavelength. The X-ray experiments of British physicist Henry Moseley, one of von Laue's students, linked the chemical behavior of elements to their atomic structure.

Von Laue's work led directly to the 1913 development of X-ray crystallography by the English physicists William and Lawrence Bragg; X-ray crystallography is a powerful tool for determining the structure of molecules. British crystallographers Dorothy Hodgkin and Kathleen Lonsdale used the technique (independently) to analyze the structure of complex biochemical molecules and organic crystals. In the 1950s the elucidation of the structure of DNA was attributable in part to the X-ray crystallography work of British physicist ROSALIND FRANKLIN.

X-rays proved to have several medical applications. They were first used for imaging bone structure, and they continue to be a valuable diagnostic tool. In the 1960s scientists introduced X-ray tomography, the viewing of soft body tissue by the combination of many X-ray images taken at various angles. The modern version of this technique, used to locate internal abnormalities in soft tissue, is known as computer-assisted tomography (CAT).

Schuyler

WORLD EVENTS		ROENTGEN'S LIFE
	1845	Wilhelm Roentgen is born
United States Civil War	1861-65	
	1869	Roentgen earns physics doctorate from Swiss Federal Institute of Technology
Germany is united	1871	
	1895	Roentgen announces discovery of X-rays
	1896	Henri Becquerel discovers radioactivity
Spanish-American War	1898	
	1901	Roentgen wins Nobel Prize for Physics
	1912	Max von Laue determines nature of X-rays
	1913	William and Lawrence Bragg develop X-ray crystallography
World War I	1914-18	
	1923	Roentgen dies

For Further Reading:

Claxton, Keith. *Wilhelm Roentgen.* London: Heron Books, 1970.

Keller, Peter A. *The Cathode-Ray Tube: Technology, History, and Applications.* New York: Palisades Press, 1991.

Michette, A., and C. Buckley. *X-ray Science and Technology.* Philadelphia, Penn.: Institute of Physics Technology, 1993.

Niske, Robert W. *The Life of Wilhelm Conrad Roentgen, Discoverer of the X-ray.* Tucson: University of Arizona Press, 1971.

Russell, Bertrand

Originator of Logicism
1872-1970

Life and Work

Bertrand Russell made great contributions to mathematics, particularly in the field of logic. His work extended into other areas, including philosophy, economics, and politics.

Russell was born in Trelleck, England, on May 18, 1872. His mother died when he was two and his father when he was three. His parents, friends of liberal philosopher John Stuart Mill, were outspoken free thinkers who favored women's suffrage and birth control. Russell's

WORLD EVENTS		RUSSELL'S LIFE
Germany is united	1871	
	1872	Bertrand Russell is born
Spanish-American War	1898	Russell joins Trinity as lecturer and fellow
	1900	Russell meets Giuseppe Peano
	1910-13	Russell and Alfred North Whitehead publish *Principia Mathematica*
World War I 1914-18		
	1920s -30s	Russell writes prolifically for lay audience
World War II 1939-45		
	1944	Russell reinstated at Trinity
	1950	Russell receives Nobel Prize for Literature
Korean War 1950-53		
	1970	Russell dies
End of Vietnam War	1975	

grandparents, who did not share such views, raised the boy in their own house.

Russell was privately educated. He read many of the books in his grandfather's library and at age 11 encountered EUCLID and became interested in the content and structure of geometry. Between the ages of 14 and 18, Russell was very interested in religion and understanding free will, immortality, and God. He discarded these one at a time, becoming an agnostic after reading Mill's explanation of the difficulty in proving God's existence.

In 1890 he entered Trinity College at Cambridge University and after graduating traveled to Germany and wrote *German Social Democracy*. In 1898 he returned to Trinity as a fellow and lecturer.

In 1900 Russell met Giuseppe Peano, considered to be the founder of symbolic logic, who was building mathematics on precise axioms, which Russell felt extended backward to philosophical vagueness. Russell insisted that the logic itself needed to be placed on stronger and more precise footing. A little over a year later, Russell disproved mathematician Georg Cantor's proof of no greatest cardinal number with what was later called Russell's Paradox, a form of logical fallacy known as a "vicious circle" in which an entity is used in the definition of the entity itself.

In 1902 he started to outline his ideas on logic in *The Principles of Mathematics* and began his work with mathematician Alfred North Whitehead on *Principia Mathematica*. Published in three volumes between 1910 and 1913, *Principia* suggested that mathematics arises solely from the rules of logic and uses these rules to explain, extend, and correct number theory, set theory, and logic as developed by Peano, Cantor, and other mathematicians.

Russell was an outspoken pacifist (a position that changed as the threats of fascism and World War II grew). The opinions he voiced during World War I got him fired from Trinity, fined, and put in prison. While incarcerated he wrote the 1919 work *Introduction to Mathematical Philosophy*. Russell's goal was to formalize his view that "Mathematics, rightly viewed, possesses not only truth, but supreme beauty ... [it is] sublimely pure, and capable of a stern perfection such as only the greatest art can show."

Russell visited and taught in Russia and China in 1920 and 1921. During the 1920s and 1930s, he wrote extensively for a lay audience on a variety of topics, including *The ABC of Atoms* in 1923, *The ABC of Relativity* in 1925, and *The Scientific Outlook* in 1931. Just before World War II he was made a philosophy professor at the City College of New York. A woman challenged his appointment on the grounds that he advocated free love, and a member of the New York State Supreme Court revoked Russell's appointment.

In 1944 Russell was re-appointed to Trinity College; in 1950 Russell received the Nobel Prize for Literature. He remained an activist and was arrested at an antinuclear demonstration when he was 89. He died on February 2, 1970, in Wales.

Legacy

Russell's work in mathematics laid the foundation for logicism, the theory that all mathematics is rooted in logic. It also provided a basis for important philosophical trends of the twentieth century.

First suggested by the German mathematician Gottlob Frege in the late 1800s, logicism greatly influenced mathematics in the first half of the twentieth century. Russell, along with Whitehead, extended and spread Frege's ideas about the fundamental nature of logic. Russell's corrections and elaborations on existing postulates of logic refined ideas on number theory, set theory, and other mathematical concepts that later influenced not only mathematics but a variety of subjects, including computer science.

In the 1920s Russell's logicism led to a broader philosophical application known as logical positivism. This doctrine suggested that the only true factual knowledge is scientific in nature, and that such knowledge can only be derived from empirical inquiry. By the middle of the twentieth century, logical positivism, in turn, led to the philosophical school known as analytical philosophy, which suggests that philosophical investigation deals largely with the constructs of language.

Russell's writings aimed at the layman helped to explain complex scientific and mathematical concepts to a wide audience.

Schuyler

For Further Reading:

Dunham, William. *Journey through Genius*. New York: John Wiley, 1990.

Kline, Morris. *Mathematical Thought from Ancient to Modern Times*. New York: Oxford University Press, 1972.

Russell, Bertrand. "My Mental Development." In *The World of Mathematics*. Edited by James R. Newman. New York: Simon & Schuster, 1956.

Rutherford, Ernest

Discoverer of the Atomic Nucleus
1871-1937

Life and Work

Ernest Rutherford was the first to discover that every atom is hollow, with a tiny but dense nucleus at its center.

The fourth of 12 children, Ernest Rutherford was born to poor parents in Spring Grove, New Zealand, on August 30, 1871. A popular and athletic boy, he won scholarships to an advanced high school and then to college in New Zealand. In 1895 he moved to the famous Cavendish Laboratory at Cambridge, England, where he immediately attracted the attention of the lab's director, J. J. Thomson, as well as the jealousy of some of the older scientists. A year later, the French physicist HENRI BECQUEREL discovered that uranium emitted rays that could expose photographic film (called radioactivity by MARIE CURIE in 1898), as did X-rays. Rutherford soon demonstrated that the uranium rays consisted of two types, which he named alpha and beta. He believed they were made of minute bits of matter, and he later showed that alpha particles were helium atoms, but without electrons.

His early success won the attention of many university leaders, and he was hired as a professor at McGill University in Montreal in 1898. By 1902 he and a young chemist, Frederick Soddy, had concluded that when an atom like uranium decays by emitting an alpha or beta particle, it converts to an atom of an entirely new element, which is also often radioactive. They identified several families of radioactive elements where each member turned into the next by radioactive decay, a discovery that won Rutherford the Nobel Prize for Chemistry in 1908.

Manchester University brought him back to England in 1907, where he continued his brilliant but simple experiments. The most famous began in 1911, when Rutherford used alpha particles from radium as bullets to probe into a thin gold foil. His student, Hans Geiger, had developed a gas-filled tube to count alpha particles, and together Rutherford, Geiger, and another student, Ernest Marsden, counted the particles penetrating the foil. If the then-current model of the atom as a spongy ball was correct, the alpha particles should have all blasted straight through and, in fact, most of them did. But the team was amazed to find that a very few alpha particles actually bounced away at large angles, sometimes even ricocheting straight back. Rutherford's explanation was that all of the positive charge and virtually all of the mass of an atom were concentrated in a tiny spot at the center, the nucleus.

In 1919 Rutherford produced the first nuclear reaction when be bombarded nitrogen atoms with alpha particles, producing oxygen and hydrogen. In that same year he returned to the Cavendish Lab to serve as the new director. He was knighted in 1914, made a baron in 1931, and died in Cambridge on October 19, 1937.

Legacy

Rutherford's discovery that atoms have nuclei was a groundbreaking step in science for the twentieth century.

Like all scientific discoveries Rutherford's produced new questions. What kept the tiny electrons so far from the nucleus? Since opposite charges attract, why didn't the positive nucleus take in the negative electrons in an instant? It could not be that the electrons move too fast to fall in, because electrons circling that rapidly should radiate away large quantities of energy, but they do not. Fifteen years passed before Louis de Broglie and Erwin Schrödinger realized that electrons are not tiny specks, but extended waves that wrap around the nucleus, and their wavelength happens to be atom-sized, large compared to the tiny nucleus. In this way, Rutherford's work helped fuel the great discoveries of quantum theory, the greatest advance in physics in the twentieth century.

Rutherford's experiments with nuclear reactions and radioactivity suggested that nuclei could be manipulated and used for other purposes. In nuclear medicine, for example, radioactive compounds are used to diagnose and treat some illnesses. Rutherford had a gift for being right in his hunches and insights, and the only time he was certainly wrong was when he claimed in the early 1930s that releasing quantities of energy from nuclei was "pure moonshine." Nuclear weapons and power plants prove that we can cause some atoms, at least, to spew out huge amounts of energy.

Perhaps Rutherford's greatest legacy was the swarm of students whose work he encouraged. He was always willing to share with students, and most scientists who contributed to nuclear physics in the first half of the twentieth century had worked with Rutherford.

Secaur

WORLD EVENTS		RUTHERFORD'S LIFE
Germany is united	1871	Ernest Rutherford is born
	1895	Rutherford begins work at Cavendish Laboratory, Cambridge, England
	1896	Rutherford separates natural radioactivity into alpha and beta types
Spanish-American War	1898	
	1902	Rutherford and Frederick Soddy produce radioactive decay series
	1908	Rutherford receives Nobel Prize for Chemistry
	1911	Rutherford, Hans Geiger, and Ernest Marsden discover atomic nucleus
World War I 1914-18		
	1919	Rutherford produces first nuclear reaction, converting nitrogen to oxygen and hydrogen
		Rutherford becomes director of Cavendish Laboratory
	1937	Rutherford dies

For Further Reading:

Crowther, James Gerald. *Ernest Rutherford*. London: Methuen, 1972.

Moon, Philip Burton. *Ernest Rutherford and the Atom*. London: Priory Press, 1974.

Snow, C. P. *The Physicists*. Boston: Little, Brown, 1981.

Sakharov, Andrey

Developer of Soviet Hydrogen Bomb
1921-1989

Life and Work

Andrey Sakharov was one of the scientists who developed the Soviet Union's hydrogen bomb. Later in his career, he became a human-rights activist, an advocate for democracy in the Soviet Union, and one of the world's most outspoken critics of the nuclear arms race.

The son of a physicist, Sakharov was born May 21, 1921, in Moscow. He received his doctorate in 1947 and became a full member of the Soviet Academy of Sciences in 1953, at age 32, relatively young for Soviet scientists in that era. From 1948 to 1956 he conducted research on controlled nuclear fusion (the combining of atomic nuclei) and developed the theory necessary for the construction of a thermonuclear, or hydrogen, bomb during the arms race with the United States. Hydrogen bombs differ from the "atomic" bombs used at the end of World War II, which were based on nuclear fission (the

splitting of an atomic nucleus) and made of plutonium and uranium. Hydrogen bombs are vastly more powerful; small hydrogen bombs have about 50 times the explosive power of the original atomic bombs.

Soviet leader Nikita Khrushchev announced in 1961 the upcoming test of a 100-megaton bomb, equal to 100 million tons of explosives. Khrushchev planned the detonation of the largest bomb ever built to show the Soviet Union's superiority in the growing Cold War with nations of the West. Sakharov came out publicly against the test after his subtler efforts to discourage its use failed, but it went ahead despite his objections.

Some of Sakharov's most important work was in the theoretical physics of elementary (subatomic) particles. In 1967 he developed some of the key ideas of what has come to be called Grand Unified Theory, an attempt to understand the particles and laws of the universe at the most fundamental level.

By 1968 he had virtually abandoned his scientific work and turned to campaigns for political and intellectual freedom. In that year he published "Progress, Coexistence, and Intellectual Freedom," an essay calling for reductions in nuclear arms, cooperation with non-Communist countries, and freedom for imprisoned Soviet dissidents. In 1971 he married Yelena Bonner, also an activist for political freedom and human rights.

He was awarded the Nobel Peace Prize in 1975 but was not allowed to travel to Norway to accept it. His outspoken criticism in 1979 of his nation's invasion of Afghanistan exceeded the limit of his government's tolerance, and in January 1980 he was stripped of all his honors and sentenced to internal exile in the prison city of Gorky. Four years later his wife was convicted of anti-Soviet activities and joined him there. In 1986 the more tolerant government of Mikhail Gorbachev released them both, and permitted them to return to Moscow, where Sakharov continued to advocate for democratic reforms as a member of the legislature, to which he was elected in 1989. He died in Moscow on December 14, 1989.

Legacy

Sakharov's efforts in particle physics moved work on unified theories a step forward.

And the legacy of his activism against the deployment of the hydrogen bomb and advocacy of disarmament and human rights outlives his contribution to the Soviet Union's development of the hydrogen bomb.

The debate over nuclear weapons and fear of their use has colored much of life in the second half of the twentieth century; for the first time in history, humankind holds the key to its own swift destruction. However, largely because of the efforts of Sakharov and other anti-nuclear activists, treaties have been signed to reduce the numbers of weapons worldwide. Continuing to control nuclear armaments and safely destroying the thousands of remaining weapons will be an important diplomatic function and an expensive burden well into the twenty-first century.

Sakharov's treatment as a political dissident in the Soviet Union became a lightening rod for human-rights activists. His continued outspoken advocacy for human rights, even as he was a victim of human-rights abuses, empowered less prominent activists across the world to continue and expand their struggle for political and intellectual freedom.

Sakharov's greatest legacy may be his work in the physics of elementary particles. His work on the Grand Unified Theory has laid the groundwork for further attempts to explain the fundamental forces and particles of nature in a single theoretical model. Testing and refining these theories will be one of the great challenges of the twenty-first century.

Secaur

World Events	Sakharov's Life
World War I 1914-18	
	1921 Andrey Sakharov is born
World War II 1939-45	
	1948-56 Sakharov develops principles of hydrogen bomb
Korean War 1950-53	
	1961 Sakharov protests testing of huge hydrogen bomb
	1967 Sakharov works on ideas in Grand Unified Theory
End of Vietnam War	1975 Sakharov wins Nobel Peace Prize
	1980 Sakharov sentenced to internal exile for criticism of Soviet government
	1986 Sakharov is released from exile and returns to Moscow
	1989 Sakharov is elected to legislature; he dies in Moscow
Dissolution of Soviet Union	1991

For Further Reading:

Babyonyshev, A. *On Sakharov.* New York: Knopf, 1982.

Bailey, George. *Galileo's Children: Science, Sakharov, and the Power of the State.* New York: Arcade, 1990.

Gell-Mann, Murray. *The Quark and the Jaguar.* New York: Freeman, 1994.

Sakharov, Andrey. *Sakharov Speaks.* New York: Knopf, 1974.

Semmelweis, Ignaz

Medical Reformer; Discoverer of
Cause of Puerperal Fever
1818-1865

Life and Work

Like his contemporary, FLORENCE NIGHTINGALE, Ignaz Semmelweis was a pioneer in the use of statistics as the basis for reform in medical practice. He found the cause for "childbed," or puerperal, fever, which was a major cause of death among women in the 1800s.

Semmelweis, a Hungarian physician, was born on July 1, 1818, in Buda (now Budapest, Hungary), in what was then the Austrian Empire. He received his medical degree from Vienna and in 1846 accepted a position as first assistant of the First Obstetrical Clinic at Vienna's General Hospital.

At the time of Semmelweis's appointment, 25 to 30% of women who delivered babies in the clinic died. They died of a disease known as puerperal fever, often referred to as "childbed fever." This infection was claiming the lives of new mothers and babies all over Europe. Most doctors at the time assumed it to be a contagious disease similar to smallpox.

Semmelweis was alarmed by the number of lives the disease was taking in his clinic and began making careful observations. He noted that medical records showed that the mortality rate was much higher in the clinic where doctors delivered the babies than in the clinic where midwives delivered babies. Semmelweis also looked into the death of his former professor, Dr. Phillip Kolletchska, who had accidentally punctured himself while performing an autopsy on a patient who died of puerperal infection. The professor died of complications identical to those of mothers who had died of the disease.

Putting his observations together, Semmelweis concluded that the disease was being passed from deceased mothers to mothers in the delivery ward on the hands of the doctors. He immediately initiated a policy of having all doctors wash their hands in a chlorine solution that would kill the agents of the disease, especially after performing autopsies. The number of deaths in Semmelweis's ward dropped from 18% to 1% in 1848.

Other doctors did not accept his findings and were infuriated by the fact that he was telling them they had to wash their hands in chlorinated water. Semmelweis was dismissed from his job. He went to other clinics and initiated his hand-washing policy. In each case the number of deaths from puerperal fever dropped dramatically.

In 1855 he accepted an appointment in obstetrics at University of Pest, in Hungary, where he married and had a family. His ideas gained acceptance in Hungary, but throughout the rest of Europe, they were denounced. Semmelweis published his findings in 1861 in *Etiology, Understanding and Preventing Childbed Fever,* which he sent to many doctors and clinics outside of Hungary. However, most doctors still refused to accept his findings and to follow the hand-washing procedure.

After many years of fighting to implement his policies, Semmelweis suffered a breakdown in 1865 and was taken to a mental hospital. He died on August 17, ironically from puerperal fever contracted through a wound resulting from surgery he performed prior to his admission.

Legacy

Semmelweis's discovery of the cause of puerperal fever had a major effect on the longevity of women, and his solution for it would help to reform medical practice and save many lives.

During his life, Semmelweis's influence was aggressively countered by the medical establishment who resented the fact that they were ordered to wash their hands after autopsies. His influence, however, did not die with him. Some doctors saw the logic in his discoveries, and two scientists were particularly influenced by him: JOSEPH LISTER and LOUIS PASTEUR. Lister (1827-1912) was a British surgeon who advocated antiseptics and acknowledged Semmelweis's example. Pasteur (1822-1895) was a French chemist and biologist who proved the germ theory of disease, founded the science of microbiology, invented the process of pasteurization, and developed vaccines for several diseases.

Semmelweis's *Etiology, Understanding and Preventing Childbed Fever,* while not accepted during his lifetime, later was regarded as a milestone in medical history.

Semmelweis's legacy, however, goes beyond the field of medicine. While some learn from his medical findings, others learn from his life. Despite his well-grounded findings, Semmelweis was rejected, often with hostility. This has become known as the "Semmelweis reflex," referring to the tendency of authorities to greet a discovery with open hostility.

Weaver

WORLD EVENTS	SEMMELWEIS'S LIFE
Napoleonic Wars 1803-15 in Europe	
	1818 Ignaz Semmelweis is born
	1846 Semmelweis accepts a position as first assistant in Vienna's General Hospital
	1848 Women's mortality rates drop in clinic after Semmelweis establishes hand-washing policy
	1855 Semmelweis accepts appointment at University of Pest in Hungary after being dismissed from other posts because of resistance to his hand-washing policies
	1861 Semmelweis publishes *Etiology, Understanding and Preventing Childbed Fever*
United States 1861-65 Civil War	
	1865 Semmelweis dies
Germany is united 1871	

For Further Reading:

Carmichael, A. G., and R. M. Ratzan, eds. *Semmelweis and Hippocrates in Medicine: A Treasury of Art and Literature.* New York: Hugh Lauter Levin, 1991.

Carter, Kay. *Childbed Fever: A Scientific Biography of Ignaz Semmelweis.* Westport, Conn.: Greenwood Press, 1994.

Sholes, Christopher

Developer of the Typewriter
1819-1890

Life and Work

Christopher Sholes made mechanical improvements to an early prototype of the modern typewriter, producing the first practical typing machine.

Sholes was born on February 14, 1819, near Mooresburg, Pennsylvania. Little is known of his early years. He received a general education but no formal higher schooling, after which he apprenticed to a printer for four years.

Sholes moved with his parents to the newly established territory of Wisconsin and became editor of Madison's newspaper, the *Wisconsin Enquirer*. He later tried his hand at politics, serving in Wisconsin's legislature.

In the early 1860s Sholes edited two Milwaukee, Wisconsin, newspapers but was soon appointed by President Abraham Lincoln to the position of Milwaukee's port collector.

The new job demanded less of Sholes's time, and he began to explore his creative and mechanical talents. He and his friend Samuel Soulé obtained a patent for a page-numbering device in 1864. Referring him to publications about earlier attempts to devise mechanical writing machines, inventor Carlos Glidden encouraged Sholes to transform his machine so that it could print letters.

The first documented typing machine had appeared in 1829 when William Burt of Detroit, Michigan, obtained a patent for a device called a typographer. A rotating semi-circular frame held the type mount, and levers depressed the frame onto paper. The result was slower than writing by hand, and the machine was barely used.

In 1868 Sholes, Soulé, and Glidden were granted a patent for a typewriter that embodied the basic mechanics of Burt's machine, but was much improved. With practice, it could produce documents at least as fast as a person could write.

After making further technical improvements, Sholes sold the patent rights in 1873 to the Remington Arms Company, a gun manufacturer with the necessary equipment, skilled labor force, and capital to build and market the machine. The first Remington Typewriter, a large, heavy model, sold for about $125, a high price that prevented its immediate commercial success.

Sholes was involved in improving the typewriter until his death on February 17, 1890, in Milwaukee.

Legacy

Sholes's development of the typewriter increased the efficiency of many communications and helped transform business practices from manual- to machine-oriented.

Before becoming popularly accepted, the typewriter had to overcome several challenges. Because of the cost of the original Remington machine, the company sold only 1,200 of the 25,000 models it built. Writers were wary because it was difficult to master the typing procedure, partly because the writer could not immediately view the typed material (in early models). Physicians accused the typewriter of causing consumption and schizophrenia in frequent users.

The typewriter, once its merits had been recognized and its mechanics improved (by the turn of the nineteenth century), offered advantages over writing by hand. It allowed people to create typed letters, stories, and official documents more quickly and legibly, with more precision, and in a standardized manner.

The typewriter was part of a twentieth-century revolution in business practices. Along with various desk-top instruments, such as dictating machines, adding machines, cash registers, and duplicating machines, the typewriter made the clerking profession vanish. Clerks, usually young men without the means for a formal education, had been essential to commerce and industry; they kept beautifully handwritten records of a company's entire operation. The advent of the typewriter also likely heralded a declining interest in the art of calligraphy.

The typewriter was improved throughout the century. In 1901 the first electric typewriter appeared, although mechanical machines dominated the popular market until the 1970s. By the 1980s typewriters were becoming archaic, as computers and computerized instruments replaced most desk-top business machines.

Schuyler

For Further Reading:
Duffy, Francis. *The Changing Workplace*. London: Phaidon, 1992.
Foulke, Arthur Toye. *Mr. Typewriter: A Biography of Christopher Latham Sholes*. Boston: Christopher Publishing House, 1961.
Pursell, Carroll W. *The Machine in America: A Social History of Technology*. Baltimore: John Hopkins University Press, 1995.

Sikorsky, Igor

Inventor of the Helicopter
1889-1972

Life and Work

Among the most significant innovators in the history of manned flight, Igor Sikorsky designed the first successful four-engine bombers, numerous versions of passenger-carrying seaplanes, and the first single-rotor helicopter.

Sikorsky was born on May 25, 1889, to a prominent family in Kiev, Russia. He was greatly influenced by his mother's interest in the work of fifteenth-century Italian artist and inventor Leonardo da Vinci. Fascinated by da Vinci's drawings of helicopter-like flying machines, at age 12 Sikorsky built a model helicopter powered by rubber bands.

In 1903 Sikorsky entered the Imperial Naval College at St. Petersburg but resigned three years later to study aeronautical engineering at Kiev's Polytechnic Institute. Inspired by news of the Wright brothers' first airplane flights, Sikorsky accepted his sister Olga's offer of money to buy a motor and start construction of his own flying machine.

Sikorsky devised two helicopter prototypes, in 1909 and 1910, both of which failed to fly; they required more powerful engines. He turned to the development of conventional aircraft instead.

Several successful designs earned him a commission to construct a bomber for the Russian Imperial Army. In 1913 Sikorsky unveiled the Ilya Mourometz (known also as *Le Grand*), a huge plane with four engines and the ability to carry heavy bomb loads. This first multi-motored airplane was used during World War I.

In 1919 Sikorsky fled to New York City to escape the political unrest following the Russian Revolution. Subsisting for several years as a schoolteacher, he acquired financial support to start an aeronautical engineering company in 1923.

In 1925 he completed the S-29, an all-metal passenger-carrying monoplane, followed by the versatile S-38 seaplane. In 1929 his successful company was purchased by the United Aircraft Corporation. The S-40 (the *American Clipper*), an advanced version of the S-38, appeared in the 1930s and became the standard craft for international flights.

In the late 1930s, Sikorsky convinced United Aircraft to invest in the development of a helicopter. In 1939 Sikorsky piloted the first flight of a practical single-rotor helicopter, the VS-300. The vehicle had an open-air cockpit and was powered by a 75-horsepower engine turning a three-bladed rotor. It could rise, float, and move in any direction.

Sikorsky continued to refine his designs until 1957, when he retired from active involvement in research and development. He died at Easton, Connecticut, on October 26, 1972.

Legacy

Sikorsky introduced many of the principles of aircraft design that remain staples of modern aeronautical engineering. He is best-known for his invention of the helicopter, a unique vehicle with a wide range of uses, and his numerous contributions to the evolution of airplanes have had a great impact on modern living and international trade.

Sikorsky's seaplanes introduced intercontinental flight as a new transportation option. The S-38s were used primarily between North and South America, creating an efficient means of transporting passengers and trade goods between the continents. The S-40 opened trans-oceanic air travel to the general public (although for several decades the high cost

restricted flight to the upper economic class).

Sikorsky's helicopters proved valuable for practical and emergency purposes in the 1940s. They carried mail and transported passengers from airports to city centers. In January 1944, helicopters carried blood plasma from New York to the New Jersey site of a steamship explosion, saving the lives of hundreds of victims.

Helicopters were first purchased by the United States Army in 1941, but they were rarely used until the 1950s, when they became a standard military vehicle. In both the Korean and Vietnam Wars, they were used to pick up and deliver forces and supplies in remote areas without airstrips.

By the 1960s helicopters were used to observe traffic, fight fires, spray crops, and lift heavy loads to the tops of buildings. They have become invaluable for rescue operations at sea and in mountainous regions, and are used to supply aid to disaster areas.

Schuyler

WORLD EVENTS		SIKORSKY'S LIFE
	1889	Igor Sikorsky is born
Spanish-American War	1898	
	1906	Sikorsky leaves Imperial Naval College to study engineering
	1910	Sikorsky's helicopter prototypes fail
	1913	Sikorsky's first bomber takes flight
World War I 1914-18		
	1919	Sikorsky flees to New York
	1925	Sikorsky launches his mono- and seaplane
	1939	Sikorsky completes the first single-rotor helicopter
World War II 1939-45		
Korean War 1950-53		
	1972	Sikorsky dies
End of Vietnam War	1975	

For Further Reading:

Delear, Frank. *Igor Sikorsky: His Three Careers in Aviation.* New York: Dodd, Mead, 1969.

Finne, N. *Igor Sikorsky: The Russian Years.* Washington, D.C.: Smithsonian Institution Press, 1987.

Sikorsky, Igor. *The Story of the Winged-S: An Autobiography.* New York: Dodd, Mead, 1952.

Somerville, Mary Fairfax

Popular Science and
Mathematics Author
1780-1872

Life and Work

Mary Somerville overcame the prejudices of her family and little formal education to become one of the preeminent writers of her time. Her books enabled others to understand recent discoveries in both mathematics and science.

Born in Jedburgh, Scotland, in 1780, Mary Fairfax seemed an unlikely candidate for academic success. Her family felt the only education necessary for a young woman entailed the simple writing, reading, and math skills needed for housekeeping. Becoming bored, she decided to teach herself Latin; a friendly uncle helped her read Virgil. After seeing several mathematical puzzles in a magazine, Fairfax became intrigued by algebra. Upon hearing her brother's tutor discuss EUCLID, she became determined to read the text herself. Her parents were appalled that she would

undertake such an unfeminine task and hid candles so that she could not read at night. Her response was to memorize the text and solve the problems in her head.

Fairfax's first marriage stifled her academic interests. It was only after she was widowed three years later that she resumed her interest in mathematics. Her next husband, Dr. William Somerville, supported her work and even aided her research. Their home became a meeting place for many of the British intellectuals of the time. Eventually, a friend persuaded her to translate the work of Pierre-Simone Laplace, a famous French mathematician who studied astronomy, as well as ISAAC NEWTON's *Principia.* English science had been stagnant for several years and out of touch with advances made in the rest of Europe. Somerville's task was to render the text so that it could be read in England. She reluctantly agreed to the challenge, unsure of her abilities and training. The resulting book, *The Mechanisms of the Heavens,* went far beyond simple translation. Published in 1831, Somerville included extensive annotations that explained the material in clear, concise language. She went on to write several other books that sought to bring together mathematical and scientific knowledge in understandable prose. These included *The Connection of the Physical Sciences* in 1834 and *Physical Geography* in 1848.

Somerville remained active in mathematics and science throughout her life. Her last book, *Molecular and Microscopic Science,* was released in 1869.

She regretted her lack of formal education and strenuously advocated for women's rights. Somerville died in Italy in 1872.

Legacy

Mary Somerville's writings enabled countless English-speaking people to understand the scientific and mathematical discoveries of her time.

During her life, Somerville's work sparked new interest in science. Her books were popular reading among the educated. The clarity of her style allowed readers to understand many technical aspects of astronomy and physics. Her *Mechanisms of the Heavens,* although not an original work, immediately rekindled interest in astronomy in England. This encouraged

her to write *The Connection of the Physical Sciences.* Not only did she construct a clear definition of physical science, she also described the interrelationships between science and mathematics. John Couch Adams, an astronomer of that time, credited his 1846 discovery of Neptune to a statement in this book. Soon after its publication, Somerville became one of the first women inducted into the Royal Astronomical Society.

As a woman she was barred from many scientific activities, but her stature caused others to question the validity of such an unequal system. She was recognized by the Royal Academy of Science, but the all-male membership did not go so far as to offer her membership.

Both directly and indirectly, her influence opened doors for other women into the halls of science. Somerville tutored Ada Byron (daughter of the poet Lord Byron), later to become Countess of LOVELACE, who is credited with writing the first computer program while an associate of CHARLES BABBAGE.

Long after Somerville stopped writing, her books inspired and educated many. Her texts were required reading for advanced students at Cambridge. A women's college at Oxford University is named after her. Without her influence, it might have been many years before England's scholars became acquainted with the advancements made by others. Through her writing, Somerville enabled many others to understand physical science and higher mathematics.

Wilson

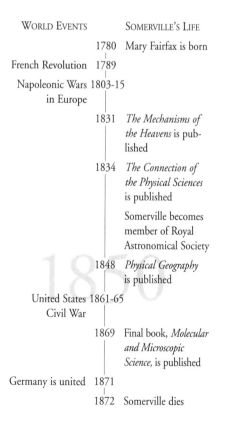

WORLD EVENTS		SOMERVILLE'S LIFE
	1780	Mary Fairfax is born
French Revolution	1789	
Napoleonic Wars in Europe	1803-15	
	1831	*The Mechanisms of the Heavens* is published
	1834	*The Connection of the Physical Sciences* is published
		Somerville becomes member of Royal Astronomical Society
	1848	*Physical Geography* is published
United States Civil War	1861-65	
	1869	Final book, *Molecular and Microscopic Science,* is published
Germany is united	1871	
	1872	Somerville dies

For Further Reading:

Osen, Lynn. *Women in Mathematics.* Cambridge, Mass.: MIT Press, 1974.

Perl, Teri. *Math Equals.* Reading, Mass.: Addison-Wesley, 1978.

Stephenson, George

Inventor of the First Practical
Steam Locomotive
1781-1848

Life and Work

George Stephenson designed and built the first practical steam locomotive; he further improved his original model and created the first public passenger train.

Stephenson was born on June 9, 1781, in Wylam, England. He followed his father into the coal-mining industry, and by age 19, with no formal schooling, he had operated Newcomen steam engines (named after THOMAS NEWCOMEN) in several mines around Newcastle. Attending night classes, he learned to read and write, and he learned mathematics by going over homework with his son Robert.

By 1812, as chief engine mechanic at Killingworth coal mine, Stephenson had built 39 stationary steam engines and several wagons that ran on rails and were drawn by steam engines.

In 1813 Stephenson observed a "steam boiler on wheels," a wagon with a built-in steam engine, designed by John Blenkinsop and utilized at a neighboring mine. The vehicle had cogwheels that fit into a third cogged rail, which made the system more expensive and complicated than one utilizing smooth rails. Several versions of steam locomotives that ran on smooth cast-iron rails had been built by other engineers, but the rails would break under the weight of the engines.

Stephenson decided to try his hand at improving existing steam locomotives. His first design, the *Blucher,* had flanged wheels, which have special rims to guide them along the rails. The *Blucher* weighed less than four tons and could pull 30 tons, a load capacity greater than previous models, and furthermore, the rails did not break under the relatively lightweight engine. The locomotive began hauling coal at Killingworth in 1814.

Stephenson later introduced the "steam blast," which directed steam up the engine's chimney, pulling air upward, creating greater draft, and thus increasing the engine's efficiency. He built several steam locomotives for coal mines over the next few years.

In 1825 engineers completed the Stockton-to-Darlington railroad on which Stephenson's *Locomotion,* the first passenger train, carried 450 people the 18-kilometer distance at 24 kilometers per hour.

In 1829 Stephenson built a 64-kilometer railroad from Liverpool to Manchester, and eight locomotives, built at his newly established Newcastle Locomotive Works, ran on it.

Stephenson died on August 12, 1848, in Chesterfield, England.

Legacy

Stephenson's innovative accomplishments in steam locomotion set the stage for the rapid evolution of this new mode of transportation.

Steam locomotives were the first land transportation faster than a horse-drawn buggy, and they changed people's lives drastically. Once their safety had been established, they spread around Europe and North America, linking distant cities and enabling people to travel farther faster.

Railroad development almost everywhere followed the practices instituted at Stephenson's Liverpool and Manchester Railway, which had two tracks, one for east-bound traffic and one for west-bound traffic. There were railway stations along the lines for the convenience of passengers, who could travel first, second, or third class, and were charged a certain amount per mile. Locomotives ran on a strict timetable, forcing the synchronization of local time from place to place.

Stephenson's son, Robert, was educated as an engineer and joined his father at the Newcastle Locomotive Works. Better trained than his father, Robert Stephenson systematically improved the steam locomotive's components; his efforts paid off in 1831 with the superb model called *Planet,* which became the prototype for most subsequent designs.

The next step in locomotive development was marked by the advent of compressed air brakes, invented by George Westinghouse in 1869. Previous brakes consisted of hand-operated shaft systems. Westinghouse's brakes represented the first fully mechanical braking system and thus ensured the safety of railway transportation.

Locomotives changed little until electric trains appeared in the twentieth century. Diesel-electric trains (using the engine invented by RUDOLF DIESEL) began to be used in the United States in 1935 for passenger trains. Requiring less fuel and less service, diesel-powered freight trains became the norm after World War II.

Schuyler

WORLD EVENTS		STEPHENSON'S LIFE
	1781	George Stephenson is born
French Revolution	1789	
Napoleonic Wars in Europe	1803-15	
	1812	Stephenson is chief engine mechanic at a coal mine
	1813	Stephenson begins designing steam locomotive
	1814	Stephenson's *Blucher* is completed
	1825	*Locomotion* runs as first passenger train
	1829	Stephenson builds railway from Liverpool to Manchester
	1831	Robert Stephenson's *Planet* is completed
	1848	Stephenson dies
United States Civil War	1861-65	

For Further Reading:

Cardwell, Donald. *Turning Points in Western Technology: A Study of Technology, Science, and History.* Canton, Mass.: Science History Publications/USA, 1991.

Rolt, L. T. C. *The Railway Revolution: George and Robert Stephenson.* New York: St. Martin's, 1962.

Swallow Richards, Ellen

Founder of Home Ecology
1842-1911

Life and Work

With her skills as a chemist and determination to assure the quality of the environment, Ellen Swallow Richards pioneered the concept of home ecology, or home economics, creating a connection between science and the home.

Ellen Swallow was born on December 3, 1842, in Dunstable, Massachusetts. Both

parents were teachers, allowing her to be schooled at home. Eventually she sought a higher education, which led her to Vassar College. In 1870 she graduated from Vassar and looked for another educational challenge. Massachusetts Institute of Technology had recently opened and she applied; she was the first woman accepted. She earned a B.S. degree from MIT: although Swallow completed the coursework for a doctoral degree in chemistry, the school refused to grant the degree to a woman.

While a student at MIT she conducted an extensive survey of the state water system. This work helped establish guidelines for the first state water-quality standards in the United States.

Swallow then studied mineralogy with Dr. Robert Richards. They married in 1875, two years after they began working together. He was an ardent supporter of her work throughout their marriage.

Richards would work at MIT for the remainder of her career. She established a women's laboratory at the school in 1876 to be used by women taking science courses. (Because women were not considered part of the regular student body, they were excluded from laboratory work.)

Richards wrote extensively about sanitation in the home, nutrition, and health issues. She began a series of lectures in 1899 to promote this new area, which she called home economics. Her own home was turned into a type of laboratory where she studied ventilation and heating systems and tested innovations. She tried to help people understand how the environment in which they lived affected their lives and, in turn, was changed by them. Thus she is one of the first pioneers in human ecology.

Richards continued to teach at MIT until she died in 1911.

Legacy

Richards was an indefatigable champion for the quality of the environment. Although she recognized the interrelationships between people and their environment, her views were not widely accepted until many years later.

During her tenure at MIT, Richards focused on women's education. Her lab provided women with a place to expand their scientific

knowledge until they finally gained full rights to an education at MIT in 1883. Some critics condemned her approach to science education for women who did not attend college as a distraction that might reduce time spent on cooking and cleaning. Richards recognized that women controlled the home environment and sought to help them understand science in that context. Her public lectures in the new field of home economics introduced women to ideas such as good nutrition, the health risks of the tight clothing that was fashionable during her time, and sanitation of air, water, and food.

Richards's water surveys indirectly affected thousands of people as they were the basis for setting government standards for water cleanliness. The first modern sewage-treatment plants built to meet the standards led to improved quality of drinking water throughout Massachusetts. Her work on environmental systems was a precursor to our current understanding of the complexity of the ecosystem.

Finally, Richards was instrumental in founding three influential organizations. She helped establish the research facility now called the Marine Biological Laboratory at Woods Hole in Massachusetts; a collegiate alumnae association that became today's American Association of University Women; and the American Home Economics Association. The laboratory at Woods Hole would eventually host important scientists and research work including BARBARA MCCLINTOCK, JAMES D. WATSON, and the Human Genome Project.

Wilson

WORLD EVENTS	SWALLOW'S LIFE
	1842 Ellen Swallow is born
United States 1861-65 Civil War	
	1870 Swallow earns B.A. degree from Vassar
Germany is united 1871	
	1873 Swallow receives B.S. in chemistry from Massachusetts Institute of Technology
	1875 Swallow and Dr. Robert Richards marry
	1876 Swallow establishes Women's Laboratory at MIT
Spanish-American 1898 War	
	1899 Swallow begins lectures on home economics
	1911 Swallow dies
World War I 1914-18	

For Further Reading:

Clarke, Robert. *Ellen Swallow: The Woman who Founded Ecology.* Chicago: Follett Publishing, 1973.

Douty, Esther M. *America's First Woman Chemist, Ellen Richards.* New York: Messner, 1961.

Kass-Simon, G., and Patricia Farnes. *Women of Science.* Indianapolis: Indiana University Press, 1990.

Takamine, Jokichi

Isolator of Adrenaline

1854-1922

Life and Work

Jokichi Takamine was the first chemist to isolate a hormone, adrenaline, from the adrenal gland.

Takamine was born November 3, 1854, in Takaoka, Japan. He was trained in the strict Samurai code, which encourages ancestral respect and strong morals and ethics. In 1879 Takamine earned a degree in chemical engineering from the College of Science and Engineering of the Imperial University of Tokyo. Takamine's education served as a strong foundation for his many positions in government service, chemical industry, and basic research. After his graduation, the Japanese government sent him to attend Anderson's College in Glasgow, Scotland, for two years (1879-1881) to improve his English.

While in Scotland Takamine was introduced to phosphate fertilizers and their production. In 1881 he returned to Tokyo to become the head chemist of the Imperial Department of Agriculture and Commerce as well as the head of the patent office. After six years he left civil service to begin his own superphosphate plant, the Tokyo Artificial Fertilizer Company. He opened this first major artificial-fertilizer plant in Japan in spite of resistance from traditional Japanese farmers.

In 1890 he married an American, Caroline Hitch, and moved permanently to the United States. He built his own laboratory in Clifton, New Jersey, to produce biochemicals in 1894. At the lab, he developed a starch-digesting enzyme from rice that he called Takadiastase. He tried without success to interest the brewing industry in the enzyme. However, the pharmaceutical firm, Parke, Davis and Company, became interested in Takadiastase, and Takamine maintained a close relationship with the firm during the rest of his life.

In 1901 Takamine isolated and crystallized adrenaline from the medulla (inner part) of the adrenal gland. He made this breakthrough while working on a joint project between Armour & Co. and John Jacob Abel, a chemist-pharmacologist of Johns Hopkins Medical School, who had called the substance epinephrine when he studied it in 1898.

Although he made his home in the United States, Takamine remained a consultant to the Japanese government on chemical and technological matters. He also continued to seek ways to foster good will between Japan and the United States. In 1912 he arranged with the Mayor of Tokyo for thousands of Japanese cherry trees to be delivered and planted near the basin of the Potomac in Washington, D.C.

For his sustained record of accomplishments in science and technology, Takamine was appointed to the Royal Academy of Sciences of Japan (1913), and for his diplomatic efforts, he was decorated with the Order of the Rising Sun (1915). Takamine died on July 22, 1922, in New York City.

Legacy

Takamine's primary legacy rests on his isolation of adrenaline, which led to significant applications in both medicine and the pharmaceutical industry. Adrenaline is used currently in the treatment of low blood pressure and in medical emergencies.

Due to his work in purifying and identifying the composition of adrenaline, it became the first natural hormone to be manufactured synthetically instead of being extracted from animal organs. Following Takamine's lead, American biochemist Edward Kendall isolated additional hormones from the adrenal gland. While Takamine had examined the medulla (inner part of the gland that acts like an independent body), Kendall isolated 28 different hormones from the cortex (the outside wrap) of the adrenal gland. Of these, cortisone has been most useful. Isolated by Kendall in 1935, its anti-inflammatory properties have led to many applications. Kendall was rewarded for his work with a Nobel Prize for Medicine or Physiology in 1950.

The enzyme Takadiastase, which breaks complex starches into alcohols and sugars, was eventually used by the brewing industry and has also been used as a medical treatment for abnormal digestion of carbohydrates.

One of Takamine's most enduring legacies is the display of cherry trees in Washington, D.C., a memorial of his desire to promote friendly relations between his native and adopted countries. The beauty of the cherry blossoms is enjoyed by all visitors to the capital of the United States during the spring. The flowering trees have been used by innumerable artists, photographers, and poets as symbols of international good will, relaying the message that Takamine intended.

Luoma

WORLD EVENTS		TAKAMINE'S LIFE
	1854	Jokichi Takamine is born
United States Civil War	1861-65	
Germany is united	1871	
	1879	Takamine graduates from College of Science and Engineering of the Imperial University of Tokyo as a chemical engineer
	1879-81	Takamine attends Anderson's College in Glasgow, Scotland
	1881-87	Takamine is head chemist at Imperial Department of Agriculture and Commerce, Tokyo
	1887	Takamine founds chemical fertilizer industry in Japan
	1894	Takamine establishes lab in Clifton, N.J., to produce biochemicals; he develops Takadiastase
Spanish-American War	1898	
	1901	Takamine isolates and purifies adrenaline (epinephrine)
	1912	Takamine negotiates with Mayor of Tokyo for cherry trees for Washington, D.C.
	1913	Takamine gains membership in Royal Academy of Sciences, Japan
World War I	1914-18	
	1922	Takamine dies

For Further Reading:

Hashizume, Yoshi. *Three Eminent Japanese in the Field of Medicine and Pharmacy.* Tokyo: Kogakusha, 1947.

Kawakami, K. K. *Jokichi Takamine: A Record of His American Achievements.* New York: W. E. Rudge, 1928.

Tesla, Nikola

Developer of Alternating
Current Electricity
1856-1943

Life and Work

Nikola Tesla, a pioneer in the field of electric power, developed the first practical system of using alternating current to transmit electricity.

Tesla was born on July 10, 1856, to Serbian parents living in the Croatian village of Smiljan. His father was a Greek Orthodox priest, and his parents expected him to enter the clergy. However, his talents were in mathematics and science, and he decided to train for an engineering career at the Technical University at Graz, Austria.

At Graz a professor demonstrated the Gramme dynamo, a direct-current motor. Tesla remarked that the dangerous sparks emitted by the dynamo could be avoided by eliminating the commutator, an attachment that reverses the current in the coil. His suggestion was met with skepticism, but the experience kindled a desire to overcome the problems of direct-current motors.

After brief electrical-engineering tenures in Budapest and Paris, in 1884 Tesla sailed for New York, where he took a position at the laboratory of THOMAS EDISON. Tesla failed to convince Edison of the need to devise a new type of motor based on alternating current (a current that continuously changes direction), which Tesla saw as the wave of the future, but which Edison claimed was dangerous. After a short time, Tesla resigned.

In 1887 Tesla founded the Tesla Electric Company and dedicated himself to designing an alternating-current motor. In his first successful model, a rotating magnetic field was created when an alternating current was supplied to the wire-taped blocks surrounding the rotor. He demonstrated that alternating current allows strong electrical currents to travel long distances, whereas direct current is limited to local use.

In 1891 Tesla devised a transformer, now called the Tesla coil, that allowed the transmission of high-frequency, high-voltage alternating current.

Inventor and manufacturer George Westinghouse bought Tesla's patent for the alternating-current motor and used the system to light the 1892-93 Columbian Exposition in Chicago, the first world's fair to have electricity. Tesla subsequently won a contract to build the first hydroelectric power plant, at Niagara Falls, New York. It was completed in 1895.

Turning to the investigation of radio waves, Tesla accomplished wireless communication across a 25-mile distance in 1897. From 1898 to 1900 he lived in Colorado, where the dry air was a good transmitter for high-frequency waves, the focus of his research during his remaining years.

In the early 1900s Tesla turned his attention to transmitting high-frequency waves across long distances without wires. His most ambitious endeavor at this time, a project for building a wireless world broadcasting tower, was defeated by financial difficulties.

His claims began to verge on the outlandish, tarnishing his public image. He was ridiculed for announcing that he had received radio signals from intelligent life on other planets and that he had invented a death ray with an enormous destructive capacity.

He became increasingly reclusive, possibly the victim of a medical disorder causing mental degeneration, and retired to a hotel room in New York City. He died there on January 7, 1943.

Legacy

Tesla's accomplishments helped shape the modern face of electrical technology.

Because Tesla's work on high-frequency waves involved general concepts associated with electricity and magnetism, it provided the basis for research in other fields. Physicist Robert Golka used Tesla's notes (housed at Belgrade after World War II) in his studies of plasma—gas composed of equal numbers of negatively and positively charged atomic particles. Plasma, formed in stars, lightning, and thermonuclear reactors, is a good conductor of electricity; plasma plays a crucial role in the development of fusion-powered energy (a theoretical future method of providing power via fusion of atomic nuclei). Soviet Pyotr Kapitsa, a 1978 physics Nobelist, used Tesla's experiments as models for his pioneering research on the electrical-conducting properties of materials placed in high-strength magnetic fields.

Tesla's alternating-current system has become the dominant power transmission method throughout the world. Power grids that supply cities with electricity use alternating current, which enables the use of transformers to control the level of power supplied to industries, businesses, and homes. The Tesla coil is an integral component of modern radio and television sets.

The design principles of the hydroelectric power plant that Tesla constructed at Niagara Falls became the model for hydroelectric plants throughout the United States.

Schuyler

WORLD EVENTS		TESLA'S LIFE
	1856	Nikola Tesla is born
United States	1861-65	
Civil War		
Germany is united	1871	
	1884	Tesla immigrates to the United States
	1887	Tesla designs alternating-current motor
	1891	Tesla designs transformer, now called a Tesla coil
	1892-93	Tesla's electrical system is used at Columbian Exposition in Chicago
	1895	Tesla's hydroelectric power plant is completed in Niagara Falls, New York
	1897	Tesla achieves radio transmission
Spanish-American	1898	Tesla moves to Colorado
War		
World War I	1914-18	
World War II	1939-45	
	1943	Tesla dies

For Further Reading:

Cheney, Margaret. *Tesla: Man Out of Time.* Englewood Cliffs, N.J.: Prentice-Hall, 1981.

O'Neill, John J. *Prodigal Genius.* London: Spearman, 1968.

Torricelli, Evangelista

Inventor of the Barometer
1608-1647

Life and Work

Evangelista Torricelli was the first person to grasp that air had weight, to measure its pressure, and to create a vacuum, a region without air.

Torricelli was born on October 15, 1608, in the northern Italian town of Faenza. An excellent student, he went to Rome to study at the Collegio di Sapienza where he heard about GALILEO GALILEI's work on gravity and motion. In 1641 Torricelli moved to Florence and served as secretary and aide to Galileo, who by that time was blind and suffering from arthritis.

Only three months later Galileo died, but not before he had given Torricelli an interesting problem to pursue—how to make a vacuum. In 1643 Torricelli filled glass tubes with mercury, a dense liquid metal, covered the open end with his finger, and inverted it into an open bowl of mercury. The mercury fell, but not all the way out. It stayed in a column about 30 inches (76 centimeters) high above the mercury in the bowl. He noted that the exact height of the mercury column varied slightly from day to day.

Torricelli explained these results in a 1644 paper by claiming that we live at the bottom of an ocean of air hundreds of miles high, with a density about 1/800 that of water. The weight of all that air pressed down on the mercury in the bowl, balancing the weight of the mercury held up in the tube. In pursuing Galileo's challenge to make a vacuum, Torricelli discovered the effects of air pressure, in effect creating the first barometer, a device to measure air pressure.

The region above the mercury in Torricelli's tube was a vacuum, as there was no way for air to get in through the mercury. No one had ever created a sustained vacuum before; many scientists had believed it was impossible to create a vacuum. Torricelli observed that light and magnetism were conducted through the vacuum as well as through the atmosphere.

Torricelli worked on a variety of other topics in mathematics, fluids, and motion. He focused on ways to calculate and describe cycloids, curves formed by points rotating on the edge of a spinning circle. He discovered that liquid flowing from the bottom of an open container leaves with the same speed as if it had fallen a distance equal to the container's height, which was later known as Torricelli's theorem. In 1644 he published *Geometric Works*, which described his conclusions on the motion of fluids and projectiles.

He died when he was only 39, in Florence, Italy, on October 25, 1647.

Legacy

Torricelli's discovery of the effects of air pressure and his development of the principles of the barometer opened investigations into the relationship between air pressure, altitude, and changes in the weather. His creation of a vacuum challenged scientists to find other ways to create vacuums and to explore their properties.

Soon after Torricelli announced his experiments with mercury-filled tubes, French mathematician and physicist BLAISE PASCAL determined that thinner air at a high altitude has a lower air pressure. Conducting experiments with a "Torricellian tube," Pascal found that mountain air supported a mercury column centimeters shorter than one at the base of a mountain. Pascal's conclusion amounted to evidence of the physical reality of the "heavens" or outer space, which was traditionally presumed to be unpredictable and unlike Earth.

Pascal also used the device, known as a barometer, to predict the weather: he observed that a drop in air pressure often precedes rain or snow. Barometric pressure is a standard part of any weather report today, and one of the units used to measure pressure, a torr, is named in honor of Torricelli.

Contemporary barometers have come a long way since Torricelli's tubes. As mercury is hazardous to handle, most barometers now are built around a vacuum chamber with flexible walls, and are called dry or aneroid barometers. Because air pressure decreases with altitude, aneroid barometers are also used in altimeters in airplanes to tell pilots how high they are above ground. Modern weather forecasting and aviation both trace back to Torricelli.

As soon as Torricelli showed that making a vacuum was possible, scientists looked for other ways to make them. German physicist Otto von Guericke (1602-1686) made a practical air pump in 1645, modeling it on a water pump. This led to other experiments showing that sound (unlike light) cannot travel in a vacuum and both fires and animals die without air. Progress in identifying air as real and having measurable mass was a first step in recognizing a diversity of gases with unique properties, fundamental concepts in modern science.

Secaur

WORLD EVENTS		TORRICELLI'S LIFE
	1608	Evangelista Torricelli is born
Thirty Years' War in Europe	1618-48	
	1641	Torricelli begins work with Galileo Galilei
	1643	Torricelli invents barometer
	1644	Torricelli publishes paper explaining air pressure
		Torricelli publishes *Geometric Works*
	1645	Otto von Guericke makes air pump, a spin-off of Torricelli's vacuum
	1647	Torricelli dies
England's Glorious Revolution	1688	

1650

For Further Reading:
Gindikin, Semen Grigorsevich. *Tales of Physicists and Mathematicians.* Boston: Birkhauser, 1988.
Middleton, William Edgar Knowles. *The History of the Barometer.* Baltimore: Johns Hopkins Press, 1964.
Schwartz, George. *Moments of Discovery.* New York: Basic Books, 1958.

Ts'ai Lun

Inventor of Paper

c.50-121

Life and Work

In the second century, Ts'ai Lun invented paper made of macerated (moistened and softened) vegetable fibers; the same process is employed by modern papermaking machines.

Ts'ai Lun was born about the year 50 in what is now the Hunan province of China. The events of his early life are unknown.

In the year 75 he was employed at the court of the Emperor Ho Ti in Beijing, and by the year 97 he had become head of a group of engineers producing swords and other instruments.

In the year 105 Ts'ai Lun announced to the Emperor the invention of paper, but it is not known whether Ts'ai Lun was the actual inventor.

The papermaking process involved macerating a combination of tree bark, hemp scraps, rags, and fish nets (made of plant matter) into a pulp of individual cellulose fibers. The pulp was soaked in water and then lifted onto a sieve-like screen, yielding a sheet of matted fiber that could be written upon. Cheaper than silk and less cumbersome than wood and bamboo, paper quickly replaced these traditional Chinese writing materials.

An outline of Ts'ai Lun's life compiled in the fifth century states that in the year 106 Ts'ai Lun was made a private counselor in the Emperor's court, and in the year 114 he gained the honorable title of Marquis for his many years of service. The story tells of Ts'ai Lun's demise: the Empress ordered Ts'ai Lun to circulate slanderous rumors about a member of the Imperial family; upon the Empress's death, Ts'ai Lun was ordered to face judgment and he poisoned himself. The year was 121.

Legacy

Few inventions have been as widely used as that of paper; it is one of the most common and versatile manufactured items in the modern world.

During the first several centuries following Ts'ai Lun's invention, paper was known only in China, where it was used as a writing surface and for crafting ornaments. In the seventh century, paper was brought to Japan, where papermaking was rapidly adopted. The first text printing followed directly from the introduction of paper. To drive out a smallpox epidemic sweeping Japan, a number of miniature pagodas were to be constructed, each with a prayer carved on the bottom. The prayer-imprints were dipped in ink and printed on paper.

Paper traveled west in the seventh century as well. First appearing in Turkestan (now Afghanistan), paper spread to Baghdad, Damascus, Egypt, and then Morocco. It made its way to southern Europe (through either Italy or Spain), but not until the twelfth or thirteenth century.

Paper was at first rejected in Europe, as it was more costly and fragile than the parchment (stretched dried animal skins) then widely used for bookmaking. Paper was eventually recognized as superior: it took up less space, it was easier both to write on and transport, and its source was more abundant.

When JOHANNES GUTENBERG devised the first printing press in the mid-fifteenth century, the design of the equipment was dictated by the properties of the paper available. Europeans were using paper made of linen and cotton fibers overlaid with a hard surface of gelatin; this hardness required heavy metal types and a forceful pressing mechanism.

Most modern paper is produced by machine, but the basic process is similar to the one developed by Ts'ai Lun. Paper has an uncountable number of everyday applications and its availability is largely taken for granted today. Demand for paper is so high that an alarming percentage of the world's forests continue to be cut down each year, as wood fiber is the most common component in the papermaking process; the practices of recycling and replanting have only partly alleviated the problem.

Schuyler

For Further Reading:

Hunter, Dard. *Papermaking: The History and Technique of an Ancient Craft*. New York: Knopf, 1967.

Munsell, Joel. *Chronology of the Origin and Progress of Paper and Paper-Making*. New York: Garland. 1980.

Rudin, Bo. *Making Paper: A Look into the History of an Ancient Craft*. New York: Rudins, 1990.

Vesalius, Andreas

Founder of Anatomy
1514-1564

Life and Work

Andreas Vesalius studied and described the anatomy of the human body through cadaver dissections, revolutionizing the field of biology and the practice of medicine.

Vesalius was born in December 1514 in Brussels, Belgium, to a family of physicians and apothecaries. From 1529 to 1536, he attended the Universities of Louvain and Paris, where he learned the techniques of animal and human dissection. He focused on learning the human skeletal system, using bones readily available from Parisian cemeteries.

In 1537 he earned his medical degree from the University of Padua in Italy and took a post there as lecturer in surgery. He gave popular human-dissection demonstrations, insisting on performing them himself.

In school he had been taught the theories of GALEN, the most accomplished physician of the Roman Empire whose books on anatomy were still considered the authority in medicine in Renaissance Europe. As Vesalius conducted more dissections and learned anatomy directly from cadavers, he grew to question Galen's views. Because Galen had dissected mostly animals, Vesalius reasoned,

Galen had simply made inferences about the human body from those dissections.

While at Padua, Vesalius worked on a textbook of human anatomy, which was published in 1543 under the title *The Seven Books on the Structure of the Human Body*. The work consisted of a complete and careful description of human anatomy, with emphases on osteology, myology, and cardiology. The refinement of its printing and illustrations and the clarity of its organization, as well as its comprehensiveness, made it a milestone in biological studies. The work earned Vesalius the position of physician to the Holy Roman Emperor, Charles V.

From 1553 to 1556, Vesalius built a flourishing medical practice in Brussels. In 1559, he moved to Madrid, Spain, to work as a physician for Phillip II, the son of Charles V. He obtained permission to go on a religious pilgrimage to Jerusalem in 1564. He died during the return trip and was buried on the Greek island of Zacynthos in June 1564.

Legacy

Vesalius' legacy lies predominantly in his comprehensive description of human anatomy based on direct observation of human cadavers. He founded the study of anatomy as a scientific discipline distinct from the fields of medicine and biology and helped to loosen the hold that Galen's theories had over medical research.

Vesalius criticized the authority of Galenic medicine during a time when other prominent physicians were leveling similar criticisms. For instance, Swiss physician PARACELSUS, who used his understanding of chemistry and new drugs to promote health and healing, directly opposed Galen's theory of the four bodily humors (a theory originating with HIPPOCRATES). While Galen's achievements in antiquity led the practice of medicine forward, his erroneous theories and methods restricted medical progress for centuries. The revolutionary ideas and methods of Vesalius and others helped to free western medicine from these limitations and to move medicine forward.

Vesalius's work inspired anatomical research well into the seventeenth century. For example, Johann Georg Wirsung described the pancreatic system in 1642, and Jean Pecquet discovered the thoracic duct in a dog in 1647.

These discoveries were the result of progressive efforts in dissection.

Vesalius's anatomy textbook ranks among the most important biological treatises in the history of western science. It marks the beginning of thorough, direct observation of internal human anatomy, and the work remained highly influential in medical curricula through the eighteenth century.

Schuyler

WORLD EVENTS		VESALIUS'S LIFE
Columbus discovers Americas	1492	
	1514	Andreas Vesalius is born
Reformation begins	1517	
	1529	Vesalius begins university study
	1537	Vesalius earns medical degree from University of Padua
	1543	*The Seven Books on the Structure of the Human Body* is published
		Vesalius is appointed physician to Charles V
	1553	Vesalius starts a medical practice in Brussels
	1559	Vesalius is appointed physician to Phillip II
	1564	Vesalius dies
Thirty Years' War in Europe	1618-48	

For Further Reading:

O'Malley, Charles Donald. *Andreas Vesalius of Brussels, 1514-1564*. Berkeley: University of California Press, 1964.

Vesalius, Andreas. *The Illustrations from the Works of Andreas Vesalius of Brussels*. Magnolia, Mass.: Peter Smith Publisher, 1979.

Volta, Alessandro

Discoverer of Electrochemical Cells
1745-1827

Life and Work

Alessandro Volta was the first to realize that steady electric currents could be produced from chemical reactions.

WORLD EVENTS	VOLTA'S LIFE
Peace of Utrecht 1713-15 settles War of Spanish Succession	
	1745 Alessandro Volta is born
	1775 Volta invents electrophorus to create and store a static charge
United States 1776 independence	
	1779 Volta appointed professor at Pavia University
French Revolution 1789	
	1800 Volta develops first electrochemical battery
Napoleonic Wars 1803-15 in Europe	
	1827 Volta dies

The descendant of a once-wealthy and powerful family, Volta was born in 1745 in the Italian city of Como. In school he wavered between interests in poetry and science, but finally decided on a science career. He began his long service as a physics professor at Pavia University in 1779.

In Volta's time the only way to generate static electricity in a laboratory was by friction from rubbing two electrical insulators together. In 1775 Volta invented an interesting device, called an electrophorus, that could store a static charge for some time after being rubbed with fur or cloth. The trouble was that all of the electrical equipment of the time held the charge in place and would release it only in high-voltage but low-power blasts. No one had a way to produce electricity smoothly or in a large quantity.

Luigi Galvani was a professor of anatomy at the University of Bologna during this period. Making a pot of soup one day, Galvani noticed that a leg from a freshly killed frog jerked and jumped when it touched against a copper hook and an iron pot. Galvani imagined that electricity was coming from the frog itself; he called it "animal magnetism."

Volta realized that the frog leg was just a source of moisture, and the electricity was actually produced by the two metal electrodes. Some metals, like copper, hold their electrons more tightly than others, such as iron. It was the combination of two different metals touching the frog leg that generated the electricity to make it twitch. Galvani did not want to give up his explanation, though, and a bitter rivalry grew between the two men because of their interpretations of the experiment.

Volta went on to develop the first electric battery in 1800. He used cups of dilute acid to provide the moisture for strips of copper and zinc. The French Emperor, Napoleon, was so impressed that he visited Volta in 1801 and awarded him a gold medal for his discovery. Volta lived the rest of his life in his hometown of Como and died on March 5, 1827.

Legacy

Volta's introduction of the electric battery extensively influenced everyday life, and the production of steady electrical currents

played a vital role in the development of nineteenth- and twentieth-century science.

All of the great scientists of the nineteenth century needed steady, reliable sources of electric current. Without Volta's batteries, ANDRÉ AMPÉRE and Hans Christian Oersted could not have investigated magnetic fields, GEORG OHM and Gustav Kirchhoff could not have mastered electric current, MICHAEL FARADAY could not have seen the way to generate electricity from magnetic fields, and JAMES CLERK MAXWELL could not have unified it all with his theories on electromagnetism.

Volta's battery and variations on it can be seen in many aspects of our daily lives. The batteries used liquid acid, so they were called wet cells; car batteries, which also contain liquid acid, are direct descendants of Volta's original models. In AA-size alkaline cells, the liquid acid has been replaced with a thick chemical paste that is far more efficient and longer lasting than anything Volta imagined, but they still operate on the same principle.

Other innovations on Volta's invention include cells based on the chemistry of lithium or silver oxide and rechargeable cells made of nickel and cadmium ore. Each has distinctions and advantages, but all are distant cousins of Galvani's frog leg and more direct descendants of Volta's cells.

The unit of measurement for electrical drive or motivation is called a volt, as a testament to Volta's influence in the development of electricity.

Secaur

For Further Reading:

Mann, A., and A. Vivian. *Famous Physicists.* John Day, 1963.

Mottelay, Paul Fleury. *Bibliographical History of Electricity and Magnetism.* New York: Arno Press, 1975.

Whittaker, E. T. *A History of The Theories of Aether and Electricity.* New York: Dover, 1989.

Von Békésy, Georg

Discoverer of the Physiology
of Hearing
1899-1972

Life and Work

Georg von Békésy discovered the physical mechanism by which the inner ear processes sound waves and detects differences in pitch.

Von Békésy was born in Budapest, Hungary, on June 3, 1899. His father was in the diplomatic service, a job that took the family to various European cities during Georg's youth. He was intrigued by gypsy music, and he later attributed his life-long interest in sound to this boyhood passion. In 1923 he earned a Ph.D. in physics from the University of Budapest for a thesis on fluid dynamics.

Upon graduation, von Békésy began a 23-year association with the laboratories of the Hungarian Telephone System. Employed as a communications engineer, he soon began investigating the physics of human hearing to help solve the practical problems of telephone systems. Physiologists had not yet determined how sound is translated by the inner ear.

During his tenure at the telephone laboratories, von Békésy also worked at various other European research institutions. For two decades, he carried out numerous experiments aimed at determining the physiology of the cochlea, the part of the inner ear involved in hearing. To do so, he designed and built various miniature tools and, with the help of tiny mirrors, observed changes that occur in the cochlea as it receives sounds.

In 1947, von Békésy immigrated to the United States to accept a research position at Harvard University's Psycho-Acoustic Laboratory, where he built a large model cochlea. The model consisted of a 30-centimeter plastic tube filled with water. He placed the tube along his forearm and stimulated one end of the tube with a sound. His forearm detected sound waves of different pitches at distinct points along the tube. From numerous experiments with the model, he concluded that sound moves through the fluid in the cochlea

in a "traveling wave" and produces a vibration at one particular point along the cochlear membrane. The wave's pitch determines the location of that point. The vibration then stimulates adjacent cells to send an auditory message to the brain, which interprets the sound as having a particular pitch.

Von Békésy's work earned him the 1961 Nobel Prize for Physiology or Medicine; he was the first physicist to win a Nobel Prize in that category.

He took a position at the University of Hawaii in 1966; he died on June 13, 1972, in Honolulu.

Legacy

Von Békésy's discovery of the basic physiology of the inner ear enhanced the understanding of hearing, the development of surgical procedures, and the diagnosis and treatment of hearing disorders.

His work influenced that of other pioneers in the study of hearing. In 1954 Ernest Glen Weaver and H. Lawrence further investigated acoustic physiology at various levels of the ear, and in 1960 H. Engström created electron microscope images of the organ of Corti, the structure inside the cochlea that contains the receptor cells that detect sound waves and their different pitches.

Von Békésy's research led directly to the design of an audiometer, a device that determines whether a person's deafness is caused by damage to the ear or to the brain. This instrument enabled physicians to prescribe the appropriate treatment at an early age for patients who were born deaf.

His research also provided the foundation for the development of instruments enabling deaf people to "hear" through tactile sensations. His model cochlea was one such instrument: placed against the forearm, the tube allowed a person to feel differences in pitch due to the vibration (caused by sound waves) of distinct points along the tube. Contemporary hearing aids now electronically amplify sound with the use of transistors, but without von Békésy's key discoveries about the physics and physiology of human hearing, modern advances in hearing aids would not have been possible.

Schuyler

WORLD EVENTS		BÉKÉSY'S LIFE
Spanish-American War	1898	
	1899	Georg von Békésy is born
World War I	1914-18	
	1923	Békésy receives Ph.D. from the University of Budapest with thesis on fluid dynamics
		Békésy begins work at Hungarian Telephone System
World War II	1939-45	
	1947	After immigrating to U.S., Békésy builds cochlear model at Harvard
Korean War	1950-53	
	1961	Békésy wins Nobel Prize for Physiology or Medicine
	1966	Békésy joins faculty at University of Hawaii
	1972	Békésy dies
End of Vietnam War	1975	

For Further Reading:

Stevens, S. S. *Sound and Hearing*. New York: Time-Life Books, 1970.

Von Békésy, Georg. *Experiments in Hearing*. Translated and edited by E. G. Weaver. Huntington, N.Y.: R.E. Krieger, 1980.

Von Humboldt, Alexander

Early Pioneer of Ecological
Concepts
1769-1859

Life and Work

Alexander von Humboldt transformed nature study from mere collection and organizing of information about plants and animals into the practice of formulating theories that would connect all natural phenomena on Earth. He was also partly responsible for the transformation of science from a hobby of the wealthy to a real profession, open to people of different classes.

WORLD EVENTS		HUMBOLDT'S LIFE
Peace of Utrecht 1713-15 settles War of Spanish Succession		
	1769	Alexander von Humboldt is born
United States independence	1776	
French Revolution	1789	Humboldt enters University of Göttingen
	1792	Humboldt accepts position in Prussia's Mining Department
	1799	Humboldt travels to the Americas to study Venezuela's grasslands and Orinoco River Basin
Napoleonic Wars 1803-15 in Europe		
	1804	Humboldt returns to Europe and begins writing *Personal Narrative of Travels...*
	1827	Humboldt returns to Berlin and begins teaching
	1834	Humboldt begins writing *Kosmos*
	1859	Humboldt dies
United States 1861-65 Civil War		

Born September 14, 1769, in Berlin, Germany, Humboldt and his brother Wilhelm were privately educated in the classics. After he became acquainted with mineralogy and geology at the University of Göttingen in 1789 to 1790, he sought further training in those areas at the School of Mines in Frieberg.

Humboldt was appointed in 1792 to the Mining Department of the Prussian government. He supervised mining activities, invented a safety lamp, and established a school for miners. But his passion was exploration. In 1797 he resigned his position in order to gain field experience in meteorological and magnetic measurements.

In 1799 Humboldt and botanist Aime Bonpland went to Central and South America to study the grasslands of Venezuela and the Orinoco River basin. Humboldt also explored the Andes, becoming the first to attribute mountain sickness to insufficient oxygen at high altitudes. He studied the oceanic Peru Current and went back to Europe with immense amounts of data on Earth's geomagnetic field, temperatures, and barometric pressures over the years of his journey; previously unknown plants; and extensive measurements for new maps.

Upon his return to Paris, Humboldt spent the years between 1804 and 1827 writing and publishing 30 volumes related to the data gathered on his trip; these volumes were called *Personal Narrative of Travels to Equinoctial Regions of the New Continent*. His descriptions included meteorological patterns that laid the foundations of comparative climatology; dynamism of Andean volcanoes as evidence of ongoing development of Earth's crust by eruptive forces; and relationships between geography and the fauna and flora of various locales.

Humboldt returned to Berlin in 1827 as tutor to the Crown Prince and also taught physical geography at the University of Berlin. In 1828 he organized one of the first international scientific conferences.

Humboldt coined the word kosmos—meaning the exquisitely ordered system of the global environment. During his last 25 years, Humboldt wrote a set of books that he named *Kosmos*. One of the most extensive scientific works ever published, the books were best sellers and were translated into nearly every European language. In these books, Humboldt emphasized unity and global factors that link diverse environments and their inhabitants. All plants and animals, he noted, were subject to the same laws of temperature that affected both the horizontal geography of latitudes and the vertical geography of altitudes. He was the first to classify forests as tropical, temperate, or boreal, terms in use today. He also coined isotherm and isobar for lines on maps connecting places with the same temperature or pressure, respectively, as he illustrated climate as a unifying global force.

He died in Berlin on May 6, 1859, while writing the fifth volume of *Kosmos*.

Legacy

Humboldt put the concept of nature's interrelationships into the mainstream of scientific thought. He did not see nature as chaotic and therefore in need of human control. He believed that the natural world was a web of harmonious interactions and tried to map some of the connections. He held up the natural world as an ideal model worthy of respect. Not until RACHEL CARSON (1907-1964) did another scientist incorporate human survival firmly within the environment instead of viewing it on or above it.

Humboldt's influence was immediate and direct. Humboldt supported young academics such as chemist Justus von Liebig and zoologist LOUIS AGASSIZ. CHARLES DARWIN was another of his disciples; Humboldt's book on the New Continent was one he read during his voyage on the Beagle. In fact, the word "scientist" first entered the English language in the 1830s to distinguish men like Humboldt and Darwin from natural historians.

The endurance of Humboldt's theories can be seen in the continued use of his classification system for forests and the use of the cartographic terms isotherm and isobar.

As environmentalists would do in the late 1900s, Humboldt worried about differentiating between natural and unnatural rates of change, the social consequences of resource exploitation, and educating people that each part of the living world is valuable. His message resonates today.

Simonis

For Further Reading:
De Terra, Helmut. *Humboldt: The Life and Times of Alexander von Humboldt, 1769-1859.* New York: Alfred A. Knopf, 1955 (reprinted 1979).
Sachs, Aaron. "Humboldt's Legacy and the Restoration of Science." *World Watch* (March/April 1995).
Zottmann, Thomas. *Alexander von Humboldt: Scientist, Explorer, Adventurer.* New York: Pantheon Books, 1960.

Von Neumann, John

Early Computer Designer;
Developer of Game Theory
1903-1957

Life and Work

In addition to contributions to logic, game theory, quantum physics, and economics, John von Neumann created much of the theory used to develop high-speed computers.

Von Neumann was born in Budapest, Hungary, on December 28, 1903. While in school his mathematical talent was recognized early and he was tutored privately. He received his Ph.D. in mathematics from the University of Budapest in 1926. He developed his reputation by publishing a series of papers on set theory, game theory, logic, and quantum mechanics. He came to Princeton University as a visiting lecturer in 1930 and three years later became one of the original professors at Princeton's Institute for Advanced Study.

He developed what are now called "Von Neumann algebras," providing new principles of mechanics needed to explain the behavior of subatomic particles. He proposed axioms to which quantum theories would comply if space were expanded from three dimensions to an infinity of dimensions. Eventually, in 1955, he was able to prove the equivalence of the two leading models of quantum theory: wave mechanics and matrix mechanics.

In 1943 von Neumann joined the Manhattan Project in Los Alamos, New Mexico, helping to

develop the atomic bombs dropped on Japan that ended World War II. He applied prior work on detonation waves to the investigation of implosions, solving difficult equations using Harvard Computational Lab's Mark I, considered the first full-scale programmable computer in the United States. Soon after von Neumann became interested in the much faster ENIAC, an electronic computer, being developed at the University of Pennsylvania.

A person of wide-ranging interests, von Neumann published many papers on game theory, beginning in 1926. Assuming rational choices, he invented mathematical strategies for winning and for analyzing social dynamics in competitive situations. His "Model of General Economic Equilibrium" (1937) became a foundation of modern economic theory in which game theory is applied. Collaborating with Oskar Morganstern led to a seminal book, *Theory of Games and Economic Behavior,* published in 1944. By this time, however, computers became both his tool and his obsession.

The ENIAC group was already planning its next computer, the EDVAC. Von Neumann joined them as leader of the team planning the computer's architecture, or logical organization, while others worked on technical problems. He wrote the influential "First Draft of a Report on the EDVAC" in 1945 advocating more reliable, faster computers. He proposed computer designs including both an arithmetical component for elementary operations and a central control organ for logical flow of information.

After World War II von Neumann continued to consult with the military and the Atomic Energy Commission at Los Alamos. He was very involved in development of thermonuclear (hydrogen) weapons and their long-range delivery systems. He opposed disarmament and favored stockpiling weapons as a defense strategy. After a decade of service on its General Advising Council, von Neumann was named Atomic Energy Commissioner in 1955. He died of bone cancer in Washington, D.C., on February 8, 1957.

Legacy

Von Neumann developed the mathematical foundations of quantum mechanics, proposed an influential theory of games, and developed the logical framework for rapid, reliable digital computers.

Von Neumann's ideas for organizing and programming computers by using specific ordered

sequences of instructions are still in use today. Although modern micro-electronics have allowed the hardware to shrink, the basic architecture of computers is inherited from von Neumann's ideas on reliability and error theory proposed more than a half century ago.

Von Neumann's work on quantum theory influenced theoretical physicists and, in general, established that traditional causality (which ALBERT EINSTEIN favored) and WERNER HEISENBERG's Uncertainty Principle could not be reconciled. The Uncertainty Principle states that an exact measurement of both the speed and position of a subatomic object is impossible given the disturbance that is caused by experimental conditions. The theory remains at the forefront of contemporary subatomic physics.

Game theory has flourished because it permits analysis of coalition formation and profitability, describing tactics beneficial in long sequences of moves. Its rigorous mathematical approach has been applied not just to games like chess and bridge but also to military strategies, business cycles, and studies of social organizations. Although von Neumann's ideas were published in the 1930s and 1940s, they were not adopted until computers were available to perform the complex mathematics involved.

Steinberg

WORLD EVENTS	VON NEUMANN'S LIFE
	1903 John von Neumann is born
World War I 1914-18	
	1926 Von Neumann completes Ph.D. in Budapest
	1933 Von Neumann becomes one of the first professors at Institute for Advanced Study at Princeton University
World War II 1939-45	
	1943 Von Neumann joins Manhattan Project
	1944 *Theory of Games and Economic Behavior* is published
	Von Neumann joins EDVAC computer development team at University of Pennsylvania
Korean War 1950-53	
	1955 Von Neumann establishes equivalence of wave mechanics and matrix mechanics
	1957 Von Neumann dies

For Further Reading:

Aspray, William. *John von Neumann and the Origins of Modern Computing.* Cambridge, Mass.: MIT Press, 1990.

Goldstine, Herman. *The Computer from Pascal to von Neumann.* Princeton, N.J.: Princeton University Press, 1972.

Heims, Steve. *John von Neumann and Norbert Wiener.* Cambridge, Mass.: MIT Press, 1980.

Watson-Watt, Robert

Developer of Radar
1892-1973

Life and Work

Robert Watson-Watt developed radar, the first system for using radio waves to detect, locate, and track moving objects.

Watson-Watt was born on April 13, 1892, in Brechin, Scotland. He won a scholarship to attend University College, Dundee, where he developed an interest in the recently established field of wireless telegraphy.

Earning a degree in electrical engineering in 1912, he taught briefly at University College. In 1914, at the beginning of World War I, he began working for the British government's Meteorological Office, researching methods of tracking storms with radio waves.

In 1919 Watson-Watt received a bachelor's degree in physics from the University of London. In 1927 the government facility at which he was posted became the Radio Research Station, and he was appointed director.

He supervised advanced research on the radio location of thunderstorms, the detection of radio signals sent from ships, and the properties of the atmosphere. He coined the term "ionosphere" for the layer of the atmosphere that reflects radio waves. (This reflection allows radio signals to be sent around the Earth despite the planet's curvature.)

In 1935 a government official asked Watson-Watt whether radio waves could be used to destroy enemy aircraft. He and his assistant A. F. Wilkins explained that, while actual destruction was not possible, radio waves would be capable of detecting the location of aircraft. A program was initiated for the development of this idea. Watson-Watt soon engineered a device capable of detecting aircraft at a range of 130 kilometers.

The term RADAR stands for Radio Detection And Ranging; it works by sending pulses of radio waves in the direction of a target. The interval of time before the waves are detected at the source after reflecting off the target indicates the target's distance, and the direction from which the reflected waves return indicates the target's position.

Watson-Watt opened a private consulting firm specializing in engineering in 1946, but little else is known of the last decades of his life. He died on December 5, 1973, in Inverness, Scotland.

Legacy

Watson-Watt's radar device became an integral part of equipment used in military operations, navigational aids, commercial air traffic, weather detection, and terrestrial and extraterrestrial mapping.

Radar systems were in place along Britain's coast by 1939, at the onset of World War II, and soon installed in fighter aircraft for the detection of targets at night and in cloudy conditions. Some observers claimed that radar was chiefly responsible for Britain's successful defense against the German air attack in the Battle of Britain in 1940. With numerous refinements and improvements, radar continues to be used in military aircraft for locating targets, measuring altitude, and detecting missiles and enemy craft.

In 1951 the first radar systems tailored for meteorology began operation. Early radar was especially effective in tropical regions and at sea and quickly became the primary instrument for tracking tropical cyclones. It was found to be invaluable as a tool for warning of approaching severe weather.

In 1969 the space probes Mariner 6 and Mariner 7 used radar to reveal mountainous structures covering the surfaces of Mars and Venus.

Modern airplanes typically have radar antennae in the nose to receive reflected signals from clouds ahead. Aircraft and ships use radar not just to monitor weather conditions, but also to regulate altitude, locate airstrips and harbor mouths, and map Earth's land contours and sea floors.

Schuyler

For Further Reading:

Brown, Jim. *Radar: How It All Began.* London: Janus, 1996.

Buderi, Robert. *The Invention That Changed the World: How a Small Group of Radar Pioneers Won the Second World War and Launched a Technological Revolution.* New York: Simon & Schuster, 1996.

Watt, James

Developer of Steam Engine
Technology
1736-1819

Life and Work

James Watt forged innovations in the steam engine that helped to fuel the Industrial Revolution.

James Watt was born at Greenock, Scotland, on January 19, 1736. Throughout his life he suffered from debilitating migraine headaches, a misfortune that hindered his education. As a boy, he worked in his father's workshop and developed great mechanical skill; from 1753 to 1757, he learned the trade of mathematical-instrument maker in Glasgow and London. He then set up a shop at the University of Glasgow for the fabrication of compasses, scales, quadrants, and other such equipment.

In 1764 Watt was asked to repair a working model of the steam engine, invented in 1705 by THOMAS NEWCOMEN. Watt began investigating the properties of steam and the operation of the engine; the following year he had the idea of adding a separate, cooled condenser to the engine. He built a model incorporating his new design, and the resulting steam engine exhibited increased thermal efficiency and reduced operational cost. He was granted a patent for his steam engine in 1769.

Following financial difficulties in 1774, Watt formed a partnership with Matthew Boulton, owner of the Soho Foundry near Birmingham, England. For the next 25 years, Watt's steam engines were manufactured at the Soho Foundry.

During that time, Watt introduced many improvements to his engine: the sun-and-planet gear wheel (which converted reciprocal motion to rotary motion); a parallel-motion mechanical linkage (which made possible a double-acting engine); an automatic centrifugal governor (which regulated steam); and a steam indicator (which measured steam pressure and degree of vacuum).

Watt coined the term "horsepower" to rate the work capabilities of his engines over time and standardized the use of the term as a unit of power.

In 1800 Watt's patent expired and he retired to Heathfield Hall, near Birmingham, where he died on August 25, 1819.

Legacy

Watt's improvements to the steam engine helped to push the Industrial Revolution forward and transformed life in industrialized countries.

Watt's engine allowed factories to be built away from rivers and streams, the primary source of power for industry in Watt's time, and closer to key transport routes, raw materials, and other resources. This enabled manufacturing to develop in the Midwest of the United States. As the steam engine changed the energy base of manufacturing from stream power to coal, it made Britain, which possessed vast reserves of coal, the world's richest industrial power during Watt's lifetime.

The Watt engine was more efficient and less costly than its predecessor and offered a practical source of power to many industries. After the incorporation of Watt's sun-and-planet gear wheel and the conversion to rotary motion, the steam engine was adopted for textile manufacture. It had previously been used only for pumping water in mines, salt works, and iron works.

Watt's work also influenced the founding of thermodynamics, the study of heat. In addition, the precision of his designs and manufacturing techniques raised engineering standards.

Watt represented a model of the connection between objective scientific research and technological development. Prior to Watt's investigation of the steam engine, JOSEPH BLACK had introduced the concept of latent heat, the heat lost or absorbed when matter changes state (for example, when liquid becomes solid or gaseous). Watt learned of Black's concept, and he discovered that significant latent heat was wasted in Newcomen's steam engine. Watt set the agenda for the development of all subsequent heat engines and demonstrated the importance of economizing heat in the operation of all engines.

Watt's steam engine was a milestone in the development and proliferation of machine-powered manufacturing and transformed the conditions of life in industrialized countries.

Schuyler

WORLD EVENTS	WATT'S LIFE
Peace of Utrecht 1713-15 settles War of Spanish Succession	
	1736 James Watt is born
	1757 Watt opens a mathematical-instrument shop at University of Glasgow
	1765 Watt designs steam engine with improvements to the Newcomen model
	1769 Watt obtains a patent for his engine
	1775 Watt joins with Matthew Boulton to manufacture steam engines
United States independence 1776	
	1770s -80s Watt engineers further improvements to the steam engine
French Revolution 1789	
	1800 Watt retires
Napoleonic Wars 1803-15 in Europe	
	1819 Watt dies

For Further Reading:

Dickinson, Henry Winram. *James Watt: Craftsman and Engineer.* New York: A. M. Kelley, 1967.

———. *James Watt and the Steam Engine: The Memorial Volume Prepared for the Committee of the Watt Centenary Commemoration at Birmingham 1919.* Ashbourne, England: Moorland, 1981.

Wegener, Alfred Lothar

Developer of Law
of Continental Drift
1880-1930

Life and Work

Alfred Wegener was the first person to develop a comprehensive view of continental drift.

Wegener was born in Berlin in 1880, the son of a clergyman who ran an orphanage. He was an energetic child who took long mountain hikes and learned Alpine skiing in order to better explore his world. He used kites and balloons to study the dynamics of the atmosphere, and in 1906 he and his brother Kurt broke the previous world record for long duration flight (52 hours) in a free balloon.

A determined student with wide-ranging curiosity, Wegener earned a Ph.D. in astronomy from the University of Berlin in 1905. Soon afterward, he made a two-year expedition to Greenland as a qualified meteorologist to study polar air circulation and glaciers.

Wegener discussed the apparent fit of continental margins of eastern Brazil and West Africa with fellow students as early as 1903. He was impressed by reports of corresponding geological formations in South America and Africa, but dismissed speculation that these areas may have been joined by a land bridge in paleohistory. Wegener was convinced that the Atlantic Ocean was not the result of sinking continental margins but of physical separation of one giant land mass, Pangaea, about 250 million years ago.

He researched facts from many specialized areas of science before writing his first papers on continental displacement in 1912. Coal deposits in Antarctica, fossil corals on the Oregon coast, remains of fossil plants on the ice-covered Arctic Ocean island of Spitsbergen, and glacial relicts in central Africa were some of the clues that Wegener used in declaring that the continents must move. His most famous work, *The Origins of the Continents and Oceans*, was first published in German in 1915 and in English in 1924.

After another Greenland trek (1912-1913), Wegener married the daughter of one of his mentors, Wladimir Koppen, a leading meteorologist in Germany. Drafted into the infantry in 1914, Wegener served in World War I. After being wounded twice, he was assigned to the German Army's weather-forecasting service. When the war ended, he was hired as a lecturer in his father-in-law's teaching position at the University and German Marine Observatory in Hamburg. But his unpopular ideas on mountain building (contradicting the prevalent "shrinking Earth" models) and on movement of continents kept him from advancing in his profession.

Moving to Austria in 1924, Wegener became professor of meteorology and geophysics at the University of Graz. He went to Greenland again in 1929-1930. He was last seen alive on his fiftieth birthday, October 30, 1930, as he headed out on the ice toward a glaciological field station.

Legacy

More than any other scientist of his time, Alfred Wegener challenged the prevailing belief in a basically static, unchanging Earth. His ideas on continental drift would provide one of the basic tenets for geological study in the twentieth century.

Wegener's hypothesis was a threat to the mosaic of complicated explanations for surface features that geologists had developed over several generations. Wegener's critics defended land bridges as explanations for fossil similarities in various parts of the world, they asked for a mechanism to push drifting masses across solid ocean floors, and they defended the permanency of such major features as entire continents. Further encouraging resistance to his ideas, Wegener could not prove that continents moved, given the limitations of 1920s technology.

Even though Wegener died knowing that his ideas about continental drift were rejected by most of his peers, the strength and comprehensiveness of his ideas persisted. Geophysicists would elaborate and corroborate his ideas 30 years after Wegener disappeared on the Arctic ice. In 1959 HARRY HAMMOND HESS proposed an elaboration of Wegener's theory of continental drift in his theory of plate tectonics, which described the outer layer of Earth as a set of plates moving across the top of Earth's mantle. Plate tectonics later helped to explain the causes for earthquakes and volcanoes.

Many contemporary scientists follow the legacy of his method, combining evidences from every area of physical and life sciences to continue to build a more complete history of the dynamic planet Earth.

Simonis

For Further Reading:

Anderson, Jr., Alan. *The Drifting Continents.* New York: Putnam, 1971.

Miller, Russell. *Continents in Collision.* Alexandria, Va.: Time-Life Books, 1983.

Sullivan, Walter. *Continents in Motion.* New York: McGraw-Hill, 1974.

Whitney, Eli

Inventor of the Cotton Gin

1765-1825

Life and Work

Eli Whitney patented the cotton gin and introduced the concept of mass-production of interchangeable parts to manufacturing.

Whitney was born on December 8, 1765, to a farming family in Westboro, Massachusetts. He demonstrated superior mechanical aptitude as a youth. He graduated from Yale College (now Yale University) in 1792 and traveled to Georgia, where he was promised a tutoring position. The job fell through, and he was left far from home without money or a job.

Catherine Greene, the owner of a cotton plantation in Georgia, befriended the jobless Whitney. He became familiar with cotton production, and Greene pointed out that the cotton seed adheres to the fiber, making the cleaning process extremely tedious. She encouraged him to use his mechanical talents to design a machine to clean the cotton more efficiently. By 1794 he had built such a device, but it didn't work well. Greene suggested that he substitute wire hooks (like crochet hooks) for the wooden teeth he had used on the rollers of the "cotton engine" or gin. With this change, Whitney patented the invention that drastically reduced time spent cleaning cotton.

Whitney and his partner Phineas Miller

(Greene's husband) opened a business in Connecticut for the manufacture and service of cotton gins. However, the gins were simple to copy, and Whitney and Miller were out of business by 1797. After the partners lost numerous patent-infringement lawsuits, several southern states reluctantly offered them monetary compensation for the invention.

Meanwhile, France had threatened war, and the United States government had solicited 40,000 muskets from private contractors. Whitney signed an agreement in 1798 to deliver 10,000 muskets in two years. He planned to break with the traditional method, whereby a skilled worker constructed a complete, unique musket. Whitney's muskets were fashioned piecemeal, each part conforming to a model and made by a worker operating a standardized machine tool. The result was a collection of muskets with interchangeable, easily replaceable parts. Because of unforeseen delays, Whitney's order was delivered eight years late in 1808. In 1812 the government requested a second installment of muskets; this one was delivered in two years.

Whitney died in New Haven, Connecticut, on January 8, 1825.

Legacy

Whitney's cotton gin increased cotton production and altered the economic and cultural future of the South. And his introduction of interchangeable parts played a key role in the industrialization of the modern world.

The cotton gin enabled a worker to produce about 23 kilograms of cleaned cotton per day, a 50-fold increase over cleaning cotton by hand. The demand for cotton from English mills was rising, and cotton plantations expanded and enjoyed prosperity in the southern United States.

The impact of the cotton gin on the South was not all positive. The cotton gin prolonged slavery and contributed to the Civil War. The increased production at cotton plantations required increased labor, and more slaves were used to fill the labor shortage. The cotton gin also helped to turn the lower South into a single-commodity economy. While cotton fueled economic growth, it left the region vulnerable to fluctuations in the cotton market and frequent devastating infestations of the boll weevil. Continued reliance on this single crop

eventually depleted the soil of its nutrients, reducing the agricultural output and economic viability of the South.

Whitney's use of milling machines to produce interchangeable, uniform parts marked the beginning of the modern era of manufacturing. Within 15 years of Whitney's impressive musket delivery, the United States government required all manufacturers of small arms to use uniform parts. It facilitated fabrication and repair. Such standards of production spread to other industries and mass-produced items increased by the mid-nineteenth century. Factory assembly lines and divisions of labor developed along with the practice of mass production. The efficiency of the new manufacturing method helped the United States grow into one of the leading industrialized nations.

Schuyler

World Events		Whitney's Life
	1765	Eli Whitney is born
United States independence	1776	
French Revolution	1789	
	1792	Whitney graduates from Yale
	1794	Whitney builds the first cotton gin
	1797	Pirating of their cotton gin design puts Whitney and Phineas Miller out of business
	1798	Whitney contracts to deliver 10,000 muskets to U.S. government
Napoleonic Wars in Europe	1803-15	
	1808	Whitney completes first musket order
	1812	Government places second order
	1825	Whitney dies
United States Civil War	1861-65	

For Further Reading:

DeCoin, Robert L. *History and Cultivation of Cotton and Tobacco.* Wilmington, Del.: Scholarly Resources, 1973.

Olmsted, Denison. *Memoir of Eli Whitney.* New York: Arno Press, 1972.

Williams, Daniel Hale

Founder of Cardiac Surgery
1858 to 1931

Life and Work

Daniel Hale Williams, the most prominent African-American physician of his time, performed the first successful open-heart surgery and helped integrate medical education and practice.

Born on January 18, 1858, in Hollidaysburg, Pennsylvania, Williams began an apprenticeship to physician Henry Palmer in 1878. He entered Chicago Medical College, received his medical degree in 1883, and opened an integrated practice on the south side of Chicago. From 1884 to 1889, he served as a physician at the Protestant Orphan Asylum, lectured in anatomy at his alma mater, and was appointed to the Illinois State Board of Health.

Most Chicago hospitals at that time did not offer internships to African-Americans or allow black physicians to practice. Williams was determined to open Chicago's medical community to blacks, and in 1891 he founded Provident Hospital, where an integrated staff focused on training black physicians and nurses.

In 1893 a man who had been stabbed in the chest was brought to Provident Hospital. After inspecting the wound, Williams decided that surgery would be the only means of saving the man's life, despite the dominant attitude among surgeons at that time that operating on the heart was too dangerous to attempt. Williams opened the victim's thoracic (chest) cavity and sutured a tear in the pericardium, the membrane that encloses the heart. The man recovered and lived for 20 years; Williams had performed the first recorded successful open-heart surgery.

Williams was appointed chief surgeon at Freedmen's Hospital in Washington, D.C., in 1894. He made vast improvements there: he reorganized the administration, upgraded the facilities, strengthened the nurse-training program, and decreased the patient death rate.

Williams accepted a professorship of clinical surgery at Meharry College in Nashville, Tennessee, in 1899. The following year, he completed research disproving the belief (held by many of his contemporaries) that black women are not at risk for ovarian cancer.

During the next decade, Williams helped establish numerous hospitals to serve black communities across the country. He died in Idlewild, Michigan, on August 4, 1931.

Legacy

Williams's emergency surgery launched the practice of cardiac surgery and his achievements represented a vital step toward ending the segregation of medical care in the United States.

The founding of Provident Hospital increased the opportunities for African-Americans both to give and to receive medical care. The hospital provided improved medical services for the black community of Chicago, and as a full-service hospital, it served as an excellent training ground for black physicians. Including the first nurse-training program open to African-Americans in Chicago, it offered black women a profession that had been closed to them previously. Williams was closely involved in the founding of integrated medical institutions in Tennessee.

Provident Hospital also served as a model for other integrated hospitals in Missouri, Kentucky, Alabama, Georgia and Texas. After black physicians and nurses began to train and practice at these hospitals, the death rate among black patients dropped significantly.

The open-heart operation Williams performed illustrated the plausibility of successful cardiac surgery. Surgeons became bolder, and the practice of open-heart surgery eventually became routine. Some of the specific techniques Williams was forced to improvise during the first operation spread quickly among surgeons across the country, leading to a greater range of surgical procedures.

Williams broke the confines of the racial inequality prevalent in late-nineteenth-century American society and eliminated the taboo against heart surgery prevalent at that time. His commitment and success elevated him to the status of hero and made him a role model not just for African-Americans, but for pioneers of medical innovation and progress.

Schuyler

For Further Reading:

Buckler, Helen. *Daniel Hale Williams: Negro Surgeon.* New York: Putnam, 1968.

Fenderson, Lewis. *Daniel Hale Williams: Open-Heart Doctor.* New York: McGraw-Hill, 1971.

Wöhler, Friedrich

Synthesizer of Urea

1800-1882

Life and Work

Friedrich Wöhler challenged the conventional wisdom of his time and demonstrated that it was possible to synthesize organic compounds from inorganic materials. His work in both organic and inorganic chemistry yielded important discoveries.

Wöhler was born in Eschersheim, Prussia (present-day Germany), on July 31, 1800. In 1820 he entered the University of Marburg to become a doctor. He graduated with a medical degree in 1823 from the University of Heidelberg. At that time, he decided to focus on another area of science, chemistry.

Wöhler worked with Jöns Berzelius, an eminent chemist in Sweden, and began a life-long correspondence with him. He soon began to study ways to isolate metals. In 1827 Wöhler heated an aluminum compound with potassium, which resulted in a nearly pure aluminum sample. Until this time, aluminum was not considered to be a metal. Wöhler's research correctly identified it. He used a similar method to isolate beryllium, another metallic element, in 1828.

Many chemists during the early 1800s believed that organic compounds contained some vital force that distinguished them from inorganic compounds. In 1828 Wöhler conducted an experiment in which he combined an ammonia solution and lead cyanate, two inorganic com-

pounds. Much to his surprise, the product resembled urea crystals, which are formed by acetic acid, an organic compound. The scientific community found it impossible to ignore his findings and the rigid distinction between inorganic and organic compounds eased. The vital force theory eventually lost support.

Wöhler investigated a wide variety of elements and compounds. He isolated or synthesized boron, silicon, and titanium. He researched the similarities between the elements carbon, silicon, and titanium. In 1831 Wöhler, in conjunction with German scientist Justus von Liebig, demonstrated that silver cyanate and silver fulminate had identical chemical compositions but very different chemical properties, a finding that would have significance for contemporary and future research. The pair later isolated the benzoyle group of radical atoms from the oil of almonds. A radical group is a group of atoms that remain together as a unit during chemical transformations.

Wöhler remained active throughout his life. He became a professor of chemistry in 1836 at the University of Göttingen and continued into his later years. He also began studying meteorites and geology. Wöhler died on September 23, 1882, in Göttingen, Germany.

Legacy

Wöhler's synthesis of urea, an organic compound, helped usher in a new understanding of organic chemistry. His research into organic compounds helped broaden the body of knowledge concerning metals and their characteristics.

Wöhler's work had an immediate impact on his contemporaries. His success with acetic acid and his discovery that organic substances could be synthesized from inorganic compounds encouraged other researchers to seek new ways of converting compounds from one chemical group to another. Wöhler's and Liebig's finding that compounds with identical chemical compositions can have distinctly different properties provided necessary experimental proof for Berzelius to propose the theory of isomerism, which states that chemical properties are dependent on both the number and types of atoms as well as the structural arrangement of those atoms. Isomerism is a fundamental concept in today's study of organic chemistry. Wöhler also helped his fellow European chemists by translating the detailed, yearly findings of Berzelius, which

included classification systems for matter and the polar theory of chemical combination.

During his time as a professor, Wöhler taught thousands of students at the prestigious teaching laboratory at the University of Göttingen. Many went on to teach chemistry at the college level themselves.

Wöhler's influence will last into the twenty-first century. The isolation of aluminum led to the discovery of many uses for this strong, light-weight metal, such as automobile parts and beverage containers. Aluminum continues to find additional applications in today's technology. The key radical groups that Wöhler and Liebig identified, in particular the benzoyle group, helped chemists understand how organic molecules behave during chemical reactions. Today, the synthesis of organic compounds affects medical research and has applications in many industries. Important research is now being done to find ways to reverse Wöhler's discovery. Now scientists seek ways to convert organic compounds, which are renewable, into inorganic compounds, such as nonrenewable petroleum products.

Wilson

WORLD EVENTS		WÖHLER'S LIFE
French Revolution	1789	
	1800	Friedrich Wöhler is born
Napoleonic Wars in Europe	1803-15	
	1823	Wöhler receives medical degree from Heidelberg
	1827	Wöhler isolates aluminum
	1828	Wöhler isolates beryllium
		Wöhler synthesizes urea
	1831	Wöhler and Justus von Liebig compare silver fulminate and silver cyanate
	1836	Wöhler becomes professor of chemistry at University of Göttingen
United States Civil War	1861-65	
Germany is united	1871	
	1882	Wöhler dies
Spanish-American War	1898	

Further Reading:

Abbott, D., ed. *The Biographical Dictionary of Scientists: Chemists.* New York: Peter Bedrick Books, 1983.

Von Hoffman, A. W. "Friedrich Wöhler." In *Great Chemists.* Edited by Eduard Farber. New York: Interscience, 1961.

Wright, Jane Cooke

Pioneer in Chemotherapy
1919-

Life and Work

Jane Cooke Wright, a physician, has devoted her career to chemotherapy cancer research and medical education.

Born in New York City on November 20, 1919, Wright comes from a family of prominent African-American physicians. In 1938 she won a four-year scholarship to Smith College in Massachusetts, performing at the top of her class and setting records in varsity swimming. Upon graduating in 1942 she entered New York Medical College and received a medical degree with honors three years later.

Wright was appointed visiting physician at Harlem Hospital in 1949 and later that year was promoted to clinician at the hospital's Cancer Foundation. Conducting research in tumor response to chemotherapy (the use of chemical agents to treat disease) in cancer treatment, she investigated the complex nature of the disease as demonstrated in test animals, tissue samples, and patient responses to chemotherapy. She became director of the foundation in 1952.

In 1955 Wright began supervising the chemotherapy research department and teaching research surgery at New York University's Medical Center. Six years later she became adjunct professor of research surgery at the university and vice president of the African Research Foundation, subsequently touring East Africa as part of a medical team.

Wright served on the 1964 President's Commission on Heart Disease, Cancer, and Stroke, advocating the development of centers for the treatment of these diseases. The following year the Albert Einstein College of Medicine honored her with its Spirit of Achievement Award.

New York Medical College appointed Wright to the positions of associate dean and professor of surgery in 1967. She also participated in the medical school's administration.

Wright earned numerous awards throughout the 1970s and 1980s, and in 1983 she presented a lecture entitled "Cancer Chemotherapy: Past, Present, and Future" at the convention of the National Medical Association. In 1987 Wright became professor emerita of surgery at New York Medical College.

She has also served as a trustee of Smith College and the American Cancer Association and has been an editor of the *Journal of the National Medical Association.*

Legacy

Wright has promoted and conducted influential research in the field of cancer chemotherapy. Along with other members of her family, she has also been an inspiration and role model in the African-American community.

Wright carried on her family's legacy in the medical field. Her father, Louis Tompkins Wright, was among the first black graduates of Harvard Medical College and was the first African-American physician to join the staff of a New York City hospital. Her grandfather was among the first graduates of Tennessee's Meharry Medical College, founded to train former slaves in medical fields. Harry D. West, another of Wright's relatives, was the first president of that institution.

Wright introduced several chemotherapeutic agents and techniques subsequently used to treat malignant tumors. Her chemotherapy research and educational efforts influenced progress in the field.

Her work directly affected public health policy both on the national and regional levels. Wright's report to the President's Commission in 1964 led to the development of a nationwide network of cancer-treatment centers. At New York Medical College, she contributed to the development of a program of study of cancer, heart disease, and stroke.

Schuyler

For Further Reading:

Sammons, Vivian O. *Blacks in Medicine and Science.* New York: Hemisphere Publishing, 1990.

Smith, Jessie Carney, ed. *Notable Black American Women.* Detroit: Gale Research, 1992.

Wright, Orville
Wright, Wilbur

Inventors of the Modern Airplane
1871-1948; 1867-1912

Life and Work

Inventors Orville and Wilbur Wright designed, built, and flew the first powered, manned airplane.

Wilbur Wright was born on April 16, 1867, near Millville, Indiana; his brother Orville was born on August 19, 1871, in Dayton, Ohio. Although they showed great mechanical aptitude as youths, neither brother finished high school nor attained any formal higher education. However, their curiosity and motivation led to a life-long career in engineering. In 1889, they began publishing a newspaper that they printed on a home-built printing press. In 1892, they opened a shop for the design, manufacture, and repair of bicycles.

The Wright brothers developed an interest in aviation while reading the works of German aviation engineer Otto Lilienthal, who had successfully performed numerous experiments with gliders before his death in 1896. That year, Orville and Wilbur began to design a flying machine. Observing buzzards in flight, they realized that an airplane must be capable of elevating and descending, turning left and right, and banking to either side. The Wrights flew their first biplane glider in 1900 at Kitty Hawk, North Carolina. The following year, they tested hundreds of miniature wings in a small wind tunnel and applied their data to the design of two improved gliders flown at nearby Kill Devil Hills in 1901 and 1902.

Turning to powered flight, the Wrights designed an efficient propeller and a light engine, and, in 1903, they completed their first airplane, *Flyer I* (now called *Kitty Hawk*). The first flight, on December 17, 1903, covered 120 feet in 12 seconds. In 1904 and 1905, the Wrights built *Flyer II* and *Flyer III*—the latter could remain airborne for half an hour and easily turn, bank, circle, and cut figure eights.

In 1908 and 1909, the brothers caused public sensations with flight demonstrations in France and New York, where Wright companies were formed for the manufacture of airplanes. The United States War Department signed a contract with the Wright brothers for the construction of the first army plane.

Wilbur died of typhoid in Dayton, Ohio, on May 30, 1912. Orville contributed to aeronautics engineering until his death in Dayton on January 30, 1948.

Legacy

The Wright brothers' airplanes revolutionized aviation engineering and forever altered the face of transportation for military and civilian travelers across the globe.

The Wrights improved aircraft technology severalfold with their wind-tunnel data, their propeller and engine designs, and their recognition of the importance of three-axis flight control. Their insight added the finishing touch to the already advanced developments in aerodynamics and structural engineering.

Following the Wrights' public flight demonstrations, an explosion of research into airplane structures and engines led to immediate advances in the progress of flight. In 1909, Louis Blériot of France flew over the English Channel, and English aviator Henri Farman completed the first flight over 100 miles. In 1910, Eugene Ely became the first pilot to take off in an airplane from the deck of a ship, demonstrating the plausibility of aircraft carriers. Russian-born American IGOR SIKORSKY built and flew the first multi-engined airplane in 1913. Captain John Alcock of Britain piloted the first non-stop flight across the Atlantic in 1919. Charles Lindbergh's famous 1927 flight was the first solo, non-stop flight across the Atlantic.

Commercial air service began in 1919 and has since become accessible to vast numbers of people. The first airplane carrying civilian passengers flew flawlessly from Paris to London, and soon a British company, Aircraft Transport and Travel, instituted regular passenger flights between those two cities. Passengers were issued leather jackets, goggles, and helmets, which they could keep as souvenirs after the flight.

Air travel has transformed modern life. It has brought people and cultures closer together, increased the speed of business communication, and escalated the flow of goods and services across the world. These profound changes resulted from the innovations of the Wright brothers, whose groundbreaking work set the field of aviation in motion.

Schuyler

WORLD EVENTS		THE WRIGHTS' LIVES
	1867	Wilbur Wright is born
Germany is united	1871	Orville Wright is born
	1892	Wrights build, repair, and sell bicycles
Spanish-American War	1898	
	1896	Wrights observe buzzards and design glider
	1900-02	Wrights fly gliders and collect wind tunnel data
	1903	Wrights pilot the first airplane flights
	1908	Wrights perform public flight demonstrations
	1909	Wrights complete the first army plane
	1912	Wilbur Wright dies
World War I	1914-18	
World War II	1939-45	
	1948	Orville Wright dies

Orville Wright

Wilbur Wright

For Further Reading:

Degan, Paula. *Wind and Sand: The Story of the Wright Brothers.* Conshohocken, Penn.: Eastern Acorn Press, 1983.

Kelly, Fred. *The Wright Brothers.* Magnolia, Mass.: Peter Smith Publisher, 1991.

Kelly, Fred C., ed. *Miracle at Kitty Hawk: The Letters of Wilbur and Orville Wright.* New York: Arno Press, 1971.

Wu, Chien-Shiung

Demonstrator of
Non-Conservation of Parity
c. 1912-

Life and Work

Chien-Shiung Wu demonstrated experimentally that parity, a fundamental property of nuclear particles related to symmetry, is not always conserved in certain weak interactions of the atomic nucleus.

Wu was born in Liu Ho (near Shanghai), China, in May 1912 (although some sources say 1915, and others 1913). Until age nine, she attended a school for girls operated by her parents and then transferred to a private high school, where she became politically active. After graduating as valedictorian in 1930, she matriculated at the National Central University in Nanking and four years later received a degree in physics.

Wu crossed the Pacific in 1936 and began graduate studies at the University of California, Berkeley, where she worked under Ernest Orlando Lawrence, inventor of the cyclotron, a device that accelerates and smashes nuclear particles. The university granted her a Ph.D. in 1940 and then hired her as a research assistant. From 1942 to 1944, she taught at Smith College and Princeton University, and finally started a long-term post at Columbia University, New York, in March 1944. Her first duties were war related; she joined the Manhattan Project, the program aimed at building the first atomic bomb.

In 1945 Wu became a research associate at Columbia; in 1952 she was promoted to associate professor, and in 1958 she attained the level of full professor.

Wu's most important work occurred in 1957. She had become proficient at experiments involving beta decay, a radioactive transformation of atoms in which beta particles are ejected from the nucleus. Because of her expertise in this area, she was approached by her colleagues Chen Ning Yang and Tsung-Dao Lee after they had formulated a revolutionary theory concerning parity.

Parity refers generally to the symmetry, or the absence of a bias in direction, of the physical laws of nature. Prior to 1956, physicists thought that parity is conserved in all nuclear reactions; with respect to particle emission, this means that the emitted particles in a given reaction are emitted in all directions equally. In 1956 Yang and Lee found theoretical evidence suggesting that parity is not conserved in certain weak interactions of the nucleus, for example, those associated with beta decay.

Wu immediately began testing this hypothesis, using the beta decay of radioactive cobalt-60. In 1957 she showed experimentally that when these cobalt atoms are placed in a magnetic field, more beta particles are emitted in the direction opposite the magnetic field than in any other direction. Thus Wu proved that parity, in this case the symmetrical emission of beta particles, is not conserved. She received numerous awards for this accomplishment.

Later in 1957, Yang and Lee were awarded the Nobel Prize for Physics for their work on parity; some observers have commented that Wu should have been included in this honor.

Wu subsequently provided experimental confirmation of two other theories of nuclear physics: the conservation of a vector current in beta decay and the release of electromagnetic radiation upon the collision of an electron and positron.

Legacy

Wu's confirmation that parity is not always conserved represented a revolutionary step in the field of nuclear physics. Her contribution was all the more important as it enforced the fact that women could make influential contributions to science, particularly in the field of nuclear physics, a field dominated by men.

By the 1950s, the conservation of physical properties was an integral element of classical physics. It had been demonstrated that important physical properties of particles, such as mass, energy, momentum, and electrical charge, are the same before and after a nuclear reaction. The nonconservation of parity forced physicists to rethink some basic assumptions and was also valuable in explaining some otherwise puzzling experimental results in nuclear physics.

Wu's work stimulated research in the progressive investigation of symmetry, pursued by influential physicists RICHARD FEYNMAN, MURRAY GELL-MANN, Robert Marshak, Ennackel Chandy, and George Sudarshan.

Schuyler

For Further Reading:

Bertsch McGrayne, Sharon. *Nobel Prize Women in Science.* Secaucus, N.J.: Carol Publishing Group, 1993.

Kass-Simon, G., and Patricia Farnes, eds. *Women of Science: Righting the Record.* Bloomington: Indiana University Press, 1990.

Yalow, Rosalyn Sussman

Developer of Radioimmunoassay
1921-

Life and Work

Rosalyn Yalow, born Rosalyn Sussman, developed radioimmunoassay, a technique that enables researchers to identify minute amounts of proteins, hormones, and other substances in body fluids. This powerful tool can be used to study how the body is affected by factors such as disease and chemicals.

Rosalyn Sussman was born on July 19, 1921, in New York City to a second-generation immigrant family in the Bronx. Her parents had little education but worked hard to provide for the family and recognized the value of school for their children. Sussman attended local schools and eventually went to Hunter College in New York City, graduating in 1941. During these years she became enthralled with science, particularly the new field of nuclear physics. With World War II beginning, many graduate fellowships held by men became available. In 1941 Sussman was offered a teaching assistant position at the University of Illinois, providing her the means to pursue her doctoral degree in physics. While there she met another young doctoral student, Aaron Yalow. They married two years

later in 1943. She received her degree in nuclear physics in 1945.

After several jobs in teaching and research, Rosalyn Yalow began working at the Bronx Veterans' Administration Hospital. There she began a long collaboration in nuclear medicine with physiologist Solomon Berson. From 1950 to 1972, this was a formidable partnership: her nuclear expertise complemented his medical background.

Together, in 1959, they developed a new technique for identifying and quantifying the presence of both biological and chemical substances in the body, a process known as radioimmunoassay. Their new technique enabled them to draw a fluid sample and subject it to radioactively-tagged antigens, foreign substances to which the human body reacts by producing antibodies to neutralize them. The degree of antibody bonding could be calculated and compared to a known sample. Using the body's immune system in this way was revolutionary and incredibly precise, allowing small traces of substances—as small as 0.000000001 grams—to be detected.

Initially Yalow and Berson used radioimmunoassay to investigate levels of insulin in adult diabetics. Later they worked long hours together developing new applications for it. Berson died unexpectedly in 1972, leaving Yalow to work on her own. Yalow was awarded the Lasker Prize in basic medical research in 1976, the first woman to be so honored.

In 1977 she was awarded the Nobel Prize for Physiology or Medicine for the development of radioimmunoassay. The prize was shared with two other researchers who had used the technique in their work. Yalow continued research until retiring in 1991.

Legacy

Rosalyn Yalow provided the medical research community with a powerful tool. Her technique of radioimmunoassay has been described as the most innovative method in physiology since the invention of the X-ray machine.

Her initial work on diabetes prompted new theories on how the body reacts to insulin in adult diabetics. It had been previously thought that the insulin molecule was too small to trigger an antibody response; radioimmunoassay's

precise and acute detection capabilities proved those ideas false.

The radioimmunoassay method has been adapted to many uses. It can detect small traces of poison in bodies that have been buried for years. Forensic science now uses this technique. Testing for exposure to chemicals or drugs can be done through blood or hair samples. One reason for the rapid spread of this invaluable technique may be that Yalow and Berson declined to patent it, sharing it with other scientists.

Yalow's contributions to medical science can not yet be fully measured because new applications continue to be developed. Nevertheless, Yalow seeks to make a difference, even during retirement. She lectures on science education and is a role model for young women, promoting the role of women in science.

Wilson

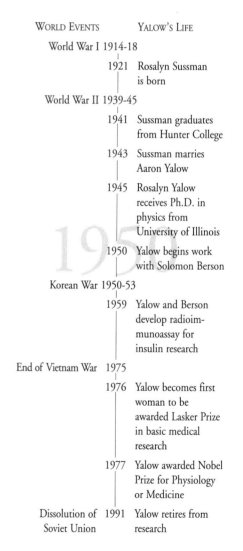

WORLD EVENTS	YALOW'S LIFE
World War I 1914-18	
	1921 Rosalyn Sussman is born
World War II 1939-45	
	1941 Sussman graduates from Hunter College
	1943 Sussman marries Aaron Yalow
	1945 Rosalyn Yalow receives Ph.D. in physics from University of Illinois
	1950 Yalow begins work with Solomon Berson
Korean War 1950-53	
	1959 Yalow and Berson develop radioimmunoassay for insulin research
End of Vietnam War 1975	
	1976 Yalow becomes first woman to be awarded Lasker Prize in basic medical research
	1977 Yalow awarded Nobel Prize for Physiology or Medicine
Dissolution of Soviet Union 1991	Yalow retires from research

For Further Reading:

Chard, T. *An Introduction to Radioimmunoassay and Related Techniques.* New York: North-Holland Pub., 1978.

Shiels, Barbara. *Winners: Women and the Nobel Prize.* Minneapolis, Minn.: Dillon Press, 1985.

Straus, Eugene. *Rosalyn Yalow, Nobel Laureate: Her Life and Work in Medicine.* New York: Plenum Trade, 1998.

Yukawa, Hideki

Founder of Meson Theory
1907-1981

Life and Work

Hideki Yukawa created meson theory to explain nuclear forces within the atom, predicting the existence of the subatomic pi-meson and the strong nuclear force associated with this elementary particle.

Hideki Ogawa was born in Tokyo, Japan, on January 23, 1907. His grandfather began acquainting the child with both Chinese and Japanese classical literature before he entered primary school in Kyoto. While still in middle school, Ogawa found Laotse and Chuangtse, early writers and interpreters of the Chinese religion Taoism, especially pleasing because their thinking centered on nature, not the social behaviors emphasized by Confucius.

Ogawa was studying physics at the University of Kyoto (1926-29) when Erwin Schrödinger introduced wave mechanics, creating a sensation among scientists. The continuity of nature that unified wave theory implied appealed to Ogawa and he read everything available on the subject.

In 1932 Ogawa married Sumi Yukawa. Following Japanese tradition, he was adopted by her family and changed his surname. That same year Hideki Yukawa became a lecturer at Osaka University while he worked toward the doctorate awarded him in 1938. He then accepted a professorship at Kyoto University.

Yukawa proposed a new theory of nuclear forces in 1935 during a period of controversy about the nature of atomic particles. JAMES CHADWICK had discovered neutrons, so named because they are neutral in electrical charge, in 1932. However, Yukawa pointed out that if the nucleus contains only neutrons and positively-charged protons, identified and described by scientists in 1920, the protons, all of the same positive charge, should repel each other and not hold together in the nucleus as they do. The Japanese student struggled for more than two years to explain the nature of nuclear forces and why the nucleus held together as it did. His conception of a nuclear force field suggested that there must be an intermediate-sized "exchange" particle between protons and neutrons for the nucleus to exist. Yukawa calculated that such a particle would have a mass about 200 times greater than that of an electron, the same negative charge as an electron, and be radioactive with a very short half-life. He predicted that such a particle, called a meson, could be found in cosmic rays.

In 1936 Carl Anderson discovered in cosmic rays the mu-meson, which possesses some properties of Yukawa's meson. By 1947 Yukawa had developed a two-meson model to accommodate this new experimental data to his theory. Cecil Powell then discovered the pi-meson, which fulfilled all the requirements Yukawa described for his imagined particles. Yukawa was awarded the Nobel Prize for Physics in 1949 for this work, the first Japanese scientist to be so honored.

Yukawa wrote several essays and books on creativity, intuition, and cultural influences on thinking. He contended that the tension between Eastern and Western cultures stimulated his thinking in ways a more traditional acceptance of one culture would not.

Yukawa worked at the Institute for Advanced Study in Princeton, New Jersey, in 1948 and taught at Columbia University from 1949 to 1953 before returning to Kyoto to direct the new Research Institute of Fundamental Physics. He also founded an international journal in 1946 called *Progress of Theoretical Physics,* published in several European languages. Yukawa was its editor until he died in Kyoto on September 8, 1981.

Legacy

Yukawa's meson theory, while later refined, provided a bold new way of thinking about questions concerning elemental forces in the nucleus.

Yukawa's prediction of the existence of the meson prompted a wave of discoveries in nuclear physics. In 1947 British physicists Clifford Butler and George Rochester discovered a new type of particle that was lighter than a proton but heavier than the meson particles previously identified. These particles behaved strangely in relation to nuclear matter and, in fact, were referred to as "strange particles." Four types of strange particles were identified by 1953. American physicist MURRAY GELL-MANN and Japanese physicist Nishijima Kazuhiko independently proposed a new theory of conservation to help understand the behavior of these various new subatomic entities. In 1964 Gell-Mann and American scientist George Zweig independently suggested that these elements were composed of tinier building blocks that Gell-Mann later termed quarks, the word still used today.

Yukawa helped to influence the direction of contemporary physics in other ways. He taught and encouraged young Japanese collaborators at Osaka University (notably Mituo Taketani, Shoichi Salzata, and Minoru Kobayashi) who later helped him formulate the two-meson theory. He also fostered global communication among theoreticians with his new journal.

As the first Japanese scientist to win a Nobel Prize, Yukawa became an inspiration for non-Western people and a spokesperson for the value of Eastern ways of thinking in science. His writing on Eastern and Western approaches toward scientific investigation are provocative and pertinent today.

Simonis

WORLD EVENTS	YUKAWA'S LIFE
	1907 Hideki Ogawa is born
World War I 1914-18	
	1926-29 Ogawa studies physics at Kyoto University in Japan
	1932 Ogawa marries Sumi Yukawa and adopts her surname
	1937 Yukawa develops meson theory
	1938 Yukawa earns doctorate from Osaka University and becomes professor at Kyoto University
World War II 1939-45	
	1946 Yukawa founds journal *Progress of Theoretical Physics*
	1948-53 Yukawa works at Princeton and Columbia Universities
	1949 Yukawa receives Nobel Prize for physics
Korean War 1950-53	
	1953 Yukawa directs Research Institute of Fundamental Physics, Kyoto University
End of Vietnam War 1975	
	1981 Yukawa dies

For Further Reading:

Newman, Harvey B., and Thomas Ypsilantis, eds. *History of Original Ideas and Basic Discoveries in Particle Physics.* New York: Plenum Press, 1996.

Yukawa, Hideki. *Creativity and Intuition.* New York: Harper & Row, 1973.

———. *Tabibito: The Traveler.* Singapore: World Scientific Publishing, 1982.

Appendices,
Bibliography, and Index

Highlights in the History of Science, Mathematics, and Invention

B.C.E.

c. 530   Pythagoras flees to Crotona, a Greek colony in the south of Italy, where he founds a secret society, dedicated to his ideas about number theory.

c. 330   The *Hippocratic Collection,* an influential medical work written by Hippocrates and his students, is finished.

c. 280   Euclid completes *Elements,* which compiles and systematizes geometry and sets the organizational standard for later mathematical texts.

c. 240   Eratosthenes, a mathematician, geographer, and librarian at Alexandria, publishes *Geography,* which gives the first mathematical basis for measuring Earth as a sphere and presents his calculation of Earth's circumference.

c. 212   Archimedes, an inventor and early pioneer in geometry, uses his cranes and catapults in the Second Punic War, during which he is killed.

C.E.

c. 105   Ts'ai Lun announces the invention of paper.

161   Galen of Pergamum, one of the leading physician of antiquity, moves to Rome, where he writes treatises that dominate medical thinking until the 1500s.

c. 400   Hypatia, the first woman mathematician, becomes head of the Neoplatonic school in Alexandria.

c. 850   Mohammed ibn Musa Al-Khwarizmi dies, having introduced Europe to algebra and the Arabic number system.

c. 894   Abu Bakr Muhammad Ar-Razi begins to study medicine; Ar-Razi eventually produces *Continens,* a basic medical text that becomes a standard reference in Middle Eastern and European medical colleges.

c. 1022   Ibn Sina (Avicenna) begins the *Canon of Medicine,* a standard encyclopedia of medicine until the eighteenth century.

1202   Leonardo Fibonacci publishes *Liber Abaci,* which introduces Arabic numerals, including zero, to Europe and helps to popularize their use.

c. 1455   Johannes Gutenberg publishes the Gutenberg Bible, introducing moveable type printing, which revolutionizes the distribution of information.

1528   Paracelsus flees the University of Basel and begins to write medical treatises, which challenge medical mysticism and introduce innovative treatments.

1537   Ambroise Paré begins work as an army surgeon, a career which leads him to numerous innovations and the development of surgical techniques.

1542   Leonhard Fuchs publishes *The Natural History of Plants,* which becomes the standard for botanical studies.

1543   Nicolaus Copernicus' *Revolutions of the Heavenly Spheres* is published; it proposes that the solar system is heliocentric and launches one of the biggest scientific revolutions in history.

1543   Andreas Vesalius publishes *The Seven Books on the Structure of the Human Body,* a comprehensive description of human anatomy.

1546   Georgius Agricola introduces a classification system for minerals in *On the Nature of Fossils.*

1569   Gerardus Mercator completes a world map using his Mercator projection method, a technique still used in cartography.

1572   Tycho Brahe discovers a new star—later identified as a supernova—in the constellation Cassiopeia; this suggests that the heavens are not static.

1618-21   Johannes Kepler introduces his three laws of planetary motion in *Epitome of Copernican Astronomy.*

1620   Francis Bacon introduces new scientific methodology, which stresses inductive logic and empirical evidence, in *Novum Organum.*

1627   Johannes Kepler completes the Rudolphine Tables, which are used to calculate the position of planets.

1628   William Harvey publishes *An Anatomical Exercise on the Motion of the Heart and Blood in Animals,* which correctly describes the circulation of blood.

1632   Galileo Galilei publishes *The Starry Messenger,* which supports Copernicus's heliocentric model for the universe and notes several telescopic discoveries.

1637   René Descartes's *Discourse on Method* is published; it establishes the modern approach to mathematics.

1638   Galileo Galilei publishes book on motion and mechanics, which introduces the idea of inertia.

1642   Blaise Pascal devises first digital calculating machine.

1643   Evangelista Torricelli invents barometer.

1648   Blaise Pascal formulates law of fluids and pressure.

1654   Pierre de Fermat and Blaise Pascal establish mathematics of probability.

1656   Christiaan Huygens invents the first accurate pendulum clock.

1661   Robert Boyle postulates particulate matter theory in *The Sceptical Chemist,* which suggests that all matter comprises smaller components.

1661   Marcello Malpighi discovers capillaries and lung structure.

1665   Pierre de Fermat dies, leaving behind the infamous "Fermat's last theorem," for which he claimed to have developed a proof, though never recorded it; the first successful proof for it is developed in 1995.

1665   Robert Hooke publishes *Micrographia*, a pioneering study of cell biology.

1665-68 Isaac Newton develops laws of motion and gravitation (published in his *Principia* in 1687).

1666   Isaac Newton develops the calculus, but does not publish his ideas.

1668   Isaac Newton invents the reflecting telescope.

1674   Antoni Van Leeuwenhoek records discoveries of microorganisms with advanced high-quality microscopes that he constructs.

1684   Gottfried Wilhelm Leibniz publishes work that introduces differential and integral calculus; his notation is easier to use than Newton's, and therefore eventually becomes the dominant version to be used.

1690   Christiaan Huygens publishes his *Treatise on Light,* which introduces the theory that light moves in waves.

1697   Jakob and Johann Bernoulli work on brachistochrone problem, initiating the calculus of variations.

1705   Edmond Halley correctly predicts the return of the comet now named after him.

1709   Abraham Darby first uses coke to smelt iron, a process which revolutionizes the iron industry.

1712   Thomas Newcomen completes the first practical steam engine.

1741   Leonhard Euler returns to Germany, where he solves numerous algebraic equations, describes transcendental numbers, and introduces standardized mathematical symbols.

1744   Benjamin Franklin invents an efficient stove for heating homes and cooking.

1748   Maria Agnesi publishes *Analytical Institutions,* a comprehensive math text used by students throughout Europe.

1749   First volumes of Georges Buffon's *Natural History* appear, a vastly influential survey of natural history.

1751   Carolus Linnaeus publishes *Philosophia Botanica,* which introduces binomial nomenclature and revolutionizes taxonomy.

1752   Benjamin Franklin investigates electricity and performs kite experiment, which proves that lightning is an electrical phenomenon and leads to Franklin's invention of the lightning rod.

c. 1760 Joseph Black defines heat capacity.

1762   John Harrison invents a practical chronometer, which greatly facilitates sea navigation.

1762   Joseph Black discovers latent heat.

1764   James Hargreaves invents the spinning jenny, which increases output and productivity in the textile industry.

1765   James Watt designs an improved steam engine.

1774   Joseph Priestley discovers oxygen.

1784   William Herschel determines that nebulae consist of stars.

1787   Antoine Lavoisier publishes *Nomenclature,* which systematizes the naming of chemical compounds.

1789   Antoine Lavoisier publishes *Elementary Treatise of Chemistry,* which introduces the law of conservation of mass.

1794   Eli Whitney invents the cotton gin, which causes sweeping agricultural reform in the southern United States.

1795   Georges Cuvier joins the Museum of Natural History in Paris, where he pioneers fossil reconstruction and develops an influential classification system for animals.

1795   Joseph Bramah builds a hydraulic press, which inspires other hydraulic technology, including jacks, lifts, and brakes.

1796   Edward Jenner develops a vaccination for smallpox.

1798   Henry Cavendish publishes his calculation of Earth's gravitational constant.

1800   Alessandro Volta develops first electrochemical battery.

1801   Carl Friedrich Gauss publishes *Disquisitiones Arithmeticae,* a milestone in number theory.

1820s   Augustin-Louis Cauchy publishes lectures on the calculus and gives the calculus a logically acceptable algebraic foundation.

1821   Friedrich Bessel begins calculating star positions, which lead to the first estimates of the size of the universe.

1824   Friedrich Bessel introduces Bessel functions, which are used in pure mathematics and wave theory, elasticity, hydrodynamics; they are also used to study the movement of heat and electricity in cylinders.

1806   Humphry Davy publishes *Some Chemical Agencies of Electricity,* which outlines the use of electrolysis.

1808   John Dalton publishes *New System of Chemical Philosophy,* which introduces modern atomic theory.

1808   Joseph-Louis Gay-Lussac publishes *Law of Combining Volumes,* the basis for understanding how molecules form and gases interact.

1810   Nicolas-François Appert develops a food canning process.

1811   Amedeo Avogadro publishes hypothesis of the number of molecules in equivalent volumes of gases, later called Avogadro's Law.

1821   Sophie Germain publishes a paper on elasticity, which describes the predictable movement of particles due to vibration.

1822   Caroline Lucretia Herschel completes *A Catalogue of the Nebulae,* a benchmark for astronomers, which charts some 2,500 nebulae.

1824   Sadi Carnot's *Reflections on the Motive Power of Fire* is published; it introduces the heat engine theory and is the starting point for later studies of thermodynamics.

1825   George Stephenson's *Locomotion,* the first practical steam locomotive, runs with passengers.

1825   Louis Braille introduces his raised-dot alphabet, allowing the blind to read.

1827   André Ampère introduces Ampère's law for measuring the flow of electric current.

1827   Robert Brown discovers Brownian motion, which is the constant movement of particles suspended in a solution.

1827   Georg Ohm discovers relationships between current, voltage, and resistance (Ohm's law).

1828 Friedrich Wöhler synthesizes urea, which reveals the relationship between inorganic and organic chemistry.

1830-33 Charles Lyell's *Principles of Geology* is published; it popularizes the principle of uniformitarianism, which states that gradual physical processes shape Earth's surface features over long periods of time.

1831 Cyrus Hall McCormick invents the mechanical reaper, a vital tool in the agricultural revolution.

1831 Michael Faraday discovers electromagnetic induction and makes the world's first dynamo, an electric generator driven by magnetism.

1833 Charles Babbage begins to design the Analytical Engine, a predecessor of mechanical calculators.

1834 Mary Fairfax Somerville defines physical science and outlines the relationship between mathematics and science in *The Connection of the Physical Sciences*.

1834 Alexander von Humboldt begins writing *Kosmos*, a multi-volume work which introduces many ecological concepts, including biome classification, the interconnectedness between organisms, and the danger of human disturbance of nature.

1837 Samuel Morse engineers the single-wire electric telegraph and formulates Morse code the next year.

1839 The first photograph, invented by Louis Daguerre and initially called a "daguerreotype," is introduced to the public.

1839 Charles Goodyear vulcanizes rubber, strengthening it and making it more useful.

1840 Louis Agassiz publishes his ice age theory, which his wife Elizabeth Cary Agassiz helps popularize after his death in 1873.

1841 Niels Henrik Abel's paper on transcendental functions is published, introducing Abel's theorem.

1842 Augusta Ada Lovelace (Countess of), considered the first computer programmer, translates and supplements Luigi Menabrea's work on the Analytical Engine, providing Howard Aiken ideas for his design of the Mark I (1944).

1846 Elias Howe patents the sewing machine.

1846 William Morton introduces ether as anesthesia in surgery.

1847 George Boole establishes and systematizes symbolic logic in *Mathematical Analysis of Logic*.

1847 Maria Mitchell discovers the first comet that cannot be seen with the naked eye.

1847 James Prescott Joule first explains the relationship between heat and electricity.

1849 Elizabeth Blackwell graduates from Geneva College; she becomes the first woman to gain a medical degree in the United States.

1851 Rudolf Clausius develops the laws of thermodynamics.

1851 Lord Kelvin (William Thomson) publishes "On the Dynamical Theory of Heat," which proves mathematically that heat is a form of mechanical motion.

1853 Elizabeth Blackwell opens a dispensary providing medical care for the poor, which becomes a model for other humanitarian medical facilities.

1855 Henry Bessemer develops an inexpensive process of making steel more versatile, later called the "Bessemer process."

1855 Florence Nightingale introduces sanitary practices to Britain's military hospital in Scutari, Turkey, during the Crimean War; her innovations reduce the death rate from 42% in 1854 to 2%.

1857 Nightingale completes *Notes on Matters Affecting the Health, Efficiency, and Hospital Administration of the British Army*, which prompts reform in military health care.

1857 August Kekulé von Stradonitz publishes theory of carbon tetravalence, which provides revolutionary insight into molecular structure of organic chemistry.

1859 Charles Darwin's *On the Origin of Species* is published, which introduces the principle of natural selection and the theory of evolution.

1859 Robert Bunsen and Gustav Kirchhoff invent the spectroscope and use spectral analysis to identify chemicals, a technique now integral to chemistry and astrophysics.

1861 Ignaz Semmelweis publishes *Etiology, Understanding and Preventing Childbed Fever*, a milestone in medical history that outlines the cause of and preventative measures against puerperal fever.

1864 Louis Pasteur introduces pasteurization.

1865 August Kekulé publishes theory of ring structure, which broadens scientific understanding of molecular structure in organic chemistry.

1866 Gregor Johann Mendel publishes *Experiments with Plant Hybrids*, which introduces the theory and science of genetics.

1867 Christopher Sholes invents the first practical typewriter.

1867 Joseph Lister announces the success of antiseptic surgery, provoking medical reform.

1871 Dmitry Mendeleyev publishes a revised periodic table of chemical elements based upon his original table published two years prior.

1872 Amanda Jones patents a new vacuum canning process.

1873 Luther Burbank introduces an improved potato (known as Burbank or Idaho), the first of over 800 plant varieties he develops.

1873 James Clerk Maxwell outlines electromagnetic theory in *Treatise on Electricity and Magnetism*.

1874 Sonya Kovalevsky receives doctorate from University of Göttingen for her influential work on partial differential equations.

1875 Ludwig Boltzmann publishes paper on statistical method of explaining the Second Law of Thermodynamics.

1876 Robert Koch demonstrates that the disease anthrax comes from bacteria, the first proof that bacteria cause specific diseases.

1876 Alexander Graham Bell invents the telephone.

1876 Nikolaus August Otto introduces a four-stroke engine, the prototype for modern internal combustion engines.

1877 Thomas Edison invents the phonograph.

1879 Thomas Edison develops the incandescent light bulb.

1880s   Élie Metchnikoff discovers phagocytosis, revealing an important component of immunology.

1881   Clara Barton founds American Association of the Red Cross.

1882   Thomas Edison builds the world's first power station.

1884   Svante Arrhenius presents electrolytic dissociation theory, which explains the behavior of ions in a solution.

1884   Henri-Louis Le Châtelier publishes his principle of chemical equilibrium, which explains how temperature and pressure affect chemical systems.

1886   Julia and Charles Hall discover electrolytic purification of aluminum, which allows aluminum to be manufactured cheaply.

1887   Nikola Tesla designs an alternating-current motor, which reduces electrical transmission problems.

1888   George Eastman introduces the Kodak camera.

1891   Nikola Tesla designs a transformer (a Tesla coil), which modulates electrical power to make it more useful in a variety of applications—from high voltage machinery to TV sets.

1892   Carrie J. Everson completes development of an ore concentration process, which makes practical the extraction of secondary grade ore.

1893   Rudolf Diesel builds an efficient compression-ignition engine, later known as a "diesel" engine.

1893   Daniel Hale Williams performs the first open-heart surgery.

1895   Auguste and Louis Lumière's *cinématographe* projects the first moving picture before an audience.

1895   Wilhelm Roentgen announces the discovery of X-rays.

1895   Henri Poincaré writes *Analysis Situ,* a pioneering work on topology, which is the study of an object's geometrical properties that do not change as the object is contorted.

1895   Guglielmo Marconi develops the first radio transmitter and receiver.

1896   Antoine-Henri Becquerel discovers the spontaneous release of energy from an ore (radioactivity).

1898   Marie Curie isolates and purifies two elements, polonium and radium, while continuing Becquerel's work on radioactivity, a term she coins.

1899   Harriet Brooks identifies radon as the product of nuclear decay, inspiring Ernest Rutherford and Frederick Soddy to recognize radioactive transmutation.

1899   Ellen Swallow Richards, the first woman accepted to study at the Massachusetts Institute of Technology, begins a series of lectures on nutrition, health issues, sanitation, and other factors in the home environment. She calls this field of study home economics; it is later referred to as home ecology.

1900   Sigmund Freud completes *The Interpretation of Dreams,* which outlines his new psychoanalytic theory.

1900   Karl Landsteiner publishes theory of blood types, which greatly increases the reliability of blood transfusions.

1900   Max Planck publishes the basis for quantum theory in a paper on blackbody radiation.

1901   Jokichi Takamine is the first to isolate a hormone (adrenaline).

1903   Wilbur and Orville Wright pilot the first powered and manned airplane flights.

1905   Albert Einstein publishes major papers on quantum theory (supporting Plank's work), atomic theory, and relativity, introducing the famous equation $E=mc^2$.

1906   William Bateson, having translated, popularized, and elaborated on Mendel's research, coins the term "genetics."

1907   Ivan Pavlov publishes *Conditioned Reflexes;* his theory of conditioned reflexes is the precursor to behavioral psychology.

1908   Henry Ford markets the Model T, a car that makes automobile transportation available to vast numbers of people.

1909   Fritz Haber develops method for synthesizing ammonia from nitrogen and hydrogen.

1910   Paul Ehrlich tests and uses Salvarsan as a cure for syphilis.

1910-13   Bertrand Russell (with Alfred North Whitehead) publishes *Principia Mathematica,* which lays the foundation for logicism, the idea that all mathematics is rooted in logic.

1911   Ernest Rutherford (with Hans Geiger and Ernest Marsden) discovers the atomic nucleus.

1911   Anna Botsford Comstock publishes *The Handbook of Nature Study,* which becomes a best-selling guide for nature lovers and naturalists; it offers an introduction to natural cycles and processes.

1912   Garrett Morgan patents the gas mask, which helps to save many lives, especially soldiers in World War I.

1913   Hideyo Noguchi proves link between syphilis and paresis, a major contribution to bacteriology.

1913   The International Astronomical Union adopts Annie Jump Cannon's system of star classification, which categorizes stars by temperature.

1913   Niels Bohr publishes his theory of atomic structure, which is the foundation of a modern understanding of atomic structure.

1914   George Washington Carver explains the importance of crop rotation to farmers in the southern United States.

1915   Alfred Lothar Wegener publishes *The Origins of Continents and Oceans,* which introduces the principle of continental drift.

1915   Ernest Everett Just shows how the developmental orientation of the fertilized egg is established, the first of his many discoveries in developmental biology.

1915   Albert Einstein publishes the general theory of relativity, revolutionizing the modern understanding of space and time.

1921   Emmy Noether publishes "General Theory of Ideals," an important work in abstract algebra that introduces a technique now fundamental to commutative algebra.

1921   Vilhelm Bjerknes publishes *On the Dynamics of the Circular Vortex with Applications to the Atmosphere and Atmospheric Vortex and Wave Motion,* which explains the characteristics of air masses.

1922 Garrett Morgan patents the traffic signal, helping to improve the safety of increasingly busy city roads throughout the United States.

1924 Satyendranath Bose sends his paper on quantum theory to Albert Einstein; the work strengthens quantum theory and introduces Bose-Einstein Statistics, which describe the behavior of the subatomic particles now known as "bosons."

1925 Werner Heisenberg develops matrix mechanics, a milestone in quantum theory that mathematically predicts the properties of atoms and simple molecules.

1926 Max Born introduces the probability interpretation, providing the foundation for Bohr's Copenhagen Interpretation of quantum mechanics.

1926 Robert H. Goddard launches the first liquid-fueled rocket.

1927 Neils Bohr presents the Copenhagen Interpretation of quantum mechanics.

1927 Werner Heisenberg develops the Uncertainty Principle, which explains that observation of subatomic particles disturbs them and therefore introduces error into measurements and calculations at the subatomic level.

1928 Margaret Mead's *Coming of Age in Samoa* is published; it introduces cultural anthropology and the theory of cultural determinism, which suggests that a culture influences the development of personality.

1928 Chandrasekhara Venkata Raman discovers the Raman effect, which describes how matter diffracts light and changes its frequency and color.

1928 Alexander Fleming discovers penicillin.

1929 Edwin Hubble formulates a law concerning the expansion of the universe, a milestone for cosmology, which helps scientists to calculate the age and size of the universe.

1931 Barbara McClintock (with Harriet Creighton) publishes a paper demonstrating how chromosomes cross over and exchange genetic information.

1932 James Chadwick discovers the neutron.

1934 Otto Hahn, Lise Meitner, and Fritz Strassmann begin investigations which lead to the discovery of fission.

1935 Percy Lavon Julian synthesizes physostigmine, which is effective treatment for glaucoma.

1935 Wallace Carothers invents nylon.

1935 Robert Watson-Watt develops the first radar system.

1937 Hideki Yukawa refines his meson theory, which describes pi-mesons and mu-mesons, previously undiscovered subatomic particles.

1939 In *The Nature of the Chemical Bond,* Linus Pauling uses quantum theory to explain chemical bonds in terms of hybridization, bond character, and resonance theory.

1939 Igor Sikorsky completes the first single-rotor helicopter.

1939 Charles Drew develops blood-plasma preservation methods, expanding the opportunities to perform life-saving transfusions.

1939 Lise Meitner announces the discovery of fission, the splitting of an atomic nucleus, which produces an enormous amount of energy and is applied to nuclear weapons.

1942 Enrico Fermi builds the first nuclear reactor and, in so doing, achieves the first sustainable nuclear (fission) chain reaction.

1943 Jacques-Yves Cousteau co-develops the Aqua-Lung, the first of his several technological developments that assist divers and oceanographers.

1944 John von Neumann publishes *Theory of Games and Economic Behavior,* a work that applies game theory to economic theory.

1944 John von Neumann joins the team to build EDVAC; he plans the computer's logical organization, which influences the design of later computers.

1944 Howard Hathaway Aiken completes the Mark I, the first full-scale programmable computer in the United States.

1945 The Manhattan Project, under J. Robert Oppenheimer's leadership, produces the world's first nuclear weapons.

1947 Georg von Békésy builds a cochlear model, with which he determines the physical mechanism of hearing.

1948 John Bardeen (with Walter Brattain and William Shockley) invents the transistor, a device which translates alternate current to direct current and amplifies it.

1948 Andrey Sakharov begins research leading to the development of the Soviet hydrogen bomb.

1951 Rosalind Franklin begins X-ray diffraction of DNA, contributing data vital to the determination of DNA's structure.

1952 Grace Murray Hopper develops the first compiler, software that translates higher-level programmed instructions to machine-level code.

1953 Francis Crick and James Watson describe the double-helical structure of DNA.

1954 Gregory Pincus (with Min-Chueh Chang) develops and tests the first oral contraceptive, now known as "The Pill."

1955 John von Neumann proves the equivalence of wave mechanics and matrix mechanics, two models of quantum theory.

1957 John Bardeen (with Leon Cooper and John Schrieffer) formulates a theory of superconductivity, the phenomenon in which some metals exhibit zero resistance to the flow of electrons at temperatures close to absolute zero.

1957 Chien-Shiung Wu proves the nonconservation of parity, undermining the notion that all nuclear reactions emit particles in all directions equally.

1959 Rosalyn Yalow and Solomon Berson develop radioimmunoassay, a technique for identifying and quantifying the presence of biological and chemical substances in the body.

1962 Rachel Carson's *Silent Spring* is published; it raises national awareness of the effects of pesticide pollution.

1962 Harry Hammond Hess publishes "History of Ocean Basins," which introduces his theory of plate tectonics, the first comprehensive explanation of mountain building, continental drift, and sea-floor spreading.

1963   Norman Ernest Borlaug begins work in India and Pakistan, where his agricultural reforms ameliorate famine and help to launch the Green Revolution.

1963   Murray Gell-Mann develops the quark model, a major breakthrough in understanding subatomic structure.

1964   Jane Cooke Wright, an innovative cancer researcher, serves on the President's Commission on Heart Disease, Cancer, and Stroke, which leads to the development of a nationwide network of cancer-treatment centers.

1965   Richard P. Feynman shares a Nobel Prize for Physics for his theory of quantum electrodynamics (QED), which explains the interaction between particles and electromagnetic radiation in terms of quantum theory.

1968   The television program, *Undersea World of Jacques Cousteau*, begins; it is one of Cousteau's many efforts to educate the public about the beauty and fragility of marine life.

1973   Stephen Hawking writes (with G. F. R. Ellis) *The Large Scale Structure of Space-Time*, which provides mathematical support for the Big Bang and black hole theories.

1975   Bertram Fraser-Reid publishes a report on his successful efforts to synthesize non-carbohydrate compounds from simple carbohydrates (sugars).

1978   Barbara McClintock's paper, which challenges the Watson-Crick model by proposing gene reorganization mechanisms, gets international attention.

1978   Mary Leakey's team finds 3.65 million-year-old footprints of upright walking hominids, one of her many important paleoanthropological discoveries.

1980   Bill Gates designs MS-DOS, which becomes the standardized operating system for personal computers.

1986   Jane Goodall publishes *The Chimpanzees of Gombe: Patterns of Behavior,* which summarizes her observations on chimpanzees; her research methods are a model for long-term ethological study.

1988   Bertram Fraser-Reid patents process to synthesize oligosaccharides, complex carbohydrates found in human cell membranes and viruses.

1990   Bill Gates's company Microsoft introduces its graphical operating system Windows.

Geographic Listing of Biographies

The listing below classifies the individuals in the book according to the country in which they conducted important work. In most instances, this is the person's place of birth. In other cases, however, individuals left their country of origin to study and work in a different place. In these instances, we have noted the birthplace and/or additional place of work parenthetically. We have also noted ethnic and/or ancestral origins of minorities in the United States parenthetically. Our decision to include this information here resulted from the selection criteria outlined in the introduction—one of which is to include people who have changed the world by breaking through the limits that society has placed on them.

Austria

Boltzmann, Ludwig (also worked in Germany)
Developer of Statistical Mechanics   1844-1906

Freud, Sigmund
Founder of Psychoanalysis   1856-1939

Kepler, Johannes (born and also worked in Germany; also worked in Prague, capital of today's Czech Republic)
Developer of Laws of Planetary Motion   1571-1630

Landsteiner, Karl (also worked in United States)
Immunologist; Identifier of Human Blood Types   1868-1943

Meitner, Lise (also worked in Germany and Sweden)
Co-discoverer of Nuclear Fission   1878-1968

Paracelsus (born and also worked in Switzerland; also worked in the Netherlands and Italy)
Revolutionary Physician and Alchemist   1493-1541

Tesla, Nikola (born in Croatia; also worked in United States)
Developer of Alternating Current Electricity   1856-1943

Wegener, Alfred Lothar (born and also worked in Germany)
Developer of Principle of Continental Drift   1880-1930

Belgium (formerly part of Flanders)

Mercator, Gerardus
Cartographer; Creator of Mercator Map Projection   1512-1594

Vesalius, Andreas (also worked in Italy)
Founder of Anatomy   1514-1564

Canada

Bell, Alexander Graham (born in Scotland; also worked in United States)
Inventor of the Telephone   1847-1922

Brooks, Harriet
Pioneering Researcher in Radioactivity   1876-1933

Fraser-Reid, Bertram (born in Jamaica; also worked in United States)
Synthesizer of Organic Compounds   1934-

Rutherford, Ernest (born in New Zealand; also worked in England)
Discoverer of the Atomic Nucleus   1871-1937

China

Ts'ai Lun
Inventor of Paper   c. 50-121

Czech Republic (formerly part of Czechoslovakia and Austria-Hungary)

Brahe, Tycho (born and also worked in Denmark)
Discoverer of Supernova   1546-1601

Kepler, Johannes (born and also worked in Germany; also worked Austria)
Developer of Laws of Planetary Motion   1571-1630

Mendel, Gregor Johann
Founder of Modern Genetics   1822-1884

Denmark

Bohr, Niels
Developer of Theory of Atomic Structure   1885-1962

Brahe, Tycho (also worked in Prague, capital of today's Czech Republic)
Discoverer of Supernova   1546-1601

England

Babbage, Charles
Grandfather of the Modern Computer;
Designer of the Analytical Engine   1791-1871

Bacon, Francis
Philosopher of Science;
Developer of Scientific Methodology   1561-1626

Bateson, William
Early Mendelian Geneticist   1861-1926

Bessemer, Henry
Inventor of Bessemer Steel-making Process   1813-1898

Blackwell, Elizabeth (born and also worked in United States)
First American Woman Physician of Modern Times   1821-1910

Boole, George (born and also worked in Ireland)
Co-founder of Symbolic Logic   1815-1864

Boyle, Robert (born and also worked in Ireland)
Founder of Chemistry   1627-1691

Bramah, Joseph
Inventor of Hydraulic Press and Bramah Lock   1748-1814

Brown, Robert (born and also worked in Scotland)
Botanist; Discoverer of Brownian Motion   1773-1858

Cavendish, Henry
    Investigator of Hydrogen; Calculator of Earth's Gravitational
    Constant   1731-1810

Chadwick, James
    Discoverer of the Neutron   1891-1974

Dalton, John
    Founder of Modern Atomic Theory   1766-1844

Darby, Abraham
    Originator of Coke for Iron Smelting   c. 1678-1717

Darwin, Charles
    Founder of Evolutionary Theory   1809-1882

Davy, Humphry
    Discoverer of Elements; Pioneer in Electrolysis   1778-1829

Faraday, Michael
    Discoverer of Electromagnetic Induction   1791-1867

Fleming, Alexander (born in Scotland)
    Bacteriologist; Discoverer of Penicillin   1881-1955

Franklin, Rosalind
    Pioneer of X-Ray Diffraction for DNA Imaging   1920-1958

Goodall, Jane
    Researcher and Ethologist of Chimpanzees   1934-

Halley, Edmond
    Pioneering Investigator of Comets   1656-1742

Hargreaves, James
    Inventor of the Spinning Jenny   c. 1720-1778

Harrison, John
    Inventor of Practical Chronometer   1693-1776

Harvey, William
    Discoverer of Blood Circulation   1578-1657

Hawking, Stephen
    Developer of Mathematical Proof for Black Hole Theory   1942-

Herschel, Caroline Lucretia (born in Germany)
    First Major Woman Astronomer of Modern Times   1750-1848

Herschel, William (born in Germany)
    Founder of Star-Related Astronomical Research   1738-1822

Hooke, Robert
    Identifier of Cells; Formulator of Laws of Elasticity   1635-1703

Jenner, Edward
    Developer of Modern Smallpox Vaccine   1749-1823

Joule, James Prescott
    Discoverer of Mechanical Equivalent of Heat   1818-1889

Kelvin, Lord (born in Ireland)
    Pioneer in Electromagnetism, Heat, and Mechanics   1824-1907

Leakey, Mary
    Paleoanthropologist;
    Researcher of Early Human Ancestors   1913-1996

Lister, Joseph
    Founder of Antiseptic Surgery   1827-1912

Lovelace (Countess of), Augusta Ada
    First Computer Programmer   1815-1852

Lyell, Charles (born in Scotland)
    Originator of Fundamental Principles of Geology   1797-1875

Maxwell, James Clerk (born and also worked in Scotland)
    Founder of Modern Electrical Theory   1831-1879

Newcomen, Thomas
    Inventor of Steam Engine   1663-1729

Newton, Isaac
    Discoverer of Laws of Motion and Gravitation   1642-1727

Nightingale, Florence
    Founder of Nursing Profession   1820-1910

Priestley, Joseph (also worked in United States)
    Discoverer of Oxygen   1733-1804

Russell, Bertrand
    Originator of Logicism   1872-1970

Rutherford, Ernest (born in New Zealand; also worked in Canada)
    Discoverer of the Atomic Nucleus   1871-1937

Somerville, Mary Fairfax (born and also worked in Scotland)
    Popular Science and Mathematics Author   1780-1872

Stephenson, George
    Inventor of Practical Steam Locomotive   1781-1848

Watson-Watt, Robert (born and also worked in Scotland)
    Developer of Radar   1892-1973

Watt, James (born and also worked in Scotland)
    Developer of Steam Engine Technology   1736-1819

France

Ampère, André
    Formulator of Laws of Electromagnetism   1775-1836

Appert, Nicolas-François
    Developer of Canned Food Process   c. 1750-1841

Becquerel, Antoine-Henri
    Discoverer of Radioactivity   1852-1908

Braille, Louis
    Inventor of Braille Alphabet   1809-1852

Buffon, Georges
    Developer of Survey of Natural World   1707-1788

Carnot, Sadi
    Developer of Heat Engine Theory   1796-1832

Cauchy, Augustin-Louis
    Early Developer of the Calculus   1789-1857

Cousteau, Jacques-Yves
    Pioneer Oceanographer;
    Developer of Scuba Equipment   1910-1997

Cuvier, Georges
    Early Pioneer in Comparative
    Anatomy and Paleontology   1769-1832

Curie, Marie (born in Poland)
    Founder of Radiation Chemistry   1867-1934

Daguerre, Louis
    Inventor of the First Photograph   1789-1851

Descartes, René
Founder of Analytic Geometry   1596-1650

Fermat, Pierre de
Pioneer in the Calculus and Number Theory   1601-1665

Gay-Lussac, Joseph- Louis
Discoverer of Law of Combining Volumes   1778-1850

Germain, Sophie
Pioneer in Number Theory and Study of Elasticity   1776-1831

Huygens, Christiaan (born and also worked in Netherlands)
Originator of Wave Theory of Light;
Inventor of Pendulum Clock   1629-1695

Lavoisier, Antoine
Originator of Law of Conservation of Mass   1743-1794

Le Châtelier, Henri-Louis
Originator of Le Châtelier's Principle of
Chemical Equilibrium   1850-1936

Lumière, Auguste
Co-inventor of Cinema   1862-1954

Lumière, Louis
Co-inventor of Cinema   1864-1948

Metchnikoff, Élie (born and also worked in Russia)
Immunologist; Discoverer of Phagocytes   1845-1916

Paré, Ambroise
Founder of Modern Surgery   1510-1590

Pascal, Blaise
Pioneer in Geometry, Probability, and Hydrostatics;
Inventor of Calculating Machine   1623-1662

Pasteur, Louis
Pioneer Microbiologist; Developer of Pasteurization   1822-1895

Poincaré, Henri
Influential Modern Mathematician   1854-1912

Von Humboldt, Alexander (born and also worked in Germany)
Early Pioneer of Ecological Concepts   1769-1859

Germany

Agricola, Georgius
Father of Mineralogy   1494-1555

Bessel, Friedrich
Astronomer; Originator of Bessel Functions   1784-1846

Boltzmann, Ludwig (born and also worked in Austria)
Developer of Statistical Mechanics   1844-1906

Born, Max
Early Pioneer in Quantum Mechanics   1882-1970

Bunsen, Robert
Developer of Chemical Spectral Analysis   1811-1899

Clausius, Rudolf
Founder of Thermodynamics   1822-1888

Diesel, Rudolf
Inventor of Diesel Engine   1858-1913

Ehrlich, Paul
Immunologist; Developer of Cure for Syphilis   1854-1915

Einstein, Albert (also worked in Switzerland and United States)
Developer of Relativity and Quantum Theories   1879-1955

Euler, Leonhard (born and also worked in Sweden; also worked in Russia)
Pioneer in Algebra and Number Theory   1707-1783

Fuchs, Leonhard
Developer of Foundations of Botany   1501-1566

Gauss, Carl Friedrich
Multi-faceted Mathematician;
Developer of Complex Number System   1777-1855

Gutenberg, Johannes
Inventor of Movable-Type Printing   c. 1400-1468

Haber, Fritz
Chemist; First to Synthesize Ammonia   1868-1934

Hahn, Otto
Co-discoverer of Nuclear Fission   1879-1968

Heisenberg, Werner
Developer of Matrix Mechanics
and the Uncertainty Principle   1901-1976

Kekulé von Stradonitz, August
Discoverer of Carbon Benzene Rings   1829-1896

Kepler, Johannes (also worked Austria and Prague, capital
of today's Czech Republic)
Developer of Laws of Planetary Motion   1571-1630

Koch, Robert
Founder of Medical Bacteriology   1843-1910

Leibniz, Gottfried Wilhelm
Inventor of Differential and Integral Calculus   1646-1716

Meitner, Lise (born and also worked in Austria; also worked in Sweden)
Co-discoverer of Nuclear Fission   1878-1968

Noether, Emmy (also worked in United States)
Pioneer in Modern Algebra   1882-1935

Ohm, Georg
Originator of Ohm's Law of Electrical Resistance   1789-1854

Otto, Nikolaus August
Inventor of Four-Stroke Internal Combustion Engine   1832-1891

Planck, Max
Developer of Foundation for Quantum Theory   1858-1947

Roentgen, Wilhelm
Discoverer of X-rays   1845-1923

Von Humboldt, Alexander (also worked in France)
Early Pioneer of Ecological Concepts   1769-1859

Wegener, Alfred Lothar (also worked in Austria)
Developer of Principle of Continental Drift   1880-1930

Wöhler, Friedrich
Synthesizer of Urea   1800-1882

Greek and Roman Empires

Archimedes (Alexandria, present-day Egypt, under Greek rule)
Inventor; Early Pioneer in Plane
and Solid Geometry   c. 290-c. 212 B.C.E.

Eratosthenes (Alexandria, present-day Egypt, under Greek rule)
 Mathematician; Geographer;
 Librarian at Alexandria   c. 284-192 B.C.E.

Euclid (Alexandria, present-day Egypt, under Greek rule)
 Organizer of Geometry;
 Author of Landmark Elements   c. 300 B.C.E.-c. 275 B.C.E..

Galen of Pergamum (Pergamum, in Asia Minor, under Roman rule)
 Leading Physician of Antiquity   129-c.200

Hippocrates (Greece and Asia Minor, under Greek rule)
 Founder of Medicine   c. 460-c. 377 B.C.E.

Hypatia of Alexandria (Alexandria, present-day Egypt, under Roman rule)
 First Recognized Woman
 Mathematician and Scientist   c. 370-415

Pythagoras (Samos and Crotona, under Greek rule)
 Early Pioneer in Mathematics   c. 580-500 B.C.E.

Hungary

Semmelweis, Ignaz
 Medical Reformer;
 Founder of Cause for Puerperal Fever   1818-1865

Von Békésy, Georg (also worked in United States)
 Discoverer of the Physiology of Hearing   1899-1972

Von Neumann, John (also worked in United States)
 Early Computer Designer;
 Developer of Game Theory   1903-1957

India

Bose, Satyendranath
 Pioneer in Quantum Mechanics   1894-1974

Raman, Chandrasekhara Venkata
 Pioneer Researcher in Modern Optics   1888-1970

Ireland

Boole, George (also worked in England)
 Co-founder of Symbolic Logic   1815-1864

Boyle, Robert (also worked in England)
 Founder of Chemistry   1627-1691

Italy

Agnesi, Maria
 First Important Woman
 Mathematician of Modern Times   1718-1799

Avogadro, Amedeo
 Originator of Concept of Molecules;
 Developer of Avogadro's Law   1776-1856

Fermi, Enrico (also worked in United States)
 Designer of First Nuclear Reactor   1901-1954

Fibonacci, Leonardo
 Popularizer of Arabic Numerals;
 Pioneer in Algebra and Geometry   c. 1170-c. 1250

Galilei, Galileo
 Pioneer in Laws of Motion and Early Telescopy 1564-1642

Malpighi, Marcello
 Early Pioneer in Medical Research   1628-1694

Marconi, Guglielmo
 Developer of Radio Communication   1874-1937

Paracelsus (born and also worked in Switzerland;
 also worked in Austria and the Netherlands)
 Revolutionary Physician and Alchemist   1493-1541

Torricelli, Evangelista
 Inventor of the Barometer   1608-1647

Vesalius, Andreas (born and also worked in Belgium)
 Founder of Anatomy   1514-1564

Volta, Alessandro
 Discoverer of Electrochemical Cells   1745-1827

Japan

Noguchi, Hideyo (also worked in United States)
 Isolator of Cause of Paralysis in Syphilis Patients 1876-1928

Takamine, Jokichi (also worked in United States)
 Isolator of Adrenaline   1854-1922

Yukawa, Hideki (also worked in United States)
 Founder of Meson Theory   1907-1981

Netherlands

Huygens, Christiaan (also worked in France)
 Originator of Wave Theory of Light;
 Inventor of Pendulum Clock   1629-1695

Leeuwenhoek, Antoni Van
 Pioneer Microscopist and Microbiologist   1632-1723

Paracelsus (born and also worked in Switzerland; also worked
 in Austria and Italy)
 Revolutionary Physician and Alchemist   1493-1541

Norway

Abel, Niels Henrik
 Advocate of Rigorous Mathematical Proofs   1802-1829

Bjerknes, Vilhelm
 Founder of Modern Meteorology   1862-1951

Persian Empire

Al-Khwarizmi, Mohammed ibn Musa (born in present-day
 Uzbekistan; worked in present-day Iraq)
 Developer of Arabic Numerals; Pioneer in Algebra c. 780-850

Ar-Razi, Abu Bakr Muhammad (born in present-day Iran;
 worked in present-day Iraq)
 Pioneering Islamic Physician of Middle Ages   c. 865-c. 935

Ibn Sina (born in present-day Uzbekistan; worked in Iran)
 Physician; Author of Canon of Medicine   980-1037

Poland

Copernicus, Nicolaus
 Father of Modern Astronomy   1473-1543

Russia and other former Soviet republics

Euler, Leonhard (born and also worked in Sweden; also worked in Germany)
Pioneer in Algebra and Number Theory — 1707-1783

Kovalevsky, Sonya (also worked in Sweden)
Developer of Partial Differential Equations — 1850-1891

Mendeleyev, Dmitry
Creator of Periodic Table of Elements — 1834-1907

Metchnikoff, Élie (also worked in France)
Immunologist; Discoverer of Phagocytes — 1845-1916

Pavlov, Ivan
Originator of Theory of Conditioned Reflexes — 1849-1936

Sakharov, Andrey
Developer of Soviet Hydrogen Bomb — 1921-1989

Sikorsky, Igor (also worked in United States)
Inventor of the Helicopter — 1889-1972

Scotland

Black, Joseph
Isolator of Carbon Dioxide;
Definer of Latent Heat and Heat Capacity — 1728-1799

Brown, Robert (also worked in England)
Botanist; Discoverer of Brownian Motion — 1773-1858

Kelvin, Lord (born in Ireland; also worked in England)
Pioneer in Electromagnetism, Heat, and Mechanics — 1824-1907

Maxwell, James Clerk (also worked in England)
Founder of Modern Electrical Theory — 1831-1879

Somerville, Mary Fairfax (also worked in England)
Popular Science and Mathematics Author — 1780-1872

Watson-Watt, Robert (also worked in England)
Developer of Radar — 1892-1973

Watt, James (also worked in England)
Developer of Steam Engine Technology — 1736-1819

Sweden

Arrhenius, Svante
Originator of Electrolytic Dissociation Theory — 1859-1927

Euler, Leonhard (also worked in Russia and Germany)
Pioneer in Algebra and Number Theory — 1707-1783

Kovalevsky, Sonya (born and also worked in Russia)
Developer of Partial Differential Equations — 1850-1891

Linnaeus, Carolus
Botanist; Originator of Taxonomic
Classification System — 1707-1778

Meitner, Lise (born and also worked in Austria; also worked in Germany)
Co-discoverer of Nuclear Fission — 1878-1968

Switzerland

Agassiz, Louis (also worked in United States)
Originator of Ice Age Theory — 1807-1873

Bernoulli, Jakob
Co-originator of the Calculus of Variations — 1654-1705

Bernoulli, Johann
Co-originator of the Calculus of Variations — 1667-1748

Einstein, Albert (born and also worked in Germany;
also worked in United States)
Developer of Relativity and Quantum Theories — 1879-1955

Euler, Leonhard (also worked in Germany and Russia)
Pioneer in Algebra and Number Theory — 1707-1783

Paracelsus (also worked in Austria, Italy, and the Netherlands)
Revolutionary Physician and Alchemist — 1493-1541

United States

Agassiz, Elizabeth Cary
Science Writer — 1822-1907

Agassiz, Louis (born and also worked in Switzerland)
Originator of Ice Age Theory — 1807-1873

Aiken, Howard Hathaway
Pioneer in Early Computer Engineering — 1900-1973

Bardeen, John
Superconductivity Theorist; Inventor of Transistor — 1908-1991

Barton, Clara
Founder of American Red Cross — 1821-1912

Bell, Alexander Graham (born in Scotland; also worked in Canada)
Inventor of the Telephone — 1847-1922

Blackwell, Elizabeth (also worked in England)
First American Woman Physician of Modern Times — 1821-1910

Borlaug, Norman Ernest
Agronomist; Founder of the Green Revolution — 1914-

Burbank, Luther
Pioneer Breeder of New Plant Varieties — 1849-1926

Cannon, Annie Jump
Developer of Star Classification System — 1863-1941

Carothers, Wallace
Inventor of Nylon — 1896-1937

Carson, Rachel
Early Advocate of Environmental Movement — 1907-1964

Carver, George Washington (African American)
Originator of Crop Rotation Practices — c. 1860-1943

Comstock, Anna Botsford
Entomologist; Pioneer in Science Education — 1854-1930

Crick, Francis
Co-discoverer of DNA Structure — 1916-

Drew, Charles (African American)
Developer of Technique for Storing Blood Plasma — 1904-1950

Eastman, George
Inventor of Hand-held Camera — 1854-1932

Edison, Thomas
Prolific Inventor; Creator of the Light Bulb — 1847-1931

Einstein, Albert (born and also worked in Germany;
also worked in Switzerland)
Developer of Relativity and Quantum Theories — 1879-1955

Everson, Carrie J.
    Inventor of Ore Concentration Process   c.1844-1913

Fermi, Enrico (born and also worked in Italy)
    Designer of First Nuclear Reactor   1901-1954

Feynman, Richard P.
    Developer of Quantum Electrodynamics   1918-1988

Ford, Henry
    Automobile Engineer; Inventor of the Model T   1863-1947

Franklin, Benjamin
    Prolific Inventor; Investigator of Electricity 1706-1790

Fraser-Reid, Bertram (born in Jamaica; also worked in Canada)
    Synthesizer of Organic Compounds   1934-

Gates, Bill
    Computer Programmer;
    Founder of Microsoft Corporation   1955-

Gell-Mann, Murray
    Developer of the Quark Model   1929-

Goddard, Robert H.
    Pioneer in Early Rocket Design   1882-1945

Goodyear, Charles
    Inventor of Vulcanized Rubber   1800-1860

Hall, Charles
    Co-developer of Aluminum Manufacturing Process 1863-1914

Hall, Julia (born in British West Indies)
    Co-developer of Aluminum Manufacturing Process 1859-1925

Hess, Harry Hammond
    Originator of Theory of Plate Tectonics   1906-1969

Hopper, Grace Murray
    Inventor of the Computer Language Compiler   1906-1992

Howe, Elias
    Inventor of the Sewing Machine   1819-1867

Hubble, Edwin
    Discoverer of Galaxies Beyond Milky Way   1889-1953

Jones, Amanda
    Inventor of Vacuum Canning   1835-1914

Julian, Percy Lavon (African American)
    Organic Chemist;
    Developer of Treatment for Glaucoma   1899-1975

Just, Ernest Everett (African American)
    Pioneer in Developmental Biology   1883-1941

Landsteiner, Karl (born and also worked in Austria)
    Immunologist; Identifier of Human Blood Types   1868-1943

McClintock, Barbara
    Pioneer in Genetic Analysis   1902-1992

McCormick, Cyrus Hall
    Inventor of Mechanical Reaper   1809-1884

Mead, Margaret
    Pioneer in Cultural Anthropology   1901-1978

Mitchell, Maria
    First American Woman Astronomer   1818-1889

Morgan, Garrett (African American)
    Inventor of Gas Mask and Traffic Light   1877-1963

Morse, Samuel
    Inventor of First Single-Wire Telegraph   1791-1872

Morton, William
    Developer of General Anesthesia   1819-1868

Noether, Emmy (born and also worked in Germany)
    Pioneer in Modern Algebra   1882-1935

Noguchi, Hideyo (born and also worked in Japan)
    Isolator of Cause of Paralysis in Syphilis Patients 1876-1928

Oppenheimer, J. Robert
    Head Scientist on First Atomic Bomb Project   1904-1967

Pauling, Linus
    Pioneer Researcher in Nuclear and Molecular Structure 1901-1994

Pincus, Gregory
    Developer of Oral Contraceptives   1903-1967

Priestley, Joseph (born and also worked in England)
    Discoverer of Oxygen   1733-1804

Sholes, Christopher
    Developer of Typewriter   1819-1890

Sikorsky, Igor (born and also worked in Russia)
    Inventor of the Helicopter   1889-1972

Swallow Richards, Ellen
    Founder of Home Ecology   1842-1911

Takamine, Jokichi (born and also worked in Japan)
    Isolator of Adrenaline   1854-1922

Tesla, Nikola (born in Croatia; also worked in Austria)
    Developer of Alternating Current Electricity   1856-1943

Von Békésy, Georg (born and also worked in Hungary)
    Discoverer of the Physiology of Hearing   1899-1972

Von Neumann, John (born and also worked in Hungary)
    Early Computer Designer; Developer of Game Theory 1903-1957

Watson, James
    Co-discoverer of DNA Structure   1928-

Whitney, Eli
    Inventor of the Cotton Gin   1765-1825

Williams, Daniel Hale (African American)
    Founder of Cardiac Surgery   1858-1931

Wright, Jane Cooke (African American)
    Pioneer in Chemotherapy   1919-

Wright, Orville
    Co-inventor of the Modern Airplane   1871-1948

Wright, Wilbur
    Co-inventor of the Modern Airplane   1867-1912

Wu, Chien-Shiung (born in China)
    Demonstrator of Non-Conservation of Parity   c. 1912-

Yalow, Rosalyn Sussman
    Developer of Radioimmunoassay   1921-

Yukawa, Hideki (born and also worked in Japan)
    Founder of Meson Theory   1907-1981

Listing of Biographies by Discipline

The listing below classifies the individuals in the book according to the discipline in which they worked. As many individuals frequently worked in several disciplines, particularly prior to twentieth-century specialization, some are listed more than once.

Anthropology

Goodall, Jane
    Researcher and Ethologist of Chimpanzees — 1934-

Leakey, Mary
    Paleoanthropologist;
    Researcher of Early Human Ancestors — 1913-1996

Mead, Margaret
    Pioneer in Cultural Anthropology — 1901-1978

Astronomy

Bessel, Friedrich
    Astronomer; Originator of Bessel Functions — 1784-1846

Brahe, Tycho
    Discoverer of Supernova — 1546-1601

Cannon, Annie Jump
    Developer of Star Classification System — 1863-1941

Copernicus, Nicolaus
    Father of Modern Astronomy — 1473-1543

Galilei, Galileo
    Pioneer in Laws of Motion and Early Telescopy — 1564-1642

Halley, Edmond
    Pioneering Investigator of Comets — 1656-1742

Herschel, Caroline Lucretia
    First Major Woman Astronomer of Modern Times — 1750-1848

Herschel, William
    Founder of Star-Related Astronomical Research — 1738-1822

Hubble, Edwin
    Discoverer of Galaxies Beyond Milky Way — 1889-1953

Huygens, Christiaan
    Originator of Wave Theory of Light;
    Inventor of Pendulum Clock — 1629-1695

Hypatia of Alexandria
    First Recognized Woman Mathematician and Scientist — c. 370-415

Kepler, Johannes
    Developer of Laws of Planetary Motion — 1571-1630

Mitchell, Maria
    First American Woman Astronomer — 1818-1889

Biology

Agassiz, Louis
    Originator of Ice Age Theory — 1807-1873

Buffon, Georges
    Developer of Survey of Natural World — 1707-1788

Comstock, Anna Botsford
    Entomologist; Pioneer in Science Education — 1854-1930

Darwin, Charles
    Founder of Evolutionary Theory — 1809-1882

Just, Ernest Everett
    Pioneer in Developmental Biology — 1883-1941

Leeuwenhoek, Antoni Van
    Pioneer Microscopist and Microbiologist — 1632-1723

Linnaeus, Carolus
    Botanist; Originator of Taxonomic
    Classification System — 1707-1778

Metchnikoff, Élie
    Immunologist; Discoverer of Phagocytes — 1845-1916

Pasteur, Louis
    Pioneer Microbiologist; Developer of Pasteurization — 1822-1895

Botany

Borlaug, Norman Ernest
    Agronomist; Founder of the Green Revolution — 1914-

Brown, Robert
    Botanist; Discoverer of Brownian Motion — 1773-1858

Burbank, Luther
    Pioneer Breeder of New Plant Varieties — 1849-1926

Carver, George Washington
    Originator of Crop Rotation Practices — c.1860-1943

Fuchs, Leonhard
    Developer of Foundations of Botany — 1501-1566

Linnaeus, Carolus
    Botanist; Originator of Taxonomic
    Classification System — 1707-1778

Von Humboldt, Alexander
    Early Pioneer of Ecological Concepts — 1769-1859

Chemistry

Arrhenius, Svante
    Originator of Electrolytic Dissociation Theory — 1859-1927

Avogadro, Amedeo
    Originator of Concept of Molecules;
    Developer of Avogadro's Law — 1776-1856

Black, Joseph
  Isolator of Carbon Dioxide;
  Definer of Latent Heat and Heat Capacity   1728-1799

Boyle, Robert
  Founder of Chemistry   1627-1691

Brooks, Harriet
  Pioneering Researcher in Radioactivity   1876-1933

Bunsen, Robert
  Developer of Chemical Spectral Analysis   1811-1899

Cavendish, Henry
  Investigator of Hydrogen;
  Calculator of Earth's Gravitational Constant   1731-1810

Curie, Marie
  Founder of Radiation Chemistry   1867-1934

Dalton, John
  Founder of Modern Atomic Theory   1766-1844

Davy, Humphry
  Discoverer of Elements; Pioneer in Electrolysis   1778-1829

Faraday, Michael
  Discoverer of Electromagnetic Induction   1791-1867

Fraser-Reid, Bertram
  Synthesizer of Organic Compounds   1934-

Gay-Lussac, Joseph- Louis
  Discoverer of Law of Combining Volumes   1778-1850

Haber, Fritz
  Chemist; First to Synthesize Ammonia   1868-1934

Hooke, Robert
  Identifier of Cells; Formulator of Laws of Elasticity   1635-1703

Julian, Percy Lavon
  Organic Chemist;
  Developer of Treatment for Glaucoma   1899-1975

Kekulé von Stradonitz, August
  Discoverer of Carbon Benzene Rings   1829-1896

Lavoisier, Antoine
  Originator of Law of Conservation of Mass   1743-1794

Le Châtelier, Henri-Louis
  Originator of Le Châtelier's
  Principle of Chemical Equilibrium   1850-1936

Mendeleyev, Dmitry
  Creator of Periodic Table of Elements   1834-1907

Pasteur, Louis
  Pioneer Microbiologist; Developer of Pasteurization   1822-1895

Pauling, Linus
  Pioneer Researcher in Nuclear and Molecular Structure   1901-1994

Priestley, Joseph
  Discoverer of Oxygen   1733-1804

Takamine, Jokichi
  Isolator of Adrenaline   1854-1922

Wöhler, Friedrich
  Synthesizer of Urea   1800-1882

Computer Science

Aiken, Howard Hathaway
  Pioneer in Early Computer Engineering   1900-1973

Babbage, Charles
  Grandfather of the Modern Computer;
  Designer of the Analytical Engine   1791-1871

Bardeen, John
  Superconductivity Theorist; Inventor of Transistor   1908-1991

Boole, George
  Co-founder of Symbolic Logic   1815-1864

Gates, Bill
  Computer Programmer;
  Founder of Microsoft Corporation   1955-

Hopper, Grace Murray
  Inventor of the Computer Language Compiler   1906-1992

Lovelace (Countess of), Augusta Ada
  First Computer Programmer   1815-1852

Pascal, Blaise
  Pioneer in Geometry, Probability, and Hydrostatics;
  Inventor of Calculating Machine   1623-1662

Von Neumann, John
  Early Computer Designer;
  Developer of Game Theory   1903-1957

Ecology

Carson, Rachel
  Early Advocate of Environmental Movement   1907-1964

Swallow Richards, Ellen
  Founder of Home Ecology   1842-1911

Von Humboldt, Alexander
  Early Pioneer of Ecological Concepts   1769-1859

Earth Sciences

Agricola, Georgius
  Father of Mineralogy   1494-1555

Agassiz, Elizabeth Cary
  Science Writer   1822-1907

Agassiz, Louis
  Originator of Ice Age Theory   1807-1873

Buffon, Georges
  Developer of Survey of Natural World   1707-1788

Bjerknes, Vilhelm
  Founder of Modern Meteorology   1862-1951

Cousteau, Jacques-Yves
  Pioneer Oceanographer;
  Developer of Scuba Equipment   1910-1997

Cuvier, Georges
  Early Pioneer in Comparative
  Anatomy and Paleontology   1769-1832

Hess, Harry Hammond
  Originator of Theory of Plate Tectonics   1906-1969

Lyell, Charles
Originator of Fundamental Principles of Geology — 1797-1875

Von Humboldt, Alexander
Early Pioneer of Ecological Concepts — 1769-1859

Wegener, Alfred Lothar
Developer of Principle of Continental Drift — 1880-1930

Genetics

Bateson, William
Early Mendelian Geneticist — 1861-1926

Crick, Francis
Co-discoverer of DNA Structure — 1916-

Franklin, Rosalind
Pioneer of X-Ray Diffraction for DNA Imaging — 1920-1958

McClintock, Barbara
Pioneer in Genetic Analysis — 1902-1992

Mendel, Gregor Johann
Founder of Genetics — 1822-1884

Watson, James
Co-discoverer of DNA Structure — 1928-

Invention

Aiken, Howard Hathaway,
Pioneer in Early Computer Engineering — 1900-1973

Appert, Nicolas-François
Developer of Canned Food Process — c.1750-1841

Babbage, Charles
Grandfather of the Modern Computer,
Designer of the Analytical Engine — 1791-1871

Bardeen, John
Superconductivity Theorist; Inventor of Transistor — 1908-1991

Bell, Alexander Graham
Inventor of the Telephone — 1847-1922

Bessemer, Henry
Inventor of Bessemer Steel-making Process — 1813-1898

Boole, George
Co-founder of Symbolic Logic — 1815-1864

Braille, Louis
Inventor of Braille Alphabet — 1809-1852

Bramah, Joseph
Inventor of Hydraulic Press and Bramah Lock — 1748-1814

Carothers, Wallace
Inventor of Nylon — 1896-1937

Cousteau, Jacques-Yves
Pioneer Oceanographer;
Developer of Scuba Equipment — 1910-1997

Daguerre, Louis
Inventor of the First Photograph — 1789-1851

Darby, Abraham
Originator of the Use of Coke for Iron Smelting — c. 1678-1717

Diesel, Rudolf
Inventor of Diesel Engine — 1858-1913

Eastman, George
Inventor of Hand-held Camera — 1854-1932

Edison, Thomas
Prolific Inventor; Creator of the Light Bulb — 1847-1931

Everson, Carrie J.
Inventor of Ore Concentration Process — c.1844-1913

Ford, Henry
Automobile Engineer; Inventor of the Model T — 1863-1947

Franklin, Benjamin
Prolific Inventor; Investigator of Electricity — 1706-1790

Galilei, Galileo
Pioneer in Laws of Motion and Early Telescopy — 1564-1642

Gates, Bill
Computer Programmer;
Founder of Microsoft Corporation — 1955-

Goddard, Robert H.
Pioneer in Early Rocket Design — 1882-1945

Goodyear, Charles
Inventor of Vulcanized Rubber — 1800-1860

Gutenberg, Johannes
Inventor of Movable-Type Printing — c. 1400-1468

Hall, Charles
Co-developer of Aluminum Manufacturing Process — 1863-1914

Hall, Julia
Co-developer of Aluminum Manufacturing Process — 1859-1925

Hargreaves, James
Inventor of the Spinning Jenny — c. 1720-1778

Harrison, John
Inventor of Practical Chronometer — 1693-1776

Hooke, Robert
Identifier of Cells; Formulator of Laws of Elasticity — 1635-1703

Hopper, Grace Murray
Inventor of the Computer Language Compiler — 1906-1992

Howe, Elias
Inventor of the Sewing Machine — 1819-1867

Huygens, Christiaan
Originator of Wave Theory of Light;
Inventor of Pendulum Clock — 1629-1695

Jones, Amanda
Inventor of Vacuum Canning — 1835-1914

Leeuwenhoek, Antoni Van
Pioneer Microscopist and Microbiologist — 1632-1723

Lovelace (Countess of), Augusta Ada
First Computer Programmer — 1815-1852

Lumière, Auguste
Co-inventor of Cinema — 1862-1954

Lumière, Louis
Co-inventor of Cinema — 1864-1948

Marconi, Guglielmo
Developer of Radio Communication — 1874-1937

McCormick, Cyrus Hall
Inventor of Mechanical Reaper — 1809-1884

Mercator, Gerardus
Cartographer; Creator of Mercator Map Projection   1512-1594

Morgan, Garrett
Inventor of Gas Mask and Traffic Light   1877-1963

Morse, Samuel
Inventor of First Single-Wire Telegraph   1791-1872

Newcomen, Thomas
Inventor of Steam Engine   1663-1729

Otto, Nikolaus August
Inventor of Four-Stroke Internal Combustion Engine   1832-1891

Pascal, Blaise
Pioneer in Geometry, Probability, and Hydrostatics;
Inventor of Calculating Machine   1623-1662

Sholes, Christopher
Developer of Typewriter   1819-1890

Sikorsky, Igor
Inventor of the Helicopter   1889-1972

Stephenson, George
Inventor of Practical Steam Locomotive   1781-1848

Tesla, Nikola
Developer of Alternating Current Electricity   1856-1943

Torricelli, Evangelista
Inventor of the Barometer   1608-1647

Ts'ai Lun
Inventor of Paper   c.50-121

Von Neumann, John
Early Computer Designer;
Developer of Game Theory   1903-1957

Watson-Watt, Robert
Developer of Radar   1892-1973

Watt, James
Developer of Steam Engine Technology   1736-1819

Whitney, Eli
Inventor of the Cotton Gin   1765-1825

Wright, Orville
Co-inventor of the Modern Airplane   1871-1948

Wright, Wilbur
Co-inventor of the Modern Airplane   1867-1912

Mathematics

Abel, Niels Henrik
Advocate of Rigorous Mathematical Proofs   1802-1829

Agnesi, Maria
First Important Woman
Mathematician of Modern Times   1718-1799

Al-Khwarizmi, Mohammed ibn Musa
Developer of Arabic Numerals; Pioneer in Algebra   c. 780-850

Archimedes
Inventor; Early Pioneer in Plane
and Solid Geometry   c. 290-c. 212 B.C.E.

Bernoulli, Jakob
Co-originator of the Calculus of Variations   1654-1705

Bernoulli, Johann
Co-originator of the Calculus of Variations   1667-1748

Bessel, Friedrich
Astronomer; Originator of Bessel Functions   1784-1846

Boole, George
Co-founder of Symbolic Logic   1815-1864

Brahe, Tycho
Discoverer of Supernova   1546-1601

Buffon, Georges
Developer of Survey of Natural World   1707-1788

Cauchy, Augustin-Louis
Early Developer of the Calculus   1789-1857

Descartes, René
Founder of Analytic Geometry   1596-1650

Eratosthenes
Mathematician; Geographer;
Librarian at Alexandria   c. 284-192 B.C.E.

Euclid
Organizer of Geometry;
Author of Landmark Elements   c. 300-c. 275 B.C.E.

Euler, Leonhard
Pioneer in Algebra and Number Theory   1707-1783

Fermat, Pierre de
Pioneer in the Calculus and Number Theory   1601-1665

Fibonacci, Leonardo
Popularizer of Arabic Numerals;
Pioneer in Algebra and Geometry   c. 1170-c. 1250

Gauss, Carl Friedrich
Multi-faceted Mathematician;
Developer of Complex Number System   1777-1855

Germain, Sophie
Pioneer in Number Theory and Study of Elasticity   1776-1831

Huygens, Christiaan
Originator of Wave Theory of Light;
Inventor of Pendulum Clock   1629-1695

Hypatia of Alexandria
First Recognized Woman Mathematician and Scientist   c. 370-415

Kovalevsky, Sonya
Developer of Partial Differential Equations   1850-1891

Leibniz, Gottfried Wilhelm
Inventor of Differential and Integral Calculus   1646-1716

Lovelace (Countess of), Augusta Ada
First Computer Programmer   1815-1852

Newton, Isaac
Discoverer of Laws of Motion and Gravitation   1642-1727

Noether, Emmy
Pioneer in Modern Algebra   1882-1935

Pascal, Blaise
Pioneer in Geometry, Probability, and
Hydrostatics; Inventor of Calculating Machine   1623-1662

Poincaré, Henri
Influential Modern Mathematician   1854-1912

Pythagoras
Early Pioneer in Mathematics   c. 580-500 B.C.E.

Russell, Bertrand
Originator of Logicism   1872-1970

Somerville, Mary Fairfax
Popular Science and Mathematics Author   1780-1872

Von Neumann, John
Early Computer Designer;
Developer of Game Theory   1903-1957

Natural History

Agassiz, Elizabeth Cary
Science Writer   1822-1907

Agassiz, Louis
Originator of Ice Age Theory   1807-1873

Buffon, Georges
Developer of Survey of Natural World   1707-1788

Cuvier, Georges
Early Pioneer in Comparative
Anatomy and Paleontology   1769-1832

Darwin, Charles
Founder of Evolutionary Theory   1809-1882

Von Humboldt, Alexander
Early Pioneer of Ecological Concepts   1769-1859

Psychology

Freud, Sigmund
Founder of Psychoanalysis   1856-1939

Pavlov, Ivan
Originator of Theory of Conditioned Reflexes   1849-1936

Philosophy of Science

Bacon, Francis
Philosopher of Science;
Developer of Scientific Methodology   1561-1626

Russell, Bertrand
Originator of Logicism   1872-1970

Physiology and Medicine

Ar-Razi, Abu Bakr Muhammad
Pioneering Islamic Physician of Middle Ages   c. 865-c. 935

Barton, Clara
Founder of American Red Cross   1821-1912

Blackwell, Elizabeth
First American Woman Physician of Modern Times   1821-1910

Drew, Charles
Developer of Technique for Storing Blood Plasma 1904-1950

Ehrlich, Paul
Immunologist; Developer of Cure for Syphilis   1854-1915

Fleming, Alexander
Bacteriologist; Discoverer of Penicillin   1881-1955

Freud, Sigmund
Founder of Psychoanalysis   1856-1939

Galen of Pergamum
Leading Physician of Antiquity   129-c.200

Harvey, William
Discoverer of Blood Circulation   1578-1657

Hippocrates
Founder of Medicine   c. 460-c. 377 B.C.E.

Ibn Sina (Avicenna)
Physician; Author of Canon of Medicine   980-1037

Jenner, Edward
Developer of Modern Smallpox Vaccine   1749-1823

Julian, Percy Lavon
Organic Chemist;
Developer of Treatment for Glaucoma   1899-1975

Koch, Robert
Founder of Medical Bacteriology   1843-1910

Landsteiner, Karl
Immunologist; Identifier of Human Blood Types   1868-1943

Lister, Joseph
Founder of Antiseptic Surgery   1827-1912

Malpighi, Marcello
Early Pioneer in Medical Research   1628-1694

Metchnikoff, Élie
Immunologist; Discoverer of Phagocytes   1845-1916

Morton, William
Developer of General Anesthesia   1819-1868

Nightingale, Florence
Founder of Nursing Profession   1820-1910

Noguchi, Hideyo
Isolator of Cause of Paralysis in Syphilis Patients 1876-1928

Paracelsus
Revolutionary Physician and Alchemist   1493-1541

Paré, Ambroise
Founder of Modern Surgery   1510-1590

Pavlov, Ivan
Originator of Theory of Conditioned Reflexes   1849-1936

Pincus, Gregory
Developer of Oral Contraceptives   1903-1967

Semmelweis, Ignaz
Medical Reformer;
Founder of Cause for Puerperal Fever   1818-1865

Vesalius, Andreas
Founder of Anatomy   1514-1564

Von Békésy, Georg
Discoverer of the Physiology of Hearing   1899-1972

Williams, Daniel Hale
Founder of Cardiac Surgery   1858-1931

Wright, Jane Cooke
Pioneer in Chemotherapy   1919-

Yalow, Rosalyn Sussman
Developer of Radioimmunoassay   1921-

Physics

Ampère, André
  Formulator of Laws of Electromagnetism   1775-1836

Bardeen, John
  Superconductivity Theorist; Inventor of Transistor   1908-1991

Becquerel, Antoine-Henri
  Discoverer of Radioactivity   1852-1908

Black, Joseph
  Isolator of Carbon Dioxide;
  Definer of Latent Heat and Heat Capacity   1728-1799

Bohr, Niels
  Developer of Theory of Atomic Structure   1885-1962

Boltzmann, Ludwig
  Developer of Statistical Mechanics   1844-1906

Born, Max
  Early Pioneer in Quantum Mechanics   1882-1970

Bose, Satyendranath
  Pioneer in Quantum Mechanics   1894-1974

Boyle, Robert
  Founder of Chemistry   1627-1691

Brooks, Harriet
  Pioneering Researcher in Radioactivity   1876-1933

Carnot, Sadi
  Developer of Heat Engine Theory   1796-1832

Cavendish, Henry
  Investigator of Hydrogen;
  Calculator of Earth's Gravitational Constant   1731-1810

Chadwick, James
  Discoverer of the Neutron   1891-1974

Clausius, Rudolf
  Founder of Thermodynamics   1822-1888

Curie, Marie
  Founder of Radiation Chemistry   1867-1934

Dalton, John
  Founder of Modern Atomic Theory   1766-1844

Descartes, René
  Founder of Analytic Geometry   1596-1650

Einstein, Albert
  Developer of Relativity and Quantum Theories   1879-1955

Faraday, Michael
  Discoverer of Electromagnetic Induction   1791-1867

Fermi, Enrico
  Designer of First Nuclear Reactor   1901-1954

Feynman, Richard P.
  Developer of Quantum Electrodynamics   1918-1988

Galilei, Galileo
  Pioneer in Laws of Motion and Early Telescopy   1564-1642

Gay-Lussac, Joseph-Louis
  Discoverer of Law of Combining Volumes   1778-1850

Gell-Mann, Murray
  Developer of the Quark Model   1929-

Hahn, Otto
  Co-discoverer of Nuclear Fission   1879-1968

Hawking, Stephen
  Developer of Mathematical Proof for Black Hole Theory   1942-

Heisenberg, Werner
  Developer of Matrix Mechanics
  and the Uncertainty Principle   1901-1976

Hooke, Robert
  Identifier of Cells; Formulator of Laws of Elasticity   1635-1703

Huygens, Christiaan
  Originator of Wave Theory of Light;
  Inventor of Pendulum Clock   1629-1695

Joule, James Prescott
  Discoverer of Mechanical Equivalent of Heat   1818-1889

Kelvin, Lord (William Thomson)
  Pioneer in Electromagnetism, Heat, and Mechanics   1824-1907

Lavoisier, Antoine
  Originator of Law of Conservation of Mass   1743-1794

Maxwell, James Clerk
  Founder of Modern Electrical Theory   1831-1879

Meitner, Lise
  Co-discoverer of Nuclear Fission   1878-1968

Newton, Isaac
  Discoverer of Laws of Motion and Gravitation   1642-1727

Ohm, Georg
  Originator of Ohm's Law of Electrical Resistance   1789-1854

Oppenheimer, J. Robert
  Head Scientist on First Atomic Bomb Project   1904-1967

Pascal, Blaise
  Pioneer in Geometry, Probability, and Hydrostatics;
  Inventor of Calculating Machine   1623-1662

Planck, Max
  Developer of Foundation for Quantum Theory   1858-1947

Raman, Chandrasekhara Venkata
  Pioneer Researcher in Modern Optics   1888-1970

Roentgen, Wilhelm
  Discoverer of X-rays   1845-1923

Rutherford, Ernest
  Discoverer of the Atomic Nucleus   1871-1937

Sakharov, Andrey
  Developer of Soviet Hydrogen Bomb   1921-1989

Somerville, Mary Fairfax
  Popular Science and Mathematics Author   1780-1872

Torricelli, Evangelista
  Inventor of the Barometer   1608-1647

Volta, Alessandro
  Discoverer of Electrochemical Cells   1745-1827

Wu, Chien-Shiung
  Demonstrator of Non-Conservation of Parity   c. 1912-

Yalow, Rosalyn Sussman
  Developer of Radioimmunoassay   1921-

Yukawa, Hideki
  Founder of Meson Theory   1907-1981

Bibliography

General Histories of Science, Mathematics, and Invention

The following sources are included to give readers a fuller context for the subjects covered in this volume, a context that is more complete than is possible through the individual biographies. They will provide researchers with encyclopedic information on the history of science, mathematics, and invention; details concerning highlights in the history of science, mathematics, and invention; and various analytical perspectives on these fields.

Brody, David Eliot and Arnold R. *The Science Class You Wish You Had: The Seven Greatest Discoveries in History and the People Who Made Them.* New York: Berkley, 1997.

Burke, James. *Connections.* Boston: Little, Brown, 1995.

Bynum, William F., et al., eds. *Dictionary of the History of Science.* Princeton, N.J.: Princeton University Press, 1981.

Cardwell, D.S.L. *Turning Points in Western Technology: A Study of Technology, Science, and History.* Canton, Mass.: Science History Publications, 1991.

Cowan, Ruth Schwartz. *A Social History of American Technology.* New York: Oxford University Press, 1997.

Crosland, Maurice. *The Science of Matter: A Historical Survey.* Philadelphia, Penn.: Gordon & Breach Science, 1992.

Crowther, J.G. *Science in Modern Society.* New York: Schocken Books, 1968.

Koestler, Arthur. *The Sleepwalkers: A History of Man's Changing Vision of the Universe.* New York: Macmillan, 1968.

Matossian, Mary Allerton Kilbourne. *Shaping World History: Breakthroughs in Ecology, Technology, Science, and Politics.* Armonk, N.Y.: M.E. Sharpe, 1997.

Pullman, Bernard. *The Atom in the History of Human Thought.* Translated by Axel Reisinger. New York: Oxford University Press, 1998.

Pursell, Carroll W. *The Machine in America: A Social History of Technology.* Baltimore: John Hopkins University Press, 1995.

Ronan, Colin A. *Science: Its History and Development Among the World Cultures.* New York: Facts On File, 1982.

Schwartz, George. *Moments of Discovery.* New York: Basic Books, 1958.

Selin, Helaine, ed. *Encyclopaedia of the History of Science, Technology, and Medicine in Non-Western Cultures.* Boston: Kluwer Academic Publishing, 1997.

Serres, Michel. *A History of Scientific Thought.* Cambridge, Mass.: Blackwell Reference, 1995.

General Collections of Biographies

Included in this section are various sources that will provide readers additional biographical information on many of the individuals included in this volume in addition to information on others not included. The sources here vary from basic biographical encyclopedias and dictionaries to biographical sources that specialize in particular fields or ideas.

Abbott, D., ed. *The Biographical Dictionary of Scientists: Chemists.* New York: Peter Bedrick Books, 1983.

Bell, E. T. *Men of Mathematics.* New York: Simon & Schuster, 1937.

Bowden, Mary Ellen. *Chemical Achievers.* Philadelphia, Penn.: Chemical Heritage Foundation, 1997.

Crowther, J. G. *British Scientists of the Twentieth Century.* London: Routledge & K. Paul, 1952.

DeFries, Amelia. *Pioneers of Science: Seven Pictures of Struggle and Victory.* Freeport, N.Y.: Books for Libraries Press, 1970.

Farber, Eduard, ed. *Great Chemists.* New York: Interscience, 1961.

Greenstein, George. *Portraits of Discovery.* New York: John Wiley, 1998.

Heathcote, N. H. *Nobel Prize Winners in Physics, 1901-1950.* Freeport, N.Y.: Books for Libraries Press, 1971.

Slater, Robert. *Portraits in Silicon.* Cambridge, Mass.: MIT Press, 1987.

Snow, C. P. *The Physicists.* Boston: Little, Brown, 1981.

Spencer, Donald. *Great Men and Women of Computing.* Ormond Beach, Fla.: Camelot Publishing, 1996.

Turner, Roland, and Steven Goulden. *Great Engineers and Pioneers in Technology.* New York: St. Martin's Press, 1981.

Wilford, John Noble. *The Mapmakers.* New York: Knopf, 1981.

Women, Minorities, and Non-Western Cultures

In keeping with one of the selection criteria used for this volume—to include women and minorities who have changed their world by breaking the confines imposed on them by society—we have separated out the biographical sources for women and minorities. These sources provide biographical information on the women and minorities in various fields throughout history. It also includes several sources on the contributions of non-Western cultures to the development of science, mathematics, and invention, especially the work of Arabic scientists and practitioners in the fields of mathematics and medicine.

Alic, Margaret. *Hypatia's Heritage: A History of Women in Science from Antiquity Through the Nineteenth Century.* Boston: Beacon Press, 1986.

Altman, Linda Jacobs. *Women Inventors.* New York: Facts On File, 1997.

Bailey, Martha J. *American Women in Science.* Denver: ABC-CLIO, 1994.

Bertsch McGrayne, Sharon. *Nobel Prize Women in Science.* Secaucus, N.J.: Carol Publishing, 1993.

Burt, McKinley. *Black Inventors of America.* Portland, Oreg.: National Book, 1989.

Elgood, Cyril. *A Medical History of Persia.* Cambridge, Mass.: Cambridge University Press, 1951.

Grinstein, Louise and Paul Campbell. *Women of Mathematics.* New York: Greenwood Press, 1987.

Haber, Louis. *Black Pioneers of Science and Invention.* San Diego: Harcourt, Brace, 1970.

The History of Women and Science, Health, and Technology: A Bibliographic Guide to the Professions and the Disciplines. Madison: University of Wisconsin System Women's Studies Librarian, 1993.

James, Portia P. *The Real McCoy: African-American Invention and Innovation, 1619-1930.* Washington, D.C.: Smithsonian Institution Press, 1989.

McDonald, Anne L. *Feminine Ingenuity: Women and Invention in America.* New York: Ballantine Books, 1992.

Morrow, Charlene, and Teri Perl, eds. *Notable Women in Mathematics.* Westport, Conn.: Greenwood Press, 1998.

Osen, Lynn W. *Women in Mathematics.* Cambridge, Mass.: MIT Press, 1975.

Perl, Teri. *Math Equals.* Reading, Mass.: Addison-Wesley, 1978.

Qurashi, M. M. and S.S.H. Rizvi. *History and Philosophy of Muslim Contribution to Science and Technology.* Islamabad: Pakistan Academy of Sciences, 1996.

Rayner-Canham, M. F., and G. W. *A Devotion to the Science: Pioneer Women of Radioactivity.* Philadelphia, Penn.: Chemical Heritage, 1997.

Sammons, Vivian Ovelton. *Blacks in Science and Medicine.* New York: Hemisphere Publishing, 1990.

Selin, Helaine, ed. *Encyclopaedia of the History of Science, Technology, and Medicine in Non-Western Cultures.* Boston: Kluwer Academic Publishing, 1997.

Shearer, Benjamin F., and Barbara S. Shearer. *Notable Women in the Physical Sciences.* Westport, Conn.: Greenwood Press, 1997.

Shiels, Barbara. *Winners: Women and the Nobel Prize.* Minneapolis, Minn.: Dillon Press, 1985.

Smith, Jessie Carney, ed. *Notable Black American Women.* Detroit: Gale Research, 1992.

Stanley, Autumn. *Mothers and Daughters of Invention: Notes for a Revised History of Technology.* New Brunswick, N.J.: Rutgers University Press, 1995.

Trescott, M. M. *Dynamos and Virgins Revisited: Women and Technological Change in History.* Metuchen, N.J.: Scarecrow Press, 1979.

Van Sertima, Ivan. *Blacks in Science: Ancient and Modern.* New Brunswick, N.J.: Transaction Books, 1983.

Vare, Ethlic and Greg Ptacek. *Mothers of Invention.* New York: William Morrow, 1988.

Yount, Lisa. *Black Scientists.* New York: Facts On File, 1991.

DISCIPLINE BY DISCIPLINE

Each section below presents several sources on various fields in science, mathematics, and invention. They include general histories of the field, analyses of major contributions and historical trends in these disciplines, the history of specialized topics within the fields, and biographies of important people who shaped these areas of endeavor. Some sources cited here may also be included in other sections of the bibliography.

Astronomy and Cosmology

Chapman, Allan. *Astronomical Instruments and Their Users: Tycho Brahe to William Lassell.* Brookfield, Vt.: Variorum, 1996.

Lightman, Alan P. *Origins: The Lives and Worlds of Modern Cosmologists.* Cambridge, Mass.: Harvard University Press, 1990.

North, J. D. *The Measure of the Universe: A History of Modern Cosmology.* New York: Dover, 1990.

Overby, Dennis. *Lonely Hearts of the Cosmos.* New York: Harper Collins, 1991.

Rosen, Edward. *Copernicus and His Successors.* London: Hambledon Press, 1995.

Thrower, Norman J., ed. *Standing on the Shoulders of Giants: A Longer View of Newton and Halley.* Berkeley: University of California Press, 1990.

Biology, Botany, and Ecology

Bajema, Carl Jay, ed. *Natural Selection Theory: From the Speculations of the Greeks to the Quantitative Measurements of the Biometricians.* New York: Van Nostrand Reinhold, 1983.

De Kruif, Paul. *Microbe Hunters.* San Diego: Harcourt Brace, 1996.

DeCoin, Robert L. *History and Cultivation of Cotton and Tobacco.* Wilmington, Del.: Scholarly Resources, 1973.

Dies, Edward J. *Titans of the Soil: Great Builders of Agriculture.* Chapel Hill: University of North Carolina Press, 1976.

Dunn, L. C. *A Short History of Genetics: The Development of Some of the Main Lines of Thought, 1864-1939.* Ames: Iowa State University Press, 1991.

Elseth, Gerald D. *Principles of Modern Genetics.* St. Paul, Minn.: West Publishing, 1995.

Grene, Marjorie, ed. *Dimensions of Darwinism: Themes and Counterthemes in Twentieth Century Evolutionary Theory.* New York: Cambridge University Press, 1983.

Hoppe, Brigitte, ed. *Biology Integrating Scientific Fundamentals: Contributions to the History of the Interrelations between Biology, Chemistry, and Physics from the 18th to the 20th Centuries.* Mèunchen: Institut fèur Greschichte der Naturwissen-Schaften, 1997.

Matossian, Mary Allerton Kilbourne. *Shaping World History: Breakthroughs in Ecology, Technology, Science, and Politics.* Armonk, N.Y.: M.E. Sharpe, 1997.

Sachs, Julius. *History of Botany, 1530-1860.* Authorized translation by Henry E. F. Garnsey. New York: Russell & Russell, 1967.

Chemistry

Bowden, Mary Ellen. *Chemical Achievers.* Philadelphia, Penn.: Chemical Heritage Foundation, 1997.

Brock, William. *The Norton History of Chemistry.* New York: Norton, 1992.

Elliott, Eric. *Polymers & People: An Informal History.* Philadelphia, Penn.: Beckman Center for the History of Chemistry, 1990.

Farber, Eduard, ed. *Great Chemists.* New York: Interscience, 1961.

Farber, Eduard. *Milestones of Modern Chemistry.* New York: Basic Books, 1966.

Higby, Gregory and Elaine C. Stroud. *The History of Pharmacy: A Selected Annotated Bibliography.* New York: Garland, 1995.

Hudson, John. *The History of Chemistry.* New York: Chapman & Hall, 1992.

Jaffe, Bernard. *Crucibles: The Story of Chemistry from Ancient Alchemy to Nuclear Fission.* New York: Simon & Schuster, 1948.

Knight, David M. *Ideas in Chemistry: A History of the Science.* New Brunswick, N.J.: Rutgers University Press, 1992.

Knight, David, et al., eds. *The Making of the Chemist: The Social History of Chemistry in Europe, 1789-1914.* New York: Cambridge University Press, 1998.

Nye, Mary Jo. *Before Big Science: The Pursuit of Modern Chemistry and Physics, 1800-1940.* London: Prentice-Hall International, 1996.

Russel, C. A. *The Origins of Organic Chemistry, 1800-1900.* London: Royal Society of Chemistry, 1992.

Salsberg, Hugh W. *From Caveman to Chemist.* Washington, D.C.: American Chemical Society, 1991.

Szabadvâary, Ferenc. *History of Analytical Chemistry.* Langhore, Penn.: Gordon & Breach Science Publishers, 1992.

Wieland, Theodor. *The World of Peptides: A Brief History of Peptide Chemistry.* Berlin, N.Y.: Springer Verlag, 1991.

Computer Science

Blohm, Hans, Stafford Beer, and David Suzuki. *Pebbles to Computers: The Thread.* Toronto: Oxford University Press, 1986.

Campbell-Kelly, Martin. *Computer: A History of the Information Machine.* New York: Basic Books, 1996.

Goldstine, Herman. *The Computer from Pascal to von Neumann.* Princeton, N.J.: Princeton University Press, 1972.

Richie, David. *The Computer Pioneers.* New York: Simon & Schuster, 1986.

Shurkin, Joel. *Engines of the Mind: A History of the Computer.* New York: Norton, 1984.

Slater, Robert. *Portraits in Silicon.* Cambridge, Mass.: MIT Press, 1987.

Spencer, Donald. *Great Men and Women of Computing.* Ormond Beach, Fla.: Camelot Publishing, 1996.

Williams, Michael R. *A History of Computing Technology.* Los Alamitos, Calif.: IEEE Computer Society Press, 1997.

Earth Sciences

Brush, Stephen G. *Transmuted Past: The Age of the Earth and the Evolution of the Elements from Lyell to Patterson.* Cambridge, Mass.: Cambridge University Press, 1996.

Dragomir, V. C., et al. *Theory of the Earth's Shape.* New York: Elsevier Scientific Publishing, 1981.

Fenton, Carroll and Mildred Fenton. *Giants of Geology.* Garden City, N.Y.: Doubleday, 1952.

Oldroyd, D. R. *Thinking About the Earth: A History of Ideas in Geology.* Boston: Harvard University Press, 1996.

Invention

Blohm, Hans, Stafford Beer, and David Suzuki. *Pebbles to Computers: The Thread.* Toronto: Oxford University Press, 1986.

Brown, Jim. *Radar: How It All Began.* London: Janus, 1996.

Buderi, Robert. *The Invention That Changed the World: How a Small Group of Radar Pioneers Won the Second World War and Launched a Technological Revolution.* New York: Simon & Schuster, 1996.

Burke, James. *Connections.* Boston: Little, Brown, 1995.

Cardwell, Donald. *The Norton History of Technology.* New York: Norton, 1995.

Carr, J. C., and W. Taplin. *History of the British Steel Industry.* Cambridge, Mass.: Harvard University Press, 1962.

Crowther, J.G. *Discoveries and Inventions of the 20th Century.* New York: Dutton, 1966.

Crone, G.R. *Maps and Their Makers: An Introduction to the History of Cartography.* Hamden, Conn.: Archon Books, 1978.

Davenport, Alma. *The History of Photography: An Overview.* Boston: Focal Press, 1991.

Garratt, G. R. M. *The Early History of Radio: From Faraday to Marconi.* London: Institution of Electrical Engineers, 1994.

Gianetti, Louis D. *Flashback: A Brief History of Film.* Englewood Cliffs, N.J.: Prentice-Hall, 1996.

Ginsburg, Madeleine, ed. *The Illustrated History of Textiles.* London: Studio Editions, 1991.

Hawke, David. *Nuts and Bolts of the Past: A History of American Technology, 1776-1860.* New York: Harper & Rowe, 1989.

Hills, Richard Leslie. *Power from Steam: A History of the Stationary Steam Engine.* New York: Cambridge University Press, 1989.

Marton, L. *Early History of the Electron Microscope.* San Francisco: San Francisco Press, 1994.

Noonan, G. Jon. *Nineteenth Century Inventors.* New York: Facts On File, 1992.

Rudin, Bo. *Making Paper: A Look into the History of an Ancient Craft.* New York: Rudins, 1990.

Steinberg, S. H. *Five Hundred Years of Printing.* New Castle, Del.: Oak Knoll Press, 1996.

Torgerson, Nancy. *Food Preservation: Before the Mason Jar.* Decatur, Ill: Glimpse of the Past, 1995.

Turner, Roland and Steven Goulden. *Great Engineers and Pioneers in Technology.* New York: St. Martin's Press, 1981.

Tylecote, R. F. *A History of Metallurgy.* Brookfield, Vt.: Institute of Metals, 1992.

Wilford, John Noble. *The Mapmakers.* New York: Knopf, 1981.

Mathematics

Bell, E. T. *Men of Mathematics.* New York: Simon & Schuster, 1937.

Calinger, Ronald. *Classics of Mathematics.* Englewood Cliffs, N.J.: Prentice-Hall, 1995.

Dunham, William. *Journey through Genius.* New York: John Wiley, 1990.

Dunham, William. *The Mathematical Universe: An Alphabetical Journey Through the Great Proofs, Problems, and Personalities.* New York: John Wiley, 1994.

Gindikin, S. *Tales of Physicists and Mathematicians.* Cambridge, Mass.: Birkhauser, 1988.

Morrow, Charlene, and Teri Perl, eds. *Notable Women in Mathematics.* Westport, Conn.: Greenwood Press, 1998.

Muir, Jane. *Of Men and Numbers: The Story of the Great Mathematicians.* New York: Dover, 1961.

Newman, James R., ed. *The World of Mathematics.* New York: Simon & Schuster, 1956.

Perl, Teri. *Math Equals.* Reading, Mass.: Addison-Wesley, 1978.

Simmons, George F. *Calculus Gems: Brief Lives and Memorable Mathematics.* New York: McGraw-Hill, 1992.

Singh, Simon. *Fermat's Enigma.* New York: Walker, 1997.

Smith, Sanderson M. *Agnesi to Zeno.* Berkeley, Calif.: Key Curriculum Press, 1996.

Struik, Dirk J. *A Concise History of Mathematics.* New York: Dover, 1987.

Psychology

Hothersall, David. *History of Psychology.* New York: McGraw-Hill, 1990.

Danziger, Kurt. *Constructing the Subject: Historical Origins of Psychological Research.* New York: Cambridge University Press, 1990.

Reisman, John M. *A History of Clinical Psychology.* New York: Hemisphere, 1991.

Brennan, James F. *History and Systems of Psychology.* Englewood Cliffs, N.J.: Prentice-Hall, 1990.

Leahey, Thomas Hardy. *A History of Psychology: Main Currents in Psychological Thought.* Englewood Cliffs, N.J.: Prentice-Hall, 1992.

Stephen Everson, ed. *Psychology.* New York: Cambridge University Press, 1991.

Viney, Wayne. *A History of Psychology: Ideas and Context.* Boston: Allyn & Bacon, 1993.

Physiology and Medicine

Dolan, John P. and William Adams-Smith. *Health and Society: A Documentary History of Medicine.* New York: Seabury Press, 1978.

Dolan, Josephine. *Nursing in Society: A Historical Perspective.* Philadelphia, Penn.: Saunders, 1983.

Donahue, M. Patricia. *Nursing: The Finest Art.* St. Louis, Mo.: Mosby, 1985.

Duin, Nancy. *A History of Medicine: From Pre-history to the Year 2000.* New York: Simon & Schuster, 1992.

Elgood, Cyril. *A Medical History of Persia.* Cambridge, Mass.: Cambridge University Press, 1951.

French, R. K. *Ancient Natural History.* London: Routledge & K. Paul, 1994.

Grainger, Thomas H. *A Guide to the History of Bacteriology.* New York: Ronald Press, 1958.

Robbin, Irving. *Giants of Medicine.* New York: Grosset & Dunlop, 1962.

Rutkow, Ira M. *Surgery: An Illustrated History.* St. Louis, Mo.: Mosby Year Book, 1993.

Sigerist, Henry E. *The Great Doctors: A Biographical History of Medicine.* Freeport, N.Y.: Books for Libraries Press, 1971.

Singer, Charles J. *A Short History of Anatomy from the Greeks to Harvey.* New York: Dover, 1957.

Physics

Albert, David. *Quantum Mechanics and Experience.* Cambridge, Mass.: Harvard University Press, 1992.

Ballentine, Leslie E. *Quantum Mechanics*. Englewood Cliffs, N.J.: Prentice-Hall, 1990.

Bordeau, Sanford P. *Volts to Hertz: The Rise of Electricity*. Minneapolis, Minn.: Burgess Publishing, 1982

Brown, Laurie M., et al., eds. *Twentieth Century Physics*. New York: American Institute of Physics Press, 1995.

Cardwell, D.S.L. *From Watt to Clausius: The Rise of Thermodynamics in the Early Industrial Age*. Ames: Iowa State University Press, 1989.

Chung, Kai Lai. *From Brownian Motion to Schrödinger's Equation*. New York: Springer-Verlag, 1995.

Gindikin, S. *Tales of Physicists and Mathematicians*. Cambridge, Mass.: Birkhauser, 1988.

Graetzer, Hans G., and David L. Anderson. *The Discovery of Nuclear Fission*. New York: Arno Press, 1981.

Heathcote, N. H. *Nobel Prize Winners in Physics, 1901-1950*. Freeport, N.Y.: Books for Libraries Press, 1971.

Newman, Harvey B. and Thomas Ypsilantis, eds. *History of Original Ideas and Basic Discoveries in Particle Physics*. New York: Plenum Press, 1996.

Nye, Mary Jo. *Before Big Science: The Pursuit of Modern Chemistry and Physics, 1800-1940*. London: Prentice-Hall International, 1996.

Purrington, Robert D. *Physics in the 19th Century*. New Brunswick, N.J.: Rutgers University Press, 1997.

Segrè, Emilio. *From Falling Bodies to Radio Waves*. New York: Freeman, 1984.

Segrè, Emilio. *From X-rays to Quarks: Modern Physicists and the Discoveries*. San Francisco: Freeman, 1983.

Snow, C. P. *The Physicists*. Boston: Little, Brown, 1981.

Thrower, Norman J., ed. *Standing on the Shoulders of Giants: A Longer View of Newton and Halley*. Berkeley: University of California Press, 1990

York, Herbert F. *The Advisors: Oppenheimer, Teller, and the Superbomb*. Stanford, Calif.: Stanford University Press, 1989.

Zukav, Gary. *The Dancing Wu Li Masters: An Overview of the New Physics*. New York: Bantam Books, 1980.

ERA BY ERA

The sources in this section offer readers a more focused analysis of specific eras in the history of science, mathematics, and invention. The books here may cover specific subject areas in a designated era, a particular idea important within an era, or a group of individuals who have shaped a specific time period. We have tried to include at least one source for each of the major eras, and some sources cited here may also be included in other sections of the bibliography.

Adams, Alexander B. *Eternal Quest: The Story of the Great Naturalists*. New York: Putnam, 1969.

Allman, George Johnston. *Greek Geometry from Thales to Euclid*. New York: Arno Press, 1976.

Aspray, William, et al. *Computing Before Computers*. Ames: Iowa State University Press, 1990.

Brown, Laurie M., et al., eds. *Twentieth Century Physics*. New York: American Institute of Physics Press, 1995.

Cardwell, D.S.L. *From Watt to Clausius: the Rise of Thermodynamics in the Early Industrial Age*. Ames: Iowa State University Press, 1989.

Chappell, Vere, ed. *Seventeenth-century Natural Scientists*. New York: Garland Publishing, 1992.

Conrad, Lawrence, et al. *The Western Medical Tradition: 800 B.C.-1800 A.D.* New York: Cambridge University Press, 1995.

Crowther, J. G. *British Scientists of the Twentieth Century*. London: Routledge & K. Paul, 1952.

Crowther, J. G. *Founders of British Science: John Wilkins, Robert Boyle, John Ray, Christopher Wren, Robert Hooke, Isaac Newton*. Westport, Conn.: Greenwood Publishing, 1982.

Crowther, J. G. *Discoveries and Inventions of the 20th Century*. New York: Dutton, 1966.

Crowther, J. G. *Scientists of the Industrial Revolution: Joseph Black, James Watt, Joseph Priestly, Henry Cavendish*. London: Cresset Press, 1962.

Debus, Allen G. *The Chemical Philosophy: Paracelsian Science and Medicine in the Sixteenth and Seventeenth Century*. New York: Science History Publications, 1977.

Devlin, Keith. *Life by the Numbers*. New York: John Wiley, 1998.

Dolan, John P. and William Adams-Smith. *Health and Society: A Documentary History of Medicine*. New York: Seabury Press, 1978.

Dolan, Josephine. *Nursing in Society: A Historical Perspective*. Philadelphia, Penn.: Saunders, 1983.

Donahue, M. Patricia. *Nursing: The Finest Art*. St. Louis, Mo.: Mosby, 1985.

French, R. K. *Ancient Natural History*. London: Routledge & K. Paul, 1994.

Grainger, Thomas H. *A Guide to the History of Bacteriology*. New York: Ronald Press, 1958.

Grene, Marjorie, ed. *Dimensions of Darwinism: Themes and Counterthemes in Twentieth Century Evolutionary Theory*. New York: Cambridge University Press, 1983.

Gullberg, Jan. *Mathematics: From the Birth of Numbers*. New York: Norton, 1997.

Hawke, David. *Nuts and Bolts of the Past: A History of American Technology, 1776-1860*. New York: Harper & Rowe, 1989.

Hoppe, Brigitte, ed. *Biology Integrating Scientific Fundamentals: Contributions to the History of the Interrelations between Biology, Chemistry, and Physics from the 18th to the 20th Centuries*. Mèunchen: Institut fèur Greschichte der Naturwissen-Schaften, 1997.

Jaffe, Bernard. *Crucibles: The Story of Chemistry from Ancient Alchemy to Nuclear Fission.* New York: Simon & Schuster, 1948.

Jenkins, Alan C. *The Naturalists.* New York: Wildflower Books, 1978.

King, Lester S. *The Road to Medical Enlightenment, 1650-1695.* New York: American Elsevier, 1970.

King, Lester S. *The Medical World of the Eighteenth Century.* Chicago: University of Chicago Press, 1958.

Knight, David M. *Ideas in Chemistry: A History of the Science.* New Brunswick, N.J.: Rutgers University Press, 1992.

Koestler, Arthur. *The Sleepwalkers: A History of Man's Changing Vision of the Universe.* New York: Macmillan, 1968.

Lemmon, Kenneth. *The Golden Age of Plant Hunters.* London: Phoenix House, 1968.

Lloyd, G. E. R. *Magic, Reason, and Experience.* New York: Cambridge University Press, 1979.

Maor, Eli. *E: The Story of a Number.* Princeton, N.J.: Princeton University Press, 1994.

Murdoch, John E. *Antiquity and the Middle Ages.* New York: Scribner's, 1984.

Newman, Harvey B. and Thomas Ypsilantis, eds. *History of Original Ideas and Basic Discoveries in Particle Physics.* New York: Plenum Press, 1996.

Noonan, G. Jon. *Nineteenth Century Inventors.* New York: Facts On File, 1992.

Nye, Mary Jo. *Before Big Science: The Pursuit of Modern Chemistry and Physics, 1800-1940.* London: Prentice-Hall International, 1996.

Qurashi, M. M. and S.S.H. Rizvi. *History and Philosophy of Muslim Contribution to Science and Technology.* Islamabad: Pakistan Academy of Sciences, 1996.

Purrington, Robert D. *Physics in the 19th Century.* New Brunswick, N.J.: Rutgers University Press, 1997.

Reiser, Stanley J. *Medicine and the Reign of Technology.* New York: Cambridge University Press, 1978.

Rocke, Alan J. *Chemical Atomism in the Nineteenth Century: From Dalton to Canizzaro.* Columbus: Ohio State University Press, 1984.

Ronan, Colin A. *Lost Discoveries: The Forgotten Science of the Ancient World.* New York: McGraw-Hill, 1973.

Russel, C. A. *The Origins of Organic Chemistry, 1800-1900.* London: Royal Society of Chemistry, 1992.

Salsberg, Hugh W. *From Caveman to Chemist.* Washington, D.C.: American Chemical Society, 1991.

Segrè, Emilio. *From X-rays to Quarks: Modern Physicists and the Discoveries.* San Francisco: Freeman, 1983.

Shurkin, Joel. *Engines of the Mind: A History of the Computer.* New York: Norton, 1984.

Singer, Charles J. *A Short History of Anatomy from the Greeks to Harvey.* New York: Dover, 1957.

Stahl, William H. *Roman Science.* Madison: University of Wisconsin Press, 1962.

Thorndike, Lynn. *Science and Thought in the Fifteenth Century.* New York: Hafner, 1963.

Zukav, Gary. *The Dancing Wu Li Masters: An Overview of the New Physics.* New York: Bantam Books, 1980.

Index

Note: Page numbers in **boldface** indicate subjects of articles.

Bonpland, Aime, 188
Book of Healing (Ibn Sina), 112
Boole, George, 1, **26**
Boolean algebra, 26
Border, Jules, 145
borium, 58
Borlaug, Norman Ernest, **27**, 93
Born, Max, **28**, 165
boron, 58, 86, 195
Bosch, Karl, 60
Bose, Satyendranath, **29**
Bose-Einstein Condensate, 29
Bose-Einstein Statistics, 29
bosons, 29, 73
botany, 35, 38
　nomenclature systems, 81, 130
　See also plants
Bothe, Walther, 38
bottom (fifth quark), 87
Boulton, Matthew, 191
Bourgeois, Louise, 23
Boyle, Robert, **30**, 106
Boyle's Law, 30, 106
brachistochrone problem, 18
Bragg, William and Lawrence, 171
Brahe, Tycho, **31**, 66, 120
Braille, Louis, **32**
braille alphabet, 32
brakes
　compressed air, 179
　hydraulic, 33, 160
Bramah, Joseph, **33**
Bramah lock, 33
Brattain, Walter, 12
breeding. *See* genetics
bridges, cast-iron, 56
Brief History of Time, A: From the Big Bang to Black Holes (Hawking), 100
Bristol Iron Company, 56
British Broadcasting Corporation (BBC), 136
British Institute of Preventive Medicine. *See* Lister Institute
British Museum, 35
British Royal Society. *See* Royal Society of London for Improving Natural Knowledge
broadcasting. *See* radio; television
Broglie, Louis de, 28, 101, 173
Brookhaven National Laboratory, 87
Brooks, Harriet, **34**
Brown, Robert, **35**
Brownian motion, 35
Brunfels, Otto, 81
Bryn Mawr College, 153
bubonic plague, 122
Buffon, Georges, **36**
Bullard, Edward, 104
Bunsen, Robert, **37**
Bunsen burner, 37
Burbank, Luther, **38**
Burbank potato (Idaho potato), 38
Burt, William, 176
business machines, 17, 63, 160, 176
Butler, Clifford, 29, 200

Byron, Augusta Ada. *See* Lovelace, Countess of
Byron, Lord (George Gordon Byron), 132

C

cacodyl compounds, 37
cadaver dissection, 185
calcium, 58
calculating machine
　Babbage, 11, 132
　Leibniz, 129
　Pascal, 129, 160
calculus
　algebraic foundation, 44
　as analytic geometry offshoot, 59
　Cauchy's systemization of, 44
　defined, 71
　differential, 71, 123, 129, 160
　Euler's textbooks, 68
　Fermat's work and beginnings of, 71
　Gauss's unpublished work in, 85
　integral, 8, 129, 160
　Leibniz-Newton invention dispute, 11, 18, 129, 151
　Leibniz's contribution, 71, 129
　Newton's contribution, 71, 151
　Pascal's influence on, 160
　of variations, 18
Calcutta University, 169
California Institute of Technology (Caltech), 39, 73, 87, 162
Calley, John, 150
calligraphy, 144, 176
calorimeter, 37
Cambridge Mathematical Journal, The, 26
Cambridge University, 90, 100, 178
　DNA double-helix discovery, 51, 78
　EDSAC computer, 5
　genetics studies, 15
　Maxwell association, 137
　Newton association, 151
　Russell association, 172
　See also Cavendish Laboratory
camera
　daguerreotype, 54
　first motion picture, 133
　hand-held, 54, 62
　Polaroid "instant," 54, 62
camera obscura, 54
cancer, 79
　chemotherapy, 196
　radiation treatment, 16, 52
candles, 86
canned food processes, 7, 114
Cannizzaro, Stanislao, 10
Cannon, Annie Jump, **39**
Canon of Medicine (Ibn Sina), 112
Cantor, Georg, 44, 85, 172
capillaries, 99, 128, 135
carbohydrates science, 79
carbolic-acid solution, 131
carbon-14, 52
carbonated beverages, 167
carbon benzene rings, 118

carbon dating, 52
carbon dioxide
　cycles, 167
　greenhouse effect, 9
　isolation of, 22
　qualities of, 167
carbon monoxide, 167
carbon tetravalence, 118
carbon-zinc cell, 37
carburetor, 157
Cardano, Girolamo, 74
cardiac surgery, 194
Carnegie, Andrew, 20
Carnegie Foundation, 21
Carnot, Sadi (Nicholas-Leonard-Sadi), **40**, 47
Carothers, Wallace, **41**
Carrion's disease. *See* Bartonellosis
Carron Ironworks, 56
cars. *See* automobiles
Carson, Rachel, **42**, 188
Cartesian coordinate system, 59
cartography, 59
　first world map of prevailing winds, 96
　isotherms and isobars, 188
　Mercator projection, 144
　radar use, 190
Carver, George Washington, **43**
Cary, Elizabeth. *See* Agassiz, Elizabeth Cary
Case, J. I., 139
Cassiopeia (constellation), 31
catalog
　natural history, 36
　nebulae, 102, 103
Catalogue of the Nebulae, A (Herschel), 102
catalysts, 37
catapults, 8
cataract extraction, 170
catastrophism (geologic), 53, 134
catechisms, 92
cathode rays, 171
Catholic church. *See* Roman Catholic church
Catholicon, 92
Cauchy, Augustin-Louis, 1, **44**, 85
causality, 24, 189
Cavendish, Henry, 22, **45**
Cavendish Laboratory, 34, 51, 156
　Cavendish family funding of, 45
　Chadwick as assistant director, 46
　design and construction, 137
　Rutherford as director, 173
CD-ROMS ("read-only" disks), 84
cell biology, 117, 128
cellophane, 41
cells
　chromosomal genetic function, 138
　DNA, 51
　first identification of, 106
　first observation of, 128
　membrane physiology, 79
　microbe-destroying, 145
　nucleus identification, 35

reproduction, 117
　staining technique, 64, 122
cell theory, 128
cellular telephone, 137
celluloid, 41
celluloid film, 133
Celsius, 119
cement, 63
census statistics, 85
centripetal force, 110
Cepheid star, 109
Ceres (asteroid), 85
cesium, 37
Chadwick, James, **46**, 72, 200
Chain, Ernst, 75
chain reaction, 72, 94, 123, 141
chain structure (organic compounds), 118
Challenger space shuttle explosion, 73
chance, laws of, 25
Chandy, Ennackel, 198
Chang, Min-Chueh, 164
charcoal, 56
Charcot, Jean, 80
Charles I, King of England, 99
Charles V, Holy Roman Emperor, 4, 185
Charles IX, King of France, 159
Charles X, King of France, 44
Charles, Jacques, 86
Charles' Law, 86
Charles R. Drew Blood Center, 61
charm (fourth quark), 87
Charpentier, Jean de, 2
chemical bonds. *See* bonding, chemical
chemical detection methods, 72
chemical equilibrium, 9, 47, 127
chemical fertilizers, 181
Chemical Institute (Berlin), 141, 165
chemical pesticides, 42
chemical reactions
　Boyle's Law application, 30
　electric currents produced by, 186
　equilibrium principle, 9, 47, 127
　gases released by, 167
　light effects, 86
　mass conservation, 125
　polymer formation, 41
　rates, 9
　specific heat capacity calculations, 22
　stoichiometry, 55
chemical warfare, 93
chemistry
　Arabic, 170
　biochemistry, 124, 162, 181
　drug development, 64
　electrical conduction, 70, 167
　electrolysis, 58, 95
　founding as separate discipline, 30
　gas laws as fundamental theoretical basis, 86
　industrial, 41, 91, 95

flea metamorphisis, 128
Fleming, Alexander, 75
Fleming, John, 63
flexible film, 62
Flexner, Simon, 154
flooding, mine, 4
Florey, Howard, 75
Flourens, Marie, 149
FLOW-MATIC (compiler), 107
fluids
    ion concentration of, 9
    motion of, 183
    Pascal's Law, 160
fluorescence, 171
fluoroscope, 63
flushable toilet, 33
Fluxions (Newton; Buffon trs.), 36
food preservation, 7, 114, 161
food production. *See* agriculture
footprints, hominid, 126
force, law of, 151
Ford, Henry, **76**
Ford Foundation, 76
Ford Motor Company, 76
forensic science
    blood typing, 124
    radioimmunoassay, 199
forest classifications, 188
Formal Logic (De Morgan), 26
formica, 41
fossils, 53, 57, 126, 134
four bodily humors belief, 82, 105,
    112, 158, 185
four-element theory, 30, 45
Fourier, Jean-Baptiste-Joseph, 119,
    155
Francis II, King of France, 159
Franco-Prussian War, 14, 131
Frankland, Edward, 37
Franklin, Benjamin, 77, 167
Franklin, Rosalind, 51, **78**, 171
Franklin stove, 77
Fraser-Reid, Bertram, **79**
Frederick II, Holy Roman Emperor,
    74
Frederick II, King of Denmark, 31
Frederick II (the Great), King of
    Prussia, 68
free-association technique, 80
Freedmen's Hospital (Washington,
    D.C.), 194
freezers, 115
freezing point, 9
Frege, Gottlob, 26, 172
French Academy of Sciences, 1, 44
    founders, 110
    Poincaré's presidency, 166
    Prix Bordin, 123
French Mint, 86
French Revolution, 6, 40, 44, 54
    execution of Lavoisier, 125
Frere, James H., 32
Freud, Sigmund, **80**, 163
friction, 186
Friedmann, Aleksandr, 100, 109
Frisch, Otto, 24, 141
Frisius, Gemma, 144
fruit fly, 138

Fuchs, Leonhard, **81**
Fuchsian function, 166
fuel cells, 9
fuel pumps, rocket, 89
functions theory, 1
furnaces
    blast, 127
    coke for iron smelting, 56
    regenerative blast, 20
fuschia, 81
fusion reaction, 141, 174, 182
Fust, Johann, 92

G

Gagnan, Emile, 50
galaxies, 19, 102
    beyond Milky Way, 103, 109
Galen of Pergamum, **82**, 105, 170
    Agricola and, 4
    blood flow theory, 99
    as Ibn Sin influence, 112
    refutations of theories of, 99,
        135, 158, 185
Galilean moons, 83
Galilei, Galileo, 30, **83**
    Archimedes as influence on, 8
    heliocentric theory, 49, 59, 83,
        120
    Huygens and, 110
    Torricelli association, 183
gallinium, 143
Galvani, Luigi, 58, 186
galvanometer, 6, 119
gametes, 136
game theory, 189
Gamow, George, 100
gangrene, 131
garment industry, 108
gas clouds, 102
gases, 183
    as air components, 125
    Bose-Einstein Condensate, 29
    Boyle's Law, 30, 106
    Cavendish experiments, 22, 45
    combining volumes, law of, 86,
        136
    discoveries of new, 143, 167
    equal volume and numbers of
        molecules, 6, 10
    expansion-related cooling, 115,
        119
    explosions, 127
    Hooke studies, 106
    inert (noble), 143
    molecular velocity distribution,
        137
    partial pressures, law of, 55
    pressure-volume relationship,
        106
    quantitative analysis of, 22
    thermal properties, 115
gas expansion, law of (Charles'
    Law), 86
gas mask, 93, 147
gasoline, 76, 157
Gasometric Methods (Bunsen), 37
Gates, Bill, **84**
Gauss, Carl Friedrich, 1, 68, **85**,
    137

advancement of Euler's work by,
    71
Gay-Lussac, Joseph-Louis, 10, **86**,
    88
Geiger, Hans, 173
Gell-Mann, Murray, 29, 46, **87**,
    198, 200
gematria, 168
general anesthesia, 149
*General Investigations of Curved
    Surfaces* (Gauss), 85
General Motors, 76
"General Theory of Ideals"
    (Noether), 153
general theory of relativity (1915).
    See relativity, general theory of
generator, 6, 70, 186
genes, 15, 52
Genesis, 36
gene therapy, 51
genetic code, 138
genetic counseling, 142
genetic diseases, 51, 142
genetic engineering, 27, 51
genetics
    acquired characteristics theory,
        38, 53, 57
    animal inheritance experiments,
        15
    blood-factors inheritance, 124
    chromosome cross-over, 138
    coining of term, 15
    developmental, 117
    DNA double-helix model, 51
    Mendelian, 15, 142
    molecular, 78
    mutations, 57, 117, 138
    plant breeding, 15, 38, 138, 142
    radioactivity uses, 52
    See also heredity
Geneva College, 23
Geneva Convention of 1864, 14
genome, 138
geocentric astronomy, 49, 83, 120
geodesy, 66, 85
geography, 65, 188
Geography (Eratosthenes), 66
*Geological Evidences of the Antiquity
    of Man, The* (Lyell), 134
geology
    Agricola's mineral codification, 4
    catastrophe theory, 53, 134
    continental drift theory, 104,
        192
    fundamental principles, 134
    ice ages, 2, 9
    ocean floor, 104
    plate tectonics, 104
    stratigraphy, 134
geomagnetism, 103
geometric series, 123
Geometric Works (Torricelli), 183
geometry, 66, 68
    analytic, 59, 71
    coordinate, 71
    curve description, 3, 85
    differential, 85
    Euclidean, 67
    Fibonacci's work, 74

non-Euclidean, 67, 85, 166
    "Pascal's mystic hexagram," 160
    projective, 160
    Pythagorean concepts, 168
    solid and plane, 8
geophysics, 96, 192
George I, King of Great Britain,
    129
George III, King of Great Britain,
    98, 102, 103
Germain, Sophie, **88**
Germain's Theorem, 88
germanium, 65, 143
German Social Democracy (Russell),
    172
germ theory, 7, 63, 122, 128, 131,
    154, 161, 175
Gibbs, J Willard, 47
Gilchrist, Percy, 20
Girard of Cremona, 121
glaciers, 2
glaucoma, 116
Glidden, Carlos, 176
Glidden Company, 116
globe construction, 144
gluons, 87
Goddard, Robert H., **89**
gold
    alchemy, 158
    mining separation process, 69
Goldstine, Herman, 189
Golka, Robert, 182
Gombe Game Reserve (Tanzania),
    90
Goodall, Jane, **90**
Goodrich, Benjamin, 91
Goodyear, Charles, **91**
Goodyear Rubber and Tire
    Company, 91
Gorbachev, Mikhail, 174
Göttingen Observatory (Germany),
    85
Gould, Stephen Jay, 15
Grace Murray Hopper Award, 107
grain
    hybrids, 27
    reaper-increased production, 139
Grand Unified Theory, 174
graphite pencil, 20
graphs, 59
gravity
    application to three-body
        problem, 68, 166
    of black holes, 100
    Cavendish calculation of Earth's,
        45
    falling motion and, 83
    Hooke's theory, 106
    Huygens's (incorrect) theory, 110
    Newton's theory, 31, 96, 106,
        120, 151
    stellar motion and, 19
    See also falling objects
grease-spot photometer, 37
Great Depression (1930s), 14
Great Exhibition of London
    (1851), 139
Great Train Robbery, The (film), 63
Greene, Catherine, 193

Yukawa, Hideki, 200
Nobel Prize for Physiology or
    Medicine
    Crick, Francis, 51, 78
    Ehrlich, Paul, 64, 145
    Fleming, Alexander, 75
    Kendall, Edward, 181
    Koch, Robert, 122
    Landsteiner, Karl, 124
    McClintock, Barbara, 138
    Metchnikoff, Élie, 64, 145
    Pavlov, Ivan, 163
    von Békésy, Georg, 187
    Watson, James, 51, 78
    Wilkins, Maurice, 51, 78
    Yalow, Rosalyn Sussman, 199
noble (inert) gases, 143
Noether, Emmy, **153**
Noguchi, Hideyo, **154**
nomenclature
    animal, 130
    botanical, 81, 130
    chemical, 125
Nomenclature (Lavoisier), 125
No More War! (Pauling), 162
noncommutative systems, theory of,
    153
non-Euclidean geometry, 67, 85,
    166
normalization technique, 153
North Atlantic ridge, 104
*Notes on Matters Affecting the
    Health, Efficiency, and Hospital
    Administration of the British
    Army* (Nightingale), 152
Notes on Nursing (Nightingale), 152
Novum Organum (Bacon), 12
nuclear energy
    Becquerel's groundwork, 16
    Dalton's groundwork, 55
    Einstein's groundwork, 65
    first reactor, 72
    first sustained chain reaction, 72
    Noether's groundwork, 153
    peaceful uses advocates, 72, 94,
        101, 156, 162, 174
    plasma, 182
    power plants, 156
    radioactive uranium fueling, 52
    See also nuclear weapons
nuclear fission
    Bohr description of, 24, 156
    chain reaction, 72, 94, 123, 141
    co-discovery of, 25, 94, 141, 156
    implications of, 141, 156
    neutron and, 46
    See also atomic bomb
nuclear fusion. *See* fusion reaction
nuclear isotopes. *See* isotopes
nuclear magnetic resonance
    (NMR), 79
nuclear medicine, 173, 199
nuclear physics
    beginnings of, 16, 24, 52
    Brooks papers, 34
    Enrico Fermi Award, 156
    fission discovery implications,
        141

infinite series and, 123
    meson theory, 200
    neutron discovery implications,
        46, 156
    nonconservation of parity, 198
    Uncertainty Principle
        implications, 101
    See also atomic theory; particle
        physics; quantum theory;
        quantum mechanics
nuclear power. *See* nuclear energy
nuclear radiation. *See* radioactivity
nuclear reactions, 73, 173, 198 (*see
    also* chain reaction)
nuclear reactors, 72, 94, 141, 156
nuclear weapons, 24, 46, 65, 94,
    123
    fission and fusion processes, 141,
        174
    German project, 101
    hydrogen bomb vs. atomic
        bomb, 174
    opposition to, 156, 162, 174
    testing of first, 156
    See also atomic bomb
nucleolus, 138
nucleotides, 51
nucleus, 101
    cell biology, 117
    as constant cellular element, 35
    discovery of, 173
    liquid-drop model, 24
    meson "exchange" particle, 200
    particles within, 46, 52, 72, 200
    radioactive disintegration of, 16,
        52
    splitting of uranium, 72, 94
Nuh ibn Mansur (Samanid ruler),
    112
numbering systems, 74, 121
numbers
    atomic, 168
    complex, 85
    mystical attributes, 168
    prime, 66, 67, 168
    Pythagorean representation, 168
number system (Gaussian), 85
number theory
    abstract, 88
    Bernoulli, 18
    Euclid, 67
    Euler, 68
    Gauss, 85
    modern methodology, 71
    Pythagoras, 168
    real numbers, 44
    Russell refinements, 172
numerology, 168
nursing, 152, 194
nutritional diet, 145
nylon, 41

O

OBAFGKM (seven star classes), 39
Oberlin College, 95
occupational hazards, 4
ocean-depth measurement, 119
ocean floors, 104, 192
ocean navigation. *See* navigation

Oceanographic Institute and
    Museum of Monaco, 50
oceanography, 42, 50, 104
Ochoa, Severo, 51
odd numbers, 168
Oedipus complex, 80
Oersted, Hans Christian, 137, 148,
    155, 186
offshore oil drilling, 50
Ogden, William, 139
ohm (unit of resistance), 155
Ohm, Georg, **155**, 186
Ohm's law, 155
oil. *See* petroleum
Olbers, Wilhelm, 19
Olduvai Gorge (Tanzania), 126
oligosaccharides, 79
omega particle, 87
On Metallurgy (Agricola), 4
"On the Art of Combination"
    (Leibniz), 129
On the Beauty of Music (Razi), 170
"On the Dynamical Theory of
    Heat" (Kelvin), 119
*On the Dynamics of the Circular
    Vortex with Applications to the
    Atmosphere and Atmospheric
    Vortex and Wave Motion*
    (Bjerknes), 21
On the Nature of Fossils (Agricola), 4
*On the Origin of Species by Means of
    Natural Selection* (Darwin), 57,
    134, 145
On the Sphere and Cylinder
    (Archimedes), 8
open-heart surgery, 194
Opera Omnia (Euler), 68
operating systems (computer), 84
Oppenheimer, J. Robert, 46, 72,
    156
optics, 71, 83, 169
    fish-eye lens, 137
    Newtonian, 129, 151
    See also light; telescopes
Optics (Euclid), 67
Optiks (Newton), 129, 151
oral contraceptives, 164
Order of Pythagoras, 168
ore concentration process, 69
Orestes (prefect of Alexandria), 111
organic chemistry, 37, 79, 116,
    118, 143, 195
organometallics, 37
*Origins of the Continents and
    Oceans, The* (Wegener), 192
orthomolecular medicine, 162
Osaka University, 200
oscillator, 136
osmotic pressure, 9
osteoporosis, 164
Otto, Nikolaus August, 60, **157**
Otto engine, 60, 157
"outer space as vacuum" concept,
    160, 183
ovum (egg), 117
Oxford University, 39, 96, 106,
    178
oxygen, 58, 86, 173
    combustibility, 125, 167

cycles, 167
discovery of, 167
high-altitude depletion of, 188
as water component, 125

P

page-numbering machine, 176
pagers, 137
paints, 116
paleoanthropology, 126
paleontology, 53, 57
Palmer, Henry, 194
pancreatic system, 185
Pangaea, 192
paper coatings, 116
paper invention, 184
Pappus of Alexandria, 66
Paracelsus, 82, 105, 135, **158**, 185
parallel postulate, 67
paralysis. *See* paresis
parasites, 64, 158
parchment, 184
Paré, Ambroise, 23, **159**
paresis, 154
Paris Exhibition (1867), 157
parity, non-conservation of, 198
Parke, Davis, and Company, 181
parthenogenesis, 117
partial differential equations, 123
partial pressures, law of, 55
particle accelerators, 87
particle physics, 29, 174
    motion laws, 151
    probability theory and, 71
    quarks and, 87
    See also subatomic particles
particle scattering, 169
Pascal (computer language), 160
Pascal, Blaise, 71, 110, 129, **160**,
    183
Pascal, Étienne, 160
Pascal's Law, 33, 160
"Pascal's mystic hexagram," 160
"Pascal triangle," 160
Pasteur, Louis, 7, 122, 131, **161**,
    175
Pasteur Institute (Paris), 145, 161
pasteurization, 161, 175
patent disputes
    aluminum manufacturing
        process, 95
    cotton gin, 193
    ether anesthesia, 149
    four-stroke engine, 157
    sewing machine, 108
paternity cases, 124
pathogens, 7, 145
pathology, 124, 145
Pauling, Linus, **162**
Pavia University, 186
Pavlov, Ivan, **163**
"Pavlov's dog," 163
PCs (personal computers), 84
Peano, Giuseppe, 172
peanuts, 43
pea plants, 15, 142
Pecquet, Jean, 185
pediatrics, 170

quantum chromodynamics (QCD), 73, 87

quantum electrodynamics (QED), 73

quantum mechanics
  Bohr's contribution, 24
  bond character, 162
  Born's formulation of, 28
  Bose's pioneer work in, 29
  complementarity principle, 24
  Copenhagen Interpretation, 24, 28
  defined, 73, 165
  hybridization, 162
  matrix mechanics, 28, 101, 165
  motion laws, 151
  Oppenheimer's studies, 156
  Pauling's work, 162
  probability theory and, 28, 71, 101
  quantum electrodynamics, 73
  resonance theory, 162
  Von Neumann's work, 189

quantum theory
  atomic model, 52
  Bohr's application of, 24, 28
  Bose-Einstein Statistics, 29
  Bose's derivation of fundamental idea of, 29
  Einstein's advancement of, 65
  equivalence of matrix and wave mechanics, 189
  Heisenberg's development of, 101
  of light, 29, 65, 169
  matrix mechanics, 28, 101, 165
  ongoing discussions, 165
  photochemistry and, 86
  Planck's foundation for, 24, 28, 29, 65, 165
  Poincaré's mathematical proof, 165
  Rutherford's foundations for, 173
  statistical mechanics and, 25
  transistor development, 65
  Uncertainty Principle, 28, 101
  wave mechanics, 28, 165, 189

quarks
  binding of, 73
  model development, 29, 46, 87, 200
  spin, 72

Quereler, Lambert, 85

quintic equations (fifth degree), 1

Quitting Time at the Lumière Factory (film), 133

R

rabies vaccine, 113, 161
radar, 190
Radcliffe College, 2, 39
radiates, 53
radiation, 171
  blackbody, 165
  black hole emission of, 100
  infrared, 103
  nuclear recoil effect, 34
  physiological effects, 16
  spectral electromagnetic, 37

thermal, 25, 100
uses of, 16
See also electromagnetism; light; radioactivity; X-rays

radiation chemistry, 52
radical atoms, 195
radio
  diode device, 63
  modulation system, 136
  ship-to-shore, 136
  Tesla-coil as set component, 182
  wartime use, 136
  wireless communication, 136
  See also radio waves
radioactive decay series, 34, 87, 141, 173, 198
radioactivity
  alpha and beta types, 173
  artificial, 52
  atomic disintegration from, 52
  atomic transmutation from, 52
  chain reaction, 72, 94, 123, 141
  coining of term, 52
  discovery of, 16, 171
  of elements, 52, 94, 141, 173
  events sequence, 34
  isotopes, 46, 72
  pioneering research in, 34
  recoil effect, 34
  scientific applications, 52
  weak nuclear force, 73
  See also X-rays
radiochemistry, 141
Radio Detection and Ranging. *See* radar
radioimmunoassay, 199
Radio Research Station, 190
radio waves, 63, 136, 137, 182
  radar, 190
radium
  alpha particles as probes, 173
  decay transmutation, 34
  discovery of, 52
radon, 34, 143
railroads
  development of, 179
  diesel locomotive, 60, 179
  steam locomotive, 56, 160, 179
Rajavski Sisters, The (Kovalevsky), 123
Raman, Chandrasekhara Venkata, **169**
Raman effect, 169
Raman-Nath theory, 169
Raman Research Institute (India), 169
Ramsay, Andrew, 2
Ramsay, William, 94
randomness. *See* entropy
Razi, Abu Bakr Muhammad, ar-, **170**
reaction. *See* action and reaction, law of; chemical reactions; chain reaction
reactor. *See* nuclear reactors
real numbers, 44, 85
reaper, 139
receiver, radio, 136

recoil effect, 34
rectangle, 168
red blood cells, 128, 162
Red Cross. *See* American Red Cross
reflecting telescope, 102, 103, 106, 151
reflection, laws of, 110
Reflections on the Motive Power of Fire (Carnot), 40
Reformation, 92
refraction, laws of, 110
refrigeration, 115
regenerative blast furnace, 20
relativity, general theory of (1915), 153
  described, 65
relativity, special theory of (1905), 25
  conservation of mass and energy, 125
  described, 65
  first English translation, 29
  Maxwell's groundwork, 137
  Poincaré's independent publication, 166
religion vs. science controversies
  Buffon's *Natural History*, 36
  creationism vs. evolution theory, 2, 57
  Sun-centered vs. Earth-centered universe, 49, 59, 120
Remarks on the Proper Mode of Administering Sulphuric Ether by Inhalation (Morton), 149
Remington Arms Company, 176
Remington Typewriter, 176
remote controls, 136
Renaissance, 92
reproduction
  asexual-sexual alternation, 145
  cellular, 117
  cloning, 51
  lower-level animal, 128
  oral contraceptives, 164
  plants, 130
  selective breeding, 142
"Researches, Chemical and Physical" (Davy), 58
Research Institute of Fundamental Physics (Hyoto), 200
resistance, electrical, 155
resonance theory, 162
Revolutions of the Heavenly Spheres (Copernicus), 49
rheumatoid arthritis, 116
Rh factor, 124
Ricardo, Harry, 60
Richards, Robert, 180
Riemann, Bernhard, 44, 85
Riemann, Georg, 67
Righi, Augusto, 136
rinderpest, 122
ring structure, 118
Rochester, George, 29, 200
Rochester Institute of Technology, 62
Rockefeller Foundation, 27
Rockefeller Institute, 124, 154
rockets, 89

rocks
  moon, 104
  sedimentation, 134
Roentgen, Wilhelm, 16, **171**
Roman Catholic church, 49, 83, 120
Roman numerals, 74, 75, 121
Roosevelt, Franklin D., 141, 156
Roswell (New Mexico), 89
rotating bodies, 110
Royal Academy of Sciences (Japan), 181
Royal Astronomical Society (G.B.), 11, 102, 178
Royal College of Physicians (G.B.), 99
Royal Institution, *See* Royal Society of London for Improving Natural Knowledge
Royal Society of London for Improving Natural Knowledge (Royal Society), 104
  Cavendish membership, 45
  Copley Medal, 155
  Davy Medal, 127
  Davy's lecture to, 58
  Faraday membership, 70
  founding of, 12, 30, 106
  Huygens's presentation to, 110
  Leibniz membership, 129
  Malpighi membership, 135
  Newton's optics lectures to, 151
  Priestley membership, 167
  publication of Leeuwenhoek's microscopic findings, 128
rubber, 41, 91
rubidium, 37
Rudolphine Tables, 31, 120
Russell, Bertrand, 26, **172**
Russell's Paradox, 172
Russo-Japanese War, 136
rust fungus, 27
Rutherford, Daniel, 22
Rutherford, Ernest, 55, **173**
  atomic model, 16, 52, 173
  Bohr studies with, 24
  Brooks association, 34
  Chadwick association, 46
  Hahn association, 94
  Oppenheimer studies with, 156

S

safety lock (Brahman), 33
safety valve, 114
Saha, M. N., 29
Saint-Hilaire, Étienne Geoffroy, 53
Sakharov, Andrey, **174**
Salk Institute for Biological Studies, 51
salmonella, 122
Salvarsan, 64, 154
Salzata, Shoichi, 200
Samanid dynasty, 112
Sandstrom, J. J., 21
San Francisco Earthquake (1906), 14
Sanger, Margaret, 164
sanitation, 180

Santa Fe Institute (New Mexico), 87

Saturn (planet), 103, 110, 146

Savery, Thomas, 4, 150

Saxon Wheel, 97

scandium, 143

Sceptical Chemist, The (Boyle), 30

Schaudinn, Fritz, 154

Scheutz, Pehr Georg, 11

Schleiden, Matthias Jakob, 35, 106, 128

Schöffer, Peter, 92

Schrieffer, John, 12

Schrödinger, Erwin, 28, 101, 165, 173, 200

Schwann, Theodor, 35, 106, 128

Schwarzschild, Karl, 100

Schweigger, Johann, 6

Schwinger, Julian S., 73

Science (journal), 17

Science and Hypothesis (Poincaré), 166

Science and Method (Poincaré), 166

science education, 48, 180

science fiction, 89

scientific method, 70, 90, 99, 110, 170

  origination of, 4, 12, 30

Scientific Outlook, The (Russell), 172

"scientist," first use as term, 188

screw pumps, 8

scuba diving equipment, 50

Sea Around Us, The (Carson), 42

sea floor spreading, 104

Seaman, Barbara, 164

seaplanes, 177

Sea Spider, 50

seawater desalination, 58

second-degree equations, 1

Second Law of Thermodynamics

  Boltzmann's statistical explanation, 25

  Carnot's work as basis of, 40

  Kelvin's contribution, 119

  statement of, 40

Second Punic War, 8

sedimentation, 134

Seiberling, Frank, 91

selective breeding programs, 142

semiconductors, 12

semi-permeable membranes, 9

Semmelweis, Ignaz, **175**

"Semmelweis reflex," 175

sepsis, 131

series

  Fibonacci sequence, 74

  infinite, 44, 123

serology, 124

serums. *See* immunology

set theory, 26, 172

Seven Books on the Structure of the Human Body, The (Vesalius), 185

sewage treatment, 8, 180

sewing machine, 108

Sex and Temperament in Three Primitive Societies (Mead), 140

sex hormones synthesis, 116

sexuality, 80

Shams ad-Dawlah (Buyid prince), 112

Shin'ichiro, Tomonaga, 73

ships

  diesel engine, 60

  wireless communication, 136

  See also navigation

Shockley, William, 12

Sholes, Christopher, **176**

short-wave radio, 136

shuttle breeding, 27

sickle-cell anemia, 162

Siemens, Friedrich and Wilhelm, 20

Siemens-Martin process, 20

Sieve of Eratosthenes, 66

Sikorsky, Igor, **177**, 197

Silent Spring (Carson), 42

Silent World, The (Cousteau), 50

silicon, 195

silicosis, 158

silver

  alchemy, 158

    mining separation process, 69

    purity determination, 86

silver cyanate, 195

silver fulminate, 195

silver oxide, 186

Simpson, James Young, 149

Singer, Isaac, 108

siphon recorder, 119

Sirius and Sirius B (stars), 19

Sitter, Willem de, 109

six-carbon ring, 118

Sixty Feet Down (film), 50

Skinner, B. F., 163

sky, mapping of, 39

Slipher, Vesto, 109

smallpox, 170

  vaccine, 113, 161

Smithsonian Institution, 89

snake venom, 154

Snell, Willebrod van Roijen, 66

Snow, John, 149

social Darwinism, 57

soda water, 167

Soddy, Frederick, 34, 173

sodium, discovery of, 58

software, 84

Soho Foundry, 191

soil conservation, 43

solar flares, 38

solar system. *See* astronomy; planetary motion; *specific components*

solenoids, 6

solutions, conductivity of, 9, 70

solvents, 35

Some Chemical Agencies of Electricity (Davy), 58

Somerville, Mary Fairfax, **178**

Somerville, William, 178

Somerville College (Oxford University), 178

sonar detection, 17

Sorbonne, 44, 52, 166

SOS (help signal), 148

Soulé, Samuel, 176

sound

  electrical transmission of, 17

  hearing mechanism, 155, 187

  light effects on, 169

  vacuum effect on, 183

sound movies, 133

South Sea Islands, 98

Soviet Academy of Sciences, 174

soybeans, 43, 116

space

  conceptualization of continuous, 67

  curvature of, 65, 85

space flight

  Challenger disaster, 73

  Hubble Telescope launching, 109

  probe flight paths, 151

  probe radar use, 190

  rocket design, 89

space-time singularities, 100

Spallanzani, Lazzaro, 7

Spanish-American War, 14

special theory of relativity (1905). *See* relativity, special theory of

species

  classification systems, 53, 130

  extinction causes, 53

  immutability theory, 57

  natural selection, 53, 57

  variations, 36

Species Plantarum (Linnaeus), 130

specific heat capacity, 22

Specificity of Serological Reactions, The (Landsteiner), 124

spectral analysis, 37, 39, 136

spectroscopy, 37, 39, 169

Spencer, Herbert, 57

sperm, 117, 128

Sperry Rand, 107

sphere, 8, 67, 146

spin, particle, 29, 72

Spingarn Medal

  Drew, Charles, 61

  Julian, Percy Lavon, 116

  Just, Ernest Everett, 117

spinning jenny, 97

spoilage. *See* food preservation

spontaneous generation theory, 7, 128

spores, 9

spring balance, 106, 110

square numbers, 168

square root, 168

Stahl, Franklin, 51

staining method, cell, 64, 122, 138

standardization

  Arabic numerals, 74

  botanical nomenclature, 81, 130

  chemical nomenclature, 125

  computer language, 107

  energy-related concepts, 155

  interchangeable parts, 76, 193

  mathematical symbols, 68

  silver purity, 86

  universal symbols concept, 129

  weights and measures, 143

Stanford Linear Accelerator Center, 87

Staphyloccocus aureus, 75

Starry Messenger, The (Galileo), 83

stars

  Cannon classification system, 39

  distances between, 103

  in distant galaxies, 109

  galaxy arrangement of, 19, 102, 103

  "Harvard" ranking system, 39

  motion of, 19, 96

  nebulae and, 102, 103, 109

  seven classes of, 39

  southern sky mapping, 103

  spectral analysis, 37

  supernova, 31

  tables of future positions, 121

static electricity, 186

statistical mechanics, 25, 166

statistics, 85

steam condensation, 22

steam engine

  coal-gas, 157

  first practical, 56, 150, 179

  heat flow, 40, 47

  high-pressure, 150

  Watt's improved model, 150, 191

steam locomotive, 56, 150, 179

steam power, 22, 97

steam sterilization, 131

steel-making processes, 20

Stefan, Josef, 25

stellar motion, 19, 96

stellar parallax, 19

stellar spectra, 37

stellar system, 103

Stephenson, George, **179**

Stephenson, Robert, 33, 179

stereochemistry, 161

stereoisomers, 161

stereoscopy, 133

Stern, Curt, 138

steroids, 164

stockings, nylon, 41

stock market prediction, 168

Stockton-to-Darlington railroad (G.B.), 179

stoichiometry, 55

Stokes, Adrian, 154

Stokes, George, 119

Story of My Childhood (Barton), 14

Story of the Red Cross, A (Barton), 14

stove, closed, 77

strain (mechanical), 106

"strange particles," 29, 200

strange quark, 87

Strassmann, Fritz, 94, 141

stratigraphy, 134

streptomycin, 75

stress (mechanical), 106

strontium, 58

Strowger, Almon B., 17

structural chemistry, 37, 117, 162, 171

Studies on Hysteria (Freud), 80

Sturgeon, William, 6

styrene, 41

X

xenon, 143

$x^n + y^n = z^n$ (Fermat's last theorem), 68, 71, 88

X-rays
chromosome damage by, 138
crystallography, 171
discovery of, 16, 171
DNA imaging, 51, 78, 171
lunar-rock composition analysis, 104
molecular structure determination, 162
X-ray tomography, 171

Y

Yale lock, 33
Yalow, Rosalyn Sussman, **199**
Yang, Chen Ning, 198
yeast cells, 128
yellow fever, 154
yogurt, 145
Yukawa, Hideki, 29, 46, **200**

Z

Z-1 computer, 5
zero
in Arabic numeral system, 74, 121
in binary system, 129
in Kelvin temperature scale, 86, 119
zinc dialkyls, 37
zoological species classification, 53
Zuse, Konrad, 5
Zweig, George, 29, 46, 200